# Lecture Notes in Mathematics

Edited by A. D




1351

I. Laine S. Rickman T. Sorvali (Eds.)

# Complex Analysis Joensuu 1987

Proceedings of the XIIIth Rolf Nevanlinna-Colloquium, held in Joensuu, Finland, Aug. 10–13, 1987


Springer-Verlag

Berlin Heidelberg New York London Paris Tokyo

**Editors**

Ilpo Laine
Tuomas Sorvali
University of Joensuu, Department of Mathematics
SF-80101 Joensuu, Finland

Seppo Rickman
University of Helsinki, Department of Mathematics
SF-00100 Helsinki, Finland

Mathematics Subject Classification (1980): 26B20, 30C45, 30C60, 30D20, 30D30, 30D45, 30D55, 30D99, 30F15, 30F20, 30F35, 30F40, 30F99, 30G05, 31B99, 31D05, 32D15, 32G15, 32H30, 46E35, 58B20, 81C35

ISBN 3-540-50370-6 Springer-Verlag Berlin Heidelberg New York
ISBN 0-387-50370-6 Springer-Verlag New York Berlin Heidelberg



Printing and binding: Druckhaus Beltz, Hemsbach/Bergstr.
2146/3140-543210

Dedicated to Professor

Lars V. Ahlfors

## PREFACE

This volume consists of papers presented at the XIII Rolf Nevanlinna – Colloquium held at the University of Joensuu, August 10–13, 1987. The program of the Colloquium concentrated on complex analysis. A substantial part of the program was devoted to the scientific work of Professor Lars V. Ahlfors to celebrate his $80^{th}$ birthday. This volume is dedicated to Lars Ahlfors, by the consent with all authors.

We wish to thank the staff of the Department of Mathematics in the University of Joensuu for their co-operation in preparing this volume. Our special thanks are due to Eero Posti and Harri Pekonen for their patient job of processing the manuscripts. Eero Posti also provided a great help for us by mastering all details of the text processing systems used to prepare these Proceedings. Finally, our gratitude is directed to Springer-Verlag for their willingness to publish this volume.

Joensuu and Helsinki, February 1988

Ilpo Laine　　Seppo Rickman　　Tuomas Sorvali

## CONTENTS

## OTHER LECTURES GIVEN AT THE COLLOQUIUM

Anderson, G. D.: Special functions of quasiconformal theory

Andrzejczak, G.: Riemannian characteristic classes for Kähler foliations

Becker, J.: Boundary values of asymptotically conformal mappings

Bertin, E. M. J.: Non-linear potential theory and Monge-Ampère equation

Brück, R.: On uniqueness theorems for entire harmonic functions of exponential type

Carleson, L.: On the work of Lars Ahlfors 1929–1955

Chen, Jixiu: Extremal problems of functionals of quasiconformal mappings

Cima, J.: The Dunford-property for function algebras

Earle, C.: Quasiconformal isotopies

Eriksson-Bique, S.-L.: A decomposition theorem for positive superharmonic functions

Essén, M.: On Beurling's theorem on harmonic measure and the rings of Saturn

Fernandez, J.: Internal distortion under conformal mappings

Gehring, F. W.: On the work of Lars Ahlfors 1954–1979

Gilman, J.: A geometric approach to the hyperbolic Jørgensen inequality

Gromov, M.: Pseudo-holomorphic curves

Haslinger, F.: Convolution equations and the problem of division in spaces of entire functions with nonradial weights

Heinonen, J.: On quasiconformal rigidity

Janson, S.: Hankel operators between weighted Bergman spaces

Jenkins, J.: On quasiconformal mappings with given boundary values

Knebusch, M.: Isoalgebraic geometry

Kopiecki, R.: A variational approach to the Newlander-Nirenberg theorem

Kra, I.: Monodromy and Schwarzian derivatives

Kruzhilin, N.: Locally biholomorphic maps of real hypersurfaces

Krzyż, J.: Harmonic analysis and boundary correspondence under quasiconformal mappings

Kuran, Ü.: NTA-conical domains

Lehtinen, M.: On majorization of quasisymmetric functions

Lesley, F. D.: On theorems of Jackson and Bernstein type in the complex plane

Maitani, F.: Ahlfors-Rauch type variational formulas on complex manifolds

Martens, H.: Mapping of closed Riemann surfaces

Martin, G. J.: Iteration of rational maps and discreteness in Kleinian groups

Martio, O.: Elliptic equations and maps of bounded length distortion

Maskit, B.: Parameters for certain Fuchsian groups

Minda, C. D.: The hyperbolic metric in k-convex regions

Mues, E.: Zum Vier-Punkte-Satz von Nevanlinna

## COLLOQUIUM PARTICIPANTS

Ahlfors, Lars, Harvard University, Cambridge, MA 02138, U.S.A.
Aikawa, Hiroaki, Gakushuin University, Tokyo 171, Japan
Anderson, Glen D., Michigan State University, East Lansing, MI 48824, U.S.A.
Anderson, J. Milne, University College, London WC1E 6BT, England, United Kingdom
Andreian Cazacu, Cabiria, University of Bucharest, Bucharest, Romania
Andrzejczak, Grzegorz, Polish Academy of Sciences, Łódź, Poland
Astala, Kari, University of Helsinki, SF–00101 Helsinki, Finland
Aulaskari, Rauno, University of Joensuu, SF–80100 Joensuu, Finland
Baker, I.N., Imperial College, London SWY 2BZ, England, United Kingdom
Bârză, Ilie, University of Bucharest, Bucharest, Romania
Becker, Jochen, Technical University of Berlin, D–1000 Berlin 12, F.R. Germany
Bertin, Emile M. J., University of Utrecht, Utrecht, The Netherlands
Bojarski, Bogdan, Polish Academy of Sciences, Warsaw, Poland
Bracalova, Melkana, Bulgarian Academy of Sciences, 1090 Sofia, Bulgaria
Brück, Reiner, University of Gießen, D–6300 Gießen, F.R. Germany
Carleson, Lennart, Royal Institute of Technology, S–10044 Stockholm, Sweden
Chen, Jixiu, Fudan University, Shanghai, China
Cima, Joseph, University of North Carolina, Chapel Hill, NC 27514, U.S.A.
Earle, Clifford, Cornell University, Ithaca, NY 14850, U.S.A.
Eriksson–Bique, Sirkka–Liisa, University of Joensuu, SF–80101 Joensuu, Finland
Erkama, Timo, University of Joensuu, SF–80101 Joensuu, Finland
Essén, Matts, University of Uppsala, S–75238 Uppsala, Sweden
Fehlmann, Richard, University of Helsinki, SF–00100 Helsinki, Finland
Fernandez, Jose, University of Maryland, College Park, MD 20742, U.S.A.
Ferrand, Jacqueline, University of Paris VI, F–75005 Paris Cedex 05, France
Frank, Günter, University of Dortmund, D–4600 Dortmund 50, F.R. Germany
Fuka, Jaroslav, Czechoslovak Academy of Science, 11567 Prague 1, ČSSR
Gehring, Frederick W., University of Michigan, Ann Arbor, MI 48109, U.S.A.
Gilman, Jane, Rutgers University, Newark, NJ 07102, U.S.A.
Godula, Janusz, University of Lublin, 20–031 Lublin, Poland
Gromov, Mikhail, IHES, F–91440 Bures–sur–Yvette, France
Haahti, Heikki, University of Oulu, SF–90100 Oulu, Finland
Hà Huy Khoái, Institute of Mathematics, Hanoi, Vietnam

Hag, Kari, Norwegian Institute of Technology, N–7034 Trondheim, Norge
Hag, Per, University of Trondheim, N–7055 Dragvoll, Norge
Hanson, Bruce, St Olaf College, Northfield, MN 550057, U.S.A.
Harmelin, Reuven, Israel Institute of Technology, Haifa, Israel
Haslinger, Friedrich, University of Vienna, A–1090 Vienna, Austria
Hayman, W.K., University of York, Heslington, YO1 5DD, England, United Kingdom
Hedberg, Lars Inge, University of Linköping, S–58183 Linköping, Sweden
Heinonen, Juha, University of Jyväskylä, SF–40100 Jyväskylä, Finland
Heins, Maurice, University of Maryland, College Park, MD 20742, U.S.A.
Hejhal, Dennis, University of Minnesota, Minneapolis, MN 55455, U.S.A.
Herron, David A., University of Cincinnati, Cincinnati, OH 45221, U.S.A.
Holopainen, Ilkka, University of Helsinki, SF–00100 Helsinki, Finland
Huber, Alfred, Federal Institute of Technology, CH–8092 Zürich, Switzerland
Hurri, Ritva, University of Jyväskylä, SF–40100 Jyväskylä, Finland
Hyvönen, Jaakko, University of Joensuu, SF–80101 Joensuu, Finland
Ikegami, Teruo, Osaka City University, Osaka 558, Japan
Ikonen, Leena, University of Joensuu, SF–80101 Joensuu, Finland
Imayoshi, Yoichi, Osaka University, Toyonaka, 560 Osaka, Japan
Jacob, Niels, University of Erlangen–Nürnberg, D–8520 Erlangen, F.R. Germany
Janson, Svante, University of Uppsala, S–75238 Uppsala, Sweden
Jenkins, James A., Washington University, St. Louis, MO 63130, U.S.A.
Jussila, Tapani, University of Helsinki, SF–00100 Helsinki, Finland
Kahramaner, Suzan, Istanbul University, Istanbul, Turkey
Karrer, Guido, University of Zürich, CH–8001 Zürich, Switzerland
Keen, Linda, Lehman College, Leonia, NJ 07605, U.S.A.
Keller Heinrich, University of Zürich, CH–8001 Zürich, Switzerland
Kilpeläinen, Tero, University of Jyväskylä, SF–40100 Jyväskylä, Finland
Kinnunen, Liisa, University of Joensuu, SF–80101 Joensuu, Finland
Kinnunen, Mervi, University of Joensuu, SF–80101 Joensuu, Finland
Kinnunen, Veli, University of Joensuu, SF–80101 Joensuu, Finland
Knebusch, Manfred, University of Regensburg, D–8400 Regensburg, F.R. Germany
Koch, Helga, Free University of Berlin, D–1000 Berlin, F.R. Germany
Koepf, Wolfram, Free University of Berlin, D–1000 Berlin, F.R. Germany
Kopiecki, Ryszard, Warsaw University, 00–901 Warsaw, Poland
Kra, Irwin, State University of New York, Stony Brook, NY 11794, U.S.A.
Królikowski, Wieslaw, Polish Academy of Sciences, Łódź, Poland
Kruzhilin, Nikolay, Steklov Mathematical Institute, Moscow, USSR

Krzyż, Jan, University of Lublin, 20–031 Lublin, Poland
Kukkurainen, Paavo, University of Joensuu, SF–80101 Joensuu, Finland
Kuran, Ülkü, University of Liverpool, Liverpool L69 3BX, England, United Kingdom
Kuusalo,Tapani, University of Jyväskylä, SF–40100 Jyväskylä, Finland
Laine, Ilpo, University of Joensuu, SF–80101 Joensuu, Finland
Lappan, Peter A., Michigan State University, East Lansing, Michigan 48824, U.S.A.
Lehtinen, Matti, University of Helsinki, SF–00100 Helsinki, Finland
Lehto, Olli, University of Helsinki, SF–00100 Helsinki, Finland
Lehtola, Pasi, University of Jyväskylä, SF–40100 Jyväskylä, Finland
Lesley, F. David, San Diego State University, San Diego, CA 92182, U.S.A.
Leutwiler, Heinz, University of Erlangen–Nürnberg, D–8520 Erlangen, F.R. Germany
Lewis, John, University of Kentucky, Lexington, KY 40506, U.S.A.
Li Zhong, Beijing University, Beijing, China
Luukkainen, Jouni, University of Helsinki, SF–00100 Helsinki, Finland
Määttä, Eljas, University of Oulu, SF–90570 Oulu, Finland
Maitani, Fumio, Kyoto Institute of Technology, Matsugasagi, Sakyo–ku, Kyoto, Japan
Marden, Albert, University of Minnesota, Minneapolis, MN 55455, U.S.A.
Martens, Henrik H., Norwegian Institute of Technology, N–7034 Trondheim, Norway
Martin, Gaven J., Yale University, New Haven, CT 06520, U.S.A.
Martio, Olli, University of Jyväskylä, SF–40100 Jyväskylä, Finland
Maskit, Bernard, State University of New York, Stony Brook, NY 11794, U.S.A.
Masur, Howard, University of Illinois at Chicago, Chicago, IL 60680, U.S.A.
McKemie, M. Jean, University of Missouri–Rolla, Rolla, MO 65410, U.S.A.
Metzger, Thomas, University of Pittsburgh, Pittsburgh, PA 15280, U.S.A.
Minda, C. David, University of Cincinnati, Cincinnati, OH 45221, U.S.A.
Miniowitz, Ruth, 60 Shoshanat Hacarmel St., Haifa 34322, Israel
Mues, Erwin, University of Hannover, D–3000 Hannover, F.R. Germany
Näätänen, Marjatta, University of Helsinki, SF–00100 Helsinki, Finland
Näkki, Raimo, University of Jyväskylä, SF–40100 Jyväskylä, Finland
Nguyen Xuan–Loc, Institute of Computer Science and Cybernetics, Hanoi, Vietnam
Nikolaus, Johannes, University of Siegen, D–5900 Siegen, F.R. Germany
Nolder, Craig A., Florida State University, Tallahassee, Florida 32306–3027, U.S.A.
Obradović, Milutin, University of Belgrade, 11000 Belgrade, Yugoslavia
Ohtake, Hiromi, Kyoto University, Kyoto, Japan
Ohtsuka, Makoto, Gakushuin University, Tokyo 171, Japan
Oikawa, Kotaro, University of Tokyo, Tokyo 180, Japan
Pansu, Pierre, Ecole Polytechnique, F–91128 Palaiseau, France

Pekonen, Osmo, SF–50100 Mikkeli, Finland

Pesonen, Martti Ensio, University of Joensuu, SF–80101 Joensuu, Finland

Pfluger, Albert, Federal Institute of Technology, CH–8092 Zürich, Switzerland

Pinkall, Ulrich, Technical University of Berlin, D–1000 Berlin, F.R. Germany

Pirinen, Aulis, Helsinki University of Technology, SF–02150 Helsinki, Finland

Posti, Eero, University of Joensuu, SF–80101 Joensuu, Finland

Pöyhönen, Juha, University of Joensuu, SF–80101 Joensuu, Finland

Puppe, Clemens, University of Karlsruhe, D–7500 Karlsruhe, F.R. Germany

Reich, Edgar, University of Minnesota, Minneapolis, MN 55455, U.S.A.

Reich, Ludwig, University of Graz, A–8010 Graz, Austria

Reimann, Martin, University of Bern, CH–3000 Bern, Switzerland

Renggli, Heinz, Kent State University, Kent, OH 44242, U.S.A.

Rickman, Seppo, University of Helsinki, SF–00100 Helsinki, Finland

Riihentaus, Juhani, University of Joensuu, SF–80101 Joensuu, Finland

Rintanen, Kirsi, University of Helsinki, SF–00100 Helsinki, Finland

Royden, Halsey, Stanford University, Stanford, CA 94305, U.S.A.

Sakan, Ken–ichi, Osaka City University, Osaka 558, Japan

Schiff, Joel Linn, University of Auckland, Auckland, New Zealand

Schwarz, Binyamin, Israel Institute of Technology, Haifa, Israel

Semmler, Klaus–Dieter, Federal Institute of Technology, CH–1015 Lausanne, Switzerland

Seppälä, Mika, University of Helsinki, SF–00100 Helsinki, Finland

Shea, Daniel, University of Wisconsin, Madison, WI 53706, U.S.A.

Shibata, Keiichi, Okayama University of Sciences, Okayama, Japan

Shiga, Hiroshige, Kyoto University, Kyoto 606, Japan

Siejka, Henryka, Łódź University, 90–238 Łódź, Poland

Sigurdsson, Ragnar, Science Institute, 107 Reykjavik, Iceland

Siu, Yum–tong, Harvard University, Cambridge, MA 02138, U.S.A.

Sorvali, Tuomas, University of Joensuu, SF–80101 Joensuu, Finland

Stegenga, David A., University of Hawaii, Honolulu, HI 96822, U.S.A.

Steinmetz, Norbert, University of Karlsruhe, D–7500 Karlsruhe, F.R. Germany

Stephenson, Kenneth, University of Tennessee, Knoxville, TN 37916, U.S.A.

Storvick, David Arne, University of Minnesota, Minneapolis MN 55455, U.S.A.

Stray, Arne, University of Bergen, Bergen, Norway

Strebel, Kurt, University of Zürich, CH–8001 Zürich, Switzerland

Suita, Nobuyuki, Tokyo Institute of Technology, Tokyo, Japan

Süss, Harry, University of Zürich, CH–8001 Zürich, Switzerland

Syčev, Anatoliĭ, Academy of Sciences, 630090 Novosibirsk 90, USSR
Tammi, Olli, University of Helsinki, SF–00100 Helsinki, Finland
Toppila, Sakari, University of Helsinki, SF–00100 Helsinki, Finland
Tukia, Pekka, University of Helsinki, SF–00100 Helsinki, Finland
Tylli, Hans–Olav, University of Helsinki, SF–00100 Helsinki, Finland
Väisälä, Jussi, University of Helsinki, SF–00100 Helsinki, Finland
Valtonen, Esko, University of Joensuu, SF–80101 Joensuu, Finland
Vamanamurthy, Mavena K., University of Auckland, Auckland, New Zealand
Vuorinen, Matti, University of Helsinki, SF–00100 Helsinki, Finland
Weill, Georges, Polytechnic Institute of New York, Brooklyn, NY 11201, U.S.A.
Weissenborn, Gerd, Technical University of Berlin, D–1000 Berlin, F.R. Germany
Winkler, Jörg, Technical University of Berlin, D–1000 Berlin, F.R. Germany
Wohlhauser, Alfred, Federal Institute of Technology, CH–1015 Lausanne, Switzerland
Wolpert, Scott, University of Maryland, College Park, MD 20742, U.S.A.
Yao, Bi–yun, Hangzhou University, Hangzhou, China
Ben Yattou, M. L., University of Tunis, Tunis, Tunisia
Zajac, Josef, Polish Academy of Sciences, Łódź, Poland

# INEQUALITIES FOR THE EXTREMAL DISTORTION FUNCTION [1]

**G. D. Anderson**
Michigan State University
East Lansing, MI 48824, U.S.A.

**M. K. Vamanamurthy**
University of Auckland
Auckland, New Zealand

**M. Vuorinen**
University of Helsinki, Department of Mathematics
Hallituskatu 15, 00100 Helsinki, Finland

## 1. Introduction

Throughout this paper $n \geq 2$ will denote a positive integer. For $s > 1$ and $n \geq 2$ let $\gamma_n(s)$ denote the conformal capacity of the unbounded Grötzsch ring, whose complementary components are the closed unit ball $\bar{B}^n$ and the ray $[se_1,\infty]$ on the $x_1$-axis. We define the function $M_n(r)$ by $\gamma_n(s) = \sigma_{n-1} M_n(r)^{1-n}$, $s = 1/r$, $0 < r < 1$, where $\sigma_{n-1}$ is the $(n-1)$-dimensional measure of the unit sphere $S^{n-1} = \partial B^n$, and denote $\lambda_n = \lim_{r\to 0}(M_n(r) + \log r)$ [AVV1]. We set $\mu(r) = M_2(r)$ [LV].

Next, for $K > 0$ and $n \geq 2$, we define a homeomorphism $\varphi_K = \varphi_{K,n} : [0,1] \to [0,1]$ by $\varphi_K(0) = 0$, $\varphi_K(1) = 1$, and

$$\varphi_{K,n}(r) = 1/\gamma_n^{-1}(K\gamma_n(1/r)) = M_n^{-1}(\alpha M_n(r)) \tag{1.1}$$

for $0 < r < 1$, where $\alpha \equiv K^{1/(1-n)}$. For $K \geq 1$ and $n \geq 2$ let $QC_K(B^n)$ denote the class of K-quasiconformal mappings of $B^n$ into itself [V, 13.1]. Then we define

$$\varphi^*_{K,n}(r) = \sup\{|f(x)| : f \in QC_K(B^n),\ |x| = r,\ f(0) = 0\} \tag{1.2}$$

for $0 < r < 1$ and $\varphi^*_{K,n}(0) = 0$, $\varphi^*_{K,n}(1) = 1$. We also define, for $K \geq 1$, $n \geq 2$,

---

[1] This paper is in final form and no version of it will be submitted for publication elsewhere.

$$\varphi^*_{1/K,n}(r) = \inf\{|f(x)| : f \in QC_K(B^n), f(0) = 0, fB^n = B^n, |x| = r\}$$

for $r \in [0,1)$, and $\varphi^*_{1/K,n}(1) = 1$. We shall refer to $\varphi^*_{K,n}$ as the *extremal distortion function* in n-space [AVV2].

These definitions are motivated by the quasiconformal Schwarz lemma ([LV, Theorem 3.1, p. 64], [Sh], [MRV, 3.1], [AVV1], [AVV2]), which says that

$$\varphi^*_{K,n}(r) \le \varphi_{K,n}(r) \tag{1.3}$$

for $n \ge 2$, $K \ge 1$, $0 \le r \le 1$, while the opposite inequality holds for $0 < K \le 1$. By [LV, Theorem 3.1, p. 64],

$$\varphi^*_{K,2}(r) = \varphi_{K,2}(r) \tag{1.4}$$

for all $K > 1$ and $r \in [0,1]$.

We are going to show that the multidimensional counterpart of (1.4) is false. Continuing our earlier work in [AVV1] and [AVV2], we shall also obtain certain functional inequalities and estimates for $\varphi^*_{K,n}(r)$, some of which are sharp even for $n = 2$ and $K = 1$.

## 2. Comparison theorems

In view of (1.3) and (1.4) it is natural to ask if $\varphi^*_{K,n} = \varphi_{K,n}$ for $n \ge 3$, $K > 1$. We shall show that this is not the case. For this we need the following lemma.

**2.1. Lemma.** *Let* $0 \le r \le 1$, $n \ge 2$, *and* $K \ge 1$. *Then*

(1) $$1 - \varphi^2_{K,n}(r) \le (1 - r^2)^K,$$

(2) $$1 - \varphi^2_{1/K,n}(r) \ge (1 - r^2)^{1/K}.$$

*Proof.* Since (2) follows from (1) by inversion, we need only prove (1). This is trivial if $r = 1$. Otherwise, from the inequality $\varphi_{K,n}(r) \ge \tanh(K \operatorname{artanh} r)$ [AVV1, (4.5)] it follows that

$$\frac{1 - \varphi^2_{K,n}(r)}{(1 - r^2)^K} \le \frac{4}{((1+r)^K + (1-r)^K)^2}.$$

Since the last denominator increases with r on [0,1], we obtain the desired bound by putting $r = 0$ there. □

**2.2. Theorem.** *For* $n \geq 3$ *and* $K \in (1,\infty)$, *let*

$$E = \{r \in [0,1] : \varphi^*_{K,n}(r) = \varphi_{K,n}(r)\}.$$

*Then* E *is a proper subset of* [0,1] *containing* 0 *and* 1. *Moreover,* 1 *is not a limit point of* E.

*Proof.* Suppose 1 is a limit point of E. Then there exists a sequence $(r_j)$ in $(0,1) \cap E$ such that $r_j \to 1$ as $j \to \infty$. Then $\varphi^*_{K,n}(r_j) = \varphi_{K,n}(r_j)$ for all j, and by [AVV2, Theorem 2.24] we would have

$$\varphi^2_{K,n}(r_j) + \varphi^2_{1/K,n}(r_j') \leq 1 \tag{2.3}$$

for each j, where $r_j' \equiv (1 - r_j^2)^{1/2}$. Then by Lemma 2.1, (2.3), and [AVV1, (4.12)],

$$1 \geq \frac{1 - \varphi^2_{K,n}(r_j)}{(1 - r_j^2)^K} \geq \left(\frac{\varphi_{1/K,n}(r_j')}{r_j'^K}\right)^2 \geq \lambda_n^{2(1-\beta)} r_j'^{2(\beta-K)},$$

for each j, where $\beta \equiv K^{1/(n-1)}$. Since $n \geq 3$ the last expression tends to $\infty$ with j, and we have a contradiction. □

**2.4. Corollary.** *For* $n \geq 3$ *and* $K > 1$, $\varphi_{K^{n-1},n} \neq \varphi_{K,2}$.

*Proof.* Otherwise Theorem 2.2 would be contradicted, since by (1.3) and [AVV1, Theorem 4.9],

$$\varphi_{K^{n-1},n}(r) \geq \varphi^*_{K^{n-1},n}(r) \geq \varphi_{K,2}(r). \quad □$$

**2.5. Remark.** We conjecture that $E = \{0,1\}$ in Theorem 2.2, for each $K > 1$.

We now obtain a pair of nontrivial bounds for $\varphi^*_{K,n}(r)$, strengthening [AVV1, (4.11)]. These bounds are of the correct order as r tends to 0. However, the proof of Theorem 2.2 shows that $\varphi^*_{K,n}(r)$ and $\varphi_{K,n}(r)$ behave quite differently as r tends to 1.

**2.6. Theorem.** *For* $n \geq 2$, $0 < r < 1$, $r' = (1 - r^2)^{1/2}$, $K > 1$, *and* $\alpha = K^{1/(1-n)}$,

$$\frac{2r^\alpha}{(1 + r')^\alpha + (1 - r')^\alpha} < \varphi^*_{K,n}(r) < \lambda_n^{1-\alpha} r^\alpha < 2^{1-\alpha} K r^\alpha.$$

*When* $K = 1$ *these inequalities reduce to identities.*

*Proof.* Since the equality statement is trivial, we may suppose that $K > 1$. First let $n = 2$, $0 < s < 1$. Since $\varphi_{K,2}(s) > s^{1/K}$ [AVV1, (4.11)], [AVV3] it follows from (1.1) that $\mu(s^{1/K}) > \mu(s)/K$. Then by [LV,(2.3), p. 60] and (1.1) we have

$$\frac{2s^{1/2K}}{1 + s^{1/K}} < \varphi_{K,2}\left(\frac{2\sqrt{s}}{1 + s}\right).$$

Now setting $r = 2\sqrt{s}/(1 + s)$ and simplifying, we obtain the lower bound when $n = 2$. If $n \geq 3$ the proof of [AVV1, Theorem 4.9] implies that

$$\varphi^*_{K^{n-1},n}(r) \geq \varphi_{K,2}(r) > \frac{2r^{1/K}}{(1 + r')^{1/K} + (1 - r')^{1/K}},$$

and the lower bound follows when we replace $K$ by $K^{1/(n-1)}$. The upper bounds follow from (1.3) and [AVV3, Theorem 3.4]. □

The hyperbolic distance between $x$ and $y$ in $B^n$ is denoted by $\rho(x,y)$ ([Be]).

**2.7. Theorem.** *For* $n \geq 2$, $1 \leq K < \infty$, $0 \leq r \leq 1$, *we have*

$$\varphi^*_{K,n}(r) \leq \frac{4r^{1/K}}{(1 + r')^{1/K} + (1 - r')^{1/K}},$$

*where* $r' \equiv (1 - r^2)^{1/2}$.

*Proof.* First, suppose $(r/(1 + r'))^{1/K} \geq 1/2$. Then

$$\frac{4r^{1/K}}{(1 + r')^{1/K} + (1 - r')^{1/K}} = \frac{4\left(\frac{r}{1 + r'}\right)^{1/K}}{1 + \left(\frac{1 - r'}{1 + r'}\right)^{1/K}} \geq \frac{2}{1 + \left(\frac{1 - r'}{1 + r'}\right)^{1/K}}$$
$$\geq 1 \geq \varphi^*_{K,n}(r).$$

On the other hand, suppose $(r/(1 + r'))^{1/K} < 1/2$, and let $f\colon B^n \to B^n$ be K-quasiconformal. For $x, y \in B^n$, denote $\rho = \rho(x,y)$, $\rho' = \rho(f(x),f(y))$. The quasi-invariance of the Ferrand invariant $\lambda_{B^n}$ ([L-F], [Vu, 2.23, 3.1(6)]) and [AVV1, (2.3)] yield

$$\gamma_n(\cosh(\rho/2)) \leq K\, \gamma_n(\cosh(\rho'/2)).$$

By [AVV3, (1.11)], [LV, (2.10), p. 61], and well-known identities for hyperbolic functions, this gives

$$\log \coth^2 \tfrac{\rho}{4} \le K\, \mu(\tanh^2 \tfrac{\rho'}{4}) \le K \log(4 \coth^2 \tfrac{\rho'}{4}),$$

or

$$\tanh \tfrac{\rho'}{4} \le 2(\tanh \tfrac{\rho}{4})^{1/K} = 2\left(\frac{r}{1+r'}\right)^{1/K} < 1.$$

Since the function $t \mapsto 2t/(1+t^2)$ is increasing on $[0,1]$, this implies that

$$\tanh \tfrac{\rho'}{2} = \frac{2\tanh \frac{\rho'}{4}}{1+\tanh^2 \frac{\rho'}{4}} \le \frac{4(\tanh \frac{\rho}{4})^{1/K}}{1+4(\tanh \frac{\rho}{4})^{2/K}} = \frac{4t^{1/K}}{1+4t^{2/K}},$$

where $t = (\tanh(\rho/2)/(1+\operatorname{sech}(\rho/2))$. Finally, since $\varphi^*_{K,n}(r) = \sup\{\tanh(\rho'/2) : \tanh(\rho/2) = r, f \in QC_K(B^n)\}$, it follows that

$$\varphi^*_{K,n}(r) \le \frac{4(r/(1+r'))^{1/K}}{1+(r/(1+r'))^{2/K}} = \frac{4r^{1/K}}{(1+r')^{1/K}+4(1-r')^{1/K}}. \quad \square$$

**2.8. Corollary.** *For* $n \ge 2$, $K \ge 1$, $r \in [0,1]$,

$$\varphi^*_{K,n}(r) \le 2^{2-1/K} r^{1/K}.$$

*Proof.* This follows from Theorem 2.7 and the well-known inequality $(1+r')^{1/K} + (1-r')^{1/K} \ge 2^{1/K}$, $0 < r' < 1$. □

**2.9. Remark.** One may give a more geometric, but longer, proof of the lower bound in 2.6 by considering the quasiconformal radial stretching [V, 16.2] composed with two Möbius transformations. This proof shows that for each $n \ge 2$, $K \ge 1$, and $r \in (0,1)$ there exists a K-quasiconformal mapping $f$ of $B^n$ onto $B^n$ such that $f(0) = 0$ and $|f(re_1)| = 2r^\alpha/[(1+r')^\alpha + (1-r')^\alpha]$, where $\alpha = K^{1/(1-n)}$.

## 3. Functional inequalities

For $n = 2$, it may be shown that $\varphi_{K,2}$ satisfies several functional identities and inequalities [AVV4]. Even though $\varphi_{K,n} \ne \varphi^*_{K,n}$ for $n \ge 3$, $K \in (1,\infty)$, we nevertheless obtain several functional inequalities for $\varphi^*_{K,n}$, $n \geq 3$, that are analogues of the functional identities for $\varphi_{K,2}$ (cf. [AVV4]).

**3.1. Theorem.** *For* $n \geq 2$, $K \geq 1$, $r, s \in (0,1)$,

$$\varphi^*_{K,n}(rs) \leq \varphi^*_{K,n}(r)\, \varphi^*_{K,n}(s).$$

*Proof.* Let $f \in QC_K(B^n)$ with $f(0) = 0$. Define $f_r : B^n \to B^n$ by $f_r(x) = f(rx)/\varphi^*_{K,n}(r)$. Then $f_r \in QC_K(B^n)$ and $f_r(0) = 0$. Hence

$$\frac{|f(rx)|}{\varphi^*_{K,n}(r)} = |f_r(x)| \leq \varphi^*_{K,n}(|x|).$$

Now set $|x| = s$ and take the supremum over all $f \in QC_K(B^n)$. □

**3.2. Theorem.** *For* $n \geq 2$, $K \geq 1$, $1/K \leq a^{n-1} \leq 1$, *and* $0 \leq r \leq 1$,

$$\varphi^*_{Ka^{n-1},n}(r^a) \leq \varphi^*_{K,n}(r) \leq (\varphi^*_{Ka^{1-n},n}(r))^{1/a}.$$

*Proof.* For convenience, for each $r \in (0,1)$ we now let $R_n(r)$ denote the ring consisting of the unit ball $B^n$ minus the slit $[0, re_1]$. The mapping $f(x) = x|x|^{a-1}$ takes $R_n(r)$ $a^{1-n}$-quasiconformally onto $R_n(r^a)$. Let $g \in QC_{Ka^{n-1}}(B^n)$ be such that $g(0) = 0$, $|g(r^a e_1)| = \varphi^*_{Ka^{n-1},n}(r^a)$. Then the composition $g \circ f$ is K-quasiconformal, and we conclude that $\varphi^*_{Ka^{n-1},n}(r^a) \leq \varphi^*_{K,n}(r)$.

Next, let $h \in QC_K(B^n)$ be such that $h(0) = 0$, $h(re_1) = \varphi^*_{K,n}(r)e_1$, and let $f$ be as above. Then the composition $f \circ h$ is $Ka^{1-n}$-quasiconformal, so that $(\varphi^*_{K,n}(r))^a \leq \varphi^*_{Ka^{1-n},n}(r)$, as desired. □

## 4. Modulus of continuity

We now recall the notion of modulus of continuity, which we shall use in obtaining further functional inequalities for $\varphi^*_{K,n}(r)$.

Let $(X,d)$, $(Y,p)$ be metric spaces and $f : (X,d) \to (Y,p)$ any function. For each $\delta \in [0,\infty)$, define

$$E_f(\delta) = \{\varepsilon \in [0,\infty) : d(x,y) \le \delta \Rightarrow p(f(x),f(y)) \le \varepsilon\}$$

and $\omega_f(\delta) = \inf E_f(\delta)$. Then $\omega_f$ is called the *modulus of continuity* of f.

If F is a family of functions $f : (X,d) \to (Y,p)$, then we let $\omega_F = \sup\{\omega_f : f \in F\}$. If $\omega_F$ is finite, we say that the family F is $\omega_F$*-equicontinuous*. In particular, if $X = Y = B^n$ with $d = p = \rho =$ the hyperbolic metric and if $F = QC_K(B^n)$, then we denote $\omega_F$ also by $\omega_K$.

**4.1. Remarks.** (1) It is easy to see that

$$\omega_f(\delta) = \sup\{p(f(x)),f(y)) : d(x,y) \le \delta\}$$

and that f is uniformly continuous if and only if $\omega_f$ is continuous at 0.

(2) Let (X,d), (Y,p) be metric spaces and $f : (X,d) \to (Y,p)$ a function. Then it is easy to see that $p(f(x),f(y)) \le \omega_f(d(x,y))$ for all $x, y \in X$.

Next, we shall say that a metric space (X,d) has the property $\mathcal{P}$ if whenever $x, y \in X$ and $0 < r < d(x,y)$ there exists $z \in X$ such that $d(x,z) = r$ and $d(x,z) + d(z,y) = d(x,y)$. In particular, this property holds in the hyperbolic space $(B^n,\rho)$.

**4.2. Lemma.** *Let* (X,d) *be a metric space having property* $\mathcal{P}$, *and let* $f : (X,d) \to (Y,p)$ *be a function. Then*

(1) $\omega_f$ *is an increasing function on* $[0,\infty)$,

(2) $\omega_f(s + t) \le \omega_f(s) + \omega_f(t)$ *for* $s, t \in [0,\infty)$,

(3) $\omega_f(kt) \le k\omega_f(t)$ *for* $t > 0$ *and* k *any positive integer.*

*Proof.* Statement (1) is clear, while (3) follows from (2) by induction. For (2), if $d(x,y) \le s$ then $p(f(x),f(y)) \le \omega_f(s)$, hence $p(f(x),f(y)) \le \omega_f(s) + \omega_f(t)$. Next, if $s < d(x,y) \le s + t$ then by property $\mathcal{P}$ there exists $z \in X$ such that $d(x,z) = s$ and $d(z,y) = d(x,y) - d(x,z) \le t$. Hence $p(f(x),f(z)) \le \omega_f(s)$ and $p(f(z),f(y)) \le \omega_f(t)$, and so by the triangle inequality $p(f(x),f(y)) \le \omega_f(s) + \omega_f(t)$. □

In the next lemma we shall use the fact that the hyperbolic distance $\rho(x,y)$ in $B^n$ satisfies

$$\tanh^2\frac{\rho(x,y)}{2} = \frac{|x-y|^2}{|x-y|^2 + (1-|x|^2)(1-|y|^2)} \tag{4.3}$$

(cf. [Be]).

**4.4. Lemma.** *For* $K \ge 1$, $QC_K(B^n)$ *is* $\omega_K$*-equicontinuous and*

$$\omega_K(t) = 2\,\mathrm{artanh}(\varphi^*_{K,n}(\tanh(t/2))).$$

*Proof.* By (1.2), (4.3), and conformal invariance,

$$\varphi^*_{K,n}(\tanh \tfrac{t}{2}) = \sup \{\tanh \frac{\rho(f(x),f(y))}{2} : \rho(x,y) \le t\} = \tanh \frac{\omega_K(t)}{2} . \ \square$$

## 5. Extremal distortion and hyperbolic distance

In our next result we use the modulus of continuity to obtain a functional inequality for $\varphi^*_{K,n}$ that is sharp for $K = 1$.

**5.1. Theorem.** *For* $n \ge 2$, $K \ge 1$, *and* $r, s \in (0,1)$,

$$\varphi^*_{K,n}\left(\frac{r+s}{1+rs}\right) \le \frac{\varphi^*_{K,n}(r) + \varphi^*_{K,n}(s)}{1 + \varphi^*_{K,n}(r)\varphi^*_{K,n}(s)} .$$

*Proof.* Since $(B^n,\rho)$ has property $\mathscr{P}$, it follows from [Be], and Lemma 4.2(2) that

$$\begin{aligned}\varphi^*_{K,n}\left(\frac{r+s}{1+rs}\right) &= \tanh \left(\frac{1}{2}\, \omega_K\left(2 \operatorname{artanh} \frac{r+s}{1+rs}\right)\right) \\ &= \tanh \left(\frac{1}{2}\, \omega_K(2(\operatorname{artanh} r + \operatorname{artanh} s))\right) \\ &\le \tanh \left(\frac{1}{2}\, \omega_K(2 \operatorname{artanh} r) + \frac{1}{2}\, \omega_K(2 \operatorname{artanh} s)\right) \\ &= \frac{\varphi^*_{K,n}(r) + \varphi^*_{K,n}(s)}{1 + \varphi^*_{K,n}(r)\varphi^*_{K,n}(s)} . \ \square\end{aligned}$$

**5.2. Theorem.** *For* $n \ge 2$, $K \ge 1$, $k$ *a positive integer, and* $r_j \in (0,1)$, $j = 1,\ldots,k$, *let* $A_k(r_1,r_2,\ldots,r_k) = r$, *where*

$$\log \frac{1+r}{1-r} \equiv \sum_{j=1}^{k} \log \frac{1+r_j}{1-r_j} .$$

*Then*

$$\varphi^*_{K,n}(A_k(r_1,r_2,\ldots,r_k)) \le A_k(\varphi^*_{K,n}(r_1),\ldots,\varphi^*_{K,n}(r_k)).$$

*Proof.* For $k = 2$, this is precisely Theorem 5.1. The general case follows by induction. $\square$

**5.3. Corollary.** *For* $n \geq 2$, $K \geq 1$, $r \in (0,1)$, $k$ *a positive integer, let*

$$A_k(r) \equiv \frac{(1+r)^k - (1-r)^k}{(1+r)^k + (1-r)^k}.$$

*Then*

$$\varphi^*_{K,n}(A_k(r)) \leq A_k(\varphi^*_{K,n}(r)).$$

*Proof.* Set $r_1 = r_2 = \ldots = r_k = r$ in Theorem 5.2. Then

$$\log \frac{1 + A_k(r)}{1 - A_k(r)} = k \log \frac{1+r}{1-r}$$

and the result follows. □

**5.4. Theorem.** *Let* $f : B^n \to fB^n \subset B^n$ *be* K-*quasiconformal. Then*

$$\tanh(\rho(f(x),f(y))/4) \leq \varphi^*_{K,n}(\tanh(\rho(x,y)/4))$$

*for all* $x, y \in B^n$. *Equality holds here for* $K = 1$ *and* $fB^n = B^n$.

*Proof.* For brevity we let $\varphi^* = \varphi^*_{K,n}$. Setting $r = s = \tanh(\rho(x,y)/4)$, $\varphi^*(\tanh(\rho(x,y)/4)) = t$ in Theorem 5.1 and employing the half-angle formula for tanh, we get

$$\varphi^*(\tanh(\rho(x,y)/2)) \leq \frac{2t}{1+t^2}. \tag{5.5}$$

Next,

$$\begin{aligned} \tanh(\rho(f(x),f(y))/4) &= \frac{\tanh(\rho(f(x),f(y))/2)}{1 + \sqrt{1 - \tanh^2(\rho(f(x),f(y))/2)}} \\ &\leq \frac{\varphi^*(\tanh(\rho(x,y)/2))}{1 + \sqrt{1 - \varphi^{*2}(\tanh(\rho(x,y)/2))}}. \end{aligned} \tag{5.6}$$

Then (5.5) and (5.6) imply that $\tanh(\rho(f(x),f(y))/4) \leq t$. The validity of the second statement is well-known. □

As a corollary we obtain an improvement of inequality (3.8) in [AVV2], replacing the former constant $2^{1-\alpha}$ by $2^{1-2\alpha}$.

**5.7. Corollary.** *Let* $f \in QC_K(B^n)$ *and let* $\alpha = K^{1/(1-n)}$. *Then*

$$|f(x) - f(y)| \leq 2^{1-2\alpha} \lambda_n^{1-\alpha} \rho(x,y)^\alpha \leq 2^{2-3\alpha} K\rho(x,y)^\alpha \tag{5.8}$$

*and*

$$|f(x) - f(y)| \le 2^{2-3/K}(\min\{K,2\})\rho(x,y)^{1/K} \tag{5.9}$$

*for all* $x, y \in B^n$.

*Proof.* Estimate (5.8) follows from [AVV2, (3.6)], Theorem 5.4, and Theorem 2.6, while (5.9) is a consequence of Theorem 5.4 and [AVV2, Corollary 2.8]. □

**5.10. Remarks.** (1) Let $f: B^n \to B^n$ be K-quasiconformal, and let $x, y \in B^n$. Then the proof of Theorem 5.4 and induction yield

$$\tanh 2^{-m}\rho(f(x),f(y)) \le \varphi^*_{K,n}(\tanh 2^{-m}\rho(x,y))$$

for each positive integer m.

(2) From the above work it follows that $\varphi^*_{K,n}(\tanh \frac{1}{2}\rho(x,y))$ is a metric on $B^n$. For let $x, y, z \in B^n$ and let $r = \tanh \frac{1}{2}\rho(x,y)$, $s = \tanh \frac{1}{2}\rho(y,z)$. Then by Theorem 5.1, since $\rho$ is a metric on $B^n$, we have

$$\varphi^*_{K,n}(\tanh(\rho(x,z)/2)) \le \frac{\varphi^*_{K,n}(r) + \varphi^*_{K,n}(s)}{1 + \varphi^*_{K,n}(r)\varphi^*_{K,n}(s)} \le \varphi^*_{K,n}(r) + \varphi^*_{K,n}(s).$$

(3) In fact, $\operatorname{artanh}(\varphi^*_{K,n}(\tanh(\rho(x,y)/2))$ is a metric on $B^n$. This follows when we apply the function artanh to the first inequality in part (2).

(4) The results in 3.1, 5.1, 5.2, 5.3 are all sharp for $K = 1$ and perhaps new even for $n = 2$.

**5.11. Open problem.** It has been conjectured by A. V. Sychev (see [AVV2] and [S, pp. 89–90]) that $\varphi^*_{K,n}(r) \le 4^{1-\alpha} r^\alpha$, $\alpha = K^{1/(1-n)}$, for $n \ge 3$, $K \ge 1$. We replace this conjecture by the following stronger one (cf. [AVV3, 3.8(5)]):

$$\varphi^*_{K,n}(r) \le \frac{2^{2-\alpha} r^\alpha}{(1-r')^\alpha + (1+r')^\alpha}$$

for $n \ge 3$, where $\alpha = K^{1/(1-n)}$ We observe that this inequality is true for $n = 2$ [AVV4].

## References

[Ah] L. Ahlfors, Möbius Transformations in Several Variables, University of Minnesota Lecture Notes, Minneapolis, 1981.

[AVV1] G. D. Anderson, M. K. Vamanamurthy, and M. Vuorinen, Dimension-free quasiconformal distortion in n-space, Trans. Amer. Math. Soc. 297 (1986), 687–706.

[AVV2] G. D. Anderson, M. K. Vamanamurthy, and M. Vuorinen, Sharp distortion theorems for quasiconformal mappings, Trans. Amer. Math. Soc. 305 (1988), 95–111.

[AVV3] G. D. Anderson, M. K. Vamanamurthy, and M. Vuorinen, Special functions of quasiconformal theory (in preparation).

[AVV4] G. D. Anderson, M. K. Vamanamurthy, and M. Vuorinen, Distortion functions for plane quasiconformal mappings (in preparation).

[Be] A. F. Beardon, The Geometry of Discrete Groups, Graduate Texts in Math., 91, Springer-Verlag, New York - Heidelberg - Berlin, 1983.

[LV] O. Lehto and K. I. Virtanen, Quasiconformal Mappings in the Plane, Die Grundlehren der math. Wissenschaften, 126, Springer-Verlag, New York - Heidelberg - Berlin, 2nd ed., 1973.

[L-F] J. Lelong-Ferrand, Invariants conformes globaux sur les varietes riemanniennes, J. Differential Geom. 8 (1973), 487–510.

[MRV] O. Martio, S. Rickman, and J. Väisälä, Distortion and singularities of quasiregular mappings, Ann. Acad. Sci. Fenn. Ser. A I Math. 465 (1970), 1–13.

[Sh] B. V. Shabat, On the theory of quasiconformal mappings in space, Soviet Math. 1 (1960), 730–733.

[S] A. V. Sychev, Moduli and n-dimensional quasiconformal mappings (Russian), Izdat. "Nauka", Sibirsk. Otdelenie, Novosibirsk, 1983.

[V] J. Väisälä, Lectures on n-Dimensional Quasiconformal Mappings, Lecture Notes in Math., 229, Springer-Verlag, Berlin - Heidelberg - New York, 1971.

[Vu] M. Vuorinen, Conformal invariants and quasiregular mappings, J. Analyse Math. 45 (1985), 69–115.

# COMPLETE KLEIN COVERINGS OF KLEIN SURFACES OF CHARACTERISTIC 0. I [1]

**Cabiria Andreian Cazacu**
Universitatea Bucureşti, Facultatea de Matematică
Str. Academiei 14, Bucureşti 1, România

## 1. Introduction

1.1. Under a *surface* we mean a connected, 2-dimensional manifold with countable basis (orientable or not, compact or non-compact, without border or with border).

A *Klein covering* is a triple (X,T,Y), where X and Y are surfaces and T is an interior transformation X → Y in the sense of Stoïlow, i.e. T is a continuous, open and 0-dimensional or light map.

Klein coverings have as their model Stoïlow's Riemann coverings [13, Ch. V, III, 4]. We choose the name of Klein coverings since the most significant example of such coverings is given by triples (X,T,Y) where X and Y are Klein surfaces and T is a morphism in the sense of Alling and Greenleaf [3].

By direct extension of the Ahlfors' and Sario's complete Riemann coverings [2, Ch. I, 21.A] we say that a Klein covering (X,T,Y) is *complete* if every point of Y has a neighbourhood V such that each component of $T^{-1}(V)$ is compact.

Our paper deals with a special case of the following general

**Problem.** Given a surface Y, determine the surfaces X for which there exist complete Klein coverings (X,T,Y) and – if such a covering does exist – give a construction for it.

The problem has its root in some classical theorems which solve it in certain particular but important cases: Y = the sphere (Stoïlow's Theorem of Riemann covering characterization [13, Ch. V, III, 4] in the constructive form given by M. Heins [8] and by L. V. Ahlfors [11, Ch. II, §4, 2.90]), Y = the projective plane (R. J. Wille's Theorem [14]), Y = the closed disc D (N. L. Alling and N. Greenleaf's Theorem 1.7.2 in [3], where X is a compact surface). In [6] we completed the last result by solving the case Y = D for non-compact surfaces X, and in [7] we treated a case where also Y is

---

[1] This paper is in final form and no version of it will be submitted for publication elsewhere.

non-compact, namely $Y = \mathbb{C}$.

Now we solve the problem in the case of a compact surface of characteristic zero without border, i.e. for $Y$ = the torus $\mathcal{T}$ and $Y$ = the Klein bottle $\mathcal{K}$, the results being formulated in Theorems 1 and 2 in Sections 2 and 3, respectively. The compact bordered cases $Y$ = the annulus and $Y$ = the Möbius strip will be treated in a forthcoming paper.

**1.2.** In addition to complete coverings we use some types of Klein coverings which extend Riemann coverings from Stoïlow's theory. Thus a Klein covering $(X,T,Y)$ is called *total* if $T$ is proper [13, Ch. VI, II] and *partially regular* [13, Ch. VI, IV] if there exists a finite family $\{\lambda\}$ of mutually disjoint Jordan curves on $Y$ such that:

$(\beta_1)$ for each component $Y_k$ of $Y \backslash \{\lambda\}$, the pre-image $T^{-1}(Y_k)$ has a finite number of components $\Delta_{kj}$, every covering $(\Delta_{kj}, T|, Y_k)$ being total, and

$(\beta_2)$ $T^{-1}(\{\lambda\})$ is either compact or empty.

We say that a Klein covering is *polyhedral* [4] if the family $\{\lambda\}$ contains besides a finite number of pairwise disjoint Jordan curves a finite number of Jordan arcs, the intersection of an arc $\lambda$ with another arc or a curve of the family consisting at most of the endpoints of the arc $\lambda$, and if the covering satisfies condition $(\beta_1)$, and a local form of condition $(\beta_2)$ at any point of the curves and arcs of $\{\lambda\}$ except for at most a finite number of points.

**Comments.** Stoïlow's definitions were given in an equivalent form. For the partial regularity, Stoïlow also assumed that every curve of $\{\lambda\}$ is the common boundary of two disjoint components of $Y \backslash \{\lambda\}$. This condition is not necessary and was omitted here. On the contrary, we formulate the definition of a polyhedral covering more restrictively than in [4], since we apply it here for compact unbordered surfaces $Y$ only. In fact the families $\{\lambda\}$, that will be considered in the following, consist of a single Jordan curve or arc $\lambda$, or of two disjoint arcs $\lambda$ and $\lambda^*$, the coverings have no exceptional points with respect to the local $(\beta_2)$, $T^{-1}(\{\lambda\})$ has a finite number of components and each of them covers totally its projection.

**1.3.** The main tools of our proofs are *Ahlfors' addition theorem for characteristics* [1] *and Hurwitz's formula for the ramification of the covering* as well as its generalizations [13, Ch. VI, III and IV], [4], [5], together with the *topological classification of the surfaces* [9, Ch. IV and V].

Let us remark that if $Y$ is of finite characteristic $\rho(Y)$ and the covering $(X,T,Y)$ is total, then Hurwitz's formula implies $\rho(X) \geq n\rho(Y)$, where $n$ is the number of sheets of the covering and $\rho(X)$ is the characteristic of $X$. However, the complete coverings are more general than the total ones, and the universal covering $(\mathbb{C},\pi,\mathcal{T})$ of the torus $\mathcal{T}$ provides an example with $\rho(X) = -1$ and $\rho(Y) = 0$.

**Notations.** We write g or $\mathfrak{g}$ for the genus of X as X is orientable or not (in the orientable case we also write $\mathfrak{g} = 2g$) and $\mu$ for the number of the borders or of the ideal boundary elements of X. Evidently g, $\mathfrak{g}$ or $\mu$ may be infinite. Similarly for g(W), $\mathfrak{g}$(W), $\mu$(W) for an arbitrary surface W. The total ramification order of a covering is denoted by r and the number of sheets of a total covering by n (n will be also the index of the general term of an exhaustion sequence).

1.4. In order to solve our problem we use again the method applied in [6]. The surface Y being now $\mathcal{T}$ or $\mathcal{K}$, the solution X has no border too.

First we treat the case of a compact surface X (2.1 and 3.1 below): We build from copies of Y by means of ramifications a surface Z, which is homeomorphic to X. Let h denote the homeomorphism $X \to Z$ and p be the natural projection $Z \to Y$, which associates to each point in a sheet of Z the same point in Y. Then the interior transformation $T = p \circ h : X \to Y$ gives the required covering (X,T,Y), which is in this case total.

For a non-compact surface X (2.3 and 3.3), we use a canonical exhaustion in Ahlfors-Sario's sense [2, Ch. I, 29.A] by a sequence of polyhedrons $\{X_n\}_{n\in\mathbb{N}}$. For every $n \in \mathbb{N}$, $\mu_n$ denotes the number of components of $\partial X_n$, which are Jordan curves $\Gamma_{nt}$, $t = 1,\ldots,\mu_n$. The genus of $X_n$ is denoted by $g_n$ or by $\mathfrak{g}_n$ as $X_n$ is orientable or non-orientable and the number of borders by $\mu_n$. The connected components $X_{nt}$ of $\overline{X_{n+1}\setminus X_n}$, $n \geq 0$, $t = 1,\ldots,\mu_n$, are polyhedrons. Every $X_{nt}$ is separated from $X_n$ by $\Gamma_{nt}$ and $\mu_{nt}$, $g_{nt}$, $\mathfrak{g}_{nt}$ have similar meanings. In contrast with the case $Y = D$, [6], a polyhedron $X_n$ or $X_{nt}$ cannot cover totally our $Y = \mathcal{T}$ or $\mathcal{K}$ so that we choose a Jordan curve or arc $\lambda$ on Y and denote by $Y^*$ the bordered surface obtained from $Y\setminus\lambda$. Next we construct – with copies of Y, $Y^*$ or another related surface which covers totally Y or $Y^*$ – a surface $Z_n$ homeomorphic to $X_n$ under $h_n : X_n \to Z_n$ such that the covering (int $Z_n$, $p_n$, Y), where $p_n : Z_n \to Y$ is the natural projection, is partially regular or polyhedral with respect to $\lambda$. With this aim we develop (in 2.2 and 3.2) a construction technique based on ramification points which will be applied to $X_0$ and to every $X_{nt}$, $n \geq 0$, $t = 1,\ldots,\mu_n$.

Let us remark that in 2.2 and 3.2, X is a polyhedron, i.e. a compact bordered surface, while in the rest of Sections 2 and 3, X is a surface without border. Since X and Z in 2.2 and 3.2 are bordered, only the coverings

$$(\text{int } Z, p, Y) \text{ and } (\text{int } X, T, Y)$$

are Klein coverings, but $h : X \to Z$, $p : Z \to Y$ and $T = p \circ h$ are well-defined on the borders too. Namely, p and T map each border so that it covers $\lambda$ totally. Therefore we consider also the (non-Klein) coverings (Z, p, Y) and (X, T, Y).

The construction of Z for a non-compact X begins (in 2.3,(i) and 3.3) with $Z_0$. Once the step $n \geq 0$ is complete, in order to obtain $Z_{n+1}$ and $h_{n+1}$, we construct (by means of the results in 2.2 and 3.2 respectively) the surfaces $Z_{nt}$ homeomorphic to $X_{nt}$ and define the homeomorphisms $h_{nt} : X_{nt} \to Z_{nt}$ such that we can glue together $Z_n$ and $Z_{nt}$ and continue $h_n$ by $h_{nt}$. Namely, for any point $a \in \Gamma_{nt}$, the condition

$$p_n \circ h_n(a) = p_{nt} \circ h_{nt}(a) \tag{1}$$

has to be fulfilled, where $p_{nt}$ is the natural projection $Z_{nt} \to Y$, so that the points $h_n(a) \in \partial Z_n$ and $h_{nt}(a) \in \partial Z_{nt}$ can be identified. The condition (1) results by modifying $h_{nt}$ in a neighbourhood of $\Gamma_{nt}$ [6, Remark 3, 1]. Thus we obtain $Z_{n+1}$ and define $h_{n+1}$ by the relations:

$$h_{n+1}|X_n = h_n, \; h_{n+1}|X_{nt} = h_{nt} \quad \text{for } t = 1,\dots,\mu_n. \tag{2}$$

The natural projection $p_{n+1} : Z_{n+1} \to Y$ satisfies similar relations

$$p_{n+1}|Z_n = p_n, \; p_{n+1}|Z_{nt} = p_{nt} \quad \text{for } t = 1,\dots,\mu_n. \tag{3}$$

The surface X is homeomorphic to $Z = \cup_n Z_n$ by the homeomorphism $h : X \to Z$ for which $h|X_n = h_n$, and (Z,p,Y), where $p|Z_n = p_n$, is a complete Klein covering. It remains to define $T = p \circ h$ in order to obtain the required covering (X,T,Y).

A similar device will be applied in 2.3, (ii), 3°, when we use two disjoint Jordan arcs on $Y = \mathcal{T}$. In 2.3, (ii), 1° and 2°, this infinite process is avoided by using the universal covering.

1.5. Concerning some recent research on Riemann coverings, connected with the problem discussed here we quote [10] and [12].

## 2. Complete Klein coverings of the torus

**Theorem 1.** *Let X be an orientable surface without border. If X is compact with $\rho(X) \geq 0$ or non-compact, then it can be represented as a complete Klein covering of the torus $\mathcal{T}$.*

*Proof.*

2.1. *The case of a compact surface* X.

The covering $(X,T,\mathcal{T})$ being total, Hurwitz's formula implies $r = 2g - 2 \geq 0$. From the different possible coverings, we choose the unramified covering with $n = 1$, if $g = 1$, and a covering with $n = 2$ and $2g - 2$ ramification points of order 1, if $g \geq 2$. The surface Z is $\mathcal{T}$ or a surface obtained by gluing together two copies of $\mathcal{T}$ by means of $2g - 2$ ramification points of order 1.

2.2. *Auxiliary coverings* $(X,T,\mathcal{T})$, *where* X *is a compact bordered orientable surface of genus* g *and with* $\mu$ *borders.*

A. $\mu \geq 2$. We choose a meridian line $\lambda$ on the torus $\mathcal{T}$ and form the bordered cylinder $\mathcal{T}^*$ by adding to $\mathcal{T}\backslash\lambda$ two borders $\lambda^+$ and $\lambda^-$ corresponding to $\lambda$. We obtain by means of the natural projection $\mathcal{T}^* \to \mathcal{T}$ a covering with a single sheet over $\mathcal{T}\backslash\lambda$ and two sheets over $\lambda$. Further we construct a surface Z homeomorphic to X consisting of copies of $\mathcal{T}$ and of $\mathcal{T}^*$ glued together successively by ramification points of order 1. In this way (int Z, p, $\mathcal{T}$) is a partially regular covering with respect to $\lambda$, each border of Z being topologically projected onto $\lambda$. The construction depends on g and $\mu$.

A.1. $g = 0$, $\mu = 2m$, $m \geq 1$. Z consists of m copies of $\mathcal{T}^*$, say $\mathcal{T}_j^*$, $j = 1,\ldots,m$, glued together as follows: we choose $\mu - 2$ distinct points $a_j$ and $b_j$, $j = 1,\ldots,m-1$, on $\mathcal{T}\backslash\lambda$ and glue $\mathcal{T}_j$ to $\mathcal{T}_{j+1}$ by means of two ramification points of order 1 projected over $a_j$ and $b_j$, respectively. The covering (int Z, p, $\mathcal{T}\backslash\lambda$) is total with m sheets and $r = 2m - 2 = \mu - 2$. The surface Z has $\mu(Z) = \mu$ and, by Hurwitz's formula, $g(Z) = g$. As it is orientable, Z is homeomorphic to X.

A.2. $g = 0$, $\mu = 2m + 3$, $m \geq 0$. First consider the case $\mu = 3$: the surface Z is formed with two copies of $\mathcal{T}^*$ glued together by a single ramification point projected in $\mathcal{T}\backslash\lambda$, so that Z has three borders: one of them covers $\lambda$ twice while each of the others covers it topologically. In what follows a border is called *simple* or *double* as it covers $\lambda$ once or twice.

If $\mu = 2m + 3 > 3$, we glue together by two ramification points of order 1 the surfaces $Z_1$ with $g(Z_1) = 0$ and $\mu(Z_1) = 3$ constructed before, and $Z_2$ with $g(Z_2) = 0$ and $\mu(Z_2) = 2m$ constructed in A.1. The resulting Z has $\mu(Z) = \mu$ and $g(Z) = 0$ since the total covering (int Z, p, $\mathcal{T}\backslash\lambda$) has $r = \mu - 2$. Hence Z is again homeomorphic to X.

A.3. $g \geq 1$, $\mu \geq 2$. The surface Z can be obtained for instance by gluing together by two ramification points of order 1 a surface $Z_1$ with $g(Z_1) = g$ and

$\mu(Z_1) = 0$ as in 2.1 and a surface $Z_2$ with $g(Z_2) = 0$ and $\mu(Z_2) = \mu$ as in A.1 if $\mu$ is even or in A.2 if $\mu$ is odd. The covering (int Z, p, $\mathcal{T}$) is partially regular and Stoïlow's formula [13, Ch. VI, IV] establishes now the homeomorphism $h : X \to Z$, since $r = 2g + \mu - 2$. If $\mu$ is odd, a double border for the covering cannot be avoided. As another possibility to build up a surface Z of A.3 type we mention the following one with a smaller number of sheets: If $\mu > 2$, we add 2g ramification points of order 1 between two of the sheets of the surface $Z_2$ constructed in A.1 or A.2 as $\mu$ is even or odd, respectively. The surface Z thus obtained from $Z_2$ is homeomorphic to X and (int Z, p, $\mathcal{T}\backslash\lambda$) is total; Z has again a double border for $\mu$ odd. If $\mu = 2$, we glue together $\mathcal{T}$ and $\mathcal{T}^*$ by 2g ramification points of order 1.

**Remark 1.** As indicated in 1.4, the results in 2.2 will be applied in 2.3 to construct $Z_0$ and $Z_{nt}$, $n \geq 0$, for $X_0$ and $X_{nt}$, where $\{X_n\}_{n\in\mathbb{N}}$ forms an exhaustion sequence of the non-compact surface X. However, double borders appear in these constructions as soon as a polyhedron has an odd number of borders. Since the condition (1) can be fulfilled only if $h_n(\Gamma_{nt})$ and $h_{nt}(\Gamma_{nt})$ are both either simple or double we have to modify A.1 and A.3 in order to construct surfaces Z with double borders for $\mu$ even, too. Namely, we replace a copy of $\mathcal{T}^*$ which enters in the composition of the surface Z already constructed in A.1 or A.3 by the unramified double cover of $\mathcal{T}^*$ which has two double borders. The new surface has the same g and $\mu$ but two double borders and will again be denoted by Z, while the corresponding covering (int Z, p, $\mathcal{T}$) has the same r as before and one sheet more over $\mathcal{T}\backslash\lambda$.

B. $\mu = 1$, *hence* $g \geq 1$.

Let now $\lambda$ be a Jordan arc on $\mathcal{T}$ and denote by $\mathcal{T}^*$ the bordered surface obtained from $\mathcal{T}\backslash\lambda$ with a border instead of $\lambda$.

B.1. $g = 1$. We choose $Z = \mathcal{T}^*$, the covering (int Z, p, $\mathcal{T}\backslash\lambda$) being total with $n = 1$.

B.2. $g \geq 2$. We take a copy of $\mathcal{T}$ and one of $\mathcal{T}^*$, and glue them together by $2g-2$ ramification points of order 1 with the projections in $\mathcal{T}\backslash\lambda$. We obtain a surface Z with $\mu(Z) = 1$ and a covering (int Z, p, $\mathcal{T}$) which is polyhedral. Formula (3) in [5] (or (2) in [4]) gives $g(Z) = g$.

In 2.3, (ii), $3^\circ$ below, we need also the following construction:

C.1. $\mu = 2$, $g = 1$.

We take two disjoint Jordan arcs $\lambda$ and $\lambda^*$ on $\mathcal{T}$ and denote by $\mathcal{T}^{**}$ the bordered surface obtained from $\mathcal{T}\backslash(\lambda\cup\lambda^*)$ by adding two borders instead of $\lambda$ and $\lambda^*$, respectively. We choose $Z = \mathcal{T}^{**}$, the covering $(\text{int } Z, p, \mathcal{T}\setminus(\lambda\cup\lambda^*))$ being total with $n = 1$.

C.2. $\mu = 2$, $g \geq 2$. We build up $Z$ with a copy of $\mathcal{T}$ and a copy of $\mathcal{T}^{**}$ glued together by $2g-2$ ramification points of order 1. Here the covering $(\text{int } Z, p, \mathcal{T})$ is polyhedral and, since $\mu(Z) = 2$, the formula (3) in [5] gives $g(Z) = g$.

In all these cases A, B, C we denote by $h$ a homeomorphism $X \to Z$ and form $(X, T = p \circ h, \mathcal{T})$.

**2.3.** *The case of a non-compact surface* X.

Let $\{X_n\}_{n\in\mathbb{N}}$ be a canonical exhaustion of X. In order to construct the surface Z homeomorphic to X and with a natural projection $Z \to \mathcal{T}$ we proceed as in 1.4. Evidently $\mu_{nt} \geq 2$ for any $n \geq 0$ (since $X \setminus \text{int } X_n$ has only non-compact components) and we distinguish two cases as $\mu_0$ can be chosen $\geq 2$ or $\mu_0 = 1$.

(i) *The surface* X *has at least two elements in its ideal boundary.* Then there exists a first index $n_0$ such that $\mu_{n_0} \geq 2$, hence $\mu_n \geq 2$ for $n \geq n_0$. Thus leaving aside, if necessary, a certain number of terms of the exhaustion sequence and renumbering the others we can suppose that $\mu_n \geq 2$ for $n \geq 0$.

By using the results in A and the method given in 1.4 we obtain $Z = \cup_n Z_n$, $h : X \to Z$, $T = h \circ p$ and the required complete covering $(X,T,\mathcal{T})$.

(ii) *The surface* X *has a single boundary element.* Clearly $\mu_n = 1$ and $\Gamma_n = \Gamma_{n1}$ for $n \geq 0$, and there are three possibilities:

1°. $g = 0$. Then X is homeomorphic to $\mathbb{C}$ and the problem has the classical solution given by the universal covering $(\mathbb{C},\pi,\mathcal{T})$.

2°. g *is finite* $\geq 1$. We may suppose $g_0 = g$ and $\mu_0 = 1$, hence $\overline{X_{n+1}\setminus X_n} = X_{n1}$ with $g_{n1} = 0$ and $\mu_{n1} = 2$. However, instead of a successive construction as in 1.4 we solve the problem in only two steps.

First we take a Jordan arc $\lambda$ in a sufficiently small closed disc $\delta$ on $\mathcal{T}$ and, according to B, we construct $Z_0$, $h_0 : X_0 \to Z_0$ and the covering $(\text{int } Z_0, p, \mathcal{T})$, which is polyhedral with respect to $\lambda$.

Next we consider the covering $(\mathbb{C},\pi,\mathcal{T})$ in 1°, choose a component $\Lambda$ of $\pi^{-1}(\lambda)$, cut $\mathbb{C}$ along $\Lambda$ and replace it by a border. Thus we obtain a bordered surface $Z^*$ homeomorphic to $X^* = X \setminus \text{int } X_0$. We extend $\pi|\mathbb{C}\setminus\Lambda$ continuously to $Z^*$ and denote it by $p^*$. Let $h^*$ be a homeomorphism $X^* \to Z^*$ such that $p^* \circ h^*|\Gamma_0 = p_0 \circ h_0|\Gamma_0$ and by identification of the points $h_0(a)$ and $h^*(a)$ for all points $a \in \Gamma_0$

we obtain by gluing of $Z_0$ and $Z^*$ the surface Z with two ramification points over the endpoints of $\lambda$. We define $h : X \to Z$ by $h|X_0 = h_0$, $h|X^* = h^*$ and $p : Z \to Y$ by $p|Z_0 = p_0$, $p|Z^* = p^*$ and obtain the complete covering $(X, T = h \circ p, \mathcal{T})$.

$3^\circ$. g *is infinite.* We take two Jordan arcs $\lambda$ and $\lambda^*$ lying in disjoint small closed discs $\delta$ and $\delta^*$ on $\mathcal{T}$ and a canonical exhaustion sequence $\{X_n\}_{n\in\mathbb{N}}$ of X so that $g_0 \geq 1$ and $g_{n1} \geq 1$ for all $n \geq 0$. Evidently $\mu_0 = 1$ and $\mu_{n1} = 2$ for $n \geq 0$. We construct $Z_0$ by B with respect to $\lambda$. Next we use C and construct $Z_{01}$ with respect to $\lambda$ and $\lambda^*$. With a convenient choice of $h_{01}$ on $\Gamma_0$ we glue together $Z_0$ and $Z_{01}$ with two ramification points of order 1 over the endpoints of $\lambda$. In this way we obtain $Z_1$ homeomorphic to $X_1$, and with the border projected by $p_1$ over $\lambda^*$. The same construction is repeated and gives $Z_2$ homeomorphic to $X_2$, and with the border projected by $p_2$ over $\lambda$, and so on. Finally, this process leads to $Z = \cup_n Z_n$, to the homeomorphism $h : X \to Z$ and to the complete coverings $(Z,p,\mathcal{T})$ and $(X, T = p \circ h, \mathcal{T})$.

### 3. Complete Klein coverings of the Klein bottle

**Theorem 2.** *Let* X *be a surface without border. If* X *is compact with* $\rho(X) \geq 0$ *or non-compact, then there exists a complete Klein covering* $(X,T,\mathcal{K})$.

*Proof.* Since the torus $\mathcal{T}$ is the orientable double of $\mathcal{K}$, Theorem 2 for an orientable surface X follows directly from Theorem 1 and it remains to consider non-orientable surfaces X.

3.1. *The case of a compact non-orientable surface* X *without border.*

Except for the case $g = 2$ when X is homeomorphic to $\mathcal{K}$ and we choose the unramified covering with $n = 1$ given by a homeomorphism $T : X \to \mathcal{K}$, we construct a total covering with $n = 2$ and $r = g - 2$ ramification points of order 1.

If $g = 2k$ it suffices to glue together two copies of $\mathcal{K}$ by $2k - 2$ ramification points.

As an example of a covering for $g = 3$, let us consider X given by a Riemann sphere of radius 1 with three cross-caps centered at $z = 0$, 1 and $-1$ (i.e. obtained by diametral identification of the points of the circles $|z| = r$, $|z-1| = r$ and $|z + 1| = r$, respectively, with $r > 0$ small enough). We choose as a ramification point $z = \infty$ and as a transformation T the mapping derived from $w = z^2$. The surface X will cover with $n = 2$ sheets the Riemann sphere with two cross-caps centered at $w = 0$ and $w = 1$, which is homeomorphic to $\mathcal{K}$. If $\mathfrak{g} = 2k + 1 > 3$, we consider a similar covering and add $2(k-1)$ ramification points of order 1 between the two sheets.

**3.2.** *Auxiliary coverings* $(X,T,\mathcal{K})$, *where* $X$ *is a compact bordered non-orientable or an orientable surface of genus* $\mathfrak{g}$ *or* $g$, *respectively, and with* $\mu$ *borders.* Here we have to consider both the non-orientable and orientable case, since we apply these results to the polyhedrons $X_0$ and $X_{nt}$, $n \geq 0$, of an exhaustion sequence $\{X_n\}_{n \in \mathbb{N}}$ of $X$ and, even if $X$ is non-orientable and if we take $X_0$ non-orientable, the presence of some orientable $X_{nt}$ cannot be avoided.

We cut $\mathcal{K}$ along a unilateral Jordan curve $\lambda$ and obtain a bordered surface $\mathcal{K}^*$ homeomorphic to the Möbius strip, and a natural projection $\mathcal{K}^* \to \mathcal{K}$, the single border of $\mathcal{K}^*$ covering $\lambda$ twice. The double unramified covering $(\mathcal{T},\pi,\mathcal{K})$ determines a double covering $(\mathcal{T}_*, \pi_*, \mathcal{K}^*)$ where $\mathcal{T}_*$ is the bordered surface corresponding to $\mathcal{T} \backslash \pi^{-1}(\lambda)$, $\pi^{-1}(\lambda)$ is regarded as a parallel line on $\mathcal{T}$, and $\mathcal{T}_*$ is homeomorphic to an annulus each of its two borders covering the border of $\mathcal{K}^*$ once, hence $\lambda$ twice.

In order to construct a partially regular covering $(\text{int } X, T, \mathcal{K})$ with respect to $\lambda$, we build $Z$ from copies of $\mathcal{K}$, $\mathcal{K}^*$ or $\mathcal{T}_*$. Since Stoïlow's formula [13, Ch. VI, IV] implies that $r = \mathfrak{g} + \mu - 2$, the construction depends on the evenness of $\mathfrak{g} + \mu$. We distinguish several cases:

A. $X$ *non-orientable,* $\mathfrak{g} + \mu$ *even.*

A.1. $\mathfrak{g} = 2k \geq 2$, $\mu = 2m \geq 2$. We construct $Z$ with two copies of $\mathcal{K}^*$ and $m-1$ copies of $\mathcal{T}_*$, which with the corresponding projections form the $n = \mu$ sheets of the covering $(Z,p,\mathcal{K})$ over $\mathcal{K} \backslash \lambda$. The copies are glued together successively by two ramification points of order 1 and apart from these $2m$ ramifications we add $2k - 2$ other ramification points of order 1 between two of the copies, so that $Z$ has the genus $\mathfrak{g}$.

In all of the following cases the construction is similar so that we indicate only the copies used to form $Z$.

A.2. $\mathfrak{g} = 2k + 1 \geq 1$, $\mu = 2m + 1 \geq 1$: a copy of $\mathcal{K}^*$ and $m$ copies of $\mathcal{T}_*$, $n = \mu$.

A.2'. $\mathfrak{g} = 2k + 1 \geq 1$, $\mu = 1$: a copy of $\mathcal{K}^*$, if $\mathfrak{g} = 1$, $n = 1$, and a copy of $\mathcal{K}^*$ and one of $\mathcal{K}$, if $\mathfrak{g} \geq 3$, $n = 2$.

B. X *non–orientable,* $\mathfrak{g} + \mu$ *odd.*

Two sheets of the covering $(Z,p,\mathcal{K})$ are glued together by an odd number of ramification points of order 1, while each other pair of sheets is glued by an even number of such points. Thus one of the borders of Z is double, i.e. it covers $\partial\mathcal{K}^*$ twice.

B.1. $\mathfrak{g} = 2k + 1 \geq 1$, $\mu = 2m \geq 2$: a copy of $\mathcal{K}^*$ and m of $\mathcal{T}_*$, $n = \mu + 1$.

B.2. $\mathfrak{g} = 2k \geq 2$, $\mu = 2m + 1 \geq 1$: two copies of $\mathcal{K}^*$ and m copies of $\mathcal{T}_*$, $n = \mu + 1$.

A'. X *orientable,* $\mathfrak{g} = 2g$, $\mu = 2m \geq 2$: m copies of $\mathcal{T}_*$, $n = \mu$.

B'. X *orientable,* $\mathfrak{g} = 2g$, $\mu = 2m + 1 \geq 3$: $m + 1$ copies of $\mathcal{T}_*$, $n = \mu + 1$; one of the borders of Z is double.

**Remark 2.** For the same reason as in 2.3, Remark 1, we need constructions with double borders in the case $\mathfrak{g} + \mu$ even, too. To this end, consider the double unramified covering $(\mathcal{T},\tau,\mathcal{T})$ where any parallel line of the covering surface $\mathcal{T}$ covers twice its projection which is a parallel line of the base surface $\mathcal{T}$. By restriction we derive the double covering $(\mathcal{T}_*, \tau_*, \mathcal{T}_*)$. We obtain thus a covering $(\mathcal{T}_*, \pi_* \circ \tau_*, \mathcal{K}^*)$ with four sheets and two double borders over the border of $\mathcal{K}^*$. Now we can replace in each of the cases A.1 with $m > 1$, A.2 and A' a double sheet of the covering $(Z,p,\mathcal{K})$, which consists of a $\mathcal{T}_*$ covering $\mathcal{K}^*$ twice, by the four sheets of the covering $(\mathcal{T}_*, \pi_* \circ \tau_*, \mathcal{K}^*)$. The new covering surface, denoted again by Z, is homeomorphic to the initial one but the covering has now two double borders, and two sheets more. It remains to discuss the case A.1 with $m = 1$: we take a copy of $\mathcal{K}$ and a copy of $\mathcal{T}_*$ covering $\mathcal{K}^*$ with four sheets under $\pi_* \circ \tau_*$ and glue them by $\mathfrak{g}$ ramification points of order 1.

**3.3.** *The case of a non–compact, non–orientable surface* X *without border.*

The general method described in 1.4 applies again. We can suppose that $X_0$ is non-orientable and we construct the surface $Z_0$ using the auxiliary results in 3.2, A or B. Further, to obtain $Z_1$ we construct $Z_{0t}$ corresponding to $X_{0t}$ using again the results in 3.2, A, B, A' or B', or in Remark 2 if $h_0(\Gamma_{0t})$ is a double border of $Z_0$, the homeomorphism $h_{0t}$ being modified in a neighbourhood of $\Gamma_{0t}$ so that (1) will be fulfilled. The surface $Z_1$ is obtained by gluing together $Z_0$ and the surfaces $Z_{0t}$, $t = 1,\dots,\mu_0$ and the procedure continues leading finally to Z, h and the complete covering $(X,T,\mathcal{K})$.

## References


[1] Ahlfors, L., Zur Theorie der Überlagerungsflächen. Acta Math. 65 (1935), 157–194.

[2] Ahlfors, L. V. and Sario, L., Riemann surfaces. Princeton University Press, 1960.

[3] Alling, N. L. and Greenleaf, N., Foundations of the theory of Klein surfaces. Lecture Notes in Math., 219, Springer-Verlag, 1971.

[4] Andreian Cazacu, C., Über eine Formel von S. Stoïlow. Rev. Roumaine Math. Pures Appl. 5, 1 (1960), 59–74.

[5] Andreian Cazacu, C., Ramification of Klein coverings. Ann. Acad. Sci. Fenn. Ser. A I Math. 10 (1985), 47–56.

[6] Andreian Cazacu, C., Morphisms of Klein surfaces and Stoïlow's topological theory of analytic functions. In: Seminar on Deformations Łódź-Lublin 1985/87, Part III: Real analytic geometry, D. Reidel Publ. Comp., Dordrecht, 235–246.

[7] Andreian Cazacu, C., Interior transformations between Klein surfaces. Rev. Roumaine Math. Pures Appl. 33, 1–2 (1988), 21–26.

[8] Heins, M., Interior mappings of an orientable surface into $S^2$. Proc. Amer. Math. Soc. 2 (1951), 951–952.

[9] Kerékjártó, B., Vorlesungen über Topologie I. Die Grundlehren der mathematischen Wissenschaften, 8, Springer-Verlag, 1923.

[10] Martens, H. H., Remarks on de Francis' theorem. In: Complex Analysis, Lecture Notes in Math., 1013, Springer-Verlag, 1983, 160–163;
Mappings of closed Riemann surfaces. Lecture at XIII R. Nevanlinna - Colloquium, Joensuu, August 10–13, 1987.

[11] Nevanlinna, R., Uniformisierung. Springer-Verlag, 1953.

[12] Srebro, U. and Wajnrub, B., Covering theorems for open surfaces. In: Geometry and Topology, Lecture Notes in Pure and Appl. Math., 105, M. Dekker, 1987, 265–275.

[13] Stoïlow, S., Leçons sur les principes topologiques de la théorie des fonctions analytiques. Gauthier-Villars, 1938, II ed. 1956, ed. russe 1964.

[14] Wille, R. J., Sur la transformation intérieure d'une surface non orientable dans le plan projectif, Indagationes Math. 56 (1953), 63–65.

# QUASICONFORMAL EXTENSION OF PLANE QUASIMÖBIUS EMBEDDINGS [1]

**Vladislav V. Aseev**
Novosibirsk Electrical Engineering Institute
Novosibirsk, USSR

**Anatolii V. Syčev**
Institute of Mathematics
Siberian Division of the USSR Academy of Sciences
Novosibirsk, USSR

## 1. Definitions

Let $\Sigma$ be a subset of $\bar{\mathbb{C}}$ and let $\omega : [0,+\infty) \to [0,+\infty)$ be a homeomorphism. A topological embedding $f : \Sigma \to \bar{\mathbb{C}}$ is said to be $\omega$-quasimöbius ($\omega$-QM) if the estimate

$$\frac{s(f(z_1),f(z_2)) \cdot s(f(z_3),f(z_4))}{s(f(z_1),f(z_3)) \cdot s(f(z_2),f(z_4))} \leq \omega\left(\frac{s(z_1,z_2) \cdot s(z_3,z_4)}{s(z_1,z_3) \cdot s(z_2,z_4)}\right)$$

holds for all distinct points $z_1,z_2,z_3,z_4$ in $\Sigma$. Here $s(\cdot,\cdot)$ denotes the chordal distance in $\bar{\mathbb{C}}$. The basic properties of QM-embeddings may be found in [1].

For a pair of continua E, F in a domain $D \subset \bar{\mathbb{C}}$ we let $M(E,F;D)$ denote the modulus of the family of all arcs joining E to F in D. If $D = \bar{\mathbb{C}}$, then we denote $M(E,F;D) = M(E,F)$. For the properties of modulii and their relation to quasiconformality see [2].

Let $\Sigma \subset \bar{\mathbb{C}}$ and $\omega : [0,+\infty) \to [0,+\infty)$ be a homeomorphism. A topological embedding $f : \Sigma \to \bar{\mathbb{C}}$ is said to be $\omega$-BMD embedding if the inequality

$$\omega^{-1}(M(E,F)) \leq M(fE,fF) \leq \omega(M(E,F))$$

holds for all continua $E, F \subset \Sigma$. The concept of BMD-embedding has been introduced by Professor P. P. Belinskiĭ (see [3]).

---

[1] This paper is in final form and no version of it will be submitted for publication elsewhere.

**1.1. Theorem** [4, Th. 2]. *Every $\omega$-quasimöbius embedding* $f : \Sigma \to \bar{\mathbb{C}}$ *is also* $\tilde{\omega}$-BMD *with a homeomorphism* $\tilde{\omega} : [0,+\infty) \to [0,+\infty)$ *depending on* $\omega$ *only.*

The proof of the theorem may be found in [5].

## 2. Quasiconformal extension

A Jordan domain $D \subset \bar{\mathbb{C}}$ is said to be of the class EXT iff for any homeomorphism $\omega : [0,+\infty) \to [0,+\infty)$ there exists $K(\omega) < \infty$ such that every sense-preserving $\omega$-quasimöbius embedding $f : D \to \bar{\mathbb{C}}$ has a $K(\omega)$-quasiconformal extension over the whole plane $\bar{\mathbb{C}}$.

**2.1.** If $D$ is a quasidisk (that is a domain bounded by a quasicircle), then $D \in$ EXT. The fact has been noted in [6] and in [7, Corollary, p. 4] in terms of BMD-embeddings.

**2.2.** The domain $D = \{z : |\mathrm{Im}\, z| < 1, \mathrm{Re}\, z > 0\}$ is of the class EXT. It follows from [7, Th. 3.1, p. 4] and the connection between the classes BMD and QM. So the class EXT does not reduce to the class of quasidisks and contains domains with degenerated angles on the boundary. On the other hand there exists an example of a Jordan domain which is not of the class EXT. The construction may be found in the next section.

## 3. The example

For $0 < \alpha < 1$ consider the closed domain $D_\alpha = \Gamma_0 \cup \Gamma_1 \cup \Gamma_2$ where $\Gamma_1 = \{0 \le \mathrm{Re}\, z \le \alpha/6,\ 0 \le \mathrm{Im}\, z \le 2\}$, $\Gamma_2 = \{11\alpha/6 \le \mathrm{Re}\, z \le 2\alpha,\ 1 \le \mathrm{Im}\, z \le 2\}$, $\Gamma_0 = \{\mathrm{Im}\, z \ge 2,\ 5\alpha/6 \le |z - \alpha - 2i| \le \alpha\}$. The mapping $w = \psi_\alpha(z)$ of $D_\alpha$ is defined by

$$\psi_\alpha(z) = \begin{cases} -x + iy & \text{for } z \in \Gamma_1 ; \\ 4\alpha - x + iy & \text{for } z \in \Gamma_2 ; \\ \alpha + 2i + \dfrac{z - \alpha - 2i}{|z - \alpha - 2i|} \cdot (2\alpha - |z - \alpha - 2i|) & \text{for } z \in \Gamma_0 , \end{cases}$$

where $z = x + iy$. The estimate

$$|dz| \le |dw| \le 7|dz|/5 \tag{1}$$

holds for the distortion of infinitesimal distance elements under the diffeomorphism $\psi_\alpha$. It is also easy to verify that the estimate

$$|a-b| \leq \rho_R(a,b) \leq (\pi+1)|a-b| \leq 5|a-b| \tag{2}$$

holds for all $a, b \in R$ where $R$ is either a circular ring or one's quarter extended by a rectangle. Here $\rho_R(\cdot,\cdot)$ denotes the inner distance in $R$.

We first show that the double inequality

$$|a-b|/7 \leq |\psi_\alpha(a) - \psi_\alpha(b)| \leq 7|a-b| \tag{3}$$

is valid for all $a, b \in D_\alpha$. Let $R$ be either $\Gamma_0$ or $D_\alpha \cap \{\operatorname{Re} z \leq \alpha\}$ or $D_\alpha \cap \{\operatorname{Re} z \geq \alpha\}$. Since (2) holds for both $R$ and $\psi_\alpha(R)$ it follows from (2) and (1) that

$$\begin{aligned} |a-b|/7 &\leq \rho_R(a,b)/7 \leq \rho_{\psi_\alpha(R)}(\psi_\alpha(a),\psi_\alpha(b))/7 \\ &\leq |\psi_\alpha(a) - \psi_\alpha(b)|\cdot 5/7 \leq |\psi_\alpha(a) - \psi_\alpha(b)| \\ &\leq \rho_{\psi_\alpha(R)}(\psi_\alpha(a),\psi_\alpha(b)) \leq 7\rho_R(a,b)/5 \leq 7|a-b|, \end{aligned}$$

so that (3) holds whenever $a, b \in R$. If this is not the case, then $|a-b| > 5\alpha/12$, $|\psi_\alpha(a) - \psi_\alpha(b)| > \alpha/2$ and

$$\begin{aligned} |a-b|/7 &\leq (|\psi_\alpha(a) - \psi_\alpha(b)| + |\psi_\alpha(a) - a| + |\psi_\alpha(b) - b|)/7 \\ &\leq (2\alpha/3 + |\psi_\alpha(a) - \psi_\alpha(b)|)/7 \leq |\psi_\alpha(a) - \psi_\alpha(b)| \\ &\leq |a-b| + |\psi_\alpha(a) - a| + |\psi_\alpha(b) - b| \\ &\leq 13|a-b|/5 \leq 7|a-b| \end{aligned}$$

so (3) holds in that situation as well. Thus the following proposition has been asserted.

**3.1.** Let $\tau(x+iy) = -x+iy$. Then the sense-preserving diffeomorphism $\varphi_\alpha = \tau \circ \psi_\alpha$ is a bilipschitz one with Lipschitz constant $k \leq 7$ and $\varphi_\alpha \equiv \mathrm{id}$ in $\Gamma_1$.

For $n = 1,2,\dots$ let $G_n = \nu_n(D_{1/2n})$ where $\nu_n(z) = z + 4(n-1)$. Consider the closed Jordan domain

$$G = (\bigcup_{n=1}^{\infty} G_n) \cup \{z : \operatorname{Im} z \leq 0\}$$

and the mapping

$$\varphi(z) = \begin{cases} z & \text{if } \operatorname{Im} z \le 0; \\ \nu_n \circ \varphi_{1/2n} \circ \nu_n^{-1}(z) & \text{if } z \in G_n\ ;\ n = 1,2,\dots . \end{cases}$$

So we have a sense-preserving topological embedding $\varphi : G \to \bar{\mathbb{C}}$ such that $\varphi \equiv \mathrm{id}$ in $\{z : \operatorname{Im} z \le 1\}$.

**3.2.** The embedding $\varphi : G \to \bar{\mathbb{C}}$ is a bilipschitz one with Lipschitz constant $k \le 7$. Thus the inequality

$$|a-b|/7 \le |\varphi(a) - \varphi(b)| \le 7|a-b| \tag{4}$$

holds for all $a, b \in G$. In particular, $\varphi$ is an $\omega$-quasimöbius embedding with $\omega(t) \le 2401 \cdot t$.

We shall prove (4) in each of the four possible cases.

CASE 1. If $a, b \in \{z : \operatorname{Im} z \le 1\}$, then $|\varphi(a) - \varphi(b)| = |a-b|$ and (4) is obvious.

CASE 2. If $a, b \in G_n$ for some $n$, then (4) follows immediately from 3.1 and the isometrical property of $\nu_n$.

CASE 3. If $a \in G_n$, $b \in G_m$ where $n \ne m$, then $3|m-n| \le |a-b| \le 5|m-n|$ and $3|m-n| \le |\varphi(a) - \varphi(b)| \le 5|m-n|$, so that

$$|a-b|/7 \le 3|m-n| \le |\varphi(a) - \varphi(b)| \le 6|m-n| \le 7|m-n|$$

as desired.

CASE 4. If $a \in G_n \cap \{\operatorname{Im} z \ge 1\}$, $b \in \{\operatorname{Im} z \le 0\}$, then $|a-b| \ge 1$, $|\varphi(a) - \varphi(b)| \ge 1$ and $|\varphi(a) - a| \le 3$. So we have the desired estimates:

$$|\varphi(a) - \varphi(b)| \le |\varphi(a) - a| + |a-b| \le 4|a-b| \le 7|a-b|$$

and

$$|a-b| \le |\varphi(a) - a| + |\varphi(b) - \varphi(a)| \le 3 + |\varphi(a) - \varphi(b)| \le 7|\varphi(a) - \varphi(b)|.$$

To complete the construction we will prove that the embedding $\varphi : G \to \bar{\mathbb{C}}$ has no quasiconformal extension over the whole plane $\bar{\mathbb{C}}$. Let $D = \bar{\mathbb{C}} \setminus G$, $D^* = \bar{\mathbb{C}} \setminus \varphi(G)$ and

$$E_n = \nu_n(\{z : \operatorname{Re} z = 0,\ 1 \le \operatorname{Im} z \le 3/2\});$$
$$F_n = \nu_n(\{z : \operatorname{Re} z = 1/n,\ 1 \le \operatorname{Im} z \le 3/2\}),$$

for $n = 1,2,\dots$ . Since any arc joining $E_n$ to $F_n$ is of euclidean diameter $\ge 1$, the metric

$$\rho(z) = \begin{cases} 1 & \text{for } z \in \nu_n(\{z : |\mathrm{Re}\, z| \le 2, 0 \le \mathrm{Im}\, z \le 3\}); \\ 0 & \text{for any other point}, \end{cases}$$

is admissible for $M(E_n,F_n;D)$. Hence $M(E_n,F_n;D) \le 12$ for all n. As regard to $E_n^* = \varphi(E_n)$, $F_n^* = \varphi(F_n)$ and $D_n^* = \nu_n(\{z : -1/n < \mathrm{Re}\, z < 0, 1 < \mathrm{Im}\, z < 3/2\})$, the estimate $M(E_n^*,F_n^*;D^*) \ge M(E_n^*,F_n^*;D_n^*) = n/2$ holds for all n. Since $M(E_n^*,F_n^*;D^*)/M(E_n,F_n;D)$ tends to $\infty$ as $n \to \infty$, it follows from [8, Th. 6.1, p. 83] that the sense-preserving homeomorphism $\varphi : \partial D \to \partial D^*$ cannot be extended quasiconformally over D.

## References


[1] Väisälä, J., Quasimöbius maps. J. Analyse Math. 44 (1985), 218–234.

[2] Väisälä, J., Lectures on n-dimensional quasiconformal mappings. Lecture Notes in Math., 229, Springer-Verlag, Berlin-Heidelberg-New York, 1971.

[3] Варисов, А. К., О продолжении пространственных квазиконформных отображений. Доклады АН СССР 234 (1977), 740–742.

[4] Aseev, V. V., Quasisymmetric embeddings and the moduli's distortion. International Conference on Complex Analysis and Applications (Varna, 1985). Summary. p.41.

[5] Асеев, В. В., Квазисимметрические вложения и отображения, ограниченно искажающие модули. Новосибирск, 1984, Деп. ВИНИТИ № 7190-84, 30 с.

[6] Gehring, F. W. and Martio, O., Quasiextremal distance domains and extension of quasiconformal mappings. J. Analyse Math. 45 (1985), 181–206.

[7] Асеев, В. В. and Журавлёв, И. В., О квазиконформном продолжении плоских гомеоморфизмов. Известия вузов. Математика, 1986, № 9(292), с. 3–6.

[8] Lehto, O. and Virtanen, K., Quasikonforme Abbildungen. Springer-Verlag, Berlin-Heidelberg-New York, 1965.

# SOME INTEGRAL CONDITIONS INVOLVING THE SPHERICAL DERIVATIVE [1]

**Rauno Aulaskari**
Department of Mathematics, University of Joensuu
SF-80101 Joensuu, Finland

**Peter Lappan**
Department of Mathematics, Michigan State University
East Lansing, Michigan 48824, U.S.A.

## 1. Notation and introduction

Let D denote the unit disk $\{z : |z| < 1\}$ in the complex plane, and let W denote the Riemann sphere. For two points $z_1$ and $z_2$ in D, let $\delta(z_1,z_2)$ denote the hyperbolic distance in D between $z_1$ and $z_2$. For $r > 0$ and $\lambda \in D$, let $D(\lambda,r) = \{z \in D : \delta(z,\lambda) < r\}$ denote the hyperbolic disk with hyperbolic center at $\lambda$ and hyperbolic radius r. For two points $w_1$ and $w_2$ in W, let $\chi(w_1,w_2)$ denote the chordal distance between $w_1$ and $w_2$. For a function f meromorphic in D and for $z \in D$, the *spherical derivative* $f^{\#}(z)$ is defined by $f^{\#}(z) = |f'(z)|/(1 + |f(z)|^2)$. A family of functions $\mathcal{F} \subset \{f : f$ meromorphic in $D\}$ is called a *normal family* if each sequence $\{f_n\}$ of functions in $\mathcal{F}$ contains a subsequence $\{f_{n_k}\}$ which converges uniformly on each compact subset of D to a function g meromorphic in D. (The "uniform convergence" here is with respect to the chordal metric, and the function g is permitted to be identically $\infty$.) The well known criterion due to Marty [4] gives a characterization of normal families in terms of the spherical derivative as follows: *a family* $\mathcal{F}$ *of functions meromorphic in* D *is a normal family if and only if, for each compact subset* K *of* D, *there exists a finite constant* $M_K$ *such that* $f^{\#}(z) \le M_K$ *whenever both* $f \in \mathcal{F}$ *and* $z \in K$. Recently, Royden [5] gave a sufficient condition for a family to be a normal family which is, in some sense, a generalization to the Marty result.

A function f meromorphic in D is called a *normal function* if the family $\mathcal{F} = \{f(\sigma(z)) : \sigma$ is a conformal mapping $D \to D\}$ is a normal family. Lehto and Virtanen [2] modified the Marty criterion for a normal family to give a criterion for a function to be

---

[1] This paper is in final form and no version of it will be submitted for publication elsewhere.

a normal function as follows: *a function* f *meromorphic in* D *is a normal function if and only if there exists a finite constant* $C_f$ *(depending on* f*) such that* $(1-|z|^2)f^{\#}(z) \le C_f$ *for each* $z \in D$. In [1], we proved the following integral version of the criterion of Lehto and Virtanen.

**Theorem AL.** *A function* f *meromorphic in* D *is a normal function if and only if, for each* $r > 0$ *and each* $p > 2$, *we have*

$$\sup\left\{\iint_{D(\lambda,r)} (1-|z|^2)^{p-2}(f^{\#}(z))^p \, dA(z) : \lambda \in D\right\} < \infty,$$

*where* dA(z) *denotes the Euclidean element of area.*

In section 2, we state and prove a number of criteria for a family of meromorphic functions to be a normal family, where the criteria involve both area integrals and contour integrals of powers of the spherical derivative.

In section 3, we improve Theorem AL by giving a number of criteria for a function to be a normal function. These criteria involve both area integrals and contour integrals of powers of the spherical derivative weighted by appropriate powers of $(1-|z|^2)$.

Finally, in section 4, we investigate the class $\mathcal{N}_0$, which is the subclass of normal functions f which satisfy the condition $\lim_{|z|\to 1}(1-|z|^2)f^{\#}(z) = 0$. We give some criteria for a function f to be in the class $\mathcal{N}_0$ involving both area and contour integrals of powers of the spherical derivative weighted by appropriate powers of $(1-|z|^2)$.

## 2. Integral criteria for normal families

In this section we prove the following criteria for a given family to be a normal family.

**Theorem 1.** *Let* $\mathcal{F}$ *be a family of functions meromorphic in* D. *The following statements are equivalent.*

(1) *The family* $\mathcal{F}$ *is a normal family.*

(2) *For each* $p > 2$ *and each* R, $0 < R < 1$, *there exists a finite constant* $C_{p,R}$ *such that, for each circle* $\gamma_r = \{z : |z| = r\}$, $0 < r \le R$,

$$\int_{\gamma_r} (f^{\#}(z))^p \, |dz| \le C_{p,R} \quad \textit{for each } f \in \mathcal{F}.$$

(3) *For each* $p > 2$ *and each* $R, 0 < R < 1$, *there exists a finite constant* $K_{p,R}$ *such that*

$$\iint_{|z|\leq R} (f^{\#}(z))^p \, dA(z) \leq K_{p,R} \quad \textit{for each } f \in \mathcal{F}.$$

(4) *There exists a number* $p_0 > 2$ *such that, for each* $R, 0 < R < 1$, *there exists a finite constant* $C_R'$ *(depending on* $R$*) such that, for each circle* $\gamma_r = \{z : |z| = r\}$, $0 < r \leq R$,

$$\int_{\gamma_r} (f^{\#}(z))^{p_0} \, |dz| \leq C_R' \quad \textit{for each } f \in \mathcal{F}.$$

(5) *There exists a number* $p_0 > 2$ *such that, for each* $R, 0 < R < 1$, *there exists a finite constant* $K_R'$ *(depending on* $R$*) such that*

$$\iint_{|z|\leq R} (f^{\#}(z))^{p_0} \, dA(z) \leq K_R' \quad \textit{for each } f \in \mathcal{F}.$$

Before proving Theorem 1, we give a lemma.

**Lemma 1.** *Let* $\mathcal{F}$ *be a normal family of functions meromorphic in* $D$. *Then for each* $p > 0$ *and each* $R$, $0 < R < 1$, *there exists a constant* $C_{p,R}$ *such that, for each circle* $\gamma_r = \{z : |z| = r\}$, $0 < r \leq R$,

$$\int_{\gamma_r} (f^{\#}(z))^p \, |dz| \leq C_{p,R} \quad \textit{for each } f \in \mathcal{F}.$$

*Proof.* Fix $p > 0$ and $R$, $0 < R < 1$. Since $\mathcal{F}$ is a normal family, by Marty's criterion there exists a constant $K_R$ such that $f^{\#}(z) \leq K_R$ for each $f \in \mathcal{F}$ and each $z$, $|z| \leq R$. Setting $C_{p,R} = 2\pi(K_R)^p$ gives the result.

Note that, in Lemma 1 we need only $p > 0$, while in Theorem 1 we have that $p > 2$ throughout. This tells us that, for $p > 0$, (1) implies (4), but (4) does not imply (1). We will give an example of this after the proof of Theorem 1.

*Proof of Theorem* 1. As noted above, Lemma 1 shows that (1) implies (2). Clearly (2) implies (4) and (3) implies (5). Thus, we need only show that (2) implies (3), (4) implies (5), and (5) implies (1).

We can use the same proof to show that (2) implies (3) and (4) implies (5). Let $p > 2$ and $R$, $0 < R < 1$, be fixed. Then, by taking the area integral in polar coordinates, we get

$$\iint_{|z|\le R} (f^{\#}(z))^p \, dA(z) = \int_0^R \left[ \int_{\gamma_r} (f^{\#}(z))^p \, |dz| \right] dr$$

so that we can set $K_{p,R} = C_{p,R}$ (and $K'_R = C'_R$) and use the obvious estimates to obtain that (2) implies (3) (and (4) implies (5) using $p = p_0$).

To complete the proof, we show that (5) implies (1) by proving the contrapositive. Let $p_0 > 2$ be given and suppose that the family $\mathcal{F}$ is not a normal family. By a result of L. Zalcman [7, Lemma, page 814] there exists a sequence of functions $\{f_n\}$ in $\mathcal{F}$, a sequence $\{z_n\}$ of points in D, a point $z_0 \in D$, and a sequence $\{q_n\}$ of positive numbers such that $z_n \to z_0$, $q_n \to 0$, and the sequence $\{g_n(t) = f_n(z_n + q_n t)\}$ of functions converges uniformly on each compact subset of the complex plane to a non-constant meromorphic function $g(t)$. Fix R, $0 < R < 1$, so that $R > |z_0|$, and fix $s > 0$. Then for n sufficiently large we have that $D_{n,s} = \{z = z_n + q_n t : |t| \le s\} \subset \{z : |z| \le R\}$. Note that $q_n f^{\#}(z_n + q_n t) = g_n^{\#}(t)$ and that, if $z = z_n + q_n t$, then $dA(z) = (q_n)^2 \, dA(t)$. Thus, making the substitution $z = z_n + q_n t$, we get

$$\iint_{D_{n,s}} (f_n^{\#}(z))^{p_0} \, dA(z) = \iint_{|t|\le s} (g_n^{\#}(t))^{p_0} (q_n)^{2-p_0} \, dA(t).$$

Thus, for n sufficiently large, we have

$$\iint_{|z|\le R} (f_n^{\#}(z))^{p_0} \, dA(z) \ge \iint_{D_{n,s}} (f_n^{\#}(z))^{p_0} \, dA(z)$$
$$= (q_n)^{2-p_0} \iint_{|t|\le s} (g_n^{\#}(t))^{p_0} \, dA(t).$$

But since the sequence of functions $\{g_n(t)\}$ converges uniformly to the non-constant meromorphic function $g(t)$ in $\{t : |t| \le s\}$, it follows that

$$\iint_{|t|\le s} (g_n^{\#}(t))^{p_0} \, dA(t) \to \iint_{|t|\le s} (g^{\#}(t))^{p_0} \, dA(t).$$

But $\iint_{|t|\le s} (g^{\#}(t))^{p_0} \, dA(t) > 0$ because g is a non-constant function. But since $q_n > 0$, $q_n \to 0$ and $p_0 > 2$, it follows that $(q_n)^{2-p_0} \to \infty$, and thus $\iint_{|z|\le R} (f_n^{\#}(z))^{p_0} dA(z) \to \infty$. This shows that the negation of (1) implies the negation of (5), and thus (5) implies (1) and the proof is complete.

We note that the "$p_0 > 2$" in (4) and (5) cannot be replaced by "$p_0 = 2$", as the following example shows.

**Example.** *There exists a family of functions* $\mathcal{F}$ *meromorphic in* D *such that, for each* $f \in \mathcal{F}$, *both* $\int_{|z|=.5} (f^{\#}(z))^2 |dz| \leq \pi$ *and* $\iint_{|z|\leq R} (f^{\#}(z))^2\, dA(z) < \pi$ *for each* $R$, $0 < R < 1$, *but* $\mathcal{F}$ *is not a normal family.*

*Proof.* For each positive integer $n$, let $f_n(z) = nz$, and let $\mathcal{F} = \{f_n\}$. It is easy to see that the pointwise limit of the sequence of functions $\{f_n(z)\}$ is the function which is identically $\infty$ at all points of $D - \{0\}$, but is $0$ at $z = 0$. Thus, no subsequence of $\{f_n\}$ can converge uniformly in a neighbourhood of $z = 0$, so $\mathcal{F}$ is not a normal family. However, $f_n^{\#}(z) = n/(1 + |nz|^2)$, and using polar coordinates it is a simple excercise to show that both

$$\int_{|z|=.5} (n^2/(1 + |nz|^2)^2)\, |dz| = 16n^2\pi/(4 + n^2)^2 < \pi, \text{ and}$$

$$\iint_{|z|\leq R} (n^2/(1 + |nz|^2)^2)\, dA(z) = \pi(1 - (1 + n^2R^2)^{-1}) < \pi$$

for each $R$, $0 < R < 1$.

## 3. Integral criteria for normal functions

Our goal in this section is to state an improved version of Theorem AL.

**Theorem 2.** *Let* f *be a function meromorphic in* D. *The following statements are equivalent.*

(1) *The function* f *is a normal function.*

(2) *For each* $p > 2$ *and each* $r > 0$ *there exists a finite constant* $K_{p,r}$ *such that, for each* $\lambda \in D$,

$$\iint_{D(\lambda,r)} (1 - |z|^2)^{p-2} (f^{\#}(z))^p\, dA(z) \leq K_{p,r}\,.$$

(3) *There exists* $p_0 > 2$ *such that, for each* $r > 0$, *there exists a finite constant* $K_r'$ *such that, for each* $\lambda \in D$,

$$\iint_{D(\lambda,r)} (1 - |z|^2)^{p_0-2} (f^{\#}(z))^{p_0}\, dA(z) \leq K_r'\,.$$

(4) *For each* $p > 1$ *and each* $r > 0$ *there exists a finite constant* $C_{p,r}$ *such that, for each* $\lambda \in D$ *and each* $s$, $0 < s < r$,

$$\int_{\partial D(\lambda,s)} (1-|z|^2)^{p-1} (f^{\#}(z))^p \, |dz| \le C_{p,r} .$$

(5) *There exists* $p_0 > 1$ *such that, for each* $r > 0$, *there exists a finite constant* $C'_r$ *such that, for each* $\lambda \in D$ *and each* $s$, $0 < s < r$,

$$\int_{\partial D(\lambda,s)} (1-|z|^2)^{p_0-1} (f^{\#}(z))^{p_0} \, |dz| \le C'_r .$$

*Proof.* Theorem AL shows the equivalence of (1) and (2). Clearly (2) implies (3). A close examination of the proof of Theorem AL in [1] shows that, in fact (3) implies (1). Thus, statements (1), (2), and (3) are equivalent.

Clearly, (4) implies (5). Thus, we need only show that (1) implies (4) and (5) implies (1). To show that (1) implies (4), we note that if $f$ is a normal function then the Lehto and Virtanen criterion for a normal function says that there exists a constant $C_f$ such that $(1-|z|^2)f^{\#}(z) \le C_f$ for each $z \in D$. Further, because of the differential relationship $d\delta(z) = |dz|/(1-|z|^2)$, we have, for a fixed $\lambda \in D$ and $s > 0$,

$$\int_{\partial D(\lambda,s)} (1-|z|^2)^{p-1} (f^{\#}(z))^p \, |dz| = \int_{\partial D(0,s)} ((1-|z|^2) f^{\#}(z))^p \, d\delta(z)$$
$$\le (C_f)^p \, \ell(\partial D(0,s)),$$

where $\ell(\partial D(0,s))$ denotes the hyperbolic circumference of $\partial D(0,s)$. Clearly, for a fixed $r > 0$ and $0 < s \le r$, we have $\ell(\partial D(0,s)) \le \ell(\partial D(0,r))$. Thus, setting $C_{p,r} = (C_f)^p \ell(\partial D(0,r))$, we have that (1) implies (4).

To show that (5) implies (1) we show the contrapositive. Suppose that $f$ is not a normal function. By a result of Lohwater and Pommerenke [3] there exists a sequence $\{z_n\}$ of points in $D$ and a sequence $\{q_n\}$ of positive real numbers such that $q_n \to 0$, $q_n/(1-|z_n|) \to 0$, and the sequence of functions $\{g_n(t) = f(z_n + q_n t)\}$ converges uniformly on each compact subset of the complex plane to a non-constant meromorphic function $g(t)$. Fix $p_0 > 1$ and let $\alpha(r)$ be defined by $\alpha(r) = \iint_{|t| \le r} (g^{\#}(t))^{p_0} \, dA(t)$, $r > 0$. Clearly $\alpha(r)$ is a positive valued increasing function of $r$ for $r > 0$. Using the polar form of the area integral, it is easy to verify that

$$\sup \{ \int_{|t|=s} (g^{\#}(t))^{p_0} \, |dt| : 0 < s < r \} \ge \alpha(r)/r.$$

Fix $r > 0$. Then there exists $r'$, $0 < r' \le r$, such that

$$\int_{|t|=r'} (g^{\#}(t))^{p_0}\,|dt| > \alpha(r)/(2r).$$

Also, since $\{g_n(t)\}$ converges uniformly to the function $g(t)$ on each compact subset of the complex plane, for $n$ sufficiently large, we have

$$\int_{|t|=r'} (g_n^{\#}(t))^{p_0}\,|dt| > \alpha(r)/(2r) > 0.$$

But $g_n^{\#}(t) = q_n f^{\#}(z_n + q_n t)$ and $|d(z_n + q_n t)| = q_n |dt|$. Thus, if $\gamma_n = \{z = z_n + q_n t : |t| = r'\}$, then

$$\int_{\gamma_n} (1 - |z|^2)^{p_0 - 1} (f^{\#}(z))^{p_0}\,|dz| =$$

$$= \int_{|t|=r'} (1 - |z_n + q_n t|^2)^{p_0-1} \left[\frac{g_n^{\#}(t)}{q_n}\right]^{p_0} q_n\,|dt|$$

$$= \int_{|t|=r'} \left[\frac{1 - |z_n + q_n t|^2}{q_n}\right]^{p_0-1} (g_n^{\#}(t))^{p_0}\,|dt|.$$

Since $q_n/(1 - |z_n|) \to 0$, we have that $u_n = \inf\{(1 - |z_n + q_n t|)/q_n : |t| \le r\} \to \infty$. Since $p_0 > 1$, we have

$$\int_{\gamma_n} (1 - |z|^2)^{p_0-1} (f^{\#}(z))^{p_0}\,|dz| \ge (u_n)^{p_0-1} \alpha(r)/(2r) \to \infty.$$

Thus, we have shown that if $f$ is not a normal function then, for each $p_0 > 1$ and each $r > 0$, there exists a sequence $\{\gamma_n\}$ of circles, where, for $n$ sufficiently large, the hyperbolic radius of each $\gamma_n$ is less than $r$, such that

$$\int_{\gamma_n} (1 - |z|^2)^{p_0-1} (f^{\#}(z))^{p_0}\,|dz| \to \infty.$$

This shows that (5) implies (1) and completes the proof.

We note that S. Yamashita [6, Lemma 3.2, Page 354] has proved that $f$ is a normal function if and only if there exists an $r > 0$ and a constant $K < \pi$ such that $\iint_{D(\lambda,r)} (f^{\#}(z))^2\,dA(z) < K$ for each $\lambda \in D$. This result is similar to, but of a different character from the equivalence of (1) and (3) in our Theorem 2. In fact, (1) is not equivalent to (3) when $p_0 = 2$ (see [1]).

## 4. Integral conditions for $\mathcal{N}_0$

The nature of functions in the class $\mathcal{N}_0$ would lead one to expect that one could characterize the class $\mathcal{N}_0$ by a result something like Theorem 2, with the finite constants replaced by a limit of 0 as the curves and regions tend to the boundary. We now prove such a result.

**Theorem 3.** *Let* f *be a function meromorphic in* D. *Then the following statements are equivalent.*

(1) *The function* f *is in the class* $\mathcal{N}_0$.

(2) *For each* $p > 0$ *and each* $r > 0$ *we have*

$$\lim_{|\lambda|\to 1} \iint_{D(\lambda,r)} (1-|z|^2)^{p-2}(f^{\#}(z))^p \, dA(z) = 0.$$

(3) *There exists a number* $p_0 \geq 2$ *such that, for each* $r > 0$,

$$\lim_{|\lambda|\to 1} \iint_{D(\lambda,r)} (1-|z|^2)^{p_0-2}(f^{\#}(z))^{p_0} \, dA(z) = 0.$$

(4) *For each* $p > 0$ *and each* $r > 0$ *we have*

$$\lim_{|\lambda|\to 1} \int_{\partial D(\lambda,r)} (1-|z|^2)^{p-1}(f^{\#}(z))^p \, |dz| = 0.$$

(5) *There exists a number* $p_0 \geq 1$ *such that, for each* $r > 0$,

$$\lim_{|\lambda|\to 1} \int_{\partial D(\lambda,r)} (1-|z|^2)^{p_0-1}(f^{\#}(z))^{p_0} \, |dz| = 0.$$

*Proof.* It is clear that (2) implies (3) and (4) implies (5). We will show that (1) implies (2), that (1) implies (4), that (3) implies (1), and that (5) implies (1).

To show that (1) implies (2), suppose that f is in the class $\mathcal{N}_0$. Then, given $\varepsilon > 0$, there exists $\beta$, $0 \leq \beta < 1$, such that $(1-|z|^2)f^{\#}(z) < \varepsilon$ whenever $\beta \leq |z| < 1$. If $dh(z)$ denotes the hyperbolic element of area, we have that $dh(z) = dA(z)/(1-|z|^2)^2$, so that, for fixed $p > 0$, and $D(\lambda,r) \subset \{z : \beta \leq |z| < 1\}$,

$$\iint_{D(\lambda,r)} (1-|z|^2)^{p-2}(f^{\#}(z))^p \, dA(z) = \iint_{D(\lambda,r)} [(1-|z|^2)f^{\#}(z)]^p \, dh(z)$$
$$< \varepsilon^p \, h(D(\lambda,r))$$

and this shows (2).

The proof that (1) implies (4) is similar, except that we use $d\delta(z) = |dz|/(1-|z|^2)$ in place of $dh(z) = dA(z)/(1-|z|^2)^2$.

To prove that (3) implies (1), we first note that, in view of Theorem 2, (3) implies that f is a normal function. If $p_0 = 2$, then (3) implies (1) by a result of S. Yamashita [6, Lemma 3.2, Page 354]. If $p_0 > 2$, we prove that (3) implies (1) by proving the contrapositive under the additional condition that f is a normal function. Suppose that f is a normal function but f is not in the class $\mathcal{N}_0$. Then there exists a sequence $\{z_n\}$ of points in D such that $|z_n| \to 1$ and $(1-|z_n|^2)f^{\#}(z_n) \geq \alpha > 0$ for each positive integer n. Then the sequence of functions $\{g_n(t) = f((t+z_n)/(1+\bar{z}_n t))\}$ contains a subsequence, which we again denote by $\{g_n(t)\}$, which converges uniformly on each of the sets $\Delta_R = \{z : |z| \leq R\}$, $0 \leq R < 1$, to a non-constant meromorphic function g(t). (The function g(t) satisfies $g^{\#}(0) \geq \alpha > 0$.) Fix $p_0 > 2$, $r > 0$, and let R, $0 < R < 1$, be such that $\delta(0,R) = r$, and let $\Omega = \iint_{\Delta_R} (g^{\#}(t))^{p_0}\, dA(t)$. Since g is not a constant function, the number $\Omega$ (which depends on both r and $p_0$) is positive. Now we have both $g_n^{\#}(t) = (1-|z_n|^2)f^{\#}((t+z_n)/(1+\bar{z}_n t))/|1+\bar{z}_n t|^2$ and if $z = (t+z_n)/(1+\bar{z}_n t)$, then both $1-|z|^2 = (1-|z_n|^2)(1-|t|^2)/|1+\bar{z}_n t|^2$ and $dA(z) = (1-|z_n|^2)^2 dA(t)/|1+\bar{z}_n t|^2$. Thus, if we set $D_n = \{z : z = (t+\bar{z}_n t)/(1+\bar{z}_n t),\ |t| \leq R\}$, we have

$$\iint_{D_n} (1-|z|^2)^{p_0-2}(f^{\#}(z))^{p_0}\, dA(z) = \iint_{\Delta_R} (1-|t|^2)^{p_0-2}(g_n^{\#}(t))^{p_0}\, dA(t)$$
$$\geq (1-R^2)^{p_0-2}\Omega/2 > 0$$

for each n sufficiently large, since $\{g_n(t)\}$ converges uniformly on $\Delta_R$ to g(t). It follows that (3) cannot happen under our assumptions. Thus, we have proved the contrapositive of (3) implies (1).

For $p_0 > 1$, the proof that (5) implies (1) follows by similar reasoning, except that where $z = (t+z_n)/(1+\bar{z}_n t)$ we use $|dz| = (1-|z_n|^2)|dt|/|1+\bar{z}_n t|$, rather than substitute for dA(z). For $p_0 = 1$, we need a different proof that (5) implies that f is a normal function (which corresponds to the first step in the proof of (3) implies (1)). Suppose that f is not a normal function. Then, as an easy consequence of the Lohwater-Pommerenke result, there exist two sequences $\{z_n\}$ and $\{w_n\}$ such that $\delta(z_n, w_n) \to 0$ and $\chi(f(z_n), f(w_n)) \to 1$. Fix $r > 0$ and, for each n, let $\lambda_n$ be a point of D such that

$\delta(z_n,\lambda_n) = \delta(w_n,\lambda_n) = r$. It follows that $\limsup \int_{\partial D(\lambda_n,r)} f^{\#}(z)|dz| \geq 2$, since the integral is the spherical arc length of $f(\partial D(\lambda_n,r))$. This shows that (5) implies that f is a normal function. Now the rest of the proof that (5) implies (1) for $p_0 = 1$ is the same as in the case for $p_0 > 1$. This completes the proof of Theorem 3.

One natural question remains associated with Theorem 3. In (3) we made the assumption that $p_0 \geq 2$. Let (3') be the following statement:

(3') *There exists* $p_0$, $0 < p_0 < 2$, *such that, for each* $r > 0$,

$$\lim_{|\lambda|\to 1} \iint_{D(\lambda,r)} (1-|z|^2)^{p_0-2}(f^{\#}(z))^{p_0}\, dA(z) = 0.$$

The question is: *Does* (3') *imply that* f *is a normal function?* We do not know the answer to this question. However, if the answer is affirmative, then we can replace statement (3) of Theorem 3 by the corresponding statement using "$p_0 > 0$" instead of "$p_0 \geq 2$". A similar question can be asked about condition (5).

## References


[1] R. Aulaskari and P. Lappan, *An integral criterion for normal functions*, Proc. Amer. Math Soc. (to appear).

[2] O. Lehto and K. I. Virtanen, *Boundary behaviour and normal meromorphic functions*, Acta Math. **97** (1957), 47–65.

[3] A. J. Lohwater and Ch. Pommerenke, *On normal meromorphic functions*, Ann. Acad. Sci. Fenn. Ser. A I, No. 550 (1972), 12 pp.

[4] F. Marty, *Recherches sur la repartition des valeurs d'une fonction meromorphe*, Ann. Fac. Sci. Univ. Toulouse (3) **23** (1931), 183–261.

[5] H. L. Royden, *A criterion for the normality of a family of meromorphic functions*, Ann. Acad. Sci. Fenn. Ser. A I Math. **10** (1985), 499–500.

[6] S. Yamashita, *Functions of uniformly bounded characteristic*, Ann. Acad. Sci. Fenn. Ser. A I Math. **7** (1982), 349–367.

[7] L. Zalcman, *A heuristic principle in complex function theory*, Amer. Math. Monthly **82** (1975), 813–817.

# THE DIANALYTIC MORPHISMS OF THE KLEIN BOTTLES [1]

**Ilie Bârză**
The University of Bucharest, Department of Mathematics
Bucharest, Romania

In this paper it is shown that a non-orientable Riemann surface whose universal covering is parabolic (that is, isomorphic to the complex plane $\mathbb{C}$) is isomorphic either to the pointed real projective plane $\mathbb{P}^2\setminus\{\hat{1}\}$ or to a Klein bottle. The set of all dianalytic morphisms from one Klein bottle to another is shown finding, in this way, the Teichmüller space of Klein bottles. Finally, the group of dianalytic automorphisms of a Klein bottle is computed.

## 1. Preliminaries

If $D_1$ and $D_2$ are open sets in $\mathbb{C}$, the function $f: D_1 \to D_2$ is dianalytic if $f$ is analytic or antianalytic on each connected component of $D_1$.

If $Y$ and $X$ are Riemann surfaces (orientable or not) and if $f: Y \to X$ is a continuous map then $f$ is a dianalytic map if expressed in terms of local uniformisers on $Y$ and $X$ one gets dianalytic maps between plane open sets.

$f: Y \to X$ is an isomorphism if it is bijective and dianalytic; then $f^{-1}$ is an isomorphism as well.

Let $\mathscr{C}(\mathbb{C})$ denote the group of conformal or anticonformal homeomorphisms of the complex plane $\mathbb{C}$ onto itself. The elements of $\mathscr{C}(\mathbb{C})$ are of the form

(i) $z \to az + b$ or

(ii) $z \to a\bar{z} + b$,

where $a, b \in \mathbb{C}$ and $a \neq 0$.

To find all non-orientable Riemann surfaces whose universal covering is (isomorphic to) $\mathbb{C}$ is an equivalent problem to the determination of the conjugacy classes of subgroups $\mathscr{K} \subset \mathscr{C}(\mathbb{C})$ having the following three properties:

---

[1] This paper is in final form and no version of it will be submitted for publication elsewhere.

(1) The orbit $\mathscr{K}z$ of each $z \in \mathbb{C}$ does not contain cluster points in $\mathbb{C}$;
(2) $\mathscr{K}$ acts without fixed points: $F(z) \neq z$ for all $F \in \mathscr{K}$ and all $z \in \mathbb{C}$;
(3) $\mathscr{K}$ contains at least one anticonformal transformation of $\mathbb{C}$.

## 2. First reduction of the problem

From now on, $\mathscr{K}$ will be a group with the properties (1)–(3).

**Proposition 1.** (i) *If* $T \in \mathscr{K}$ *and* $Tz = az + b$, *then* $a = 1$;

(ii) *if* $S \in \mathscr{K}$ *and* $Sz = \alpha\bar{z} + \beta$, *then* $|\alpha| = 1$ *and* $\omega := \alpha\bar{\beta} + \beta \neq 0$.

*Proof.* (i) is trivial. If $|\alpha| \neq 1$ or $\omega = 0$, S will have a fixed point in $\mathbb{C}$. Hence (ii) follows.

**Proposition 2.** *If* $S_k \in \mathscr{K}$, $S_k z = \alpha_k \bar{z} + \beta_k$, $k = 1,2$, *then* $\alpha_1 = \alpha_2$.

*Proof.* $S_1 \circ S_2(z) = \alpha_1 \bar{\alpha}_2 z + \alpha_1 \bar{\beta}_2 + \beta_1$. Now everything follows from Proposition 1.

**Corollary.** *There exists a single* $\alpha \in \mathbb{C}$, $|\alpha| = 1$, *so that all the anticonformal elements of* $\mathscr{K}$ *have the form* $z \mapsto \alpha\bar{z} + \beta$.

**Proposition 3.** $\mathscr{K}$ *is conjugated in* $\mathscr{C}(\mathbb{C})$ *with a group* $\mathscr{K}_0$ *of the same kind and whose corresponding number* $\alpha$ *is* 1.

*Proof.* If $\alpha = \exp(i\theta)$ and $Tz = z\exp(i\theta/2)$ then $\mathscr{K}_0 = T^{-1}\mathscr{K}T$ has the properties (1)–(3) and $\alpha_0 = 1$.

According to Proposition 3, (because the non-orientable Riemann surfaces $\mathbb{C}/\mathscr{K}$ and $\mathbb{C}/\mathscr{K}_0$ are isomorphic) one can suppose from now on that the group $\mathscr{K}$ has the number $\alpha = 1$; the antianalytic elements of $\mathscr{K}$ have the form $z \mapsto \bar{z} + b$, $b \in \mathbb{C}$, and $\omega = b + \bar{b} = 2\,\mathrm{Re}\, b \neq 0$.

## 3. Group $\mathscr{K}$ can be in two (and only two) situations

Let $\mathscr{K}_1$ be the subgroup of conformal elements of $\mathscr{K}$. Obviously the index of $\mathscr{K}_1$ in $\mathscr{K}$ is 2 and if $S \in \mathscr{K}\setminus\mathscr{K}_1$, then $\mathscr{K} = \mathscr{K}_1 \cup S\mathscr{K}_1$ and $\mathscr{K}_1 \cap S\mathscr{K}_1 = \emptyset$.

Let $A = \{|V(0)| \mid V \in \mathscr{K}\setminus\mathscr{K}_1\}$ and $r = \inf A$. From the property (1) of $\mathscr{K}$ it results that there exists $S \in \mathscr{K}\setminus\mathscr{K}_1$, $Sz = \bar{z} + \beta$ and $|\beta| = r$. If $\beta = u + iv$, because $S^{-1}z = \bar{z} - \bar{\beta}$ and $|-\bar{\beta}| = r$, we can suppose $\mathrm{Re}\,\beta = u \geq 0$. The number

$\omega \neq 0$ from Proposition 1 is $\omega = 2u$. Hence $u > 0$.

From now on $\beta$ and $\omega$ will be the preceding numbers and S will be the transformation $Sz = \bar{z} + \beta$. Obviously $S^2 z = z + \omega$.

Let $\mathcal{G}_S = \{S^k \mid k \in \mathbb{Z}\}$ be the subgroup generated by S. We shall consider the following two cases separately:

$(C_1)$ $\mathcal{K} = \mathcal{G}_S$;

$(C_2)$ $\mathcal{K} \setminus \mathcal{G}_S \neq \emptyset$.

## 4. Case $(C_1)$

**Proposition 4.** *The set* $D = \{z \mid 0 \le \operatorname{Re} z < u\}$ *is a fundamental set for* $\mathcal{K}$.

*Proof.* Obviously, $\mathcal{K}_1 = \{z \mapsto z + n\omega \mid n \in \mathbb{Z}\}$. Every $z \in \mathbb{C}$ has precisely one $\mathcal{K}_1$-equivalent point $z_1 \in D_1 = \{w \in \mathbb{C} \mid 0 \le \operatorname{Re} w < \omega\}$. If $z \in D$, $z = x + iy$, $0 < x < u$, then $Sz = x + u + i(v - y) \in D_2 = D_1 \setminus D$. If $z = iy$ then $Sz = -iy + \beta$. Hence every $z \in \mathbb{C}$ has a $\mathcal{K}$-equivalent point $z_1 \in D$. Obviously D does not contain distinct $\mathcal{K}$-equivalent points.

**Proposition 5.** *The orbit space* $\mathbb{C}/\mathcal{K}$ *is a Möbius strip.*

*Proof.* Because $S(0) = \beta$, the image of the line segment $[0,\beta]$ in $\mathbb{C}/\mathcal{K}$ is a simple closed curve. For $z = iy$, the equality $z + Sz = \beta$ states that z and Sz are symmetric points with respect to $\beta/2$. Hence the points of the border $\partial D$ are crosswise identified with respect to $\beta/2$; $\mathbb{C}/\mathcal{K}$ is then a Möbius strip.

**Theorem 1.** $\mathbb{C}/\mathcal{K}$ *with the dianalytic structure given by the canonical projection of* $\mathbb{C}$ *on* $\mathbb{C}/\mathcal{K}$ *is a non-orientable Riemann surface isomorphic to the pointed real projective plane* $\mathbb{P}^2 \setminus \{\hat{1}\}$.

*Proof.* We shall consider as a model for $\mathbb{P}^2$ the closed disc $\{z \mid |z| \le 1\}$ with the border points $e^{i\theta}$ and $-e^{i\theta}$ identified ([4], p. 28). Denote

$$A_1 = \{z \mid |z| \le 1\} \setminus \{z \mid |z| = 1, \operatorname{Im} z \ge 0\};$$
$$A_2 = \{z \mid \operatorname{Im} z \ge 0\} \setminus \{z \mid z \in \mathbb{R}, z \le 0\};$$
$$A_3 = \{z \mid 0 \le \operatorname{Re} z < 1/2\}.$$

Let us consider the following sequence of maps:

$$D \xrightarrow{g_3} A_3 \xrightarrow{g_2} A_2 \xrightarrow{g_1} A_1 ,$$

where

$$g_3(z) = z/\omega;$$
$$g_2(z) = \exp(2\pi i z);$$
$$g_1(z) = \frac{(z-i)\exp(-\pi s)}{(z+i)\exp(-\pi s)} \quad \text{with } s = \operatorname{Im} \beta/\omega.$$

Then the map induced by $g = g_1 \circ g_2 \circ g_3$ gives a dianalytic isomorphism of $\mathbb{C}/\mathscr{K}$ on $\mathbb{P}^2\backslash\{\hat{1}\}$ where $\hat{1} = \{1; -1\}$.

**Remark.** The pointed real projective plane is the single Möbius strip having a parabolic universal covering. All the other Möbius strips have hyperbolic universal coverings.

### 5. Case $(C_2)$

**Proposition 6.** $(\mathscr{K}\backslash\mathscr{G}_S) \cap \mathscr{K}_1 \neq \emptyset$.

*Proof.* Let $T \in \mathscr{K}\backslash\mathscr{G}_S$. If $T \in \mathscr{K}\backslash\mathscr{K}_1$, then $T \circ S \in \mathscr{K}_1$ and $\notin \mathscr{G}_S$.

**Proposition 7.** *If* $T \in \mathscr{K}_1$, $Tz = z + \gamma$ *and* $\operatorname{Im} \gamma = 0$, *there exists* $q \in \mathbb{Z}$ *so that* $\gamma = q\omega$.

*Proof.* If $r = \gamma/\omega$ is an irrational number, according to a theorem of Dirichlet, the set $A = \{m\gamma + n\omega \mid m, n \in \mathbb{Z}\}$ is dense in $\mathbb{R}$. This is in contradiction with the property (1) of $\mathscr{K}$. Hence $r \in \mathbb{Q}$ and there exist $p_1, q_1 \in \mathbb{Z}$ so that $p_1\omega - q_1\gamma = 0$. In this case the net $A$ is reducible: there exists $\tilde{\omega}$ so that $A = \{k\tilde{\omega} \mid k \in \mathbb{Z}\}$. Hence there exist $p, q \in \mathbb{Z}$ so that $\omega = p\tilde{\omega}$, $\gamma = q\tilde{\omega}$ and there exist $m, n \in \mathbb{Z}$ so that $\tilde{\omega} = m\omega + n\gamma$. Let $\tilde{T}z = z + \tilde{\omega}$; hence $\tilde{T} = S^{2m} \circ T^m \in \mathscr{K}_1$. Replacing (if necessary) $S$ by $S^{-1}$ we can suppose $p, q \geq 1$.

Let $V = S \circ \tilde{T}^{-1}$; $Vz = \bar{z} + \beta - \tilde{\omega}$. Let $\beta_V = \beta - \tilde{\omega}$. One obtains now

$$|\beta_V|^2 = |\beta|^2 + \frac{|\omega|^2}{p^2}(1-p) \geq |\beta|^2$$

from the minimality of $|\beta|$. Hence $p \leq 1$, that is $p = 1$. Finally, $\gamma = q\omega$.

We shall denote the $\mathscr{K}_1$-orbit of 0 by $\Sigma$. $\mathscr{K}_1$ being a discontinuous group, from Propositions 6 and 7 we get the following.

**Corollary.** $\Sigma$ *is a lattice in* $\mathbb{C}$.

We now denote

$$\hat{z} = \Sigma + z = \mathscr{K}_1\text{-orbit of } z \in \mathbb{C};$$

$$\tilde{z} = \{V(z) \mid V \in \mathcal{K}\} = \mathcal{K}\text{-orbit of } z.$$

**Proposition 8.** $\overline{\Sigma} = \Sigma$ *(symmetry with respect to the real axis).*

*Proof.* Let $\gamma \in \Sigma$ and $Vz = z + \gamma$. Obviously $S \circ V \circ S^{-1} \in \mathcal{K}_1$. But $(S \circ V \circ S^{-1})(z) = z + \overline{\gamma}$. Hence $\overline{\gamma} \in \Sigma$, $\overline{\Sigma} \subseteq \Sigma$, $\Sigma = \overline{\overline{\Sigma}} \subseteq \overline{\Sigma} \subseteq \Sigma$ and finally $\overline{\Sigma} = \Sigma$.

Let $\omega_1 \in \Sigma$ so that $\operatorname{Im} \omega_1 > 0$ and

$$|\omega_1| = \inf \{|\gamma| \mid \gamma \in \Sigma,\ \operatorname{Im} \gamma > 0\}. \qquad (*)$$

Obviously, $\Sigma = \{m\omega_1 + n\omega \mid m, n \in \mathbb{Z}\} = \mathbb{Z}\omega_1 + \mathbb{Z}\omega$.

If $\omega_1 = a + ib$, $b > 0$, $\omega_1 + \overline{\omega}_1 = 2a \in \Sigma$. From Proposition 7, there exists $k \in \mathbb{Z}$ such that $2a = k\omega$, hence $a = ku$. From $(*)$,

$$|\omega_1| \le |\omega + \omega_1|, \quad |\omega_1| \le |\omega - \omega_1|,$$

that is $a \in [-u, u]$. Hence $k \in \{-1, 0, 1\}$.

In the case $k = -1$ we shall replace $\omega_1$ by $\omega_1 + \omega$. The property $(*)$ remains unchanged because of $|\omega_1| = |\omega + \omega_1|$ in this case. Hence we may suppose $k \in \{0;1\}$. For $k = 1$, $\omega_1 = u + ib$, $b > 0$. Let $Vz = z + \omega_1$ and $S_1 = V^{-1} \circ S \in \mathcal{K}$. Then $S_1 z = \overline{z} + \beta - \omega_1 = \overline{z} + i(v-b)$. For $z_0 = i(v - b)/2$ one has $S_1 z_0 = z_0$, a contradiction with the property (2) of $\mathcal{K}$. Therefore $k = 0$. Hence

$$\Sigma = \mathbb{Z}\omega_1 + \mathbb{Z}\omega$$

where $\omega_1 = ib$, $b > 0$, verifies $(*)$.

If $Vz = z + \omega_1$, $V \in \mathcal{K}_1$ and $V^{-1} \circ S$, then $V \circ S \in \mathcal{K} \setminus \mathcal{K}_1$ and $V^{-1} \circ S(z) = \overline{z} + \beta - \omega_1$, $V \circ S(z) = \overline{z} + \beta + \omega_1$. From the definition of $\beta$ one gets:

$$|\beta| \le |\beta - \omega_1|, \quad |\beta| \le |\beta + \omega_1|.$$

These inequalities are equivalent to $v \in [-b/2, b/2]$.

**Proposition 9.** *Let* $\beta = u + iv$, $u > 0$, $Sz = \overline{z} + \beta$, $\omega_1 = ib$, $b > 0$, $v \in [-b/2, b/2]$ *and* $Vz = z + \omega_1$. *Let* $\mathcal{K} = \{S; V\}$ *be the group generated by* S *and* V. *Then* $\mathcal{K}$ *has the properties* (1)–(3).

*Proof.* Trivial.

As a consequence of the facts discussed above we obtain

**Theorem 2.** *A group* $\mathcal{K} \subset \mathcal{C}(\mathbb{C})$ *has the properties* (1)–(3) *if and only if* $\mathcal{K}$ *is* *conjugated in* $\mathscr{C}(\mathbb{C})$ *to either a group* $\mathscr{G}_S$ *mentioned in* 3 *or a group of the type mentioned in Proposition* 9.

From now on, $\mathscr{K}$ will be the group given in Proposition 9.

## 6. The Klein bottles

**Theorem 3.** *The orbit space* $\mathbb{C}/\mathscr{K}$ *is a Klein bottle.*

*Proof.* Because of $\mathscr{K}_1 = \{S^2; V\}$, the group generated by $S^2$ and V, each $z \in \mathbb{C}$ has precisely one $\mathscr{K}_1$-equivalent point $z_1$ in the rectangle $D = \{\lambda\omega + \mu\omega_1 \mid \lambda, \mu \in [0,1)\}$.

The interior of the hexagon H with vertices at $0, \beta, \omega, \omega_1 + \omega, \omega_1 + \beta, \omega_1$ together with the line segments $[0,\omega_1)$, $[0,\beta]$ and $[\beta,\omega)$ is a fundamental set for $\mathscr{K}_1$. Let $D_1$ be the interior of the parallelogram with vertices at $0, \beta, \omega_1 + \beta, \omega_1$ together with the line segments $[0,\beta)$ and $[0,\omega_1)$. Let $D_2 = H \backslash D_1$. If $z \in \mathring{D}_1$, then $(V \circ S)(z) \in \mathring{D}_2$. If $z \in [0,\omega_1)$, then $(V \circ S)(z) \in (\beta, \omega_1 + \beta]$ and if $z \in [0,\beta)$, then $(V \circ S)(z) \in [\beta + \omega_1, \omega + \omega_1)$. Hence the points of $D_1$ and $D_2$ are pairwise equivalent; $\mathscr{K}_1$ being of index 2 in $\mathscr{K}$, $D_1$ will be a fundamental set for $\mathscr{K}$.

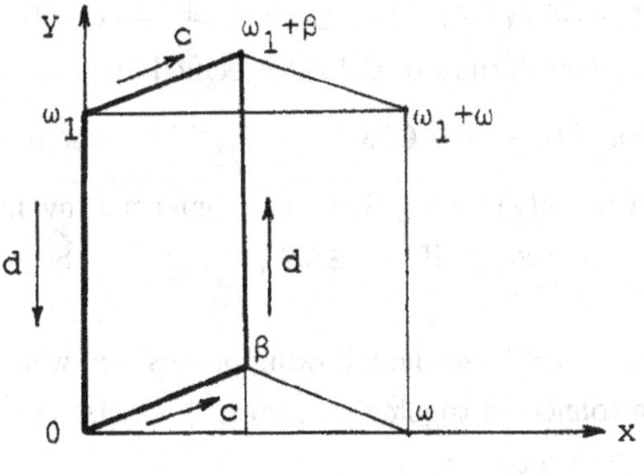


Let $z \in [0,\omega_1)$. The equality

$$\tfrac{1}{2}[z + (V \circ S)(z)] = \tfrac{1}{2}[\beta + \omega_1] \tag{$*$}$$

states that z and $(V \circ S)(z)$ are symmetric points with respect to the midpoint of the line segment $[0, \omega_1 + \beta]$. Hence, denoting by d the *oriented* line segment $[\omega_1,0]$ and by c the oriented segment $[0,\beta]$, the symbol of the border of $D_1$ is $cdc^{-1}d$. Hence $\mathbb{C}/\mathscr{K}$ is a Klein bottle.

## 7. Morphisms of Klein bottles

Let $\mathscr{K}' = \{S'; V'\}$ be another group of the same kind as $\mathscr{K}$ is:

$$S'z = \bar{z} + \beta', \ \beta' = u' + iv', \ \omega' = 2u' > 0;$$
$$V'z = z + \omega_1', \ \omega_1' = ib', \ b' > 0, \ v' \in [-b'/2, b'/2].$$

Let $\mathscr{K}_1'$ be the subgroup of conformal elements of $\mathscr{K}'$. The meanings of $\Sigma'$, $\hat{z}'$, $\tilde{z}'$ are clear. We introduce now the following notations:

a) $\mathbb{C}/\Sigma$, $\mathbb{C}/\Sigma'$ will be the tori $\mathbb{C}/\mathscr{K}_1$ and $\mathbb{C}/\mathscr{K}_1'$;

b) If $\mathscr{H}$ is a subgroup of $\mathscr{C}(\mathbb{C})$, $\sim_{\mathscr{H}}$ will be the equivalence relation given by $\mathscr{H}$ in $\mathbb{C}$: $z_1 \sim_{\mathscr{H}} z_2$ if and only if there exists $h \in \mathscr{H}$ so that $z_1 = hz_2$;

c) $\pi$, $\pi'$ will be the projections of $\mathbb{C}$ on $\mathbb{C}/\Sigma$ and $\mathbb{C}/\Sigma'$;

d) $p : \mathbb{C}/\Sigma \to \mathbb{C}/\mathscr{K}$, $p(\hat{z}) = \tilde{z}$, is the unbranched dianalytic two-sheeted covering of $\mathbb{C}/\mathscr{K}$ by $\mathbb{C}/\Sigma$. Similarly $p' : \mathbb{C}/\Sigma' \to \mathbb{C}/\mathscr{K}'$.

e) $$\mathrm{Deck}\,(\mathbb{C} \xrightarrow{\pi} \mathbb{C}/\Sigma) = \mathscr{K}_1, \ \mathrm{Deck}\,(\mathbb{C} \xrightarrow{\pi'} \mathbb{C}/\Sigma') = \mathscr{K}_1',$$
$$\mathrm{Deck}\,(\mathbb{C} \xrightarrow{p\circ\pi} \mathbb{C}/\mathscr{K}) = \mathscr{K}, \ \mathrm{Deck}\,(\mathbb{C} \xrightarrow{p'\circ\pi'} \mathbb{C}/\mathscr{K}') = \mathscr{K}',$$

are the groups of covering transformations of the specified maps.

Obviously, $\mathrm{Deck}\,(\mathbb{C}/\Sigma \xrightarrow{p} \mathbb{C}/\mathscr{K}) = \{1_{\mathbb{C}/\Sigma}; k\}$, where $1_{\mathbb{C}/\Sigma}$ is the identity of $\mathbb{C}/\Sigma$ and $k : \mathbb{C}/\Sigma \to \mathbb{C}/\Sigma$, $k(\hat{z}) := \widehat{Sz}$, is an anticonformal involution without a fixed point of the torus $\mathbb{C}/\Sigma$. Moreover, if $S_1 \in \mathscr{K} \setminus \mathscr{K}_1$, $\widehat{S_1 z} = \widehat{Sz}$, then $k$ depends of $\mathscr{K} \setminus \mathscr{K}_1$ only.

In what follows, we shall use the following proposition which we shall refer to as "property D": If $X$ is a connected topological space, $Y$ a discrete space and $h : X \to Y$ a continuous map, then $h$ is a constant.

The following theorem gives the complete list of dianalytic nonconstant morphisms between Klein bottles.

**Theorem 4.** *The following statements are equivalent:*

1) *For* $a \in \mathbb{C}$, *there exists a dianalytic nonconstant morphism* $f : \mathbb{C}/\mathscr{K} \to \mathbb{C}/\mathscr{K}'$, *so that* $f(\tilde{0}) = \tilde{a}'$;

2) *There exists* $\alpha \in \mathbb{R}^*$ *with one of the following two properties:*

(i) $\alpha\Sigma \subseteq \Sigma'$ *and* $\alpha\beta + 2i\,\mathrm{Im}\,a \in \Sigma' + \beta'$;

(ii) $\alpha\Sigma \subseteq \Sigma'$ *and* $\alpha\bar{\beta} + 2i \operatorname{Im} a \in \Sigma' + \beta'$.

*Here we may assume that* a *belongs to a fundamental set of* $\mathscr{K}'$.

*Proof.* 2) ⇒ 1). Let $\alpha \in \mathbb{R}^*$ and $a \in \mathbb{C}$ with the property (i). One easily verifies that $z_1 \sim_{\mathscr{K}} z_2 \Rightarrow \alpha z_1 + a \sim_{\mathscr{K}'} \alpha z_2 + a$. Hence one can define $f : \mathbb{C}/\mathscr{K} \to \mathbb{C}/\mathscr{K}'$ by

$$f(\tilde{z}) := \widetilde{\alpha z + a}',$$

that is $f \circ p \circ \pi = p' \circ \pi' \circ F$, $F(z) = \alpha z + a$. Obviously f is a dianalytic covering map and hence a morphism of Klein bottles, with $f(\tilde{0}) = \tilde{a}'$.

In the same way one proves that in the case 2) (ii) the map $f : \mathbb{C}/\mathscr{K} \to \mathbb{C}/\mathscr{K}'$, $f(\tilde{z}) := \widetilde{\alpha\bar{z} + a}'$, is a well defined map that is a dianalytic covering map (with the property $f(\tilde{0}) = \tilde{a}'$).

1) ⇒ 2). Let $a \in \mathbb{C}$ and $f : \mathbb{C}/\mathscr{K} \to \mathbb{C}/\mathscr{K}'$, be a nonconstant dianalytic morphism so that $f(\tilde{0}) = \tilde{a}'$. Hence

$$f \circ p \circ \pi\,(0) = p \circ \pi'\,(a).$$

Since $p' \circ \pi' : \mathbb{C} \to \mathbb{C}/\mathscr{K}'$ is a universal covering, $f \circ p \circ \pi$ a continuous map and $\mathbb{C}$ simply connected, there exists precisely one continuous (nonconstant) function $F : \mathbb{C} \to \mathbb{C}$ such that $f \circ p \circ \pi = p' \circ \pi' \circ F$ and $F(0) = a$, see [3], p. 26. Hence

$$f(\tilde{z}) = \widetilde{F(z)}' \quad \text{for all } z \in \mathbb{C}. \tag{$*$}$$

f being dianalytic and $p \circ \pi$, $p' \circ \pi'$ being dianalytic covering maps, F will be dianalytic too. Hence F is either analytic or antianalytic ($\mathbb{C}$ being connected). We shall consider these cases separately.

*Case* 1. F analytic. We shall show that the situation 2) (i) is obtained.

Now $F(z) = a + \Sigma_{n=1}^{\infty} a_n z^n$ and there exists $n \geq 1$ so that $a_n \neq 0$. For $z \in \mathbb{C}$ and $T \in \mathscr{K}_1$, $\tilde{z} = \widetilde{Tz}$. From (*) one gets $\widetilde{F(Tz)}' = \widetilde{F(z)}'$, that is $F(Tz) \sim_{\mathscr{K}'} F(z)$. There exists $G \in \mathscr{K}_1'$ (from the property D) so that $F(T(z)) = G(F(z))$, $Gw = w + \gamma'$, $\gamma' \in \Sigma'$.

Hence $F(Tz) - F(z) = \gamma' \in \Sigma'$ is a discrete space. Hence the continuous function $F \circ T - F$ takes discrete values. The property D once again gives $F(Tz) = F(z) + \gamma'$ for all $z \in \mathbb{C}$ (that is $\gamma'$ does not depend on z); it results that $\frac{dF}{dz}$ is a holomorphic doubly periodic function and hence constant. In consequence,

$$F(z) = \alpha z + a, \text{ for all } z \in \mathbb{C},\ \alpha \neq 0.$$

Now $F(T0) = F(0) + \gamma'$ and $Tz = z + \gamma'$ implies $\alpha\gamma = \gamma' \in \Sigma'$. Hence $\alpha\Sigma \subsetneq \Sigma'$ and $\alpha \in \mathbb{C}^*$.

Let now $T \in \mathcal{K}\setminus\mathcal{K}_1$, $Tz = \bar{z} + \gamma + \beta$, $\gamma \in \Sigma$. Again $F(Tz) \sim_{\mathcal{K}'} F(z)$. There exists $G \in \mathcal{K}'\setminus\mathcal{K}_1'$ (from the property D) such that $F(Tz) = GF(z)$ for all $z \in \mathbb{C}$. Now $Gw = \bar{w} + \gamma' + \beta'$, $\gamma' \in \Sigma'$. Hence $\alpha(\bar{z} + \gamma + \beta) + a = \bar{\alpha}\bar{z} + \bar{a} + \gamma' + \beta'$ for all $z \in \mathbb{C}$. Hence $\alpha \in \mathbb{R}^*$ and $\alpha\beta + 2i\,\mathrm{Im}\, a = \gamma' - \alpha\gamma + \beta' \in \Sigma' + \beta'$.

*Case* 2. $F$ antianalytic. In the same way as in the previous case one shows that $F(z) = \alpha\bar{z} + a$ where $\alpha \in \mathbb{R}^*$, $\alpha\Sigma \subsetneq \Sigma'$ and $\alpha\bar{\beta} + 2i\,\mathrm{Im}\, a \in \Sigma' + \beta'$ and we get the situation in 2) (ii).

**Remark.** One can see from Theorem 4 that a morphism is either constant or unbranched. In the second case, it is a covering map.

## 8. The connection with the orientable unbranched two-sheeted covering

We have seen in Theorem 4 that a morphism $f : \mathbb{C}/\mathcal{K} \to \mathbb{C}/\mathcal{K}'$ has one of the following two forms:

$(t_1)$ $f(\tilde{z}) = \widetilde{\alpha z + a}'$, that is $f \circ p \circ \pi = p' \circ \pi' \circ F$, $F(z) = \alpha z + a$;

$(t_2)$ $f(\tilde{z}) = \widetilde{\alpha\bar{z} + a}'$, that is $f \circ p \circ \pi = p' \circ \pi' \circ F$, $F(z) = \alpha\bar{z} + a$.

Let us consider each case separately.

*Case* $(t_1)$. If $z_1, z_2 \in \mathbb{C}$ and $z_1 \sim_{\mathcal{K}_1} z_2$, then $\alpha z_1 + a \sim_{\mathcal{K}_1'} \alpha z_2 + a$. Hence one can define $\hat{f}_1 : \mathbb{C}/\Sigma \to \mathbb{C}/\Sigma'$ by $\hat{f}_1(\hat{z}) = \widehat{\alpha z + a}'$, that is $\hat{f}_1 \circ \pi = \pi' \circ F$.

$$\begin{array}{ccc}
\mathbb{C} & \xrightarrow{\ F\ } & \mathbb{C} \\
{\scriptstyle \pi}\downarrow & & \downarrow{\scriptstyle \pi'} \\
\mathbb{C}/\Sigma & \xrightarrow{\ \hat{f}_1\ } & \mathbb{C}/\Sigma' \\
{\scriptstyle p}\downarrow & & \downarrow{\scriptstyle p'} \\
\mathbb{C}/\mathcal{K} & \xrightarrow{\ f\ } & \mathbb{C}/\mathcal{K}'
\end{array}$$

The maps $\pi$ and $\pi'$ being analytic covering maps and $F$ being a conformal automorphism of $\mathbb{C}$, $\hat{f}_1$ will be an analytic covering map. The map $\pi$ being surjective, $p' \circ \hat{f}_1 \circ \pi = p' \circ \pi' \circ F = f \circ p \circ \pi$ and one gets

$$p' \circ \hat{f}_1 = f \circ p. \tag{1}$$

Let $\hat{f}_2 := f_1 \circ k$. One obtains $p' \circ \hat{f}_2 = p' \circ \hat{f}_1 \circ k = f \circ p \circ k = f \circ p$. Obviously $\hat{f}_2$ is an antianalytic morphism of a torus. Hence

$$p' \circ \hat{f}_1 = p' \circ \hat{f}_2 = f \circ p. \tag{2}$$

Because $p'$ is a covering map of degree 2, $\hat{f}_1$ and $\hat{f}_2$ are the only continuous liftings $\varphi : \mathbb{C}/\Sigma \to \mathbb{C}/\Sigma'$ of $f \circ p$.

*Case* $(t_2)$. Again if $z_1 \sim_{\mathcal{K}_1} z_2$, $\alpha z_1 + a \sim_{\mathcal{K}_1'} \alpha z_2 + a$. Hence one can define $\hat{f}_1 : \mathbb{C}/\Sigma \to \mathbb{C}/\Sigma'$ by

$$\hat{f}_1(\hat{z}) := \widehat{\alpha\bar{z} + a}', \text{ that is } \hat{f}_1 \circ \pi = \pi' \circ F. \tag{3}$$

This time $f_1$ is an antianalytic covering map. The map $\pi$ being surjective,

$$p' \circ f_1 \circ \pi = p' \circ \pi' \circ F = f \circ p$$

and one has

$$p' \circ \hat{f}_1 = f \circ p. \tag{4}$$

Let $\hat{f}_2 := \hat{f}_1 \circ k$. Then $f_2$ is an analytic covering map and $p' \circ \hat{f}_2 = f \circ p$. Once again, $\hat{f}_1$ and $\hat{f}_2$ are the only continuous liftings $\varphi : \mathbb{C}/\Sigma \to \mathbb{C}/\Sigma'$ of $f \circ p$.

A consequence of the preceding facts is the following (see [5], Lemma 3.1)

**Theorem 5.** *If* $f : \mathbb{C}/\mathcal{K} \to \mathbb{C}/\mathcal{K}'$, *is a nonconstant morphism, then there exists a single analytic morphism* $\varphi$ *and a single antianalytic morphism* $\psi$, $\varphi, \psi : \mathbb{C}/\Sigma \to \mathbb{C}/\Sigma'$, *such that*

$$p' \circ \varphi = p' \circ \psi = f \circ p.$$

**9. Isomorphic Klein bottles**

**Proposition 10.** *Let* $f: \mathbb{C}/\mathcal{K} \to \mathbb{C}/\mathcal{K}'$, *be a nonconstant morphism. Then, if* f *is an isomorphism, the liftings* $\hat{f}_1$ *and* $\hat{f}_2$ *of* $f \circ p$ *are isomorphisms as well.*

*Proof.* Every nonconstant morphism of Klein bottles or of tori being a covering map, it is an open map.

The surfaces $\mathbb{C}/\mathcal{K}$ and $\mathbb{C}/\Sigma$ being compact, f, $\hat{f}_1$, $\hat{f}_2$ are surjective. Hence f (resp. $\hat{f}_1$) is isomorphic if and only if it is injective.

Let us suppose $f_1$ is noninjective. Let $z_1, z_2 \in \mathbb{C}$, $\hat{z}_1 \neq \hat{z}_2$ and $\hat{f}_1(\hat{z}_1) = \hat{f}_1(\hat{z}_2)$. Then $(p' \circ \hat{f}_1)(\hat{z}_1) = (p' \circ \hat{f}_2)(\hat{z}_2)$, that is $f \circ p(\hat{z}_1) = f \circ p(\hat{z}_2)$. But f is supposed to be injective, hence $p(\hat{z}_1) = p(\hat{z}_2)$.

Because $p^{-1}(\tilde{z}) = \{\hat{z}; k\hat{z}\}$ for all $\tilde{z} \in \mathbb{C}/\mathcal{K}$, one has $\hat{z}_1 = k\hat{z}_2$. Hence $\hat{f}_1(\hat{z}_1) = \hat{f}_1(\hat{z}_2) = \hat{f}_2(\hat{z}_1)$, a contradiction because of $\hat{f}_1(\hat{z}) \neq \hat{f}_2(\hat{z})$ for all $\hat{z} \in \mathbb{C}/\Sigma$, $\hat{f}_1$ and $\hat{f}_2$ being liftings of the same map $f \circ p$.

**Proposition 11.** *Let* $a \in \mathbb{C}$ *and* $\alpha \in \mathbb{R}^*$ *satisfy the condition* 2) (i) *of Theorem* 4 *and* $\varphi: \mathbb{C}/\Sigma \to \mathbb{C}/\Sigma'$ *be defined by* $\varphi(\hat{z}) := \widehat{\alpha z + a}'$. *Then, if* $\varphi$ *is injective,* $\alpha\Sigma = \Sigma'$.

*Proof.* Let us suppose $\Sigma' \setminus \alpha\Sigma \neq \emptyset$. If $\theta \in \Sigma' \setminus \alpha\Sigma$, $\theta_1 = \theta/\alpha \notin \Sigma$ and $\hat{\theta}_1 \neq \hat{0}$. Now $\varphi(\hat{\theta}_1) = \widehat{\alpha\theta_1 + a}' = \widehat{\theta + a}' = \hat{\theta}' + \hat{a}' = \hat{a}' = \varphi(\hat{0})$. Hence $\varphi$ is not injective, a contradiction.

**Corollary.** *In the context of Theorem* 4, *if* f *is an isomorphism then* $\alpha\Sigma = \Sigma'$.

**Theorem 6.** *The map* $f: \mathbb{C}/\mathcal{K} \to \mathbb{C}/\mathcal{K}'$, *is an isomorphism if and only if there exist* $a \in \mathbb{C}$ *and* $\alpha \in \mathbb{R}^*$ *such that one of the following two situations holds:*

(i) $\alpha\Sigma = \Sigma'$, $\alpha\beta + 2i \operatorname{Im} a \in \Sigma' + \beta'$ *and* $f(\tilde{z}) = \widetilde{\alpha z + a}'$;

(ii) $\alpha\Sigma = \Sigma'$, $\alpha\bar{\beta} + 2i \operatorname{Im} a \in \Sigma' + \beta'$ *and* $f(\tilde{z}) = \widetilde{\alpha\bar{z} + a}'$.

*Proof.* The necessity is a direct consequence of Theorem 4 and of the preceding corollary.

Sufficiency. *Case* (i). Clearly $\alpha\Sigma = \Sigma'$ if and only if $\frac{1}{\alpha}\Sigma' = \Sigma$. There exists $\gamma' \in \Sigma'$, so that $\alpha\beta + 2i \operatorname{Im} a = \gamma' + \beta'$. Hence $\frac{1}{\alpha}\beta' + 2i \operatorname{Im} \frac{-a}{\alpha} = \frac{1}{\alpha}(-\gamma') + \beta \in \Sigma + \beta$. According to Theorem 4, we may define $g: \mathbb{C}/\mathcal{K}' \to \mathbb{C}/\mathcal{K}$ by

$$g(\tilde{w}') := \widetilde{\frac{1}{\alpha} w - \frac{a}{\alpha}}.$$

One verifies at once that $g = f^{-1}$; hence f is an isomorphism.

*Case* (ii). In this case the function $g : \mathbb{C}/\mathcal{K}' \to \mathbb{C}/\mathcal{K}$ defined by

$$g(\tilde{w}') := \widetilde{\frac{1}{\alpha}\overline{w} - \frac{\overline{a}}{\alpha}},$$

is well defined and one verifies that $g = f^{-1}$.

**Proposition 12.** *Let* $\mathcal{K} = \{S; V\}$, $Sz = \overline{z} + \beta$, $\beta = u + iv$, $u > 0$, $Vz = z + ib$, $b > 0$ *and* $v \in [-b/2, b/2]$. *Let* $S^*z = \overline{z} + \overline{\beta}$ *and* $\mathcal{K}^* = \{S^*; V\}$. *In this context the Klein bottles* $\mathbb{C}/\mathcal{K}$ *and* $\mathbb{C}/\mathcal{K}^*$ *are isomorphic.*

*Proof.* In Theorem 6, (ii), one considers $\alpha = 1$, $\beta' = \overline{\beta}$ and $a = 0$. Obviously $\Sigma = \Sigma'$ and $\overline{\beta} \in \Sigma + \overline{\beta}$. Hence, one can define $h : \mathbb{C}/\mathcal{K} \to \mathbb{C}/\mathcal{K}^*$ by $h(\tilde{z}) := \tilde{\overline{z}}'$. Then h is an isomorphism.

As a consequence, to find the Teichmüller space of the Klein bottles it is sufficient to consider the situation $\beta = u + iv$, $u > 0$ and $v \in [0, b/2]$ only; we shall consider this situation from now on.

**Theorem 7.** *Let* $\mathcal{K}_0 = \{S_0; V\}$, $S_0z = \overline{z} + u$, $\beta = u + iv$, $v \in [0, b/2]$. *Then the Klein bottles* $\mathbb{C}/\mathcal{K}$ *and* $\mathbb{C}/\mathcal{K}_0$ *are isomorphic.*

*Proof.* Obviously, $\beta_0 = u$, $\Sigma = \Sigma_0$. In Theorem 6 one takes $\alpha = 1$ and $a = a_1 - iv/2$, $a_1 \in \mathbb{R}$. Then $\alpha\beta + 2i \operatorname{Im} a = \beta_0 \in \Sigma_0 + \beta_0$. Hence $h: \mathbb{C}/\mathcal{K} \to \mathbb{C}/\mathcal{K}_0$, $h(\tilde{z}) := \widetilde{z + a_1 - iv/2}'$, is an isomorphism.

Again we may simplify our finding of the Teichmüller space of the Klein bottles supposing $\beta = u > 0$, that is, $v = 0$.

Let $\mathcal{K}_0 = \{S_0; V_0\}$, $S_0z = \overline{z} + 1/2$ $(\beta_0 = 1/2)$ and $V_0z = z + ib/\omega$. Then $\omega_0 = 2\beta_0 = 1$, $\Sigma_0 = \mathbb{Z} + \mathbb{Z}ib/\omega$, $\Sigma = \mathbb{Z}\omega + \mathbb{Z}ib = \omega\Sigma_0$.

With $a = 0$ and $\alpha = 1/\omega > 0$, from Theorem 6, one has $\alpha\Sigma = \Sigma_0$ and $\alpha\beta + 2i \operatorname{Im} a = u/\omega = 1/2 = \beta_0 \in \Sigma_0 + \beta_0$. Hence $f : \mathbb{C}/\mathcal{K} \to \mathbb{C}/\mathcal{K}_0$, $f(\tilde{z}) := \widetilde{z/\omega}'$, is an isomorphism of Klein bottles.

Thus we have obtained the final reduction of the problem of finding the Teichmüller space of the Klein bottles:

Every Klein bottle is dianalytically equivalent to a Klein bottle of the type $\mathbb{C}/\mathcal{K}$ where $\mathcal{K} = \{S; V_b\}$, $Sz = \overline{z} + 1/2$, $V_bz = z + ib$, $b > 0$. From now on, $\mathcal{K}$ is supposed to be of this form.

Let $\mathcal{K}' = \{S; V_{b'}\}$, $V_{b'}z = z + ib'$, $b' > 0$.

**Theorem 8.** *The Klein bottles* $\mathbb{C}/\mathscr{K}$ *and* $\mathbb{C}/\mathscr{K}'$ *are isomorphic if and only if* $b = b'$.

*Proof.* The necessity is trivial. To prove the sufficiency, let $f : \mathbb{C}/\mathscr{K} \to \mathbb{C}/\mathscr{K}'$ be an isomorphism. There exist $\alpha \in \mathbb{R}^*$ and $a \in \mathbb{C}$ with the properties $\alpha\Sigma = \Sigma'$, $\alpha/2 + 2i \operatorname{Im} a \in \Sigma' + 1/2$ and $f(\tilde{z}) = \widetilde{\alpha z + a}'$ or $\widetilde{\alpha \bar{z} + a}'$. For both lattices $\Sigma$ and $\Sigma'$, $\omega = 1$ and then $\alpha\Sigma = \Sigma'$ if and only if $\alpha = 1$ or $\alpha = -1$. Since $\alpha = \pm 1$, $\Sigma = \Sigma'$ and $b = b'$.

**Corollary.** *The Teichmüller space of the Klein bottles is*

$$\mathscr{T}\mathscr{K} = \{ib \mid b \in \mathbb{R}, b > 0\}.$$

*Proof.* This is a consequence of Theorem 8 and the remark after Theorem 7.

## 10. The group of dianalytic automorphisms of a Klein bottle

Let $b > 0$, $Sz = \bar{z} + 1/2$, $Vz = z + ib$ and $\mathscr{K} = \{S; V\}$. By means of Theorem 6 we shall describe now the group of dianalytic automorphisms of $\mathbb{C}/\mathscr{K}$. In this case, $\Sigma = \mathbb{Z} + \mathbb{Z}ib$. If $\alpha \in \mathbb{R}^*$, then $\alpha\Sigma = \Sigma$ if and only if $\alpha = \pm 1$. Let $a = a_1 + ia_2 \in \{\lambda/2 + ib\mu \mid 0 \le \lambda,\mu < 1\}$. For $\alpha = \pm 1$, the conditions (i) and (ii) of Theorem 6 become a single one, namely

$$2ia_2 \in \Sigma. \tag{*}$$

This holds if and only if $a_2 = 0$ or $a_2 = b/2$. With this remark, we get

**Theorem 9.** *The map* $f : \mathbb{C}/\mathscr{K} \to \mathbb{C}/\mathscr{K}$ *is an automorphism if and only if* $f$ *has one of the following forms:*

(1) $f(\tilde{z}) = \widetilde{\alpha z + a}$;

(2) $f(\tilde{z}) = \widetilde{\alpha \bar{z} + a}$;

(3) $f(\tilde{z}) = \widetilde{\alpha z + a + ib/2}$;

(4) $f(\tilde{z}) = \widetilde{\alpha \bar{z} + a + ib/2}$,

*where* $\alpha = \pm 1$ *and* $a \in [0, 1/2)$.

**Remark.** This is the complete list of the dianalytic automorphism of $\mathbb{C}/\mathscr{K}$.

The proof of the theorem is very simple and will be omitted.

**Corollary.** *The group of dianalytic automorphisms of a Klein bottle is a continuous group depending on a single real parameter.*

## References

[1] Alling, N. L., Greenleaf, N., Foundations of the theory of Klein surfaces, Lecture Notes in Math., 219, Springer-Verlag, Berlin - Heidelberg - New York, 1971.

[2] Bârză, I., Surfaces de Riemann non orientables, Preprint Series in Mathematics, 21/1981, INCREST, Bucharest.

[3] Forster, O., Lectures on Riemann surfaces, Springer-Verlag, New York - Heidelberg - Berlin, 1981.

[4] Schiffer, M., Spencer, D., Functionals of finite Riemann surfaces, Princeton University Press, Princeton, N.J., 1954.

[5] Seppälä, M., Teichmüller spaces of Klein surfaces, Ann. Acad. Sci. Fenn. Ser. A I Math. Dissertationes 15 (1978).

# REMARKS ON SOBOLEV IMBEDDING INEQUALITIES [1]

B. Bojarski [2]
Department of Mathematics, University of Warsaw
PKIN, IX p., 00901 Warszawa, Poland

This paper is a continuation and extension of some topics discussed in [6]. In particular we present here a detailed proof of the theorem, only sketched in [6]. In §6 we also give another, rather direct proof of Sobolev imbedding inequalities for John domains (def. below) based on §3 of [6] and on a recent characterization of John domains given by O. Martio in [19].

We shall use the usual notations of Sobolev space theory, see e.g. [21]. For a bounded domain $\Omega \subset R^n$, $W^{l,p}_{loc}(\Omega)$ is the space of functions $f(x)$, $x \in \Omega$, with distributional derivatives $D^\alpha f$ of order $\leq l$ in $L^p(\Omega')$ for every compact subset $\Omega'$ of $\Omega$.

The measure of a Lebesgue measurable set $E$ is denoted by $|E|$. The average value of an integrable function $f$ on $E$ is denoted by the barred integral

$$\fint_E f\,dx = \frac{1}{|E|}\int_E f\,dx = f_E .$$

$\chi_E$ is the characteristic function of $E$. $B$ and $Q$ will be general symbols for open balls and cubes; $r(Q)$ is the sidelength of $Q$ and $\sigma Q$, $\sigma > 0$, is the cube with the same center as $Q$, extended homothetically $\sigma$ times. $P_k$ will denote the space of polynomials of degree $\leq k$ in $n$ variables $(x_1,\ldots,x_n) \in R^n$.

Our starting point is the local Sobolev inequality for the functions $f(x)$ in the Sobolev space $W^{l,p}(\Omega)$, $1 \leq lp < n$, $p \geq 1$. The case $l = 1$ is of special interest because of direct geometric interpretation. Local Sobolev inequalities in this case have the form

---

[1] This paper is in final form and no version of it will be submitted for publication elsewhere.

[2] This article was prepared during the author's stay in autumn 1987 at the Centre for Mathematical Analysis in Canberra.

$$\Big(\int_{R^n} |f|^{p^*}dx\Big)^{1/p^*} \le C_1(n,p)\Big(\int_{R^n} |\nabla f|^p\,dx\Big)^{1/p},\quad p^* = \frac{p\cdot n}{n-p} > p \tag{1.1}$$

for f compactly supported, say $f \in C_0^\infty(R^n)$ and

$$\Big(\fint_{B_R} |f - f_{B_R}|^{p^*}dx\Big)^{1/p^*} \le C_2(n,p)\,R\,\Big(\fint_{B_R} |\nabla f|^p\,dx\Big)^{1/p} \tag{1.2}$$

for functions without restrictions on the support. Here $B_R$ is a ball with radius R and $\nabla f$ is the gradient of f. The optimal constants $C_1$ and $C_2$ in (1.1) and (1.2) depend on n and p only. (1.2) implies (1.1) with $C_1 \le C_2$. For $p = 1$, (1.1) is known to be essentially equivalent to the isoperimetric inequality (see e.g. [21]). For general bounded domains $\Omega$ the Sobolev inequality for $l = 1$ should have the form

$$\Big(\fint_{\Omega} |f - f_\Omega|^{p^*}dx\Big)^{1/p^*} \le C(n,p,\Omega)\Big(\fint_{\Omega} |\nabla f|^p\,dx\Big)^{1/p} \tag{1.3}$$

or the form

$$\Big(\int_{\Omega} |f - f_\Omega|^{p^*}dx\Big)^{1/p^*} \le \hat{C}(n,p,\Omega)\Big(\int_{\Omega} |\nabla f|^p\,dx\Big)^{1/p} \tag{1.4}$$

for functions in $W^{1,p}(\Omega)$. While (1.3) and (1.4) are true for various classes of domains $\Omega$, e.g. convex domains, domains satisfying Sobolev cone condition etc. ([30], [20], [10]) they are not true in general. Counterexamples to (1.4) go back to O. Nikodym [24], see also [21]. Inequalities (1.4) and (1.3) are closely related with the essentially weaker Poincaré inequality

$$\Big(\int_{\Omega} |f - f_\Omega|^{p}\,dx\Big)^{1/p} \le P(n,p,\Omega)\Big(\int_{\Omega} |\nabla f|^p\,dx\Big)^{1/p} \tag{1.5}$$

which follows from (1.3) by the monotonicity of the arithmetic means, [13], with $P(n,p,\Omega) \le C(n,p,\Omega)$. We remark that (1.5) is not sufficient to obtain (1.1) and is thus not equivalent to the isoperimetric inequality. We remark also, and that is in fact very essential for all what follows, that (1.3) and (1.4) are conformally invariant, in the sense that $C(n,p,\lambda\Omega) = \lambda C(n,p,\Omega)$, $\lambda > 0$ $(\hat{C}(\lambda\Omega) = \hat{C}(\Omega))$, where $\lambda\Omega$ denotes the domain obtained from $\Omega$ by homotety with dilatation $\lambda$. The Poincaré inequality (1.5) is not conformally invariant. Later we shall use (1.2) for cubes Q. The only difference is that then (1.2) holds with, in general, another constant $C_2(n,p)$. For $l > 1$ the local Sobolev inequality takes the form

$$\Big(\fint_{B_R} |f-P_B(f)|^{p^*}dx\Big)^{1/p^*} \le C(n,l,p)\, R^l \Big(\fint_{B_R} |\nabla^l f|^p\Big)^{1/p} \tag{1.6}$$

with $p^* = \frac{np}{n-lp}$, where $\nabla^l f$ denotes the gradient of order $l$ of f i.e. $\nabla^l f = \{D^\alpha f, |\alpha| = l\}$ and $P_B(f)$ is a polynomial of degree $\le l-1$. This polynomial can be chosen so that for $x \in B_R$

$$|P_B(x)| \le C \fint_\omega |f| \tag{1.7}$$

where $\omega$ is some subset of positive measure of $B_R$. Then the constant C in (1.6) will depend on the ratio $|\omega|/|B_R|$. Classically $P_B(f)$ is obtained by averaging the formal Taylor expansion of f in $B_R$ over $\omega$. The proof of (1.6) goes back to the Sobolev papers from 1936. In the form (1.6) with the condition of type (1.7) it is contained in [10] (compare also [20], [21], [3], [4]). The local inequality (1.6) gives immediately the imbedding $W^{l,p}_{loc} \to L^{p^*}_{loc}(\Omega)$ for any domain $\Omega$. The main result here is that for some rather broad classes of domains the local inequalities (1.6) can be "integrated" to the global inequality

$$\Big(\int_\Omega |f-P_\Omega(f)|^{p^*}dx\Big)^{1/p^*} \le A\Big(\int_\Omega |\nabla^l f|^p\, dx\Big)^{1/p} \tag{1.8}$$

with some control, of type (1.7), for the polynomial $P_\Omega(f)$, and the coefficient A which describes the imbedding $W^{l,p}(\Omega) \to L^{p^*}(\Omega)$.

The phenomenon that in some situations the local integral estimates can be "integrated" to global estimates was probably first observed by J. Boman in [8]. He considered the case $p^* = p$, $P_\Omega(f) =$ constant. See also [4]. The novelty here is to show that the Markov type inequalities for polynomials and the scale invariance of the Sobolev inequalities make it possible to obtain a proof of a Sobolev imbedding theorem. An analogous phenomenon for Lipschitz classes was discovered by F. Gehring and O. Martio in [12]. Their result has been also applied to the Sobolev imbedding theorem for the case $p > n$: $W^{1,p}(\Omega) \to Lip_\alpha(\Omega)$, see [16].

The general imbeddings $W^{l,p}(\Omega) \to C^{k,\alpha}(\Omega)$ for $lp > n$ could be also discussed via the integral estimates. This would require to consider the local (and global) inequalities of the form

$$\sum_{j=0}^{k} \|\nabla^j(f - Pf)\|_{L^q} \leq C\|\nabla^l f\|_{L^p} \tag{1.9}$$

for various values of the parameters $k, j, q, l, p$ in agreement with the general case of the Sobolev imbedding theory [21]. For this general case the full form of the Markov inequality described in §2 would be relevant, while for the case $lp < n$ and $j = 0$ only the special case $\alpha = 0$ is needed. However, even for $lp < n$, the case of admissible positive values of $j$, see [21], requires the Markov inequality for $|\alpha| > 0$. Details of these questions are not discussed here.

We remark also that the use of balls and cubes in our considerations is not essential. We could as well work with rectangles, triangles or even curvilinear simplexes $\Delta$ and their n-dimensional analogues with some restriction on their form so that their linear dimension is controlled by their volume in some uniform way, $|\Delta| \sim (\text{diam}\,\Delta)^n$, and a uniform bound on the overlap after a homothetical dilatation $\Delta \to \sigma\Delta$, $\sigma > 1$. In particular for $n = 2$, "thin and long" rectangles are excluded. Generally speaking, the scale or conformal invariance of the Sobolev estimates makes it possible to use instead of cubes and balls the images of these under rather general classes of bilipschizian or even K-quasiconformal mappings, satisfying some general conditions. The stability theory of quasiconformal mappings and various global estimates of the distortion of shape and volume under these mappings, see [33], [29], [26], [25], [6], make them a good tool in studying the behaviour of various averaging processes under coordinate transformations. In particular, in [25] the distortion estimates for quasiconformal mappings have been applied to show that the BMO-class of F. John and L. Nirenberg is invariant under quasiconformal transformations.

The theory proposed here so far leaves aside the Sobolev trace theorems. There are no essential geometric reasons for that and it is possible to extend the discussed approach also to this case. The work on these problems is in progress.

## 2. The averaged Markov inequality

In the proof below essential use is made of the averaged Markov inequality. Various forms of Markov type inequalities can be found in the literature. The strongest and most general results in this direction seem to be the theorems of Brudnyj and Gansburg [9].

We formulate the Markov inequality in the following general form [5].

**Theorem 2.1.** *Let* $Q$ *be an arbitrary cube in* $R^n$ *and* $\omega$ *a measurable subset of* $Q$ *of positive measure:* $|\omega| = \sigma|Q|$, $0 < \sigma \leq 1$. *Then for any polynomial* $P \in P_k$ *the inequality*

$$\Big(\fint_Q |D^\alpha P|^p\, dx\Big)^{1/p} \leq C|Q|^{-|\alpha|/n}\, \sigma^{-k}\Big(\fint_\omega |P - S|^q\, dx\Big)^{1/q} \tag{2.1}$$

*holds, with the constant* $C$ *depending on* $n, k, \alpha, p, q$ *only.* $S$ *in* (2.1) *is an arbitrary polynomial of degree strictly less than* $k = |\alpha|$. *The constants* $p$ *and* $q$ *are arbitrary positive numbers, the case* $p = +\infty$ *is also included.*

As shown in [5] the inequality (2.1) in its most general form can be deduced from the results of Brudnyj and Gansburg [9], who considered the case $\alpha = 0$. We remark that for our purposes it is enough to apply (2.1) to the special case when $\omega$ is a parallepiped. Then, as shown in [5], the Markov inequality (2.1) can be deduced by elementary considerations, not appealing to the results of [9]. In §4 we apply the Markov inequality for $\alpha = 0$. It then has the simpler form

$$\Big(\fint_Q |P(x)|^p\, dx\Big)^{1/p} \leq C\sigma^{-k}\Big(\fint_\omega |P(x)|^q\, dx\Big)^{1/q}, \quad \sigma = |\omega|/|Q| \tag{2.2}$$

and expresses the fact that the averages of polynomials over measurable subsets are controlled by the ratio of the measures of the corresponding sets.

**Remark.** Weighted averaged Markov's inequalities hold also with respect to weights satisfying the general conditions of $A_p$-type of Muckenhoupt [22]. See also [6].

## 3. Domains satisfying Boman's chain condition

Let $\sigma \geq 1$, $C > 1$ and $B > 1$ be fixed numbers ($B$ integer).

**Definition.** An open set $\Omega \subset R^n$ satisfies the chain condition $F(\sigma,B,C)$ if there exist a covering $F$ of $\Omega$ by open cubes such that

i) $\Sigma_{Q \in F}\, \chi_{\sigma Q}(x) \leq B\chi_\Omega(x)$, $x \in R^n$,

ii) for some fixed cube $Q_0$ in $F$, called the central cube, and for every $Q \in F$ there exists a chain $Q_0, Q_1, \ldots, Q_N = Q$ of cubes from $F$ such that for each $\nu = 0,1,\ldots,N-1$,

$$Q \subset BQ_\nu .$$

Moreover it is required that the consecutive cubes of the connecting chain are comparable in size and overlap in some uniform way:

iii) $$C^{-1} \le \frac{|Q_\nu|}{|Q_{\nu+1}|} \le C,$$
$$|Q_\nu \cap Q_{\nu+1}| \ge C^{-1} \max(|Q_\nu|,|Q_{\nu+1}|), \quad \nu = 0,1,\ldots,N-1.$$

It is important that the lengths $N = N(Q)$ of the connecting chain may depend on the particular cube in $F$ and are not assumed to be (uniformly) bounded in the covering family $F$.

Many important classes of domains in $R^n$ satisfy the chain condition. These include bounded Lipschitz domains, bounded domains satisfying the Sobolev cone condition etc. More generally, J. Boman proves in [8] the following lemma:

**Lemma 3.1.** *Bounded John domains satisfy the chain condition (for some choice of* $B$ *and* $C$*).*

We recall that a domain $\Omega$ in $R^n$ is called an $(\alpha,\beta)$-John domain, $0 < \alpha \le \beta$, if for some point $p_0 \in \Omega$, called a John center of $\Omega$, each point $p \in \Omega$ can be joined to $p_0$ by a rectifiable (or, what is equivalent, piecewise linear) path $\gamma(t)$ of length $l(\gamma)$, $\gamma : [0, l(\gamma)] \to \Omega$, parametrized by the arc length, such that

$$l(\gamma) \le \beta,$$
$$d(\gamma(t),\partial\Omega) \ge \frac{\alpha t}{l(\gamma)}, \quad t \in [0, l(\gamma)].$$

Here $d(p,\partial\Omega)$ is the distance of the point $p \in \Omega$ to the boundary $\partial\Omega$.

A domain $\Omega$ is a John domain if it is an $(\alpha,\beta)$-John domain for some $\alpha$ and $\beta$. For an $(\alpha,\beta)$-John domain the constants $B$ and $C$ depend on $\alpha$, $\beta$ and the dimension $n$. It is known that John domains may have a very irregular non-rectifiable boundary. However, important classes of domains, e.g. some irregular fractals [8], [7], are John domains. In particular, the snow-flake domains, domains bounded by a Koch curve [7], bounded quasi-discs of two-dimensional quasiconformal theory etc. are John-domains. In general domains with external cusps are not John-domains.

An important characterisation of John domains has been recently given by O. Martio [19].

## 4. The basic lemma

We say that the function $f(x)$, $x \in \Omega$, is in the local class $\mathrm{loc}_\sigma L^r(\Omega)$ if the integral $\int_{\sigma Q} |f|^r\, dx$ is finite for each cube $Q \subset \Omega$ such that the extended cube $\sigma Q \subset \Omega$ $(\sigma \geq 1)$. For $\sigma = 1$ we write $\mathrm{loc}_1 L^r(\Omega) = \mathrm{loc}\, L^r(\Omega)$.

For a real $p$, $1 \leq p < n/l$, $n = \dim \Omega$, $l \geq 1$, $l$ integral, let $p^* = np/(n - lp) > p$ be the Sobolev conjugate to $p$.

**Lemma 4.1.** *Assume that the domain* $\Omega$ *satisfies the chain condition* $F(\sigma,B,C)$. *Let* $g$ *be a non-negative function in the local class* $\mathrm{loc}_\sigma L^p(\Omega)$. *Assume that the function* $f \in \mathrm{loc}\, L^{p^*}(\Omega)$ *satisfies the following condition: for each cube* $Q \in F$ *there exists a polynomial* $P_Q(f) \equiv P_Q$ *of order at most* $l-1$ *such that the inequality*

$$\Big( \fint_Q |f - P_Q|^{p^*} dx \Big)^{1/p^*} \leq A\, r^l(Q) \Big( \fint_{\sigma Q} g^p\, dx \Big)^{1/p} \tag{4.1}$$

*holds with a constant* $A$ *not depending on the cube* $Q$. *Then there exists a constant* $\hat{A}$, *depending on* $n, p, \sigma, B, C, l$ *and* $A$ *only, and a polynomial* $P_\Omega(f)$ *of order* $l-1$, *such that the global inequality*

$$\Big( \int_\Omega |f - P_\Omega(f)|^{p^*} dx \Big)^{1/p^*} \leq \hat{A} \Big( \int_\Omega g^p\, dx \Big)^{1/p} \tag{4.2}$$

*holds if* $g \in L^p(\Omega)$ *i.e.* $\int_\Omega g^p\, dx < \infty$.

In a rather detailed proof of the lemma we use in an essential way the ideas of the proof of Theorem 3.1 in the J. Boman's paper [8]. See also [14]. However, Sobolev type inequalities are not considered in [8].

We need also the following lemma from [8] the proof of which is given here for the sake of completeness. This lemma, as pointed in [8], is related with some ideas of O. Strömberg and A. Torchinsky in [32].

**Lemma 4.2.** *Let* $F = \{Q_\alpha\}$ *be an arbitrary family of cubes in* $R^n$ *(indexed by some parameter* $\alpha \in I$*). Assume that for each* $Q_\alpha$ *we are given a non-negative number* $a_\alpha$. *Then for* $1 \leq p < \infty$ *and* $B \geq 1$ *we have*

$$\Big\| \sum_\alpha a_\alpha \chi_{BQ_\alpha} \Big\|_{L^p} \leq D \Big\| \sum_\alpha a_\alpha \chi_{Q_\alpha} \Big\|_{L^p}$$

*where the constant* D *depends on* n, p *and* B *only.*

*Proof.* Let $\varphi \in L^p(R^n)$ with $1/p + 1/p' = 1$, $1 < p' \leq \infty$. We will denote by $M\varphi(x) = \sup_Q ⨍_Q |\varphi(t)|\,dt$ the Hardy-Littlewood maximal function of $\varphi$. Here the supremum is taken over all cubes $Q \subset R^n$ which contain x. We will use the Hardy-Littlewood inequality [31]:

$$\|M\varphi\|_{L^{p'}} \leq H_{p'}\|\varphi\|_{L^{p'}} \tag{4.3}$$

with the constant $H_{p'}$ depending only on p'. Now we have, by (4.3),

$$\begin{aligned}
\Big|\int_{R^n} \sum_\alpha a_\alpha \chi_{BQ_\alpha} \varphi(x)\,dx\Big| &= \Big|\sum_\alpha a_\alpha |BQ_\alpha| ⨍_{BQ_\alpha} \varphi\,dx\Big| \\
&\leq B^n \sum_\alpha a_\alpha \int_{Q_\alpha} M\varphi\,dx = B^n \int_{R^n} \sum_\alpha a_\alpha \chi_{a_\alpha} M\varphi\,dx \\
&\leq B^n \Big\|\sum_\alpha a_\alpha \chi_{Q_\alpha}\Big\|_{L^p} \|M\varphi\|_{L^{p'}} \\
&\leq B^n H_{p'} \Big\|\sum_\alpha a_\alpha \chi_{Q_\alpha}\Big\|_{L^p} \|\varphi\|_{L^{p'}}
\end{aligned}$$

and the lemma follows with $D = B^n H_{p'}$.

*Proof of Lemma* 4.1. In what follows we shall use only the cubes Q belonging to the family F in the definition (3.1). We shall estimate

$$\int_\Omega |f - P_{Q_0} f|^{p^*} dx.$$

We have

$$\begin{aligned}
\int_\Omega |f - P_{Q_0} f|^{p^*} dx &\leq \sum_{Q \in F} \int_Q |f - P_{Q_0} f|^{p^*} dx \\
&\leq \sum_{Q \in F} 2^{p^*}\Big(\int_Q |f - P_Q f|^{p^*} dx + \int_Q |P_Q f - P_{Q_0} f|^{p^*} dx\Big) \\
&\leq \tilde{A} \sum_{Q \in F} \Big(\int_{\sigma Q} g^p\,dx\Big)^{p^*/p} + 2^{p^*} \sum_{Q \in F} \int_Q |P_Q f - P_{Q_0} f|^{p^*} dx \\
&\equiv I_1 + 2^{p^*} I_2, \quad \tilde{A} = 2^{p^*} A^{p^*} \sigma^{-np^*/p}.
\end{aligned}$$

We estimate $I_1$ and $I_2$. Obviously

$$I_1 \le A\Big(\sum_{Q\in F}\int_{\sigma Q} g^p\,dx\Big)^{p^*/p} \le \tilde{A}B^{p^*/p}\Big(\int_{\Omega} g^p\,dx\Big)^{p^*/p}$$

where B is the constant from the definition (3.1).

To estimate $I_2$ for each pair $(Q,Q_0)$ we use the connecting chain $\{Q_\nu\}$ $\nu = 0,1,\ldots,s$ of cubes from the family F. Here $s = s(Q)$ – the length of the connecting chain – is not a priori bounded for all cubes Q in the family F. We have $Q_{s(Q)} = Q$. For each term in $I_2$ we have then

$$\Big(\int_Q |P_Q f - P_{Q_0}|^{p^*}dx\Big)^{1/p^*} \le \sum_{\nu=0}^{s-1}\Big(\int_Q |P_{Q_\nu} f - P_{Q_{\nu+1}} f|^{p^*}dx\Big)^{1/p^*} \tag{4.4}$$

Since $Q \subset BQ_\nu$ and $Q \subset BQ_{\nu+1}$ by the condition, we have

$$\begin{aligned}\int_Q |P_{Q_\nu} f - P_{Q_{\nu+1}} f|^{p^*}dx &\le |BQ_\nu| \fint_{BQ_\nu} |P_{Q_\nu} f - P_{Q_{\nu+1}} f|^{p^*}dx\\ &\le C\,B^n|Q_\nu| \fint_{Q_\nu} |P_{Q_\nu} f - P_{Q_{\nu+1}} f|^{p^*}dx\end{aligned}$$

where the Markov inequality (2.2) has been applied to the polynomial $P_{Q_\nu} f - P_{Q_{\nu+1}} f$ of order $\le l - 1$.

Applying the Markov inequality once more we get

$$\int_Q |P_{Q_\nu} f - P_{Q_{\nu+1}} f|^{p^*}dx \le B^n|Q_\nu| \int_{Q_\nu\cap Q_{\nu+1}} |P_{Q_\nu} f - P_{Q_{\nu+1}} f|^{p^*}dx.$$

Now we have

$$\begin{aligned}&\int_{Q_\nu\cap Q_{\nu+1}} |P_{Q_\nu} f - P_{Q_{\nu+1}} f|^{p^*}dx\\ &\quad\le 2^{p^*-1}\Big(\int_{Q_\nu\cap Q_{\nu+1}} |P_{Q_\nu} f - f|^{p^*}dx + \int_{Q_\nu\cap Q_{\nu+1}} |P_{Q_{\nu+1}} f - f|^{p^*}dx\Big)\\ &\quad\le 2^{p^*-1}\Big(\int_{Q_\nu} |P_{Q_\nu} f - f|^{p^*}dx + \int_{Q_{\nu+1}} |P_{Q_{\nu+1}} f - f|^{p}\,dx\Big)\end{aligned}$$

$$\le \tilde{A} 2^{p^*-1}\Big[\Big(\int_{\sigma Q_\nu} g^p\,dx\Big)^{p^*/p} + \Big(\int_{\sigma Q_{\nu+1}} g^p\,dx\Big)^{p^*/p}\Big].$$

Finally we get

$$\Big(\int_Q |P_{Q_\nu} f - P_{Q_{\nu+1}} f|^{p^*} dx\Big)^{1/p^*} \le G\Big[\Big(\int_{\sigma Q_\nu} g^p\,dx\Big)^{1/p} + \Big(\int_{\sigma Q_{\nu+1}} g^p\,dx\Big)^{1/p}\Big]$$

with $G = B^n C \tilde{A} 2^{p^*-1}$.

The right hand side of (4.4) is then estimated by

$$2G \sum_{B\tilde{Q} \supset Q} \Big(\int_{\sigma\tilde{Q}} g^p\,dx\Big)^{1/p}$$

where the summation is extended over all cubes $\tilde{Q}$ in the family $F$ such that $B\tilde{Q} \supset Q$. Put

$$a_Q = \frac{1}{|Q|^{1/p^*}} \Big(\int_{\sigma Q} g^p\,dx\Big)^{1/p}$$

and observe that for $x \in Q$,

$$\sum_{B\tilde{Q} \supset Q} a_{\tilde{Q}} \le \sum_{\tilde{Q} \in F} a_{\tilde{Q}} \chi_{B\tilde{Q}}(x)$$

where in the sum on the right all cubes in $F$ are present. Therefore the term $\int_Q |P_Q f - P_{Q_0} f|^{p^*} dx$ in $I_2$ is estimated by

$$\int_Q |P_Q f - P_{Q_0} f|^{p^*} dx \le (2G)^{p^*} \Big(\sum_{B\tilde{Q} \supset Q} a_Q\Big)^{p^*} |Q| \le (2G)^{p^*} \int_Q \Big|\sum_{\tilde{Q} \in F} a_{\tilde{Q}} \chi_{B\tilde{Q}}(x)\Big|^{p^*} dx.$$

For $I_2$ we get then, in view of Lemma 4.2.,

$$I_2 \le (2G)^{p^*} \int_{R^n} \Big|\sum_{\tilde{Q} \in F} a_{\tilde{Q}} \chi_{B\tilde{Q}}(x)\Big|^{p^*} dx$$

$$\le (2G)^{p^*} D_{p^*}^{p^*} \int_{R^n} \Big|\sum_{\tilde{Q} \in F} a_{\tilde{Q}} \chi_{\tilde{Q}}(x)\Big|^{p^*} dx$$

$$\leq (2G)^{p^*} D_{p^*}^{p^*} B^{p^*-1} \int_{R^n} \sum_{\tilde{Q} \in F} a_{\tilde{Q}}^{p^*} \chi_{\tilde{Q}}(x)\, dx$$

$$= (2G)^{p^*} D_{p^*}^{p^*} B^{p^*-1} \sum_{\tilde{Q} \in F} a_{\tilde{Q}}^{p^*} |\tilde{Q}|$$

$$= (2G)^{p^*} D_{p^*}^{p^*} B^{p^*-1} \sum_{\tilde{Q} \in F} \Big( \int_{\sigma\tilde{Q}} g^p\, dx \Big)^{p^*/p}$$

since in the sum $\Sigma_{\tilde{Q} \in F} a_{\tilde{Q}} \chi_{\tilde{Q}}(x)$ for each $x \in \Omega$ at most $B$ terms are not zero. Thus

$$I_2 \leq K\Big( \sum_{\tilde{Q} \in F} \int_{\sigma\tilde{Q}} g^p\, dx \Big)^{p^*/p} \leq KB^{p^*/p}\Big( \int_{\Omega} g^p\, dx \Big)^{p^*/p}$$

with $K_1 = (2G)^{p^*} D_{p^*}^{p^*} B^{p^*-1} B^{p^*/p}$ and we get (4.2) with $\hat{A} = 2 \max(A^{1/p^*} B, K^{1/p^*})$.

**Remark 1.** The order of the local approximating polynomials $P_Q(f)$ in the Lemma 4.1. does not need to be $l-1$. Actually the orders of the local approximating polynomials in (4.1) can be completely arbitrary (but fixed for all $Q$). In particular, if in the local estimates (4.1) we let the order of the polynomials be higher than $l-1$, and this is typically the case in various local approximations of solutions of partial differential equations, the finite element method etc., achieving by this some more precise information on the constants $A$ e.g. presenting $A$ in the form, say, $A = Dh^{\gamma}$, where $h$ is a parameter measuring the typical linear dimension (the diameter) of the mesh of "cubes" $Q$ (simplices $Q$) used in the covering $F_h$, and $D$ is a factor depending on other parameters (which are dilatation invariant) describing the geometry of the covering $\{F_h\}$ as used in the proof of Lemma 4.1., then the estimate of the form $\hat{A} = Dh^{\gamma'}$ will be preserved in the global form (4.2) with some $\gamma'$.

**Remark 2.** A weighted version of Lemma 4.1, i.e. if on both sides of (4.1) we take the local norms in $L^{p^*}(Q, d\mu)$, and the measure $d\mu = w\, dx$ satisfies conditions of the type of $A_q$-conditions of Muckenhoupt [22], holds also. See [14], [23].

## 5. The global Sobolev inequality

The global Sobolev inequalities (1.2) and (1.6) and the Lemma 4.1 imply the general Sobolev inequality for domains satisfying the chain condition. We state the Sobolev inequality in the following form ($n > lp$):

**Theorem 5.1.** *Let the bounded domain* $\Omega \subset R^n$ *satisfy the chain condition. If the* $L_p$ *norm of the* $l$*–th gradient* $\nabla^l f$ *of a function* $f \in W^{l,p}_{loc}(\Omega)$ *is finite then the global Sobolev inequality*

$$\left(\int_\Omega |f - P_\Omega f|^{p^*} dx\right)^{1/p^*} \le C(\Omega) \left(\int_\Omega |\nabla^l f|^p \, dx\right)^{1/p}, \quad p^* = \frac{pn}{n - lp} \tag{5.1}$$

*holds with the constant* $C(\Omega)$ *depending only on the parameters* $C, B, \sigma, p$ *and the dimension* $n$ *only.*

Moreover the polynomial $P_\Omega f$ (of order $\le l - 1$) can be chosen in such a way that it satisfies the inequality

$$|P_\Omega(f)(x)| \le \tilde{C}(\Omega) \int_\Omega |f| \, dx, \quad x \in \Omega \tag{5.2}$$

with the constant $\tilde{C}$ not depending on $f$.

*Proof.* If $f \in W^{l,p}_{loc}(\Omega)$ then by the local Sobolev inequality (1.6), $f$ satisfies the assumptions of Lemma 4.1 with $g(x) = |\nabla^l f(x)|$, $\sigma = 1$. Therefore we conclude (5.1) with the polynomial $P_\Omega(f)$ equal e.g. to $P_{\Omega_0}(f)$, where $Q_0$ is the "central" cube of the covering $F$.

The choice of the polynomial $P_\Omega f$ in (5.1) allows a considerable amount of freedom. If $f$ satisfies some additional conditions, say, if $f \equiv 0$ on an open subset $\omega$ of $\Omega$ then $P_\Omega(f)$ can be chosen $\equiv 0$. More precisely the polynomial $P_\Omega(f)$ can be always chosen in such a way that

$$|P_\Omega(f)(x)| \le K \int_\omega |f| \, dx \tag{5.3}$$

with $K$ depending on $\omega$ (and $\Omega$), but not on $f$.

Obviously, the way of choice of the polynomial $P_\Omega(f)$ may influence the optimal value of the constant $C(\Omega)$ in (5.1) so we should write $C = C(\Omega,\omega)$, but for the domains satisfying the chain condition, the estimates of the proof of Lemma 4.1 give (in general very crude, but finite) estimates for $C(\Omega)$. $P_\Omega(f)$ in (5.1) is always a projection operator on the polynomials of degree $\le l - 1$: if $f \in P_{l-1}(R^n)$ then $P_\Omega(f) \equiv f$. If $f$ satisfies some "orthogonality conditions" then also $P_\Omega(f) \equiv 0$. The theorem 5.1 and the inequalities (5.1) and (5.2) imply the usual Sobolev inequality

$$\|f\|_{L^{p^*}(\Omega)} \le \tilde{C}\Big(\int_\Omega |f|\,dx + \Big(\int_\Omega |\nabla^l f(x)|^p\,dx\Big)^{1/p}\Big) \le \tilde{C}\|f\|_{W^{l,p}(\Omega)} \tag{5.4}$$

with the norm $\|f\|_{W^{l,p}(\Omega)}$ defined as usual.

## 6. The Sobolev inequality for John domains

We give here a direct proof of general Sobolev imbedding inequalities for John domains. This proof relies on some ideas discussed in [6], §3, combined with a new characterisation of John domains given by O. Martio in his interesting recent paper [18]. We first recall the precise formulation of O. Martio's result. Let $A$ be an arbitrary subset of $R^n$. A mapping $f: A \to R^n$ is called an L-bilipschitz mapping if

$$|x-y|/L \le |f(x)-f(y)| \le L|x-y| \tag{6.1}$$

for all $x, y \in A$.

The open ball with center $x$ and radius $r$ will be denoted by $B(x,r)$, $B(0,r) = B(r)$.

Let $x_0$ and $x$ be two points of the domain $\Omega \subset R^n$. We say that $x$ lies in an $(\Omega,L)$-bilipschitz ball $B(r)$ centered at $x_0$ if there exists an L-bilipschitz mapping $T$ of $B(r)$ into $\Omega$ such that $T(0) = x_0$ and $x \in T(B(r))$. As $\Omega$ is fixed in what follows, we just use the term: L-bilipschitz ball.

The following theorem characterizes John domains.

**Theorem** (Martio). *If $\Omega$ is an $(\alpha,\beta)$-John domain with center $x_0$, then each point $x \in \Omega$ is contained in an L-bilipschitz ball $B(\alpha)$ centered at $x_0$ with $L = c(\frac{\beta}{\alpha})^4$ where $c$ is a constant depending on dimension $n$ only. Conversely, if $\Omega$ is a domain in $R^n$ and, for some $x_0 \in \Omega$ each point $x \in \Omega$ lies in an L-bilipschitz ball $B(\alpha)$ centered at $x_0$, then $\Omega$ is an $(\alpha/L^3,\alpha L)$-John domain with center $x_0$.*

Thus we see that John-domains $\Omega$ can be covered by a family of $(\Omega,L)$-bilipschitzian balls $T_x(B(\alpha))$ with fixed center $x_0$ in $\Omega$ and fixed "radius" $r$, equal to the parameter $\alpha$ in the definition of $(\alpha,\beta)$-John-domains. We should rather say that the L-bilipschitzian ball $T(B(r))$ has an L-radius, or quasiradius, equal to $r$. Notice that for any $(\alpha,\beta)$-John domain with center $x_0$ the ball $B(x_0,\alpha) \subset \Omega$: moreover the ball $B(x_0,\alpha/L) = E$ is contained in the L-bilipschitzian ball $T_x(B(\alpha))$ for each $x \in \Omega$. Averages over the ball $E$ can be used as good approximations of functions in

Sobolev classes $W^{1,l}(\Omega)$. In particular, imitating the usual Sobolev procedure to estimate the averaged oscillation $|⨍_E [f(x) - f(y)]\, dy| = |f(x) - f_E|$ by the $(n-1)$ order Riesz potentials in starshaped or convex domains, applied to the composition $f \circ T^{-1}(y)$, changing the variables in the resulting formulas, and using the estimates (6.1) the following estimate is readily obtained

$$|f(x) - f_E| \le CL^{4n} \int_A |x-y|^{1-n} |\nabla f(y)|\, dy \tag{6.2}$$

for $C \in A = T(B(R))$, i.e. for any L-bilipschitz ball.

The constant $C$ in (6.2) depends on the dimension $n$ only. According to the definition of $W_1(\varphi)$-domains, given in §3 of [6], the inequality (6.2) means that any L-bilipschitzian ball $A = T(B(r))$ is a $W_1(\varphi)$ domain for $\varphi = \chi_E/|E|$ equal to the normalized characteristic function of the ball $E$. Moreover the constant $C$ for all these domains is uniformly bounded from above. Since obviously the union of $W_1(\varphi)$ domains, with fixed $\varphi$ and fixed constant $C$, is also a $W_1(\varphi)$ domain, we obtain, in view of the Martio's theorem, that for any $(\alpha,\beta)$-John domain $\Omega$ the estimate

$$|f(x) - f_E| \le C(n)L^{4n} \int_\Omega |x-y|^{1-n} |\nabla f(y)|\, dy \tag{6.3}$$

holds a.e. in $\Omega$ for all functions $f \in W^{1,1}(\Omega)$.

Thus any $(\alpha,\beta)$-John domain is $W_1(\varphi)$-domain for $\varphi = \chi_E/|E|$ with the constant bounded from above by $L = C(n)(\frac{\beta}{\alpha})^4$. By the theorem in §3 of [6] we get then the inequality

$$|f(x) - P_{f,E}(x)| \le \hat{C} \int_\Omega \frac{|\nabla^l f|\, dy}{|x-y|^{n-1}} \quad \text{a.e. in } \Omega. \tag{6.4}$$

Thus in any $(\alpha,\beta)$-John domain a function $f \in W^{l,1}(\Omega)$ can be approximated by a polynomial of order $\le l-1$ with the error controlled pointwise by the $l$-th order Riesz potentials of the $l$-th gradient $|\nabla^l f|$. Moreover the constant $\hat{C}$ in (6.4) depends on the dimension $n$, the diameter of the domain $\Omega$, $(\operatorname{diam} \Omega)$, and the Lipschitz constant $L = C(n)(\frac{\beta}{\alpha})^4$. This dependence is estimated from above by very explicite and simple formulas in [6]. The polynomial $P_{f,E}(x)$ is controlled by the estimate

$$|P_{f,E}(x)| \le \hat{C}\|f\|_{W^{l-1,1}_{(E)}} \quad \text{for } x \in \Omega \tag{6.5}$$

where $\tilde{C}$ depends on n, $l$ and the diam $\Omega$. In view of the fundamental Hardy-Littlewood-Sobolev theorem on fractional integration as a bounded operation in the Lebesgue $L_p$ spaces, [2], [31], the global Sobolev imbedding inequality

$$\Big(\int_{\Omega} |f(x) - P_{f,E}(x)|^{p^*} dx\Big)^{1/p^*} \le K\Big(\int_{\Omega} |\nabla^l f|^p \, dy\Big)^{1/p} \tag{6.6}$$

for John domains follows. The constant K in (6.6) can be estimated by the inequality

$$K \le \hat{C}\, C(n,p,l) \tag{6.7}$$

where $C(n,p,l)$ is the $L_p$-norm of the Riesz $R_l(f)$ transform of order $l$, as an operator $R_l : L_p(R^n) \to L_{p^*}(R^n)$, $p^* = \frac{np}{n-lp}$, $n > lp$, and $\hat{C}$ is the constant from (6.4). The described procedure to obtain the Sobolev inequality (6.6) can be suitably modified to obtain estimates for the polynomials $P_{f,E}$ of the form

$$|P_{f,E}(x)| \le \tilde{\tilde{C}} \int_E |f| \, dx.$$

with some constants $\tilde{\tilde{C}}$ not depending on f.

Also more subtle estimates of the constant $\hat{C}$ in terms of the ratio $\beta/\alpha$ are possible (with exponent 2n instead 4n).

We remark that in the sketched proof of (6.6) above and the related discussion in §3 of [6], an idea of the paper [28] of J. Rešetniak is utilized. Also cf. [26], [27].

If more refined versions of Hardy-Littlewood Sobolev theorem of $L_p$ bounds for Riesz potentials e.g. various known weighted versions of these theorems are used, [1], [22], [23], then the corresponding weighted Sobolev inequalities follow.

## References

[1] D. R. Adams, A trace inequality for generalized potentials. Studia Math. 48 (1973), 99–105.

[2] O. V. Besov, V. P. Ilin, S. M. Nikolskii, Integral representation of functions and imbedding theorems. (Russian) Izdat. Nauka, Moscow, 1975.

[3] B. Bojarski, Pointwise differentiability of weak solutions of elliptic divergence type equations. Bull. Polish Acad. Sci. Math. 33:1–2 (1985), 1–6.

[4] B. Bojarski, Sharp maximal operator of fractional order and Sobolev imbedding inequalities. ibid. 33:1–2 (1985), 7–16.

[5] B. Bojarski, Remarks on Markov's inequalities and some properties of polynomials. ibid. 33:7–8 (1985), 355–365.

[6] B. Bojarski, Remarks on local function spaces. Proceedings of the Function Space Conference (Lund, 1986), Lecture Notes in Math., Springer-Verlag (to appear).

[7] B. Bojarski, T. Iwaniec, Analytic foundations of the theory of quasiconformal mappings in $R^n$. Ann. Acad. Sci. Fenn. Ser. A I Math. 8 (1983), 257–324.

[8] J. Boman, $L^p$-estimates for very strongly elliptic systems. J. Analyse Math. (to appear).

[9] Ju. A. Brudnyj, I. M. Gansburg, On an extremal problem for polynomials of n variables. (Russian) Izv. Akad. Nauk SSSR Ser. Mat. 37 (1973), 344–355.

[10] V. I. Burenkov, Sobolev's integral representation and Taylor's formula. (Russian) V. Trudy Mat. Inst. Steklov. 131 (1974), 33–38.

[11] R. Coifman, C. Fefferman, Weighted norm inequalities for maximal functions and singular integrals. Studia Math. 51 (1974), 241–250.

[12] F. Gehring, O. Martio, Lipschitz classes and quasiconformal mappings. Ann. Acad. Sci. Fenn. Ser. A I Math. 10 (1985), 203–219.

[13] G. H. Hardy, J. E. Littlewood and G. Polya, Inequalities. Cambridge, 1934.

[14] T. Iwaniec, C. A. Nolder, Hardy-Littlewood inequality for quasiregular mappings in certain domains in $R^n$. Ann. Acad. Sci. Fenn. Ser. A I Math. 10 (1985), 267–282.

[15] F. John, On quasi-isometric mappings I, II. Comm. Pure Appl. Math. 21 (1968), 77–100, ibid. 22 (1969), 265–278.

[16] A. Lehtonen, Embedding of Sobolev spaces into Lipschitz spaces. Preprint (University of Jyväskylä, 1986).

[17] B. Mandelbrot, The fractal geometry of nature. Freeman, San Fransisco, 1982.

[18] O. Martio, Definitions for uniform domains. Ann. Acad. Sci. Fenn. Ser. A I Math. 5 (1980), 179–205.

[19] O. Martio, John domains, bilipschitz balls and Poincaré inequality. Rev. Roum. Math. Pures Appl. (to appear).

[20] V. G. Mazja, Einbettungssätze für Sobolewsche Räume. Teil 1. Teubner, Leipzig, 1979.

[21] V. G. Mazja, Prostranstva S. L. Soboleva. Izdat. Leningr. Univ., Leningrad, 1985. English translation: Sobolev Spaces. Springer-Verlag, 1986.

[22] B. Muckenhoupt, Weighted norm inequalities for Hardy maximal function. Trans. Amer. Math. Soc. 165 (1972), 207–226.

[23] B. Muckenhoupt, R. L. Wheeden, Weighted norm inequalities for fractional integrals. Trans. Amer. Math. Soc. 192 (1974), 261–274.

[24] O. Nikodym, Sur une classe de fonctions considerée dans le problème de Dirichlet. Fund. Math. 21 (1933), 129–150.

[25] H. Reimann, Functions of bounded mean oscillation and quasiconformal mappings. Comment. Math. Helv. 49 (1974), 260–276.

[26] Ju. G. Rešetniak, Stability theorems in geometry and analysis. (Russian) Nauka Sibirsk. Otdel., Novosibirsk, 1982.

[27] Ju. G. Rešetniak, Integral representations of differentiable functions in domains with non-smooth boundary. (Russian) Sibirsk. Mat. Zh. 21:6 (1980), 108–116.

[28] Ju. G. Rešetniak, Remarks on integral representations of differentiable functions in several variables. (Russian) Sibirsk. Mat. Zh. 25:5 (1984), 198–200.

[29] Ju. G. Rešetniak, Mappings with bounded distortion. Science, Moscow, 1981.

[30] S. L. Sobolev, Some applications of functional analysis in mathematical physics. (Russian) Izdat. Leningrad. Gos. Univ., Leningrad, 1950.

[31] E. Stein, Singular integrals and differentiability properties of functions. Princeton University Press, 1970.

[32] J. O. Strömberg, A. Torchinsky, Weights, sharp maximal functions and Hardy spaces. Bull. Amer. Math. Soc. 3 (1980), 1053–1056.

[33] J. Väisälä, Lectures on n-dimensional quasiconformal mappings. Lecture Notes in Math., 229, Springer-Verlag, 1971.

# A GENERALIZATION OF THE TEICHMÜLLER THEOREM

**M. A. Bracalova**
Institute of Mathematics, Bulgarian Academy of Sciences
Sofia 1090, P.O.Box 373, Bulgaria

The Teichmüller theorem [6], [8], [4] on the behavior of a quasiconformal mapping at a point is well-known in the theory of quasiconformal mappings and can be stated as follows. Let

$$g(z) = u(z) + iv(z) \tag{1}$$

be a quasiconformal mapping (q.c. mapping) defined in the exterior of $|z| \leq 1$, such that $\lim_{z\to\infty} g(z) = \infty$, with dilatation $p(z)$ and complex dilatation $k(z)$.

**Theorem** 1 (Teichmüller). *If the q.c. mapping* (1) *satisfies the condition*

$$\iint_{|z|>1} \frac{p(z)-1}{|z|^2}\, dxdy < \infty, \tag{2}$$

*then*

$$|g(z)| \sim A|z|, \quad z \to \infty, \tag{3}$$

*where* $0 < A < \infty$.

Theorem 1 was first established by O. Teichmüller [6] and later proved in another way by H. Wittich [7].

P. P. Belinskii [1] proved the very interesting result that (2) implies also the existence of the limit

$$\lim_{z\to\infty} \arg \frac{g(z)}{z} = \varphi$$

and therefore, if the q.c. mapping (1) satisfies (2) then it holds:

$$\lim_{z\to\infty} \frac{g(z)}{z} = A,$$

where $A \neq 0, \infty$ is a complex number. This last result, known as the Teichmüller-Belinskii theorem (or the Teichmüller-Wittich-Belinskii theorem) is an important achievement in the theory of quasiconformal mappings with many applications especially in the theory of meromorphic functions ([3], [2]).

In the present paper (Theorem 3), a quite general condition is considered, under which an asymptotic formula more general than (3) for the behavior of the q.c. mapping (1) is given. We also show that Theorem 3 implies Theorem 1.

Further we shall suppose that (1) is continuously differentiable. The general case of a q.c. mapping can be obtained from the one considered.

Let $z = \rho e^{i\varphi}$ and let $g_\rho(\rho e^{i\varphi}) = u_\rho + iv_\rho$, $g_\varphi(\rho e^{i\varphi}) = u_\varphi + iv_\varphi$ be the partial derivatives of $g$ with respect to $\rho$ and $\varphi$. Later on, we shall mainly use polar coordinates.

For $\rho, r_1, r_2 \in [0,\infty]$ and $r_2 > r_1$ we denote by

$$\left.\begin{aligned} N(\rho) &= \left[\frac{1}{2\pi}\int_0^{2\pi} \frac{u_\varphi^2 + v_\varphi^2}{u_\rho v_\varphi - u_\varphi v_\rho}\, d\varphi\right]^{-1}, \\ M(\rho) &= \frac{1}{2\pi}\int_0^{2\pi} \frac{u_\rho^2 + v_\rho^2}{u_\rho v_\varphi - u_\varphi v_\rho}\, d\varphi, \\ e(r_1, r_2) &= \int_{r_1}^{r_2} [M(\rho) - N(\rho)]\, d\rho . \end{aligned}\right\} \tag{4}$$

We denote by $V_r$ the closed curve which is the image of the circle $|z| = r$ under $g$. Let $V = \{V_r\}_{r\geq 1}$ be the family of all these disjoint curves. By

$$\omega_V(r) = \max_{w\in V_r} \log|w| - \min_{w\in V_r} \log|w|$$

we denote the circular oscillation of $V_r$, $V_r \in V$. For a fixed $r$, $\omega_V(r) = 0$ if and only if $V_r$ is a circle with center at the origin.

For $r_2 > r_1$ we denote the ring with boundary components $V_{r_1}$ and $V_{r_2}$ by $Q(r_1,r_2)$ and its modulus by $\lambda(r_1,r_2)$.

We introduce the following definition.

**Definition 1.** The q.c. mapping (1) is a *Kr-q.c. mapping* if the following relation holds:

$$\lambda(r,r_2) - \lambda(r,r_1) - \lambda(r_1,r_2) = o(1), \quad r \to \infty, \tag{5}$$

where $r_2 > r_1 > r$.

The geometrical aspect of this definition can be seen from the following theorem:

**Theorem 2.** *The q.c. mapping* (1) *is a Kr–q.c. mapping if and only if*

$$\omega_V(r) = o(1), \quad r \to \infty. \tag{6}$$

Theorem 2 is a consequence of a result of O. Teichmüller [6, §5.2], called there "Bedingung für Kreisähnlichkeit". The condition (6) means that if $r$ is big enough, then $V_r$ is "very close" to a circle with center at the origin.

**Lemma 1.** *If* (1) *is a Kr–q.c.mapping then for every fixed* $r_0$, $r_0 > 1$, *the next asymptotic equality is true:*

$$\log|w| - \lambda(r_0,r) = C + o(1), \quad r \to \infty, \tag{7}$$

*where* $w \in V_r$ *and* $C = C(r_0)$ *is a real number.*

*Proof.* Let $z_1 = F(w)$ be a conformal mapping of the exterior of $V_{r_0}$ onto the exterior of $|z_1| \le 1$, such that $F(\infty) = \infty$.

In a neighbourhood of the infinity $F(w)$ has the following expansion:

$$F(w) = C_1 w(1 + o(1)), \quad w \to \infty, \tag{8}$$

where $C_1 \neq 0$ is a complex number. We denote by $T_r$ the image of $V_r$ under $F$. According to Theorem 2, (6) is fulfilled and therefore it follows from (8) that for the circular oscillation of $T_r$ we have

$$\omega_T(r) = o(1), \quad r \to \infty.$$

Using this asymptotic formula, it can be proved that

$$\log|z_1| - \lambda(r_0,r) = o(1), \quad r \to \infty,$$

for $z_1 \in T_r$. Combining this with the asymptotic expansion of $z_1 = F(w)$ from (8), we obtain (7) with $C = -\log|C_1|$.

**Lemma 2.** *If* (1) *is a Kr-q.c. mapping then for every two fixed real numbers* $r^*$ *and* $\hat{r}$, $r^* > \hat{r} > 1$ *it holds:*

$$\lambda(\hat{r},r) - \lambda(r^*,r) = C + o(1), \quad r \to \infty, \tag{9}$$

*where* $C = C(\hat{r},r^*)$ *is a real number.*

*Also for every fixed number* $r_0 > 1$ *we have*

$$\lambda(r_0,r_2) - \lambda(r_0,r_1) - \lambda(r_1,r_2) = o(1), \ r_1 \to \infty \tag{10}$$

*where* $r_2 > r_1$.

*Proof.* According to Lemma 1 there exists $C_1 = C_1(\hat{r})$ and $C_2 = C_2(r^*)$, such that for $w \in V_r$ we have

$$\log|w| - \lambda(\hat{r},r) = C_1 + o(1), \ r \to \infty$$

and

$$\log|w| - \lambda(r^*,r) = C_2 + o(1), \ r \to \infty,$$

from which (9) follows with $C = C_2 - C_1$.

Let $\varepsilon > 0$ be given. According to Definition 1 there exists $r^* = r^*(\varepsilon)$, $r^* > r_0$, such that for all $r_1$ and $r_2$, $r_2 > r_1 > r^*$ we have

$$|\lambda(r^*,r_2) - \lambda(r^*,r_1) - \lambda(r_1,r_2)| < \varepsilon/3.$$

According to (9), for a sufficiently large $N_0 > r^*$ and for all $r > N_0$ it holds:

$$|\lambda(r_0,r) - \lambda(r^*,r) - C^*| < \varepsilon/3,$$

where $C^* = C^*(r_0,r^*)$ is a real number. From the last two inequalities we obtain for all $r_1$ and $r_2$, $r_2 > r_1 > N_0$,

$$\begin{aligned}|\lambda(r_0,r_2) - \lambda(r_0,r_1) - \lambda(r_1,r_2)| &\le |\lambda(r_0,r_2) - \lambda(r^*,r_2) - C^*| + \\ &+ |\lambda(r_0,r_1) - \lambda(r^*,r_1) - C^*| + |\lambda(r^*,r_2) - \lambda(r^*,r_1) - \lambda(r_1,r_2)| \\ &\le 3(\varepsilon/3) = \varepsilon.\end{aligned}$$

This implies (10) and completes the proof.

Let us consider the following expressions in polar coordinates. The formal complex derivatives $g_z$ and $g_{\bar{z}}$ of the function $g(z)$ can be calculated as follows:

$$g_z(\rho e^{i\varphi}) = e^{-i\varphi}(\rho g_\rho - i g_\varphi)/2\rho,$$
$$g_{\bar{z}}(\rho e^{i\varphi}) = e^{i\varphi}(\rho g_\rho + i g_\varphi)/2\rho.$$

Using this we obtain:

$$p(\rho e^{i\varphi}) + 1/p(\rho e^{i\varphi}) = [\rho^2(u_\rho^2 + v_\rho^2) + u_\varphi^2 + v_\varphi^2)]/\rho(u_\rho v_\varphi - u_\varphi v_\rho) \tag{11}$$

and

$$\left.\begin{aligned} \frac{|1-e^{-2i\varphi}k|^2}{1-|k|^2} &= \frac{u_\varphi^2+v_\varphi^2}{\rho(u_\rho v_\varphi - u_\varphi v_\rho)}, \\ \frac{|1+e^{-2i\varphi}k|^2}{1-|k|^2} &= \frac{\rho(u_\rho^2+v_\rho^2)}{u_\rho v_\varphi - u_\varphi v_\rho}. \end{aligned}\right\} \tag{12}$$

Here p and k are the real and complex dilatations of g, respectively.

In [5, §2], two lemmas on moduli of quadrilaterals and rings are obtained. From Lemma 2.2. it follows that the modulus of the ring domain $Q(r,R)$ satisfies the inequalities $I(r,R) \le \lambda(r,R) \le J(r,R)$, where

$$(2\pi)^{-1} I(r,R) = \int_r^R \left[\rho \int_0^{2\pi} \frac{|1-e^{-2i\varphi}k|^2}{1-|k|^2}\, d\varphi\right]^{-1} d\rho,$$

$$(2\pi)^{-1} J(r,R) = \left\{\int_0^{2\pi} \left[\int_r^R \frac{|1+e^{-2i\varphi}k|^2}{1-|k|^2} \frac{d\rho}{\rho}\right]^{-1} d\varphi\right\}^{-1}.$$

From these estimates for $\lambda(r,R)$, (12), and the Schwarz inequality we obtain the following inequalities important for our investigation:

$$\int_r^R N(\rho)\, d\rho \le \lambda(r,R) \le \int_r^R M(\rho)\, d\rho, \tag{13}$$

where $M(\rho)$ and $N(\rho)$ are defined in (4).

**Lemma 3.** *If a q.c. mapping* (1) *satisfies the condition*

$$e(r_1,r_2) \to 0, \quad r_2 > r_1, \quad r_1 \to \infty \tag{14}$$

*then* (1) *is a Kr–q.c. mapping and the following asymptotic representation holds:*

$$\lambda(r_0,r) = C + \int_{r_0}^{r} N(\rho)\, d\rho + o(1), \quad r \to \infty \tag{15}$$

*where* $r_0 > 1$ *and* $C = C(r_0) \in \mathbb{R}$.

*Proof.* Using (13) we obtain for $r_2 > r > r_1$,

$$\lambda(r_1,r_2) - \lambda(r_1,r) - \lambda(r,r_2) \le \int_{r_1}^{r_2} M(\rho)\, d\rho - \int_{r_1}^{r} N(\rho)\, d\rho - \int_{r}^{r_2} N(\rho)\, d\rho = e(r_1,r_2).$$

According to (14), (5) is fulfilled and (1) is a Kr-q.c. mapping by Definition 1. Hence we can apply Lemma 2 to obtain (10). Combining (10) with condition (14) we deduce that the function

$$h(r) = \lambda(r_0,r) - \int_{r_0}^{r} N(\rho)\, d\rho$$

satisfies the Cauchy condition, and the desired formula (15) follows.

The main result of the present paper is the next

**Theorem 3.** *If a q.c. mapping* (1) *satisfies condition* (14) *then for every fixed* $r_0 > 1$

$$|g(z)| \sim A \exp\left\{\int_{r_0}^{r} N(\rho)\, d\rho\right\}, \quad r \to \infty \quad (z = re^{i\varphi}), \tag{16}$$

*uniformly with respect to* $\varphi$, *where* $A = A(r_0)$ *is a positive real number.*

*Proof.* By Lemma 3, (1) is a Kr-q.c. mapping. Therefore, applying Lemma 1, we obtain

$$\lim_{r\to\infty} \{\log|g(z)| - \lambda(r_0,r)\} = C_1 \quad (z = re^{i\varphi}),$$

uniformly with respect to $\varphi$, where $C_1 = C_1(r_0)$ is a real number. From this and (15), which holds according to Lemma 3, (16) follows. This completes the proof.

Condition (14) is weaker than (2), as one can see from the relation between $p(z)$ and $e(r,R)$ and the example given below.

**Lemma 4.** *The dilatation* $p(z)$ *of* (1) *satisfies the equality:*

$$\frac{1}{2\pi} \iint_{r<|z|<R} \frac{(p(z)-1)^2}{p(z)|z|^2}\, dxdy = e(r,R) + \int_{r}^{R} \frac{(\rho N(\rho)-1)^2}{\rho^2 N(\rho)}\, d\rho, \tag{17}$$

*where* $e(r,R)$ *and* $N(\rho)$ *are as in* (4).

*Proof.* Using (11) we calculate

$$\frac{1}{2\pi}\iint\limits_{r<|z|<R}\frac{(p(z)-1)^2}{p(z)|z|^2}\,dxdy=\frac{1}{2\pi}\int_r^R\int_0^{2\pi}\frac{(p(\rho e^{i\varphi})-1)^2}{\rho p(re^{i\varphi})}\,d\varphi d\rho$$

$$=\frac{1}{2\pi}\int_r^R\int_0^{2\pi}\left[\frac{\rho^2(u_\rho^2+v_\rho^2)+u_\varphi^2+v_\varphi^2}{\rho^2(u_\rho v_\varphi-u_\varphi v_\rho)}-\frac{2}{\rho}\right]d\varphi d\rho$$

$$=\int_r^R[M(\rho)-N(\rho)]\,d\rho+\int_r^R\left[\frac{(N(\rho))^{-1}}{\rho^2}-\frac{2}{\rho}+N(\rho)\right]d\rho$$

$$=e(r,R)+\int_r^R\frac{(\rho N(\rho)-1)^2}{\rho^2N(\rho)}\,d\rho.$$

This completes the proof.

Since it is always true that

$$\frac{(p(z)-1)^2}{p(z)}\le p(z)-1,$$

we obtain

**Corollary 1.** *If* (2) *is fulfilled, then*

$$\frac{1}{2\pi}\iint\limits_{|z|>1}\frac{(p(z)-1)^2}{p(z)|z|^2}\,dxdy<\infty. \tag{18}$$

Using (17) and Corollary 1 we obtain:

**Corollary 2.** *If* (2) *or* (18) *is fulfilled, then* (14) *holds and*

$$\int_1^\infty\frac{(\rho N(\rho)-1)^2}{\rho^2N(\rho)}\,d\rho<\infty. \tag{19}$$

From this and Lemma 3 we conclude

**Corollary 3.** *If* (1) *satisfies* (2), *then* (1) *is a Kr–q.c. mapping.*

Let us consider the following mapping, which is defined in the exterior of $|z|\le 1$:

$$g(z)=g(\rho e^{i\varphi})=\frac{z}{\log|z|+2},\quad |z|>1. \tag{20}$$

Evidently (20) is a q.c. mapping, for which $\lim_{z\to\infty}g(z)=\infty$. Its dilatation p is

$$p(z) = p(\rho e^{i\varphi}) = 1 + \frac{1}{\log \rho + 1}, \quad \rho > 1. \tag{21}$$

As one can verify, (18) is fulfilled for (21), while (2) is not. Therefore, Theorem 3 is applicable for the mapping (20), while Theorem 1 is not.

Comparing the asymptotic representations (3) and (16) and using Theorems 1 and 3, we conclude

$$r \sim A \exp\left\{ \int_{r_0}^{r} N(\rho)\, d\rho \right\}, \quad r \to \infty, \tag{22}$$

where $A \neq 0, \infty$, $r = |z|$ and $N(\rho)$ is as in (4), provided (2) is fulfilled (whence (14) also holds).

We further show independently from Theorem 1 and Theorem 3 that (2) is a sufficient condition for the validity of (22).

**Lemma 5.** *If a q.c. mapping* (1) *satisfies the condition* (2), *then* (22) *holds.*

*Proof.* We set

$$s(\rho,\varphi) = (u_\varphi^2 + v_\varphi^2)/\rho(u_\rho v_\varphi - u_\varphi v_\rho).$$

Then

$$\frac{(p-1)^2}{p} - \frac{(s-1)^2}{s} = \frac{(\rho u_\rho - v_\varphi s^{-1})^2 + (\rho v_\rho + u_\varphi s^{-1})^2}{\rho(u_\rho v_\varphi - u_\varphi v_\rho)} \geq 0$$

($p$ denotes the dilatation of (1)) and

$$|s - 1| < p - 1. \tag{23}$$

Let

$$E_1 = \{\rho > r_0 : \rho N(\rho) \geq 2\}, \; E_2 = \{\rho > r_0 : \rho N(\rho) \leq 2\}.$$

By Corollary 2, (19) holds and we have

$$\int_{E_1} \frac{|\rho N(\rho) - 1|}{\rho} \leq 2 \int_{E_1} \frac{(\rho N(\rho) - 1)^2}{\rho^2 N(\rho)}\, d\rho < \infty.$$

Using (23) we also obtain

$$\int_{E_2} \frac{|\rho N(\rho) - 1|}{\rho} \leq \frac{1}{2\pi} \int_{E_2} \int_0^{2\pi} \rho N(\rho) \frac{|s-1|}{\rho}\, d\varphi d\rho \leq \frac{1}{\pi} \int_{E_2} \int_0^{2\pi} \frac{p-1}{\rho}\, d\varphi d\rho < \infty.$$

From the inequalities obtained above the absolute convergence of the integral

$$\int_{r_0}^{\infty} \frac{\rho N(\rho) - 1}{\rho} \, d\rho$$

follows and therefore (22) is valid.

Lemma 5 shows that the Teichmüller theorem (Theorem 1), very important in the theory of quasiconformal mappings, follows from Theorem 3. In fact, let the q.c. mapping (1) satisfy (2). Then, according to Corollary 2, (14) is fulfilled. Applying Theorem 3 we obtain the asymptotic representation (16). Since, according to Lemma 5, (22) holds, then (3) follows from (16) and (22).

The author would like to express her deep gratitude to A. A. Gol'dberg from the L'vov University for useful discussions.

## References

[1] Белинский, П. П., Поведение квазиконформного отображения в изолированнои точке. Докл. АН СССР **91** (1953), 997–998.

[2] Drasin, D., On the Teichmüller-Wittich-Belinskii theorem. Results in Math. **10** (1986), 54–65.

[3] Гольдберг, А. А. и Островский, И. В., Распределение значений мероморфных функций. Наука, Москва, 1970.

[4] Lehto, O. und Virtanen, K. I., Quasikonforme Abbildungen. Springer-Verlag, Berlin-Heidelberg-New York, 1965.

[5] Reich, E. and Walczak, H. R., On the behavior of a quasiconformal mapping at a point. Trans. Amer. Math. Soc. **117** (1965), 338–351.

[6] Teichmüller, O., Untersuchungen über konforme und quasikonforme Abbildung. Deutsche Math. **3** (1938), 621–678.

[7] Wittich, H., Zum Beweis eines Satzes über quasikonforme Abbildungen. Math. Z. **51** (1948), 275–288.

[8] Виттих, Г., Новейшие исследования по однозначным аналитическим функциям. ГИФМЛ, Москва, 1960.

## UNBOUNDED FEYNMAN INTEGRABLE FUNCTIONALS DEFINED IN TERMS OF ANALYTIC FUNCTIONS

**R. H. Cameron and D. A. Storvick**
School of Mathematics, University of Minnesota
Minneapolis, Minnesota 55455, U.S.A.

### 1. Introduction

In his paper, *Space–time approach to non–relativistic quantum mechanics*, [11], R. P. Feynman assumed the existence of an integral over a space of paths, and he used his integral in a formal way in his approach to quantum mechanics. Various mathematicians have undertaken to give rigorously meaningful definitions of the Feynman integral with appropriate existence theorems and have expressed solutions of the Schroedinger equation in terms of their integrals. One of these approaches was based on the similarity between the Wiener and the Feynman integral, and procedures were set up by the present authors and others to obtain Feynman integrals from Wiener integrals by analytic extension from the real axis to the imaginary axis. It is this procedure of analytic continuation that we will use in this paper to define the Feynman integral that we will use. Classes of functionals based on Fourier transforms in function space have been found to be Feynman integrable in this sense. In particular we'd mention the influential monograph of Albeverio and Høegh-Krohn, [1]. Their Fresnel integral, which is their interpretation of Feynman's concept, was defined on a space of path functions (functionals) which were given in terms of Fourier transforms.

This approach to the Feynman integral via transform assumptions has been carried further in the works of Johnson and Skoug, Elworthy and Truman, Kallianpur, the present authors and others, cf. [5], [6], [7], [8], [10], [12], [13], [14], [15] and [16].

In the application of the Feynman integral to quantum theory, functionals of the type

$$\exp\left\{ \int_a^b \theta(s,\vec{x}(s))\, ds \right\} \Psi(\vec{x}(b)) \tag{1.1}$$

are often employed, with $\Psi$ corresponding to the initial condition associated with

Schroedinger's equation and $\theta$ corresponding to the potential function. It has been shown that such functionals are Feynman integrable when $\theta$ and $\Psi$ are Fourier transforms of bounded measures. However this condition requires $\theta$ and $\Psi$ to be bounded and continuous; so that they could be applied in connection with the Schroedinger equation only when $\theta$ and $\Psi$ are bounded and continuous. Recently, in [7] and [8], the authors have proved existence theorems for Feynman integrals for a wider class of functionals which includes functionals of the type (1.1) for which $\Psi$ need not be bounded or continuous.

In this paper the authors establish the existence of the Feynman integral for a large class of unbounded functionals. This class includes functionals of the form (1.1) where both $\theta$ and $\Psi$ are unbounded.

In this paper we shall use the analytic Feynman integral which is defined as follows:

**Definition.** Let $F(\vec{x}(\cdot)) \equiv F(x_1(\cdot), x_2(\cdot), \ldots, x_\nu(\cdot))$ be a functional defined on $C^\nu \equiv C^\nu[a,b]$, ($\nu$-dimensional Wiener space), so that each $x_j(t)$ is continuous for $a \le t \le b$, and $x_j(a) = 0$.

**Definition.** Let $F$ be a functional such that the $\nu$-dimensional Wiener integral

$$J(\lambda) \equiv \int_{C^\nu[a,b]} F(\lambda^{-1/2}\vec{x})\, d\vec{x} \tag{1.2}$$

exists for all real $\lambda > 0$. If there exists a function $J^*(\lambda)$ analytic in the half-plane $\mathrm{Re}\,\lambda > 0$ such that $J^*(\lambda) = J(\lambda)$ for all real $\lambda > 0$, then we define $J^*$ to be the analytic Wiener integral of $F$ over $C^\nu[a,b]$ with parameter $\lambda$, and for $\mathrm{Re}\,\lambda > 0$ we write

$$\int_{C^\nu[a,b]}^{\mathrm{anw}_\lambda} F(\vec{x})\, d\vec{x} \equiv J^*(\lambda). \tag{1.3}$$

**Definition.** Let $q$ be a real parameter ($q \neq 0$) and let $F$ be a functional whose analytic Wiener integral exists for $\mathrm{Re}\,\lambda > 0$. Then if the following limit exists, we call it the analytic Feynman integral of $F$ over $C^\nu[a,b]$ with parameter $q$, and we write

$$\int_{C^{\nu}[a,b]}^{\mathrm{anf}_q} F(\vec{x})\, d\vec{x} = \lim_{\substack{\lambda \to -iq \\ \mathrm{Re}\lambda > 0}} \int_{C^{\nu}[a,b]}^{\mathrm{anw}_{\lambda}} F(\vec{x})\, d\vec{x}. \tag{1.4}$$

As a simple example of an unbounded functional to which our first existence theorem applies, we observe that if $\nu = 1$ and $f(z)$ is an entire function of order less than two and $g(s) \in L_2[a,b]$, then $F(x) \equiv f[\int_a^b g(s)x(s)\, ds]$ is analytic Feynman integrable. We begin with several lemmas concerning Wiener integrability.

## 2. Preliminary Lemmas

**Notation.** We shall use $\|x\|$ to denote the $L_2$ norm of $x$.

**Lemma 1.** *The functional*

$$F(x) \equiv \exp\{B\|x\|^{\alpha}\} \tag{2.1}$$

*is Wiener integrable over* $C[a,b]$ *for* $B > 0$ *and* $0 < \alpha < 2$. *Moreover* $F$ *is Wiener integrable for* $\alpha = 2$ *if* $B < \dfrac{\pi^2}{8(b-a)^2}$.

P r o o f. Equation (4.4) on page 83 of [4] can be written

$$\int_{C[a,b]} \exp\left\{B \int_a^b x^2(t)\, dt\right\} dx = [\cos((b-a)\sqrt{2B})]^{-1/2} \tag{2.2}$$

when $(b-a)\sqrt{2B} < \frac{\pi}{2}$. Here we have replaced $C[0,1]$ by $C[a,b]$ using the Cuthill thesis [9], and have used the modern form of the Wiener integral where the variance over a unit time interval is unity rather than the original Wiener integral in which the variance was one half. Thus

$$\int_{C[a,b]} \exp\left\{B \int_a^b x^2(t)\, dt\right\} dx$$

is finite when $B < \dfrac{\pi^2}{8(b-a)^2}$. It follows immediately that if $0 < \alpha < 2$

$$\int_{C[a,b]} \exp\{B\|x\|^{\alpha}\}\, dx$$

is finite for all $B > 0$ and the Lemma is proved.

**Lemma 2.** *The functional*

$$F(x) \equiv \exp\{B\|x\|^{-\alpha}\} \tag{2.3}$$

*is Wiener integrable over* $C[a,b]$ *for all* $B > 0$ *and* $0 < \alpha < 2$. *Moreover* $F$ *is Wiener integrable for* $\alpha = 2$ *if* $B < \frac{(b-a)^2}{8}$.

P r o o f. Equation (1.14) page 198 of [3] when written in terms of $C[a,b]$ and the modern form of the Wiener integral is

$$\int_{C[a,b]} F\Big[\int_a^b x^2(t)\, dt\Big]\, dx = \frac{1}{\sqrt{2\pi}} \int_0^\infty F\Big(\frac{2(b-a)^2}{s^2}\Big) \sum_0^\infty (-1)^n \frac{1\cdot 3\cdots(2n-1)}{2\cdot 4\cdots\,(2n)}\,(4n+1)\, e^{-(n+\frac{1}{4})^2 s^2}\, ds. \tag{2.4}$$

Let us first assume $\alpha = 2$ and $B < \frac{(b-a)^2}{8}$.

Let

$$F(u) = \begin{cases} \exp\{Bu^{-1}\} & \text{for } 0 < u \le (\frac{b-a}{4})^2 \\ 0 & \text{for } (\frac{b-a}{4})^2 < u \end{cases}.$$

Then by equation (2.4)

$$Q \equiv \int_{C[a,b]} F\Big[\int_a^b x^2(t)\, dt\Big]\, dx = \int_{\|x\|<\frac{b-a}{4}} \exp\Big\{B\Big[\int_a^b x^2(t)\, dt\Big]^{-1}\Big\}\, dx$$

$$= \frac{1}{\sqrt{2\pi}} \int_{s=2\sqrt{2}}^{s=\infty} \exp\Big\{B\Big[\frac{s^2}{2(b-a)^2}\Big]\Big\} \sum_{n=0}^{\infty} (-1)^n \frac{1\cdot 3\cdots(2n-1)}{2\cdot 4\cdots\,(2n)}\,(4n+1)\, e^{-(n+\frac{1}{4})^2 s^2}\, ds.$$

To establish the convergence of the integral we consider

$$\varphi(v) = ve^{-v^2s^2},$$

so that

$$\varphi(n+\tfrac{1}{4}) = (n+\tfrac{1}{4})\,e^{-(n+\frac{1}{4})^2s^2}$$

and

$$\varphi'(v) = e^{-v^2s^2}(1-2v^2s^2) \le 0$$

when $2v^2s^2 \ge 1$ or $s \ge \frac{1}{\sqrt{2}v}$. Thus $\varphi(v)$ is a decreasing function of $v$ when $s \ge \frac{1}{\sqrt{2}v}$ and $\varphi(n+\frac{1}{4})$ is a decreasing function of $n$ for non-negative integers $n$ when $s \ge \frac{1}{\sqrt{2}\,\frac{1}{4}} = 2\sqrt{2}$. Consequently when $s > 2\sqrt{2}$ the infinite series appearing in $Q$ is an alternating decreasing series whose general term approaches zero as $n \to \infty$. Thus the series converges and its value is positive and less than the value of its first term, and the integrand of $Q$ is dominated by the first term of the sum so that

$$|Q| \le \frac{1}{\sqrt{2\pi}} \int_{s=2\sqrt{2}}^{s=\infty} \exp\left\{B\left[\frac{s^2}{2(b-a)^2}\right]\right\} e^{-\frac{1}{16}s^2}\,ds$$

when $B < \frac{(b-a)^2}{8}$. Thus

$$\int_{\|x\|<\frac{b-a}{4}} \exp\left\{B\left[\int_a^b x^2(t)\,dt\right]^{-1}\right\} dx < +\infty.$$

But the integrand of this integral is bounded for $\|x\| \ge \frac{b-a}{4}$ and consequently the integral

$$\int_{C[a,b]} \exp\left\{B\left[\int_a^b x^2(t)\,dt\right]^{-1}\right\} dx < +\infty,$$

and Lemma 2 is proved in the case $\alpha = 2$, $B < \frac{(b-a)^2}{8}$. Clearly the integral of $F$ will also be finite if $\alpha < 2$ for all $B > 0$.

**Lemma 3.** *The functional*

$$F(x) \equiv \exp\{B_1\|x\|^{\alpha} + B_2\|x\|^{-\alpha}\}$$

*is Wiener integrable over* C[a,b] *for* $B_1, B_2 > 0$ *and* $0 < \alpha < 2$. *Moreover* F *is Wiener integrable for* $\alpha = 2$ *if* $B_1 < \dfrac{\pi^2}{8(b-a)^2}$ *and* $B_2 < \dfrac{(b-a)^2}{8}$.

Proof. It follows from Lemma 1 that

$$\int_{\|x\|\geq 1} \exp\{B_1\|x\|^{\alpha} + B_2\|x\|^{-\alpha}\}\, dx \leq \int_{\|x\|\geq 1} \exp\{B_1\|x\|^{\alpha} + B_2\}\, dx < +\infty,$$

and from Lemma 2 we see that

$$\int_{\|x\|\leq 1} \exp\{B_1\|x\|^{\alpha} + B_2\|x\|^{-\alpha}\}\, dx \leq \int_{\|x\|\leq 1} \exp\{B_1 + B_2\|x\|^{-\alpha}\}\, dx < +\infty$$

and consequently

$$\int_{C[a,b]} \exp\{B_1\|x\|^{\alpha} + B_2\|x\|^{-\alpha}\}\, dx < +\infty$$

and Lemma 3 is proved.

**Lemma 4.** *The functional*

$$F(x) \equiv \exp\{B_1\|x\|^{\alpha} + B_2\|x\|^{-\alpha} + B_3|x(b)|^{\alpha}\}$$

*is Wiener integrable over* C[a,b] *for* $B_1, B_2, B_3 \geq 0$ *and* $0 < \alpha < 2$.

Proof. Case I: $B_1 = B_2 = 0$. Then

$$\int_{C[a,b]} F(x)\, dx = \int_{C[a,b]} \exp\{B_3|x(b)|^{\alpha}\}\, dx$$

$$= \frac{1}{\sqrt{2\pi(b-a)}} \int_{-\infty}^{\infty} \exp\{B_3|u|^{\alpha}\} \exp\left\{\frac{-u^2}{2(b-a)}\right\} du < +\infty.$$

Case II: $B_3 = 0$. This is given by Lemma 3.

Case III: General Case, $B_1, B_2, B_3 \geq 0$. By Case I and II with $B_1, B_2, B_3$

replaced by $2B_1$, $2B_2$, $2B_3$ we have

$$\int_{C[a,b]} [\exp\{2B_1\|x\|^{\alpha} + 2B_2\|x\|^{-\alpha}\}]^2 \, dx < +\infty$$

and

$$\int_{C[a,b]} [\exp\{2B_3|x(b)|^{\alpha}\}]^2 \, dx < +\infty.$$

Therefore by the Schwarz inequality for Wiener integrals, we have

$$\int_{C[a,b]} \exp\{B_1\|x\|^{\alpha} + B_2\|x\|^{-\alpha}\} \exp\{B_3|x(b)|^{\alpha}\} \, dx < +\infty.$$

**Lemma 5.** *The functional*

$$F(\vec{x}) \equiv \exp\Big\{\sum_{j=1}^{\nu} \Big[b_j\|x_j\|^{\alpha_j} + b_j'\|x_j\|^{-\alpha_j} + b_j''|x_j(b)|^{\alpha_j}\Big]\Big\}$$

*is Wiener integrable over* $C^{\nu}[a,b]$ *for* $b_j, b_j', b_j'' > 0$ *and* $0 < \alpha_j < 2$.

P r o o f. Clearly

$$F(x) = \prod_{j=1}^{\nu} F_j(x_j)$$

where

$$F_j(x_j) = \exp\{b_j\|x_j\|^{\alpha_j} + b_j'\|x_j\|^{-\alpha_j} + b_j''|x_j(b)|^{\alpha_j}\}.$$

By Lemma 4, $F_j(x_j)$ is Wiener integrable over $C[a,b]$ and hence $F(x)$ is Wiener integrable over $C^{\nu}[a,b]$.

## 3. Existence theorems for the Feynman integral of functionals on $C^{\nu}[a,b]$

**Theorem 1.** *Let* $\Omega = \{z \mid z \neq 0, \ |\arg z| \leq \pi/4\}$. *Let* $F(\vec{x})$ *be a complex valued functional defined on* $C^{\nu} \equiv C^{\nu}[a,b]$.

(3.1) *Let* $f(\rho,\vec{x}) \equiv F(\rho\vec{x})$ *for real* $\rho > 0$ *and* $\vec{x} \in C^{\nu}$.

(3.2) *Let* $f(\rho,\vec{x})$ *have an extension* $f(z,\vec{x})$ *so that* $f(z,\vec{x})$ *is defined for* $(z,\vec{x}) \in \Omega \times C^{\nu}$.

(3.3) *For each* $\vec{x} \in C^{\nu}$ *let* $f(z,\vec{x})$ *be continuous in* $z$ *on* $\Omega$.

(3.4) *For each* $\vec{x} \in C^{\nu}$ *let* $f(z,\vec{x})$ *be analytic in* $z$ *on* $\mathrm{Int}(\Omega)$.

(3.5) *For each* $z \in \Omega$ *let* $f(z,\vec{x})$ *be Wiener measurable in* $\vec{x}$ *on* $C^{\nu}$.

(3.6) *For each* $\theta$, $|\theta| < \pi/4$, *let* $f(z,\vec{x}) \equiv f(re^{i\theta},\vec{x})$ *be measurable in* $(r,\vec{x})$ *on* $R^{+} \times C^{\nu}$.

(3.7) *For each* $r$, $0 < r < \infty$, *let* $f(z,\vec{x}) \equiv f(re^{i\theta},\vec{x})$ *be measurable in* $(\theta,\vec{x})$ *on* $[-\pi/4,\pi/4] \times C^{\nu}$.

(3.8) *For* $0 < \alpha_j < 2$, *for* $(j = 1,\ldots,\nu)$ *and for each* $R > 0$ *let there exist* $A \equiv A(R) > 0$ *and* $b_j \equiv B_j(R) > 0$, $b'_j \equiv b'(R) \geq 0$, *and* $b''_j \equiv b''_j(R) \geq 0$ *such that*

$$|f(z,\vec{x})| \leq A \exp\left\{ \sum_{j=1}^{\nu} \left[ b_j \|x_j\|^{\alpha_j} + b'_j \|x_j\|^{-\alpha_j} + b''_j |x_j(b)|^{\alpha_j} \right] \right\}$$

*for* $(z,\vec{x}) \in \Omega \times C^{\nu}[a,b]$ *with* $|z| < R$.

*Then* $F$ *is analytic Feynman integrable for each real* $q$, $q \neq 0$, *and equals the Wiener integral of* $f(\sqrt{i/q},\vec{x})$:

$$\int_{C^{\nu}[a,b]}^{\mathrm{anf}_q} F(\vec{x})\, d\vec{x} = \int_{C^{\nu}[a,b]} f(\sqrt{i/q},\vec{x})\, d\vec{x}$$

*where we choose* $\mathrm{Re}(\sqrt{i/q}) > 0$.

P r o o f. We show that $f(z,\vec{x})$ is Wiener integrable in $\vec{x}$ over $C^{\nu}[a,b]$ for $z \in \Omega$. By hypothesis (3.5), $f(z,\vec{x})$ is Wiener measurable and by hypothesis (3.8),

$$|f(z,\vec{x})| \leq A \exp\left\{ \sum_{j=1}^{\nu} \left[ b_j \|x_j\|^{\alpha_j} + b'_j \|x_j\|^{-\alpha_j} + b''_j |x_j(b)|^{\alpha_j} \right] \right\}.$$

Thus by Lemma 5 we see that $f(z,\vec{x})$ is Wiener integrable in $\vec{x}$ over $C^{\nu}[a,b]$ and

$$I(z) \equiv \int_{C^{\nu}[a,b]} f(z,\vec{x})\, d\vec{x} \text{ exists for } z \in \Omega.$$

To prove that $I(z)$ is continuous on $\Omega$, fix $z_0 \in \Omega$ and let $\varepsilon > 0$ be given and consider $\{z \mid |z - z_0| < \varepsilon, |z| < R\} \cap \Omega \equiv U(z_0)$. Then for $z \in U(z_0)$,

$$\int_{C^{\nu}[a,b]} |f(z,\vec{x}) - f(z_0,\vec{x})|\, d\vec{x} < +\infty,$$

and by hypothesis (3.8),

$$|f(z,\vec{x}) - f(z_0,\vec{x})| \leq 2A \exp\Big\{ \sum_{j=1}^{\nu} \Big[ b_j \|x_j\|^{\alpha_j} + b_j' \|x_j\|^{-\alpha_j} + b_j'' |x_j(b)|^{\alpha_j} \Big] \Big\},$$

and since the right hand side is independent of $z$, we see by hypothesis (3.3) and dominated convergence that $I(z)$ is continuous for $z \in \Omega$.

To prove that

$$\int_{C^{\nu}[a,b]} f(z,\vec{x})\, d\vec{x}$$

is analytic for $z \in \Omega$, we let $S \equiv S(\alpha,\beta,r_1,r_2) \subset \Omega$ be a portion of a sector defined by $\alpha \leq \theta \leq \beta$, $r_1 \leq |z| \leq r_2$, and let $\Gamma$ be its boundary. Then $f(z,\vec{x})$ is measurable on $\Gamma \times C^{\nu}[a,b]$ by hypotheses (3.6) and (3.7) and

$$\int_{\Gamma} \int_{C^{\nu}[a,b]} |f(z,\vec{x})|\, d\vec{x} |dz| \leq \Lambda \cdot L_{\Gamma}$$

where

$$\Lambda = \max_{z \in S} \int_{C^{\nu}[a,b]} |f(z,\vec{x})|\, d\vec{x}$$

$$\leq \int_{C^{\nu}[a,b]} A \exp\Big\{ \sum_{j=1}^{\nu} \Big[ b_j \|x_j\|^{\alpha_j} + b_j' \|x_j\|^{-\alpha_j} + b_j'' |x_j(b)|^{\alpha_j} \Big] \Big\}\, d\vec{x} < +\infty$$

and $L_{\Gamma}$ is the length of $\Gamma$. By the Fubini theorem and the analyticity of $f(z,\vec{x})$ we see that

$$\int_{\Gamma}\int_{C^{\nu}[a,b]} f(z,\vec{x})\, d\vec{x}dz = \int_{C^{\nu}[a,b]}\int_{\Gamma} f(z,\vec{x})\, dzd\vec{x} = \int_{C^{\nu}[a,b]} 0\, d\vec{x} = 0.$$

Any simple closed rectifiable path $\gamma$ can be approximated uniformly by a path composed of radial segments and arcs of circles centered at the origin. Since we proved that

$$\int_{\Gamma}\int_{C^{\nu}[a,b]} f(z,\vec{x})\, d\vec{x}dz = 0$$

we conclude that

$$\int_{\gamma}\int_{C^{\nu}[a,b]} f(z,\vec{x})\, d\vec{x}dz = 0$$

and by Morera's Theorem we see that

$$\int_{C^{\nu}[a,b]} f(z,\vec{x})\, d\vec{x}$$

is analytic in $\Omega$.

Thus

$$G(\lambda) \equiv \int_{C^{\nu}[a,b]} f(\lambda^{-1/2},\vec{x})\, d\vec{x}$$

is an analytic function in the right half plane, $\text{Re}\,\lambda > 0$, and is continuous for $\text{Re}\,\lambda \geq 0$, $\lambda \neq 0$. Moreover by hypothesis (3.1) when $\lambda$ is real and positive

$$G(\lambda) = \int_{C^{\nu}[a,b]} F(\lambda^{-1/2}\vec{x})\, d\vec{x}.$$

Thus $G(\lambda)$ is the analytic extension of the right hand member of the above equation to $\text{Re}\,\lambda > 0$, hence

$$G(\lambda) = \int_{C^{\nu}[a,b]}^{\text{anw}_{\lambda}} F(\vec{x})\, d\vec{x}.$$

Since $I(z)$ is continuous for $z \in \Omega$, it follows that $G(\lambda)$ is continuous for $\text{Re}\,\lambda \geq 0$, $\lambda \neq 0$, consequently for real $q \neq 0$, the integrals and limits below exist and

$$\int_{C^\nu[a,b]}^{\mathrm{anf}_q} F(\vec{x})\, d\vec{x} = \lim_{\substack{\lambda\to -iq \\ \mathrm{Re}\lambda>0}} \int_{C^\nu[a,b]}^{\mathrm{anw}_\lambda} F(\vec{x})\, d\vec{x} = \lim_{\substack{\lambda\to -iq \\ \mathrm{Re}\lambda>0}} G(\lambda)$$

$$= \lim_{\substack{\lambda\to -iq \\ \mathrm{Re}\lambda>0}} \int_{C^\nu[a,b]} f(\lambda^{-1/2},\vec{x})\, d\vec{x} = \int_{C^\nu[a,b]} f(\sqrt{\tfrac{i}{q}},\vec{x})\, d\vec{x}.$$

Thus the theorem is proved.

**Note.** The sum and product of two functions $F_1$ and $F_2$ that satisfy the hypotheses of Theorem 1 also satisfy the hypotheses of Theorem 1 and are therefore analytic Feynman integrable.

Let $\Omega^* = \{u \mid u \in \Omega \text{ or } -u \in \Omega\}$.

**Corollary 1.** *If* $\Psi(u)$ *is analytic on* $\mathrm{Int}(\Omega^*)$ *and is continuous and satisfies* $|\Psi(u)| < A\exp\{B|u|^\alpha\}$ *on* $\Omega^*$ *for* $A, B > 0$, $0 < \alpha < 2$, *then* $F(x) \equiv \Psi(x(b))$ *satisfies the hypothesis of Theorem* 1 *and is analytic Feynman integrable.*

**Corollary 2.** *If* $F(x)$ *satisfies the hypothesis of Theorem* 1 *and* $\Psi(x)$ *satisfies the hypothesis of Corollary* 1, *then* $F(x)\Psi(x(b))$ *is analytic Feynman integrable.*

As a consequence of our Theorem 1 we obtain immediately

**Theorem 2.** *If* $\varphi(z)$ *is an entire function of order less than* 2, *and if* $g(s)$ *is a complex valued function in* $L_2[a,b]$, *then* $\varphi[\int_a^b g(s)x(s)\,ds]$, $\varphi[\int_a^b g(s)|x(s)|\,ds]$, $\varphi[|\int_a^b g(s)x(s)\,ds|]$, *and* $\varphi[|\int_a^b g(s)|x(s)|\,ds|]$ *are analytic Feynman integrable on* $C[a,b]$ *for all real values of the parameter* $q$, $q \neq 0$.

Moreover:

$$\int_{C[a,b]}^{\mathrm{anf}_q} \varphi\left(\int_a^b g(s)x(s)\,ds\right) dx = \int_{C[a,b]} \varphi\left(\sqrt{i/q}\int_a^b g(s)x(s)\,ds\right) dx,$$

$$\int_{C[a,b]}^{\text{anf}_q} \varphi\left(\int_a^b g(s)|x(s)|\,ds\right) dx = \int_{C[a,b]} \varphi\left(\sqrt{i/q}\int_a^b g(s)|x(s)|\,ds\right) dx,$$

$$\int_{C[a,b]}^{\text{anf}_q} \varphi\left(\left|\int_a^b g(s)x(s)\,ds\right|\right) dx = \int_{C[a,b]} \varphi\left(\sqrt{i/q}\left|\int_a^b g(s)x(s)\,ds\right|\right) dx,$$

and

$$\int_{C[a,b]}^{\text{anf}_q} \varphi\left(\left|\int_a^b g(s)|x(s)|\,ds\right|\right) dx = \int_{C[a,b]} \varphi\left(\sqrt{i/q}\left|\int_a^b g(s)|x(s)|\,ds\right|\right) dx.$$

Some further examples of functionals to which our theorems apply are the following:

$$F(x) = \exp\left\{\int_a^b |x(s)|\,ds\right\}|x(b)|^{100}$$

$$F(x) = [\max_{s\in[a,b]} \{x(s)\}]^{-n}, \quad n > 0$$

$$F(x) = \left[\int_a^b \log(1 + |x(s)|)\,ds\right]^n, \quad n > 0$$

$$F(x) = \left[\int_a^b x(s)x(a+b-s)\,ds\right]\|x\|.$$

**Definition.** We say $F$ defined on $C^\nu[a,b]$ is a positive-homogeneous functional of $\vec{x}$ of degree $\delta$ (where $\delta$ is a real number) if for each $\rho > 0$, we have

$$F(\rho\vec{x}) = \rho^\delta F(\vec{x}) \text{ for all } \vec{x} \in C^\nu[a,b]. \tag{3.9}$$

For such functionals, $F$ is Feynman integrable whenever it is Wiener integrable, thus:

**Theorem 3.** *If* $F(\vec{x})$ *is a positive–homogeneous function of* $\vec{x}$ *of degree* $\delta$ *on* $C^\nu[a,b]$, *and if* $F$ *is Wiener integrable on* $C^\nu[a,b]$, *then for each real* $q \neq 0$, $F$ *is analytic*

*Feynman integrable on* $C^\nu[a,b]$ *and we have*

$$\int_{C^\nu[a,b]}^{\mathrm{anf}_q} F(\vec{x})\, d\vec{x} = |q|^{-\delta/2}\, e^{\pi\delta i\, \mathrm{sgn}(q)/4} \int_{C^\nu[a,b]} F(\vec{x})\, d\vec{x}. \tag{3.10}$$

P r o o f. By (3.9), for $\lambda = |q| e^{-i\theta}$,

$$\int_{C^\nu[a,b]} F(\lambda^{-1/2}\vec{x})\, d\vec{x} = \lambda^{-\delta/2} \int_{C^\nu[a,b]} F(\vec{x})\, d\vec{x} = |q|^{-\delta/2}\, e^{i\theta\delta/2} \int_{C^\nu} F(\vec{x})\, d\vec{x} \tag{3.11}$$

and the second member has an analytic extension to a (perhaps multiple valued) analytic function of $\lambda$ on the entire complex plane (except possibly for $\lambda = 0$). In particular, it has an analytic extension to the right half plane which is single valued and this function of $\lambda$ is continuous for $\mathrm{Re}\,\lambda \geq 0$, $\lambda \neq 0$. It follows that as $\lambda \to -iq$, $\mathrm{Re}\,\lambda > 0$, both outer members of (3.11) approach the corresponding members of (3.10) and $F$ is analytic Feynman integrable for real $q \neq 0$, and (3.10) holds.

**Theorem 4.** *Let* $g(\vec{x})$ *be a complex-valued positive-homogeneous functional of degree* $\delta > 0$ *on* $C^\nu[a,b]$, *and let* $g(\vec{x})$ *be Wiener measurable on* $C^\nu[a,b]$ *and let* $|g(\vec{x})| \leq K\|\vec{x}\|^\delta \equiv K[\Sigma_{j=1}^{\nu}\|x_j\|]^\delta$ *on* $C^\nu[a,b]$ *for some constant* K. *Let* $G(z)$ *be an entire function of order* $\beta$ *less than* $2/\delta$. *Then the functional* $G[g(x)]$, *defined on* $C^\nu[a,b]$, *is analytic Feynman integrable for each real* q, $q \neq 0$, *and with* $(\frac{i}{q})^{\delta/2}$ *interpreted as in* (3.10), *we have*

$$\int_{C^\nu[a,b]}^{\mathrm{anf}_q} G(g(\vec{x}))\, d\vec{x} = \int_{C^\nu[a,b]} G[(\tfrac{i}{q})^{\delta/2} g(\vec{x})]\, d\vec{x}.$$

P r o o f. If we set $f(\rho,\vec{x}) \equiv G(g(\rho\vec{x})) = G(\rho^\delta g(\vec{x}))$ for real $\rho > 0$ and $\vec{x} \in C^\nu[a,b]$, we observe that $f(\rho,\vec{x})$ satisfies all the hypotheses of Theorem 1 and the conclusion of our Theorem follows.

We now present three examples of functionals which satisfy the conditions of Theorem 1:

$$F_1(\vec{x}) \equiv J_0(\|\vec{x}\|),$$

$\vec{x} \in C^{\nu}[a,b]$, where $J_0$ is the Bessel function of zero order,

$$F_2(\vec{x}) \equiv \begin{cases} \exp\left\{\left[\dfrac{B\|x_2\|}{\max\limits_{s\in[a,b]} |x_1(s)|}\right]^{2/3}\right\} & \text{if } x_1(s) \not\equiv 0 \\ 0 & \text{if } x_1(s) \equiv 0 \end{cases}$$

for $\vec{x} \in C^2[a,b]$ and

$$F_3(\vec{x}) \equiv \exp\left\{B\left[\int_a^b x_1(s)x_2(s)\, ds\right]^{2/3}\right\}$$

for $\vec{x} = (x_1,x_2) \in C^2[a,b]$ .

By Theorem 1 or by Theorem 4 we see that

$$\int_{C^{\nu}[a,b]}^{\text{anf}_q} F_1(\vec{x})\, dx = \int_{C^{\nu}[a,b]} J_0(\sqrt{i/q}\, \|\vec{x}\|)\, d\vec{x}.$$

By Theorem 1 or by Theorem 3 we see that

$$\int_{C^2[a,b]}^{\text{anf}_q} F_2(\vec{x})\, d\vec{x} = \int_{C^2[a,b]} F_2(\vec{x})\, d\vec{x},$$

which is independent of q.

By Theorem 1 or by Theorem 4, we see that

$$\int_{C^2[a,b]}^{\text{anf}_q} F_3(\vec{x})\, d\vec{x} = \int_{C^2[a,b]} \exp\left\{B\left(\frac{i}{q}\int_a^b x_1(s)x_2(s)\, ds\right)^{2/3}\right\} d\vec{x}.$$

## References


[1] S. Albeverio and R. Høegh-Krohn, *Mathematical theory of Feynman path integrals*, Lecture Notes in Math., 523, Springer-Verlag, Berlin, 1976.

[2] R. H. Cameron, *A family of integrals serving to connect the Wiener and Feynman integrals*, J. of Math. and Phys. vol. XXXIX (1960), 126–140.

[3] R. H. Cameron and W. T. Martin, *The Wiener measure of Hilbert neighborhoods in the space of real continuous functions*, J. of Math. and Phys. vol XXIII (1944), 195–209.

[4] R. H. Cameron and W. T. Martin, *Evaluation of various Wiener integrals by use of certain Sturm–Liouville differential equations*, Bull. Amer. Math. Soc. 51 (1945), 73–89.

[5] R. H. Cameron and D. A. Storvick, *Some Banach algebras of analytic Feynman integrable functionals*, Analytic functions (Kozubnik, 1979), 18–67, Lecture Notes in Math., 798, Springer-Verlag, Berlin, 1980.

[6] R. H. Cameron and D. A. Storvick, *Analytic Feynman integral solutions of an integral equation related to the Schroedinger equation*, J. Analyse Math. 38 (1980), 34–66.

[7] R. H. Cameron and D. A. Storvick, *New existence theorems and evaluation formulas for sequential Feynman integrals*, Proc. London Math. Soc. (3) 52 (1986), 557–581.

[8] R. H. Cameron and D. A. Storvick, *New existence theorems and evaluation formulas for analytic Feynman integrals*, accepted for publication.

[9] E. H. Cuthill, *Integrals on spaces of functions which are real and continuous on finite and infinite intervals*, Thesis, University of Minnesota, 1951.

[10] D. Elworthy and A. Truman, *A Cameron–Martin formula for Feynman integrals*, Mathematical problems in theoretical physics (Berlin, 1981), 288–294, Lecture Notes in Phys., 153, Springer-Verlag, Berlin, 1982.

[11] R. P. Feynman, *Space–time approach to non–relativistic quantum mechanics*, Rev. Modern Phys. 20 (1948), 115–142.

[12] G. W. Johnson, *The equivalence of two approaches to the Feynman integral*, J. Math. Phys. 23 (11) (1982), 2090–2096.

[13] G. W. Johnson and D. L. Skoug, *Notes on the Feynman integral*, III: *the Schroedinger equation*, Pacific J. Math. 105 (1983), 321–358.

[14] G. Kallianpur and C. Bromley, *Generalized Feynman integrals using analytic continuation in several complex variables*, Stochastic analysis, M. Pinsky, ed., 217–267, Marcel Dekker, 1984.

[15] G. Kallianpur, D. Kannan, and R. L. Karandikar, *Analytic and sequential Feynman integrals on abstract Wiener and Hilbert spaces and a Cameron–Martin formula*, Ann. Inst. Henri Poincaré 21:4 (1985), 323–361.

[16] A. Truman, *The Feynman maps and the Wiener integral*, J. Math. Phys. 19 (1978), 1742–1750.

# DYNAMICS OF MAPS WITH CONSTANT SCHWARZIAN DERIVATIVE

ROBERT L. DEVANEY*
Department of Mathematics, Boston University
Boston, Mass. 02215, U.S.A.

LINDA KEEN**
Department of Mathematics, Herbert H. Lehman College, CUNY
Bronx, N.Y., U.S.A.

There has been a resurgence of interest in the past decade in the dynamics of complex analytic functions. Most of this work has centered around the dynamics of polynomials or rational maps [Bl, DH, Ma, S] or entire transcendental functions [DK, DT, GK, BR]. Our goal is to extend some of this work to the meromorphic case, pointing out along the way some of the similarities and principal differencies between this case and the other classes of maps.

One of the principal differencies arising in the iteration of meromorphic (non-rational) functions is the fact that, strictly speaking, iteration of these maps does not lead to a dynamical system. Infinity is an essential singularity for such a map, so the map cannot be extended continuously to infinity. Hence the forward orbit of any pole terminates, and, moreover, any preimage of a pole also has a finite orbit. All other points have well defined forward orbits.

Despite the fact that certain orbits of a meromorphic map are finite, the iteration of such maps is important. For example, the iterative processes associated to Newton's method applied to entire functions often yields a meromorphic function as the root-finder. See [CGS].

In this paper we will deal exclusively with a very special class of meromorphic functions, namely, those whose Schwarzian derivative is a constant. This class of maps includes a number of dynamically important families of maps, including $\lambda \tan z$ and $\lambda \exp z$. More general results and details can be found in [DKn].

*Partially supported by NSF and DARPA.
**Partially supported by NSF.

## 1. Background

To study the dynamics of a family of maps, the family should have two properties. First, it should be topologically closed—that is, if $f$ belongs to the family and $g$ is topologically conjugate to $f$ then $g$ belongs to the family; and second, there should be parameters for the family which are easy to work with. A theorem of Nevanlinna [N] says that the meromorphic maps with polynomial Schwarzian derivative of fixed degree (and in particular degree 0) have these properties. The theorem states that meromorphic maps with exactly $p$ asymptotic values and no critical values are precisely those functions whose Schwarzian derivative is a polynomial of degree $p-2$. Moreover, it describes the asymptotic behavior of these maps which makes it easy to analyze the dynamics.

These functions are quotients of a pair of linearly independent solutions of the second order linear differential equation associated to the third order non-linear Schwarzian differential equation

$$\{F, z\} = \left(\frac{F''}{F'}\right)' - \frac{1}{2}\left(\frac{F''}{F'}\right)^2.$$

Since the Schwarzian is polynomial, the solutions to the linear differential equation are entire. Parameters for the family can either be chosen as the asymptotic values or as the coefficients of the Schwarzian polynomial and parameters which determine the linearly independent solutions in terms of some basis.

Below we look at two 1-dimensional subfamilies of the case $p = 2$. In these examples we see dynamic phenomena similar to those encountered for certain rational functions as well as those of certain entire transcendental functions. In addition we see something new: functions whose Julia sets exhibit an inhomogeneity which has not been observed before.

Before we can proceed we need the following. Let $F$ be a meromorphic function whose Schwarzian derivative is a constant $k$.

We call a point $z$ *stable* for $F$ if there is a neighborhood $U$ of $z$ such that the iterates $F^n$ are uniformly bounded on $U$. The metric here is the standard Euclidean metric. The Julia set of $F$, denoted by $J(F)$, is the complement of the stable set. In [DKn] it is proved that the Julia set of a meromorphic function with polynomial Schwarzian derivative has two additional equivalent formulations:

1. $J(F)$ is the closure of the set of repelling periodic orbits.

2. If $F$ is not entire, $J(F)$ is the closure of the set consisting of the poles of $F^n$ for $n > 0$.

We will use both of these characterizations.

## 2. Asymptotic values which are poles

It is known that entire transcendental functions of finite type often have Julia sets which contain smooth curves. Indeed, for a wide class of these maps (see [DT]), all repelling periodic orbits lie at the endpoints of invariant curves which connect the orbit to the essential singularity at $\infty$.

In this section we give an example of a family of maps with constant Schwarzian derivative for which certain of the repelling fixed points lie on analytic curves in the Julia set, but for which many of the other periodic points do not. This lack of homogeneity in the Julia set is caused by the fact that one of the asymptotic values is also a pole.

Consider the family of maps

$$F_\lambda(z) = \frac{\lambda e^z}{e^z - e^{-z}}$$

with $\lambda > 0$. Clearly, $\{F_\lambda, z\} = -2$ and $F_\lambda$ is periodic with period $\pi i$. These maps have asymptotic values at 0 and $\lambda$, and 0 is also a pole.

The graph of $F_\lambda$ restricted to $\mathbf{R}$ shows that $F_\lambda$ has two fixed points on $\mathbf{R}$ at $p$ and $q$ with $p < 0 < q$. We note that $F_\lambda(z) = L_\lambda \circ E(z)$ where $E(z) = \exp(-2z)$ and $L_\lambda$ is the linear fractional transformation

$$L_\lambda(z) = \frac{\lambda}{1-z}.$$

$F_\lambda$ has poles at $k\pi i$ where $k \in \mathbf{Z}$. $F_\lambda$ has the following mapping properties:

1. $F_\lambda$ preserves $\mathbf{R}^+$ and $\mathbf{R}^-$.
2. $F_\lambda$ maps the horizontal lines $\operatorname{Im} z = \frac{1}{2}(2k+1)\pi$ onto the interval $(0, \lambda)$ in $\mathbf{R}$.
3. $F_\lambda$ maps the imaginary axis onto the line $\operatorname{Re} z = \lambda/2$, with the points $k\pi i$ mapped to $\infty$.
4. $F_\lambda$ maps horizontal lines onto circular arcs passing through both 0 and $\lambda$.
5. $F_\lambda$ maps vertical lines with $\operatorname{Re} z > 0$ to a family of circles orthogonal to those in 4 which are contained in the plane $\operatorname{Re} z > \lambda/2$.
6. $F_\lambda$ maps vertical lines with $\operatorname{Re} z < 0$ to a family of circles orthogonal to those in 4 which are contained in the plane $\operatorname{Re} z < \lambda/2$.

As a consequence of these properties, we have

PROPOSITION 1. *If $\lambda > 0$, then the fixed point $q$ is attracting. Moreover, if $\operatorname{Re} z > 0$, then $F_\lambda^n(z) \to q$ as $n \to \infty$. Hence $J(F_\lambda)$ is contained in the half plane $\operatorname{Re} z \leq 0$.*

PROOF. $|(F_\lambda)'(q)| < 1$ follows from property 5 above and the Schwarz Lemma.

PROPOSITION 2. *$J(F_\lambda)$ contains $\mathbf{R}^- \cup \{0\}$.*

PROOF. The fixed point $p$ is repelling: if $|(F_\lambda)'(p)| \leq 1$, then it would follow that $p$ would have to attract a critical point or asymptotic value of $F_\lambda$ on $\mathbf{R}$; this does not occur since $q$ attracts $\lambda$ and 0 is a pole.

Let $x \in (-\infty, p)$. One may check easily that $|(F_\lambda^2)'(x)| > 1$. Moreover, $|(F_\lambda^{2n})'(x)| \to \infty$ as $n \to \infty$. This again follows from the fact that $F_\lambda^2$ has negative Schwarzian derivative on $\mathbf{R}^-$. Let $U$ be a neighborhood of $x$ in $\mathbf{C}$. Note that $F_\lambda^{2n}$ expands $U$ until some image overlaps the horizontal lines $y = \pm\pi/2$. By the above properties, these points are in the basin of $q$. Hence the family $\{F_\lambda^{2n}\}$ is not normal at $x$, and so $(-\infty, p) \subset J(F_\lambda)$. The image of this interval under $F_\lambda$ is $(p, 0)$, so $\mathbf{R}^- \subset J(F_\lambda)$. q.e.d.

Thus some points in the Julia set lie on analytic curves, for example, $\mathbf{R}^-$ and all of its preimages. But not all points in the Julia set lie on smooth invariant curves:

PROPOSITION 3. *There is a unique repelling fixed point $p_1$ in the half strip*

$$\pi/2 < \operatorname{Im} z < 3\pi/2$$

*and this point does not lie on any smooth invariant curve in $J(F_\lambda)$.*

PROOF. Let $R$ be the rectangle

$$\pi/2 < \operatorname{Im} z < 3\pi/2$$
$$\nu < \operatorname{Re} z < 0$$

where $\nu$ is chosen far enough to the left in $\mathbf{R}$ so that

$$|F_\lambda(\nu + iy)| < \pi/4.$$

Then $F_\lambda(R)$ is a "disk" which covers $R$ and $F_\lambda|R$ is 1-1. So, $F_\lambda^{-1}$ has a unique attracting fixed point $p_1$ in $R$. Since this argument is independent of $\nu$ for $\nu$ large enough negative, the first part of the Proposition follows.

Now suppose that $p_1$ lies on a smooth invariant curve $\gamma$ in $J(F_\lambda)$; $\gamma$ is unbounded since it is in $J(F_\lambda)$. Since $J(F_\lambda)$ is invariant under $F_\lambda^{-1}$, we may assume that $\gamma$ accumulates on the boundary of the strip

$$\pi/2 < \operatorname{Im} z < 3\pi/2$$
$$\operatorname{Re} z < 0.$$

To see this choose branches of $F_\lambda^{-1}$ which fix $\gamma$; $\gamma$ cannot leave the strip since the upper and lower boundaries of the strip are stable by property 2 and so are points on the line

$x = 0$ (except possibly at $i\pi$). So $\gamma$ can only accumulate at $\infty$ or $i\pi$. If $\gamma$ accumulates at $\infty$, then $\gamma$ must also accumulate at $i\pi$, since $F_\lambda(i\pi) = \infty$. Since all points on $\gamma$ leave the strip under iteration, it follows that $\gamma$ must contain $i\pi$. Now $\gamma$ cannot have a tangent vector at $i\pi$, for if so, $\gamma$ would enter the region $\text{Re}\, z \geq 0$, $\text{Im}\, z \neq i\pi$, which lies in the stable set. q.e.d.

REMARKS. 1. There is a continuous invariant curve which lies in the Julia set and accumulates on $p_1$. Indeed, the horizontal line $\ell_0$ given by $y = \pi$, $x \leq 0$ lies in $J(F_\lambda)$ since it is mapped onto $\mathbf{R}^-$ by $F_\lambda$. Consider the successive preimages $\ell_n = F_\lambda^{-n}(\ell_0)$, where $F_\lambda^{-1}$ is the branch of the inverse of $F_\lambda$ whose image is $\pi/2 < \text{Im}\, z < 3\pi/2$. Then $\ell_1$ meets $\ell_0$ at $i\pi$, $\ell_2$ meets $\ell_1$ at $F_\lambda^{-1}(i\pi)$, and so forth. Since $p_1$ is an attracting fixed point for $F_\lambda^{-1}$, the curve $\ell$ formed by concatenating the $\ell_i$ is invariant and accumulates on $p_i$ as $i \to \infty$. Note that this curve is considerably different from a dynamical point of view from the invariant curve $\mathbf{R}^-$ through $p$.

2. $J(F_\lambda)$ also possesses a collection of invariant curves which are quite different from $\ell$. Since $F_\lambda$ has a pole at the asymptotic value 0, $F_\lambda^2$ has an asymptotic value at $\infty$. Using techniques developed in [DT] to study entire functions with an isolated asymptotic value at $\infty$ we can prove that these curves lie in a Cantor bouquet, and all points (except the endpoints) on these curves tend toward $\infty$ or 0 under iteration of $F_\lambda^2$. (See [DT] for a complete discussion of Cantor bouquets.)

## 3. Bifurcation to an entire function

Most maps with polynomial Schwarzian derivatives are bona fide meromorphic functions, but occasionally they are entire functions. In this section we describe an "explosion" in the Julia set which occurs when a meromorphic family suddenly encounters an entire function. An *explosion* occurs at a parameter value for a family of functions whenever the Julia sets of the functions in the family change suddenly, when the parameter is reached, from a nowhere dense subset $\hat{\mathbf{C}}$ to all of $\hat{\mathbf{C}}$.

Consider the family

$$F_\lambda(z) = \frac{e^z}{\lambda e^z + e^{-z}} = \frac{1}{\lambda + e^{-2z}}.$$

When $\lambda = 0$, the corresponding element of this family is the entire function $F_0(z) = \exp(2z)$ whose dynamics are well understood. It is known that $J(F_0) = \mathbf{C}$, since the orbit of the asymptotic value 0 tends to $\infty$. See [D] or [Mi].

When $\lambda > 0$, $J(F_\lambda) \neq \mathbf{C}$. This follows since $F_\lambda$ has a unique attracting fixed point $p_\lambda$ on the real line. The graph of $F_\lambda$ is depicted in Fig. 1.

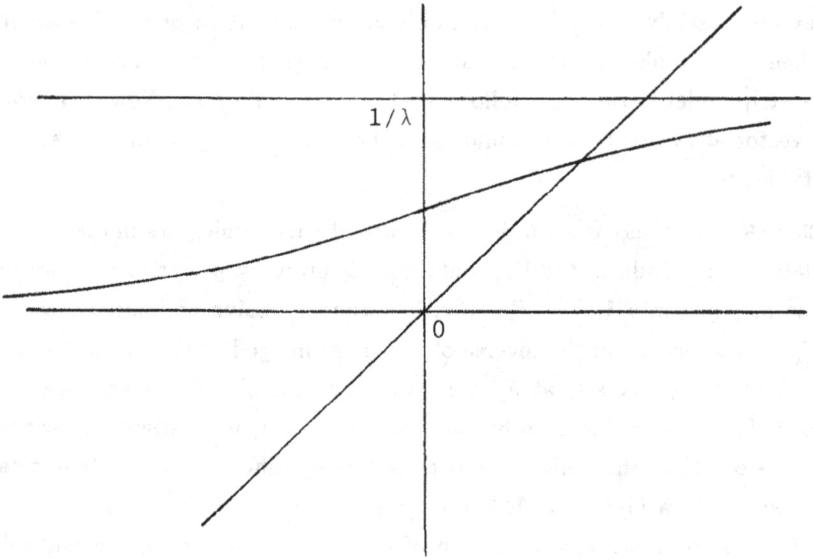


Fig. 1. The graph of $F_\lambda$.

In fact, we can say much more about $J(F_\lambda)$. Below we prove that the dynamics of $F_\lambda$ are unlike any possible for entire transcendental functions but are similar to certain polynomial and rational functions.

PROPOSITION 4. *For all $\lambda > 0$, the Julia set of $F_\lambda$ is a Cantor set in $\hat{\mathbf{C}}$ and $F_\lambda|J(F_\lambda)$ is the shift map on infinitely many symbols.*

PROOF. First note that the entire real axis lies in the basin of attraction of $p_\lambda$. This follows since $F_\lambda$ has negative Schwarzian derivative and maps $\mathbf{R}$ diffeomorphically onto the interval bounded by the asymptotic values, $(0, 1/\lambda)$. In particular, both asymptotic values lie in the immediate basin of $p_\lambda$ and so there are disks about these points which lie in the basin. Taking preimages of these disks, it follows that there are half planes of the form $\operatorname{Re} z < \nu_1$ and $\operatorname{Re} z > \nu_2$ with $\nu_1 < p_\lambda < \nu_2$ which lie in the immediate basin of $p_\lambda$.

We may find a strip $S_\mu$ surrounding the interval $[\nu_1, \nu_2]$ of the form

$$\{ z \mid |\operatorname{Im} z| < \mu,\ \nu_1 \le \operatorname{Re} z \le \nu_2 \}$$

which is mapped inside itself. Now let $B$ denote the "ladder-shaped" region consisting of the two half planes together with $S_\mu$ and all of its $\pi i$ translates. See Fig. 2. Clearly, $F_\lambda$ maps $B$ inside itself as long as the $\nu_i$ are chosen large enough.

The complement of $B$ consists of infinitely many congruent rectangles $R_j$ where $j \in \mathbf{Z}$ and the $R_j$ are indexed according to increasing imaginary part. $F_\lambda$ maps each

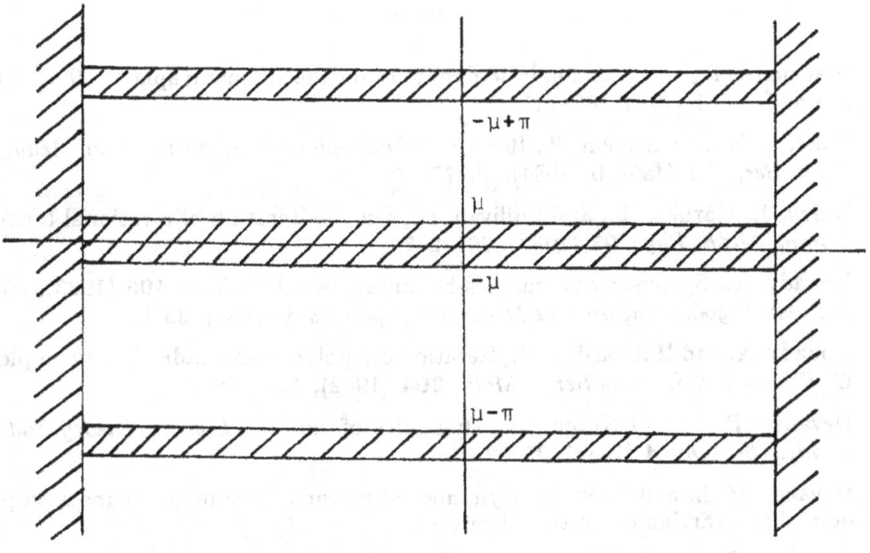


Fig. 2. The region $B$.

$R_j$ diffeomorphically onto $\hat{\mathbb{C}} - F_\lambda(B)$. In particular, $F_\lambda(R_j)$ covers each $R_k$ and $\infty$. It follows that there exists at least one point $z$ corresponding to any sequence $(s_0, s_1, \ldots)$ in the sequence space $\Gamma$.

$\Gamma$ consists of all infinite sequences $(s_0, s_1, s_2, \ldots)$ where $s_j \in \mathbf{Z}$ and all finite sequences of the form $(s_0, s_1, \ldots, s_j, \infty)$ where $s_i \in \mathbf{Z}$. There is a topology on $\Gamma$ originally defined by Moser [Mo] and a natural map $\sigma : \Gamma \to \Gamma$ called the shift automorphism defined by $\sigma(s_0, s_1, s_2, \ldots) = (s_1, s_2, \ldots)$. Note that $\sigma(\infty)$ is not defined. In Moser's topology, $\sigma$ is continuous and $\Gamma$ is a Cantor set. The pair $(\Gamma, \sigma)$ is often called the shift on infinitely many symbols; it has the property that $F_\lambda^n(z) \in R_{s_n}$ for each $n$.

We claim that the point with the prescribed itinerary $(s_0, s_1, \ldots)$ is in the Julia set; first note that $B$ is contained in the immediate attracting basin of the attractive fixed point. Moreover, $B$ intersects its preimages, so the immediate basin is completely invariant. Since it contains both asymptotic values, it is the whole stable set. Therefore, the entire Julia set is contained in the rectangles. Points whose orbits remain in the rectangles are in the Julia set, so the points corresponding to sequences are in the Julia set.

We also claim the point is uniquely determined by its itinerary. The argument of [Mc, Prop. 6.1] shows that $|(F_\lambda)'(z)| \to \infty$ for all $z \in J(F_\lambda)$, $z$ not a pole; at the poles $|(F_\lambda)'(z)| = \infty$. Therefore $F_\lambda$ is expanding on its Julia set. Since two points corresponding to the same sequence must remain a bounded distance apart, it follows that two points in the Julia set cannot have the same itinerary. We therefore have a topological conjugacy between $F_\lambda|J_\lambda$ and $\sigma|\Gamma$. q.e.d.

## References

[Bl] Blanchard, P., Complex analytic dynamics on the Riemann sphere. *Bull. Amer. Math. Soc.* **11** (1984), 85–141.

[BR] Baker, I. N. and Rippon, P., Iteration of exponential functions. *Ann. Acad. Sci. Fenn. Ser. A I Math.* **9** (1984), 49–77.

[CGS] Curry, J., Garnett, L., and Sullivan, D., On the iteration of a rational function. *Comm. Math. Phys.* **91** (1983), 267–277.

[Do] Douady, A., Systèmes dynamiques holomorphes. *Astérisque* **105** (1983), 39–63. See also *Ergodic Theory and Dynamical Systems* **4** (1984), 35–52.

[DH] Douady, A. and Hubbard, J. H., Itération des polynômes quadratiques complexes. *C. R. Acad. Sci. Paris Sér. I Math.* **294** (1982), 123–126.

[DK] Devaney, R. L. and Krych, M., Dynamics of exp($z$). *Ergodic Theory and Dynamical Systems* **4** (1984), 35–52.

[DKn] Devaney, R. L. and Keen, L., Dynamics of meromorphic maps: maps with polynomial Schwarzian derivative. Preprint.

[DT] Devaney, R. L. and Tangerman, F., Dynamics of entire functions near the essential singularity. *Ergodic Theory and Dynamical Systems* **6** (1986), 489–503.

[GK] Goldberg, L. R. and Keen, L., Finiteness theorem for a dynamical class of entire functions. *Ergodic Theory and Dynamical Systems* **6** (1986), 183–192.

[Ma] Mandelbrot, B., *The Fractal Geometry of Nature.* San Francisco: Freeman, 1982.

[Mc] McMullen, C., Area and Hausdorff dimension of Julia sets of entire functions. *Trans. Amer. Math. Soc.* **300** (1987), 329–342.

[Mi] Misiurewicz, M., On iterates of $e^z$. *Ergodic Theory and Dynamical Systems* **1** (1981), 103–106.

[Mo] Moser, J. K., *Stable and Random Motions in Dynamical Systems.* Princeton: Princeton University Press, 1973.

[N] Nevanlinna, R., Über Riemannsche Flächen mit endlich vielen Windungspunkten. *Acta Math.* **58** (1932), 295–373.

[S] Sullivan, D., Quasiconformal homeomorphisms and dynamics II, *Acta Math.* **155** (1985), 243–260.

# RATIONAL RIEMANN MAPS [1]

**Timo Erkama** [2]
Department of Mathematics, University of Joensuu
P.O.Box 111, SF-80101 Joensuu, Finland

## 1. Introduction

Let D be a domain on the Riemann sphere $\mathbb{C} \cup \{\infty\}$. We say that a biholomorphic automorphism of D is *rational* if it is the restriction of a rational transformation. The set of rational biholomorphic automorphisms of D is a semigroup under composition and will be denoted by Rat(D). The group Möb(D) of invertible elements of Rat(D) consists of sense-preserving Möbius transformations mapping D onto itself.

Let f be a biholomorphic map from a disc onto D. We say that f is a *rational Riemann map of degree* n if either f or the inverse of f is the restriction of a rational map R with $\deg R = n$. It turns out that in this situation Rat(D) = Möb(D). Moreover, D is a disc if and only if $n = 1$. In this paper we show that for $n > 1$ the semigroup Rat(D) is in general a finite cyclic group of Möbius transformations. The only exceptions occur when D is either the complement of a circular arc or the intersection of two discs.

The proofs depend on the solution of some functional equations and the classification of domains D with a non-discrete group Möb(D).

## 2. Preliminary results

A fundamental property of rational Riemann maps is contained in the following lemma.

**Lemma 1.** *Let* $\Delta$ *be a disc on the Riemann sphere, and let* $f : \Delta \to D$ *be a rational Riemann map. Then* Rat(D) = Möb(D).

---

[1] This paper is in final form and no version of it will be submitted for publication elsewhere.

[2] This research was supported by grants from the Mittag-Leffler Institute and the Emil Aaltonen Foundation.

*Proof.* Suppose $S \in \mathrm{Rat}(D)$. Then $T = f^{-1} \circ S \circ f$ maps $\Delta$ conformally onto itself; hence $T$ is the restriction of a Möbius transformation. If $f$ [resp. $f^{-1}$] is the restriction of a rational transformation $R$, we have $R \circ T = S \circ R$ [resp. $T \circ R = R \circ S$]. By analytic continuation the identity holds on the whole Riemann sphere. Moreover, $\deg(R \circ T) = \deg(T \circ R) = \deg R$, because $T$ is a Möbius transformation. It follows that either $\deg(S \circ R) = \deg R$ or $\deg(R \circ S) = \deg R$. This is possible only if $\deg S = 1$, and we conclude that $S \in \mathrm{Möb}(D)$.

Examples of domains with $\mathrm{Rat}(D) \neq \mathrm{Möb}(D)$ appear in the theory of iterations of rational functions. We say that $D$ is a *Siegel domain* if there is $\zeta \in D$ such that the stabilizer $\mathrm{Rat}_\zeta(D) = \{S \in \mathrm{Rat}(D)\ ;\ S(\zeta) = \zeta\}$ is not a group. Then $\mathrm{Rat}_\zeta(D)$ contains elements of infinite order, because every element of finite order has an inverse in $\mathrm{Rat}(D)$. The existence of Siegel domains follows from a theorem of Siegel [5]. The next theorem shows that for Siegel domains the group $\mathrm{Möb}_\zeta(D) = \mathrm{Möb}(D) \cap \mathrm{Rat}_\zeta(D)$ is always finite.

**Theorem 1.** *Let* $D$ *be a domain with at least two boundary points, and let* $\zeta \in D$. *Then either* $D$ *is a disc or* $\mathrm{Möb}_\zeta(D)$ *is a finite cyclic group.*

*Proof.* Let $h$ be a Möbius transformation such that $h(\infty) = \zeta$, and suppose that $\mathrm{Möb}_\zeta(D)$ is not a finite cyclic group. Then $G = \{h^{-1} \circ S \circ h\ ;\ S \in \mathrm{Möb}_\zeta(D)\}$ can be identified with a subgroup of the group $\mathrm{Möb}(\mathbb{C})$ of affine Möbius transformations. Since $G$ is not a finite cyclic group, the corresponding subgroup of $\mathrm{Möb}(\mathbb{C})$ has a limit point at infinity. It follows that $\zeta$ is a limit point of $\mathrm{Möb}(D)$, and the action of $\mathrm{Möb}(D)$ in $D$ is not discontinuous. Domains with this property have been classified in [3]. The classification shows that if $D$ satisfies the hypotheses of Theorem 1 and if $\mathrm{Möb}_\zeta(D)$ is not a finite cyclic group, then $D$ is a disc.

If $D$ is the whole Riemann sphere, then every element of $\mathrm{Rat}(D)$ has a fixed point in $D$. The following theorem contains information about other domains with a similar property.

**Theorem 2.** *Let* $D$ *be a proper subdomain of the Riemann sphere such that every element of* $\mathrm{Rat}(D)$ *has a fixed point in* $D$. *Then either* $D$ *is a Siegel domain or* $\mathrm{Rat}(D)$ *is a finite group. Moreover, if* $D$ *is simply connected, then* $\mathrm{Möb}(D) = \mathrm{Möb}_\zeta(D)$ *for some* $\zeta \in D$.

*Proof.* Suppose that $D$ is not a Siegel domain. Then $\mathrm{Rat}_\zeta(D)$ is a group for each $\zeta \in D$ so that $\mathrm{Rat}_\zeta(D) = \mathrm{Möb}_\zeta(D)$. Since every element of $\mathrm{Rat}(D)$ has a fixed point in $D$, it follows that $\mathrm{Rat}(D) = \mathrm{Möb}(D)$.

The hypothesis implies that $D$ is not a disc and that $D$ has at least two boundary points. Therefore, by Theorem 1 $\mathrm{Möb}_\zeta(D)$ is a finite cyclic group for each

$\zeta \in D$. In particular, every element of Möb(D) is of finite order.

Since every element of Möb(D) has a fixed point in D, the action of Möb(D) in D is discontinuous. This follows from the fact that none of the domains appearing in the classification in [3] satisfies the hypotheses of Theorem 2. Since in addition every element of Möb(D) is of finite order, we conclude that Möb(D) is a finite group (see [4, p. 91]).

Suppose next that D is simply connected. Since D has at least two boundary points, there exists a biholomorphic map f from the upper half plane $H = \{z \in \mathbb{C};\ \mathrm{Im}\, z > 0\}$ onto D. Let $G = \{f^{-1} \circ S \circ f;\ S \in \text{Möb}(D)\}$. Then G is a subgroup of Möb(H), and every element of G is of finite order. It follows that all elements of G have a common fixed point $z \in H$ [2, Lemma 3.2]. Then $\zeta = f(z)$ is a common fixed point for every element of Möb(D). The proof of Theorem 2 is now complete.

## 3. Functional equations

The following two lemmas deal with some functional equations. These will be needed in the proof of our main results in Section 4.

**Lemma 2.** *Let* R *be a rational transformation, and let* a, b *be complex constants such that* $a \neq 0$.

(a) *If* R *satisfies the functional equation* $R(az) = R(z) + b$, *then* $b = 0$.

(b) *If* R *satisfies the functional equation* $R(az) = bR(z)$, *then either* a *is a root of unity or there is a constant* $\gamma \in \mathbb{C}$ *and an integer* k *such that* $R(z) = \gamma z^k$. *If in addition* R *is conformal and finite at* $z = 0$, *then* $a = b$.

*Proof.* Since R is rational, there exists $r > 0$ such that R has a Laurent series

$$R(z) = \sum_{\nu=-N}^{\infty} c_\nu z^\nu$$

converging for $0 < |z| < r$.

(a) If $R(az) = R(z) + b$, then

$$\sum_{\nu=-N}^{\infty} c_\nu a^\nu z^\nu = \sum_{\nu=-N}^{\infty} c_\nu z^\nu + b$$

in a punctured neighborhood of the origin. Equating the constant terms of these series yields $c_0 = c_0 + b$, so that $b = 0$.

(b) If $R(az) = bR(z)$, we have

$$\sum_{\nu=-N}^{\infty} c_\nu a^\nu z^\nu = \sum_{\nu=-N}^{\infty} c_\nu b z^\nu$$

in a punctured neighborhood of the origin. Equating the coefficients yields

$$c_\nu(a^\nu - b) = 0 \qquad (\nu \geq -N).$$

If $a$ is not a root of unity, it follows that $c_\nu \neq 0$ for at most one value of $\nu$. In addition, if $R$ is conformal and finite at $z = 0$, then $c_1 \neq 0$, so that $a = b$.

The equation of Lemma 2 (b) needs to be studied also if $a$ is a root of unity:

**Lemma 3.** *Let* $R$ *be a rational transformation with* $\deg R \geq 2$. *Suppose that* $R$ *satisfies the functional equation*

$$R(\omega z) = \omega R(z)$$

*where* $\omega$ *is a primitive* n:*th root of unity. Then* $\deg R \geq n$. *Moreover, if* $R$ *is a polynomial, then* $\deg R \geq n + 1$.

*Proof.* Let $S(z) = R(z)/z$; then $S$ is rational and satisfies $S(\omega z) = S(z)$. Since $\omega^n = 1$, it follows that $S(z) = S_1(z^n)$ where $S_1$ is rational. Therefore $R(z) = zS(z) = zS_1(z^n)$. Because $S_1$ is not constant, we conclude that $\deg R \geq n$. In addition, if $R$ is a polynomial, then $S_1$ is also a polynomial, so that $\deg R \geq n + 1$.

## 4. Rational Riemann maps

Before stating our main results we recall some basic facts about Möbius transformations.

Let $R \in \text{Möb}(\mathbb{C} \cup \{\infty\})$ be a Möbius transformation, that is, a rational transformation of degree one. If $R$ is not the identity, $R$ has either one or two fixed points on the Riemann sphere. If $R$ has only one fixed point, $R$ is *parabolic*. If $R$ has two distinct fixed points, $R$ is *elliptic* or *loxodromic* according as $R$ is or is not conjugate in the group $\text{Möb}(\mathbb{C} \cup \{\infty\})$ to a rotation of the Riemann sphere. A loxodromic element $R$ is called *hyperbolic* if there exists a disc $\Delta$ such that $R(\Delta) = \Delta$.

The classes of elliptic, parabolic and loxodromic elements are invariant under conjugation in $\text{Möb}(\mathbb{C} \cup \{\infty\})$. Elements of $\text{Möb}(D)$ are classified in a corresponding way by identifying $\text{Möb}(D)$ with a subgroup of $\text{Möb}(\mathbb{C} \cup \{\infty\})$. Sometimes we identify elements of $\text{Möb}(D)$ with their representatives in $\text{Möb}(\mathbb{C} \cup \{\infty\})$.

We shall need the following lemma which follows from a similar result in [1].

**Lemma 4.** *Let* $f$ *be a rational Riemann map from a disc* $\Delta$ *onto* $D$. *Let* $\varphi \in \text{Möb}(D)$ *and* $\rho = f^{-1} \circ \varphi \circ f$. *Then* $f$ *has a continuous extension to the closure of* $\Delta$ *which maps every fixed point of* $\rho$ *to a fixed point of* $\varphi$. *Moreover,*

(a) *if* $\varphi$ *is elliptic, then* $\rho$ *is elliptic;*

(b) *if* $\varphi$ *is loxodromic, then* $\rho$ *is hyperbolic;*

(c) *if* $\varphi$ *is parabolic, then* $\rho$ *is parabolic or hyperbolic.*

Note that the extension of $f$ is even piecewise analytic on the boundary of $\Delta$, because $D$ is bounded by a finite number of analytic arcs.

In the proof of the following theorem it is convenient to consider a rational Riemann map defined in the open unit disc $U = \{z \in \mathbb{C}\ ;\ |z| < 1\}$. The result holds, of course, for arbitrary rational Riemann maps.

**Theorem 3.** *Let* $f : U \to D$ *be a rational Riemann map of degree* $n$, *and suppose that* $\text{Möb}(D)$ *does not contain any parabolic or loxodromic elements. Then* $\text{Möb}(D)$ *is a finite cyclic group of order* $k \leq n$ *and agrees with* $\text{Möb}_\zeta(D)$ *for some* $\zeta \in D$.

*Proof.* Given $\varphi \in \text{Möb}(D)$ let $\rho = f^{-1} \circ \varphi \circ f$; then Lemma 4 implies that $\rho$ is not parabolic or hyperbolic, so that $\rho$ has a fixed point in $U$. It follows that every element of $\text{Möb}(D)$ has a fixed point in $D$.

Since $\text{Rat}(D) = \text{Möb}(D)$ by Lemma 1, $D$ is not a Siegel domain. Thus by Theorem 2 $\text{Möb}(D)$ is a finite group and agrees with $\text{Möb}_\zeta(D)$ for some $\zeta \in D$. By Theorem 1 $\text{Möb}_\zeta(D)$ is a finite cyclic group.

Composing $f$ with a Möbius transformation we may assume that $f(0) = \zeta$. Let $\varphi_0$ be a generator of $\text{Möb}_\zeta(D)$; then $\rho_0 = f^{-1} \circ \varphi_0 \circ f$ is a rotation $z \mapsto \omega z$ where $\omega$ is a primitive k:th root of unity. We may assume $k \geq 2$ for otherwise there is nothing to prove. Let $h$ be a Möbius transformation such that $h(\zeta) = 0$ and $h$ maps the second fixed point of $\varphi_0$ to $\infty$. Then the rational Riemann map $g = h \circ f$ satisfies

$$g \circ \rho_0 = (h \circ f) \circ (f^{-1} \circ \varphi_0 \circ f) = (h \circ \varphi_0 \circ h^{-1}) \circ g = \psi_0 \circ g \qquad (1)$$

where $\psi_0 = h \circ \varphi_0 \circ h^{-1}$ is a Möbius transformation having fixed points at $0$ and $\infty$. Thus there is a constant $b \neq 0$ such that $\psi_0(w) = bw$, and (1) can be written

$$g(\omega z) = bg(z) \qquad (z \in U).$$

Replacing z with $g^{-1}(w)$ we see that $g^{-1}$ satisfies a similar functional equality

$$g^{-1}(bw) = \omega g^{-1}(w) \qquad (w \in g(U)).$$

Since either g or $g^{-1}$ is the restriction of a rational transformation which is conformal and finite at the origin, we conclude by Lemma 2 (b) that $b = \omega$. Finally, $k \leq n$ by Lemma 3.

The proof of Theorem 3 is now complete.

The condition of Theorem 3 is necessary and sufficient. In fact, the next theorem implies that if Möb(D) contains a parabolic or loxodromic element, then Möb(D) is a non-discrete Lie group. Moreover, the boundary of D consists of one or two circular arcs.

This time it is more convenient to consider maps defined in the upper half plane H.

**Theorem 4.** *Let* $f : H \to D$ *be a rational Riemann map of degree* n.

(a) *If* Möb(D) *contains a parabolic element, then* $n = 1$ *and* D *is a disc.*

(b) *If* Möb(D) *contains a loxodromic element, then* D *can be mapped by means of a Möbius transformation onto the domain* $\{z \in \mathbb{C}\setminus\{0\} \; ; 0 < \arg z < \theta\}$ *where either* $\theta = \pi/n$ *or* $\theta = 2\pi$.

*Proof.* Let $\varphi \in$ Möb(D) be either parabolic or loxodromic, and let $\rho = f^{-1} \circ \varphi \circ f$. Then in virtue of Lemma 4 $\rho$ is either parabolic or hyperbolic, and composing f with a Möbius transformation we may assume that $\rho(z) = az + b$ where $b = 0$ unless $a = 1$.

Let $\zeta = \lim_{|z|\to\infty} f(z)$; then by Lemma 4 $\zeta$ is a fixed point of $\varphi$. Let h be a Möbius transformation such that $h(\zeta) = \infty$; if $\varphi$ is loxodromic, we choose h so that $h^{-1}(0)$ is also a fixed point of $\varphi$. Then $g = h \circ f$ is a rational Riemann map mapping H onto h(D).

The Möbius transformation $\psi = h \circ \varphi \circ h^{-1}$ satisfies $\psi(\infty) = \infty$; therefore $\psi(w) = \alpha w + \beta$ where $\alpha, \beta \in \mathbb{C}$. Since $f \circ \rho = \varphi \circ f$ in H, we have

$$h \circ f \circ \rho = h \circ \varphi \circ f = \psi \circ h \circ f,$$

so that

$$g \circ \rho = \psi \circ g. \tag{2}$$

If $\varphi$ is parabolic, then $\psi$ is also parabolic, so that $\alpha = 1$ and

$$g(\rho(z)) = g(z) + \beta. \qquad (3)$$

In addition, the substitution $z = g^{-1}(w)$ shows that the inverse map $g^{-1} : h(D) \to H$ satisfies

$$\rho(g^{-1}(w)) = g^{-1}(w + \beta). \qquad (4)$$

If $g$ is the restriction of a rational map $R$, then (3) holds by Lemma 2 (a) only if $a = 1$, because $\beta \neq 0$. In this case differentiation of (3) yields

$$g'(z + b) = g'(z),$$

and we conclude that $g'$ is a periodic rational function with period $b$. Thus $g'$ is a constant, and it follows that $n = \deg R = 1$. Similarly, if $g^{-1}$ is the restriction of a rational map $R$, by differentiating (4) we see that either $R'$ or $R'/R$ has period $\beta$, according as $a = 1$ or $b = 0$. Therefore, either $R'$ or $R'/R$ is constant, and because $R$ is rational, this implies $n = \deg R = 1$. The proof of (a) is now complete.

To prove (b), suppose that $\varphi$ is loxodromic. Then $\rho$ is hyperbolic by Lemma 4, so that $a \neq 1$ and $b = 0$. Since in this case $\psi(0) = 0$ and $\psi$ is loxodromic, we have $\beta = 0$, $|\alpha| \neq 1$, and (2) can be written

$$g(az) = \alpha g(z) \qquad (z \in H).$$

In addition, the substitution $z = g^{-1}(w)$ yields

$$g^{-1}(\alpha w) = a g^{-1}(w) \qquad (w \in g(H)).$$

Since $g$ is a rational Riemann map of degree $n$ and $\lim_{|z| \to \infty} g(z) = \infty$, by Lemma 2 (b) either $g$ or $g^{-1}$ agrees with the polynomial $z \mapsto \gamma z^n$ for some constant $\gamma \in \mathbb{C}\backslash\{0\}$.

If $g(z) = \gamma z^n$, then $n \leq 2$ because $g$ is one-to-one in $H$. In this case $D = h^{-1}(g(H))$ is either a disc or the complement of an arc. On the other hand, if $g^{-1}(w) = \gamma w^n$, then $g(z) = (z/\gamma)^{1/n}$, and we conclude that $D = h^{-1}(g(H))$ can be mapped by means of the Möbius transformation $w \mapsto \gamma^{1/n} h(w)$ onto the domain $\{z \in \mathbb{C}\backslash\{0\} ; 0 < \arg z < \pi/n\}$. This completes the proof of Theorem 4.

## 5. Examples

The upper bound n in Theorem 3 for the order of Möb(D) is sharp. As an example we consider a rational map f defined by

$$f(z) = \frac{z}{1 + \beta z^n}$$

where $\beta$ is a constant. If $|\beta|$ is sufficiently small, f defines a rational Riemann map of degree n mapping U conformally onto a domain $D = f(U)$. Moreover, D is invariant under the rotation $w \mapsto \omega w$ where $\omega$ is a primitive n:th root of unity. If $n > 2$, it follows that D cannot be any of the domains appearing in Theorem 4. Therefore, Möb(D) does not contain any parabolic or loxodromic elements, and in view of Theorem 3 Möb(D) consists of n rotations around the origin. If $n = 2$ and $|\beta| < 1$, D is bounded by an ellipse and Möb(D) contains exactly two elements.

In this example it is essential that f(z) is not a polynomial. In fact, for polynomial maps Theorem 3 can be sharpened as follows.

**Theorem 5.** *Suppose that a rational Riemann map* $f : U \to D$ *is the restriction of a polynomial of degree* $n \geq 3$. *Then* Möb(D) *is a finite cyclic group of order* $k \leq n-1$. *Moreover, either* $k = 2$ *or every element of* Möb(D) *has a fixed point at infinity.*

*Proof.* If f is the restriction of a polynomial P with $\deg P = n \geq 3$, the image $D = P(U)$ cannot be any of the domains appearing in Theorem 4. It follows that the hypotheses of Theorem 3 are satisfied, and that Möb(D) is a finite cyclic group of order $k \leq n$.

Let S be a generator of Möb(D), and let $T = f^{-1} \circ S \circ f$; then $P \circ T = S \circ P$, because $f \circ T = S \circ f$. It follows that the topological degree of P at $T^{-1}(\infty)$ is equal to n, the degree of P at infinity.

If $T^{-1}(\infty) \neq \infty$, we conclude that P' has a zero of order $n-1$ at $T^{-1}(\infty)$. Replacing S and T by $S^{-1}$ and $T^{-1}$, respectively, we see that P' has another zero of order $n-1$ at $T(\infty)$. Since a polynomial of degree $n-1$ has at most one zero of order $n-1$, it follows that $T(\infty) = T^{-1}(\infty)$, and that $T \circ T$ is the identity. Hence $k = 2$.

It remains to consider the case $T^{-1}(\infty) = \infty$. Now S is affine, because S has a fixed point at infinity, and composing P with a translation we may assume $P(0) = 0$. Then there exists a constant $b \neq 0$ such that $S(w) = bw$. Moreover, since T maps U onto itself, T is a rotation $z \mapsto \omega z$ where $\omega$ is a primitive k:th root of unity. It follows that P satisfies the functional equation

$$P(\omega z) = bP(z).$$

Therefore $b = \omega$ by Lemma 2 (b), and we conclude by Lemma 3 that $k \leq n-1$.

The bound $n-1$ in Theorem 5 is sharp. In fact, if $|\beta|$ is sufficiently small, the polynomial

$$P(z) = z(1 + \beta z^{n-1})$$

defines a rational Riemann map of degree n from U onto a domain D. Then Möb(D) is generated by the rotation $w \mapsto \omega w$, where $\omega$ is a primitive (n–1):th root of unity.

Theorem 5 does not hold if $n = 2$. For example, the polynomial $P(z) = (z-2i)^2$ maps U conformally onto a domain D which is invariant under the transformation $w \mapsto 9/w$. Thus Möb(D) is a group of order $k = 2$. Examples of higher degree are obtained by considering the polynomial $(z-ni)^n$ which maps U conformally onto a domain invariant under the transformation $w \mapsto (1-n^2)^n/w$.

## References


[1] L. Bers, *On boundaries of Teichmüller spaces and on kleinian groups: I.* Ann. of Math. (2) **91** (1970), 570–600.

[2] T. Erkama, *Group actions and extension problems for maps of balls.* Ann. Acad. Sci. Fenn. Ser. A I Math. **556** (1973), 1–31.

[3] T. Erkama, *Möbius automorphisms of plane domains.* Ann. Acad. Sci. Fenn. Ser. A I Math. **10** (1985), 155–162.

[4] J. Lehner, *Discontinuous Groups and Automorphic Functions.* American Mathematical Society, Providence, R.I., 1964.

[5] C. L. Siegel, *Iteration of analytic functions.* Ann. of Math. (2) **43** (1942), 607–612.

# A CHARACTERIZATION OF QUASICONFORMAL MAPPINGS BY THE BEHAVIOUR OF A FUNCTION OF THREE POINTS [1]

**Jacqueline Ferrand**
University of Paris VI – U.E.R. 48
Laboratoire de Mathématiques fondamentales
4, Place Jussieu, 75252 PARIS CEDEX 05, France

## 1. Introduction

The quasiconformal mappings of $\mathbb{R}^n$ can be characterized by the following property, called *strong quasisymmetry*, which involves a function of three points ([Vä 2], [TV], [F3]).

1.1. *A map* $f : \mathbb{R}^n \to \mathbb{R}^n$ *is quasiconformal if, and only if, there exists a homeomorphism* $\eta$ *of* $\mathbb{R}^+ = [0,+\infty[$ *satisfying for any* $a, b, c \in \mathbb{R}^n$

$$|fc - fa| / |fb - fa| \leq \eta(|c - a| / |b - a|). \tag{1.2}$$

If $D, \Delta$ are any domains of $\mathbb{R}^n$, we know that the quasisymmetric maps of $D$ into $\Delta$ are quasiconformal, but that the converse property is not true unless $D$ satisfies some condition of regularity [TV]. In fact, if $f : D \to \Delta$ is quasiconformal, then $f$ is $\eta$-quasisymmetric on every compact $F \subset D$, but, $f$ given, the function $\eta$ depends on $F$ [Vä 2].

On the other hand the quasiconformal homeomorphisms of the Möbius space $\overline{\mathbb{R}}^n$ are characterized by the property of being quasimöbius ([Vä 3], [F3]) which involves the absolute ratio of four points.

The purpose of this paper is the construction, for any domain $D$ of $\mathbb{R}^n$, of a conformal invariant $j_D(a,b,c)$ associated with any three points $a, b, c$ of $D$ such that the following assertion $(P_j)$ holds

$(P_j)$ $D, \Delta$ *being any domains of* $\mathbb{R}^n$, *a map* $f : D \to \Delta$ *is quasiconformal if,*

---

[1] This paper is in final form and no version of it will be submitted for publication elsewhere.

*and only if, there exists a homeomorphism* $\varphi$ *of* $\mathbb{R}^+$ *satisfying for any* a, b, c ∈ D:

$$j_\Delta(fa,fb,fc) \le \varphi \circ j_D(a,b,c). \tag{1.3}$$

In fact, we shall construct a function $j_D$ satisfying a more precise property: the map f being not a priori supposed to be continuous neither injective nor surjective, the condition (1.3) will insure that f is a homeomorphism of D onto $\Delta$.

For convenience, we shall denote by $P'_j$ the statement obtained from $P_j$ by changing (1.3) into

$$j_D(a,b,c) \le \varphi \circ j_\Delta(fa,fb,fc). \tag{1.4}$$

We observe that, if h is a decreasing homeomorphism of $[0,+\infty]$ onto $[+\infty,0]$, $P'_j$ is equivalent to $P_{h\circ j}$ (the associated function $\varphi$ being changed into $h \circ \varphi^{-1} \circ h^{-1}$).

When $D = \Delta = \mathbb{R}^n$, $P_j$ is true with $j_{\mathbb{R}^n}(a,b,c) = |c-a|/|b-a|$. However in order to prepare a generalization, we shall construct another conformal invariant $\gamma(a,b,c)$ on $\mathbb{R}^n$ satisfying the same property. The extension to any domains will afterwards be easier.

Our main tool will be some inequalities obtained by F. W. Gehring [G1] and M. Vuorinen [Vu 3]. The methods will be similar to those used by F. W. Gehring in [G2].

## 2. The case of $\mathbb{R}^n$

**2.1. Definition.** For any three points a, b, c ∈ $\mathbb{R}^n$ $(b \ne a \ne c)$ we set

$$\gamma(a,b,c) = \inf_{C_0, C_1} \operatorname{Cap} \Gamma(C_0,C_1)$$

where $\Gamma(C_0,C_1)$ is the condenser defined by two disjoint continua $C_0$, $C_1$ such that $C_0$ joins a to $\infty$ and that b, c ∈ $C_1$.

We have also

$$\gamma(a,b,c) = \inf_u \int_{\mathbb{R}^n} |du|^n \, dm \tag{2.2}$$

the infimum being relative to the set of all continuous functions $u \in W_n^1(\mathbb{R}^n)$ such that $u^{-1}(0)$ joins $a$ to $\infty$ and that $u^{-1}(1)$ joins $b$ to $c$.

By a classical argument we have at first

**2.3.** *If* $f$ *is a* $K$*-quasiconformal homeomorphism of* $\mathbb{R}^n$

$$K^{-1}\gamma(a,b,c) \le \gamma(fa,fb,fc) \le K\ \gamma(a,b,c).$$

This assertion gives a necessary condition for $f$ to be quasiconformal. We shall obtain a sufficient condition by using inequalities of [G1] and [Vu 3] which in fact give an estimation of $\gamma(a,b,c)$ by means of

$$\sigma(a,b,c) = |c-b|^{-1} \inf(|b-a|,|c-a|). \tag{2.4}$$

By a similiraty of ratio $|c-b|^{-1}$ (which preserves $\gamma(a,b,c)$) we can indeed reduce to the case $b = 0$, $c = e_1$; hence, from inequalities (1.2) and (1.3) of [Vu 3]

$$\tau \circ \sigma(a,b,c) \le \gamma(a,b,c) \le 2^{n+1}\tau \circ \sigma(a,b,c) \tag{2.5}$$

where $\tau(s) = \gamma(se_1, -e_1, 0)$ is the capacity of the Teichmüller condenser. From (2.5) we obtain easily

**2.6. Lemma.** If $f : \mathbb{R}^n \to \mathbb{R}^n$ satisfies for any $a$, $b$, $c$ in $\mathbb{R}^n$

$$\gamma(fa,fb,fc) \le \varphi \circ \gamma(a,b,c) \tag{2.7}$$

$$[\text{resp. } \gamma(a,b,c) \le \varphi \circ \gamma(fa,fb,fc)] \tag{2.8}$$

*where* $\varphi$ *is a homeomorphism of* $\mathbb{R}^+$, *then*

$$\sigma(a,b,c) \le \psi \circ \sigma(fa,fb,fc) \tag{2.9}$$

$$[\text{resp. } \sigma(fa,fb,fc) \le \psi \circ \sigma(a,b,c)] \tag{2.10}$$

*where* $\psi(t) = \tau^{-1}(2^{-n-1}\varphi^{-1} \circ \tau(t))$ *is a homeomorphism of* $\mathbb{R}^+$.

We are thus led to study the maps $f : \mathbb{R}^n \to \mathbb{R}^n$ satisfying the inequality (2.9) or (2.10), and to give a variation of the property of quasisymmetry.

More generally we shall prove the following two theorems 2.11. and 2.12, where $D$ is any domain of $\mathbb{R}^n$, which complete our preceding results [F3]. We emphasize the fact that in these statements, $f$ is not supposed to be continuous.

**2.11. Theorem.** *Let* $f : D \to \mathbb{R}^n$ *be a non-constant map satisfying* (2.10) *for some homeomorphism of* $\mathbb{R}^+$. *Then* f *is a locally weakly* H*-quasisymmetric imbedding of* D *into* $\mathbb{R}^n$, *with* $H = (1 + \psi(1))^2$, *hence an* $H^{n-1}$*-quasiconformal mapping of* D *onto* fD. *If* $D = \mathbb{R}^n$, $fD = \mathbb{R}^n$.

*Proof.* a) f is injective. If not, there would exist a, b, c ∈ D distinct such that $fb = fc \neq fa$, in contradiction with $\sigma(fa,fb,fc) \leq \psi \circ \sigma(a,b,c) < \infty$.

b) f and $f^{-1}$ are continuous. The points a, b ∈ D ($a \neq b$) being fixed let $(a_n)$ be a sequence converging to a in D. Then $\sigma(a,b,a_n) \to 0$ and from (2.10) $\sigma(fa,fb,fa_n) \to 0$, which implies that $fa_n \to fa$. Similarly if $fa_n \to fa$, then $\sigma(fb,fa,fa_n)$ and also $\sigma(b,a,a_n)$ tends to $+\infty$, which implies that $a_n \to a$.

c) If f were not surjective when $D = \mathbb{R}^n$, there would exist a sequence $(b_n)$ tending to infinity such that $(fb_n)$ tends to a finite limit $\beta$. Then a, c being fixed such that $c \neq a$ and $fa \neq \beta$, $\sigma(a,b_n,c)$ and also (from (2.10)) $\sigma(fa,fb_n,fc)$ would tend to zero, in contradiction with $\lim \sigma(fa,fb_n,fc) = \sigma(fa,\beta,fc) > 0$.

d) f is locally H-quasisymmetric. Since f is an embedding we just have to majorize the ratio $|fc - fa|/|fb - fa|$ when a, b, c satisfy $|b - a| = |c - a|$ and belong to a suitable neighborhood $V_x$ of an arbitrary point x of D. We choose $V_x = B(x,r)$ such that $B(x,3r) \subset D$.

As a preliminary remark, we observe that for any a, b, c ∈ $\mathbb{R}^n$, the mutual ratios of the three numbers $|b - c|$, $|c - a|$, $|a - b|$ are bounded by $1 + \bar{\sigma}(a,b,c)$, with

$$\bar{\sigma}(a,b,c) = \sup(\sigma(a,b,c),\ \sigma(b,c,a),\ \sigma(c,a,b)).$$

Now, under the above hypothesis, two cases are possible.

i) If $|b - c| \geq |a - b| = |a - c|$ then $\bar{\sigma}(a,b,c) = 1$, hence $\bar{\sigma}(fa,fb,fc) \leq \psi(1)$ and $|fc - fa|/|fb - fa| \leq 1 + \psi(1)$.

ii) If $|b - c| < |a - b| = |a - c|$ there exists d ∈ D, on the prolongation of the bissectrix of $\widehat{bac}$, such that $|d - b| = |d - c| > |d - a| = |b - a|$. By applying the preceding argument to (a,b,d) and (a,d,c) we obtain

$$|fd - fa|/|fb - fa| < 1 + \psi(1) \text{ and } |fc - fa|/|fd - fa| < 1 + \psi(1).$$

In both cases i) and ii) we have for any a, b, c ∈ $V_x$:

$$|fc - fa|/|fb - fa| \leq (1 + \psi(1))^2.$$

Consequently f is weakly H-quasisymmetric on $V_x$, with $H = (1 + \psi(1))^2$; hence f is $H^{n-1}$-quasiconformal [Vä 1].

**2.12. Theorem.** *Let* $f : D \to \mathbb{R}^n$ *be a non-constant map satisfying* (2.9) *for some homeomorphism* $\psi$ *of* $\mathbb{R}^+$. *Then* f *is a locally weakly* H*-quasisymmetric embedding of* D *into* $\mathbb{R}^n$ *with* $H = (\psi^{-1}(1))^{-2}$ *hence an* $H^{n-1}$*-quasiconformal mapping of* D *onto* fD. *If* $D = \mathbb{R}^n$, $fD = \mathbb{R}^n$.

*Proof.* a) f is injective. If not there would exist a, b, c $\in$ D distinct such that fa = fb $\neq$ fc, in contradiction with (2.9).

b) f and $f^{-1}$ are continuous: the proofs are respectively similar to those given for the continuity of $f^{-1}$ and f in 2.11 b).

c) If f were not surjective when $D = \mathbb{R}^n$, there would exist a sequence $(a_n)$ such that $a_n \to \infty$ and $(fa_n) \to \alpha \neq \infty$. Then b, c being fixed (b $\neq$ c), $\sigma(a_n,b,c)$ would tend to $+\infty$ and $\sigma(fa_n,fb,fc)$ would be bounded, in contradiction with (2.9).

d) f is locally H-symmetric. As in 2.11 b) we majorize $|fc - fa|/|fb - fa|$ when a, b, c satisfy $|b - a| = |c - a|$ and belong to a ball $V_x = B(x,r)$ such that $B(x,3r) \subset D$.

i) If $|b - c| \geq |a - b| = |a - c|$ we have $\sigma(b,a,c) = 1$, hence $|fb - fa|/|fc - fa| \geq \sigma(fb,fa,fc) \geq \psi^{-1}(1)$. Similarly we have $|fc - fa|/|fb - fa| \geq \psi^{-1}(1)$, hence $\psi^{-1}(1) \leq 1$.

ii) If $|b - c| < |a - b| = |a - c|$ there exists d $\in$ D such that $|d - b| = |d - c| > |a - d| = |a - b|$. By applying the preceding argument to (a,b,d) and (a,d,c) we obtain $|fb - fa|/|fd - fa| \geq \psi^{-1}(1)$ and $|fd - fa|/|fc - fa| \geq \psi^{-1}(1)$.

In both cases i) and ii) we have, as wanted

$$|fc - fa|/|fb - fa| \leq [\psi^{-1}(1)]^{-2}.$$

By combining these results with lemma 2.6 we obtain:

**2.13. Theorem.** *Let* $f : \mathbb{R}^n \to \mathbb{R}^n$ *be a non-constant map such that there exists a homeomorphism* $\varphi$ *of* $\mathbb{R}^+$ *satisfying*

i) $\gamma(fa,fb,fc) \leq \varphi \circ \gamma(a,b,c)$ $(\forall$ a, b, c $\in \mathbb{R}^n)$ *or*

ii) $\gamma(a,b,c) \le \varphi \circ \gamma(fa,fb,fc)$ $(\forall\, a, b, c \in \mathbb{R}^n)$.

*Then* f *is a* K*−quasiconformal homeomorphism of* $\mathbb{R}^n$ *with*

$$K = [\tau^{-1} \circ \varphi(2^{n+1} c_n)]^{-2(n-1)}$$

*in case* i) *and*

$$K = [1 + \tau^{-1}(2^{-n-1} \varphi^{-1}(c_n)]^{2(n-1)}$$

*in case* ii) $(c_n = \tau(1))$.

We shall sum up the results relative to $\mathbb{R}^n$ by saying that *both* $P_j$ *and* $P'_j$ (see § 1) *are true when* $D = \Delta = \mathbb{R}^n$ *if we set* $j_{\mathbb{R}^n} = \sigma$ *or* $j_{\mathbb{R}^n} = \gamma$.

**Remark.** By using the inequality (3.11) of [AVV1] we can obtain a dimension-free variant of Theorems 2.11 and 2.12, with a constant H depending only on $\psi$, and not on n.

### 3. Extension to domains of $\mathbb{R}^n$. Preliminary results

In order to have a well adapted conformal invariant for domains of $\mathbb{R}^n$ we must modify the definition given for $\mathbb{R}^n$, and use some notions closely related to those introduced in [F1].

**3.1. Definition.** A *generalized continuum* of a domain D is a closed subset C of D which does not admit any partition $C = F \cup K$, F being closed and K compact $(F, K \ne \emptyset, F \cap K = \emptyset)$.

In other terms: if $\hat{D}$ is the Alexandrov compact extension of D, either C is a continuum of D, or $C \cup \{\infty\}$ is a continuum of $\hat{D} = D \cup \{\infty\}$.

**3.2. Definition.** For any $a, b, c \in D$ $(b \ne a \ne c)$ we set

$$\gamma_D(a,b,c) = \inf_{C_0, C_1} \operatorname{Cap} \Gamma(C_0,C_1)$$

where $\Gamma(C_0,C_1)$ is the condenser defined in D by two disjoint generalized continua $C_0$, $C_1$ of D such that $a \in C_0$, $C_0$ being not compact, and $b, c \in C_1$.

We have $\gamma_D(a,b,c) = \gamma_D(a,c,b)$ and also

$$\gamma_D(a,b,c) = \inf \int_D |du|^n \, dm \tag{3.3}$$

the infimum being relative to the set of continuous functions $u \in W_n^1(D)$ such that there exist two generalized continua $C_0$, $C_1$, respectively contained in $u^{-1}(0)$, $u^{-1}(1)$, satisfying the preceding conditions. For brevity such functions will be called ***admissible*** for a, b, c, D.

**Remarks.** We do not change the value of $\gamma_D$ if we add the condition that $C_0$ is an ordinary continuum (joining a to $\partial D$ or to $\infty$). On the other hand, when $D = \mathbb{R}^n$, all the condensers $\Gamma(C_0,C_1)$ such that $C_1$ is not compact, have an infinite capacity. From these remarks we infer that $\gamma_{\mathbb{R}^n} = \gamma$, the function defined in 2.1.

**3.4. Topological properties of $\gamma_D$.** By using the same methods as in [F1] we can easily prove:

i) $\gamma_D(a,b,c) = 0$ if, and only if, $b = c$.

ii) The points a, b $(a \neq b)$ being fixed, $\gamma_D(a,b,c)$ tends to zero [resp. $+\infty$] if and only if $c \to b$ [resp. $c \to a$].

iii) The points b, c $(b \neq c)$ being fixed, $\gamma_D(a,b,c)$ tends to zero if, and only if, a tends to $\partial D$ or to infinity.

When $D = \mathbb{R}^n$ these properties follow from the comparison of $\gamma$ with $\sigma$.

By extension, we can set $\gamma_D(a,b,c) = +\infty$ if $a = b \neq c$ or if $a = c \neq b$.

**3.5. Lemma.** *If* D, $\Delta$ *are two domains of* $\mathbb{R}^n$ *such that* $D \subset \Delta$, *then for any* $a, b, c \in D$,

$$\gamma_D(a,b,c) \le \gamma_\Delta(a,b,c).$$

*Proof.* We observe that the trace on D of a generalized continuum of $\Delta$ is a generalized continuum of D (this would not always be true for ordinary continua, as it can be checked on a figure.) Then if u is admissible for a, b, c, $\Delta$ with $a, b, c \in D$, $u|D$ is admissible for a, b, c, D and the announced inequality follows from (3.3).

In particular for any domain D,

$$\gamma_D(a,b,c) \le \gamma_{\mathbb{R}^n}(a,b,c) = \gamma(a,b,c) \le 2^{n+1} \tau \circ \sigma(a,b,c). \tag{3.6}$$

Now we shall obtain a minorant of $\gamma_D(a,b,c)$ by using balls contained in D.

**3.7. Lemma.** *Let* $a \in D$, $r \le \frac{1}{2} d(a,\partial D)$ *and* $B = B(a,2r)$. *Then for any* $b, c \in B(a,r)$

$$\gamma_D(a,b,c) \geq \gamma_B(a,b,c) \geq \theta \circ \sigma(a,b,c)$$

*where* $\theta$ *is a decreasing homeomorphism of* $[0,+\infty]$ *onto* $[+\infty,0]$, *depending only on* n.

*Proof.* Let u be an admissible function for a, b, c, B, and $C_0 \subset u^{-1}(0)$, $C_1 \subset u^{-1}(1)$ two generalized continua of B satisfying the conditions set in 3.2:

a) *First case*: $C_1$ is compact. For any $t \in ]0,2r[$ we define a function $v_t$ on $\mathbb{R}^n$ by $v_t(x) = u(x)$ if $|x-a| \leq t$ and

$$v_t(x) = a + t^2(x-a)/|x-a|^2$$

if $|x-a| > t$. From a property of inversions

$$\int_{\mathbb{R}^n} |dv_t|^n \, dm = 2\int_{B(a,t)} |du|^n \, dm. \tag{3.8}$$

Now $v_t^{-1}(0)$ contains a continuum joining a to infinity and, if $t \geq \sup_{x \in C_1} |x-a|$, $v_t^{-1}(1)$ contains $C_1$. Consequently $v_t$ is admissible for a, b, c, $\mathbb{R}^n$. Letting t tend to 2r we have, from 3.5 and (3.8)

$$\gamma_D(a,b,c) \geq \gamma_B(a,b,c) \geq \tfrac{1}{2}\,\gamma(a,b,c) \geq \tfrac{1}{2}\,\tau \circ \sigma(a,b,c).$$

b) *Second case:* $C_1$ is not compact. Then both $C_0$ and $C_1$ intersect every sphere $\partial B(a,t)$ with $2r > t \geq \rho = \inf(|b-a|, |c-a|)$. The oscillation $\delta(t)$ of u on such a sphere being not less than 1, we have from a classical inequality [G2], [F1]:

$$\mathrm{Log}\,\frac{2r}{\rho} \leq \int_\rho^{2r} \delta^n(t)\,\frac{dt}{t} \leq A_n \int_B |du|^n$$

where $A_n$ is a constant depending only on n. Hence

$$\gamma_B(a,b,c) \geq A_n^{-1}\,\mathrm{Log}\,(2r/\rho).$$

Now we have $\rho \leq r$ and $\rho = |b-c|\sigma(a,b,c) \leq 2r\sigma(a,b,c)$. We define a decreasing homeomorphism $\tau_1$ of $[0,+\infty]$ onto $[+\infty,0]$ by setting $\tau_1(0) = +\infty$, $\tau_1(+\infty) = 0$, $\tau_1(s) = \mathrm{Log}\,1/s$ if $s \leq 1/2$ and $\tau_1(s) = (\mathrm{Log}\,2)/(2s)$ if $s > 1/2$. Then we have

$$A_n \gamma_B(a,b,c) \geq \mathrm{Log}\,2r/\rho \geq \tau_1 \circ \sigma(a,b,c).$$

In both cases a) and b) we have

$$\gamma_B(a,b,c) \geq \theta \circ \sigma(a,b,c) \ \text{ with } \ \theta = \inf(\tau, A_n^{-1}\tau_1).$$

**4. Characterization of quasiconformal mappings**

From a classical argument we have at first

**4.1.** *If* f *is a* K*-quasiconformal homeomorphism of* D *onto* $\Delta$, *then, for any* a, b, c $\in$ D

$$K^{-1}\,\gamma_D(a,b,c) \leq \gamma_\Delta(fa,fb,fc) \leq K\,\gamma_D(a,b,c).$$

From 3.5 the left inequality is still true when f is a K-quasiconformal embedding of D into $\Delta$.

Conversely the following theorems will provide a sufficient condition for f to be a quasiconformal embedding [resp. homeomorphism] of D into [resp. onto] $\Delta$. *As in Theorems* 2.12 *and* 2.13, *the map* f *is not assumed to be continuous, neither injective nor surjective.*

**4.2. Theorem.** *Let* $f : D \to \Delta$ *be a non-constant map satisfying for any* a, b, c $\in$ D

$$\gamma_D(a,b,c) \leq \varphi \circ \gamma_\Delta(fa,fb,fc)$$

*where* $\varphi$ *is a homeomorphism of* $\mathbb{R}^+$. *Then* f *is a quasiconformal embedding of* D *into* $\Delta$.

*Proof.* a) f is injective. If not, there would exist a, b, c $\in$ D distinct such that fb = fc $\neq$ fa. We should have $\gamma_D(a,b,c) > 0$ and $\gamma_\Delta(fa,fb,fc) = 0$, in contradiction with the hypothesis.

b) f and $f^{-1}$ are continuous. The points a, b $\in$ D (a $\neq$ b) being fixed, let $(a_n)$ be a sequence converging to a. Then $\gamma_D(a,b,a_n)$, and also $\gamma_\Delta(fa,fb,fa_n)$ tends to $+\infty$, which implies that $fa_n \to fa$. Similarly let $(a_n)$ be a sequence such that $fa_n \to fa$. Then $\gamma_\Delta(fb,fa,fa_n)$ and also $\gamma_D(b,a,a_n)$ tends to zero, which implies that $a_n \to a$.

c) f is locally weakly quasisymmetric. For every x $\in$ D we set r = (2/5)d(x,$\partial$D) and $V_x = B(x,r/2)$. Then for any a, b, c $\in V_x$ the hypothesis of Lemma 3.7 are satisfied. Thus we have, with the same function $\theta$ :

$$2^{n+1}\tau \circ \sigma(fa,fb,fc) \geq \gamma_\Delta(fa,fb,fc) \geq \varphi^{-1} \circ \gamma_D(a,b,c) \geq \varphi^{-1} \circ \theta \circ \sigma(a,b,c).$$

Hence, since $\tau$ and $\theta$ are decreasing:

$$\sigma(fa,fb,fc) \leq \psi \circ \sigma(a,b,c) \text{ with } \psi(t) = \tau^{-1}(2^{-n-1}\varphi^{-1} \circ \theta(t)).$$

From Theorem 2.11, f is weakly H-quasisymmetric on every $V_x$, with $H = (1 + \psi(1))^2$ depending only on n and $\varphi$. Hence f is $H^{n-1}$-quasiconformal on D.

**4.3. Theorem.** *Let* $f : D \to \Delta$ *be a non-constant map satisfying for any* a, b, $c \in D$

$$\gamma_\Delta(fa,fb,fc) \leq \varphi \circ \gamma_D(a,b,c)$$

*where* $\varphi$ *is a homeomorphism of* $\mathbb{R}^+$. *Then* f *is a quasiconformal homeomorphism of* D *onto* $\Delta$.

*Proof.* a) f is injective. If not there would exist a, b, $c \in D$ distinct such that $fa = fb \neq fc$. We should have $\gamma_D(a,b,c) < \infty$ and $\gamma_D(fa,fb,fc) = \infty$ in contradiction with the hypothesis.

b) f and $f^{-1}$ are continuous. The proofs are respectively similar to those given for the continuity of $f^{-1}$ and f in 4.2.

c) f is surjective. If not, there would exist a sequence $(a_n)$ tending to $\partial D$ or to infinity such that $(fa_n)$ tends to a limit $\alpha \in \Delta$. The points b, $c \in D$ $(b \neq c)$ being fixed, $\gamma_D(a_n,b,c)$ and also $\gamma_\Delta(fa_n,fb,bc)$ would tend to zero, in contradiction with (3.4).

d) f is locally weakly quasisymmetric. For every $x \in D$, we set $r = (2/5)d(fx,\partial\Delta)$; then $V_x = f^{-1}B(fx,r/2)$ is a neighborhood of x, and for any a, b, $c \in V_x$ we can apply Lemma 3.7 to fa, fb, fc in $\Delta$. We obtain

$$\theta \circ \sigma(fa,fb,fc) \leq \gamma_\Delta(fa,fb,fc) \leq \varphi \circ \gamma_D(a,b,c) \leq \varphi(2^{n+1}\tau \circ \sigma(a,b,c)).$$

Hence, since $\tau$ and $\theta$ are decreasing

$$\sigma(a,b,c) \leq \psi \circ \sigma(fa,fb,fc) \text{ with } \psi(t) = \tau^{-1}(2^{-n-1}\varphi^{-1} \circ \theta(t)).$$

From Theorem 2.12, f is weakly H-quasisymmetric on every $V_x$ with $H = (\psi^{-1}(1))^2$ depending only on n and $\varphi$. Hence f is $H^{n-1}$-quasiconformal on D.

**Conclusion.** With the notations of §1 we can say that the property $P_j$ is true with $j_D = \gamma_D$. On the other hand the property $P'_j$ is not exactly true when $j_D = \gamma_D$ but we have the slightly weaker result

($P''_\gamma$) *A map* $f: D \to \Delta$ *is a quasiconformal embedding of* D *into* $\Delta$ *if, and only if, there exists an increasing homeomorphism* $\varphi$ *of* $\mathbb{R}^+$ *satisfying for any* a, b, $c \in D$

$$\gamma_D(a,b,c) \le \varphi \circ \gamma_\Delta(fa,fb,fc).$$

## 5. Comparison with the conformal invariant $\lambda_D$

In [F1], [F2] we introduced the conformal invariant

$$\lambda_D(x,y) = \inf_{C_0, C_1} \operatorname{Cap} \Gamma(C_0,C_1)$$

where $C_0$, $C_1$ are disjoint non compact generalized continua of D such that $x \in C_0$, $y \in C_1$; and we proved that $\delta_D(x,y) = [\lambda_D(x,y)]^{-1/n}$ is a distance on D. It is easy to check that every K-quasiconformal map $f: D \to \Delta$ satisfies

$$K^{-1/n}\, \delta_D(x,y) \le \delta_\Delta(fx,fy) \le K^{1/n}\, \delta_D(x,y) \tag{5.1}$$

and in [F2, §6] we conjectured that conversely the inequality (5.1) could imply the quasiconformality of f. The following counterexample shows that this conjecture is false.

**A non-quasiconformal $\delta$-bilipschitzian map of the unit ball onto itself.** Let B be the unit ball of $\mathbb{R}^n$ and f be the homeomorphism of B onto B defined by $f(ru) = g(r)u$ ($|u| = 1, 0 \le r < 1$) with

$$g(r) = 1/2 + 4(r - 1/2)^3 = 4r^3 - 6r^2 + 3r.$$

It is easy to check that f is not quasiconformal, its linear dilatation being infinite at any point x such that $|x| = 1/2$.

On the other hand, we have here [F1], [Vu 1], [AVV2, §7] $2\lambda_B(x,y) = \tau \circ q(x,y)$ with

$$q(x,y) = |x-y|^2(1-|x|^2)^{-1}(1-|y|^2)^{-1}.$$

By elementary calculations we obtain, for any $r, r' \in [0,1]$: $|g(r') - g(r)| \leq 3(r' - r)$, $3/4 \leq g(r)/r \leq 3$ and

$$\begin{cases} \text{if } \sup(r,r') \geq 3/4 & |g(r') - g(r)| \geq (3/16)|r' - r|, \\ \text{if } \sup(r,r') \leq 3/4 & |g(r') - g(r)| \geq (r' - r)^3. \end{cases}$$

Moreover we have the inequality

$$(3/8)(1 - r^2) \leq 1 - g^2(r) \leq 6(1 - r^2).$$

For any $x, x' \in B$ with $|x| = r$, $|x'| = r'$ and $\langle x|x'\rangle = r \cos\theta$ $(0 \leq \theta \leq \pi)$ we have

$$|x' - x|^2 = (r' - r)^2 + 4\, rr' \sin^2(\theta/2),$$
$$|f(x') - f(x)|^2 = (g(r') - g(r))^2 + 4\, g(r)\, g(r') \sin^2(\theta/2).$$

By using all these formulae, we can easily prove the existence of positive constants A, B, C such that $q(f(x),f(x')) \leq A\, q(x,x')$ and

$$\begin{cases} \text{if } \sup(|x|,|x'|) \geq 3/4 & q(f(x),f(x')) \geq B\, q(x,x'), \\ \text{if } \sup(|x|,|x'|) \leq 3/4 & q(f(x),f(x')) \geq C(q(x,x'))^3. \end{cases}$$

On the other hand, the inequalities (2.10) of [Vu 3] imply the existence of positive functions $\alpha$, $\beta$, $\gamma$ such that, for any $A, B, C, s \in \mathbb{R}^+$, $\tau(As) \geq \alpha(A)\tau(s)$, $\tau(Bs) \leq \beta(B)\tau(s)$, $\tau(Cs^3) \leq \gamma(C)\tau(s)$. (We obtain such functions by comparing As and Bs to $\sqrt{s}$ and $s^2$, and comparing $Cs^3$ to $s^2$ and $s^4$).

Consequently since $\tau$ is decreasing, we have

$$\alpha(A) \leq \lambda_B(f(x),f(y))/\lambda_B(x,y) \leq \sup(\beta(B), \gamma(C))$$

and f satisfies (5.1) for a suitable value of K, with $D = \Delta = B$, as wanted.

## 6. Additional note

The quasihyperbolic distance $k_D$ was introduced by F. W. Gehring and B. P. Palka [GP] and is now a very useful tool in quasiconformal theory. However it is not exactly a conformal invariant. In order to answer a question asked by M. Vuorinen we prove that we obtain a conformally invariant distance $\delta_D$ equivalent with $k_D$ by taking the density

$$\varphi_D(x) = \sup_{a,b\in\partial D} (|b-a|\,|x-a|^{-1}|x-b|^{-1})$$

instead of $\rho_D(x) = 1/d(x,\partial D)$ $(x \in D,\ D \subset \bar{\mathbb{R}}^n)$.

Starting with the invariance of $|y-x|\,|b-a|\,|x-a|^{-1}|x-b|^{-1}$ under Möbius maps and letting $y \to x$ we check that $|b-a|\,|x-a|^{-1}|x-b|^{-1}|dx|$ is a conformal invariant. It is the same for $\varphi_D(x)|dx|$ and for

$$\delta_D(a,b) = \inf_\gamma \int_\gamma \varphi_D(x)\ |dx|$$

where $\gamma$ is any rectifiable curve joining $a$ to $b$ in $D$.

Now we have easily

$$\varphi_D(x) \le \sup_{a,b\in\partial D} (|x-a|^{-1} + |x-b|^{-1}) \le 2/d(x,\partial D)$$

and, if we choose $a, b \in \partial D$ such that $|x-a| = d(x,\partial D)$ and $\langle x-a|x-b\rangle = 0$ (possibly $b = \infty$)

$$\varphi_D(x) \ge |b-a|\,|x-a|^{-1}|x-b|^{-1} \ge |x-a|^{-1} = 1/d(x,\partial D).$$

Hence, for any $x, y \in D$ the inequalities

$$k_D(x,y) \le \varphi_D(x,y) \le 2\, k_D(x,y)$$

hold, and a new proof of the behaviour of $k_D$ under Möbius maps [GP] follows.

We easily check that when $D$ is a ball or the Poincaré half space, $\delta_D$ coincides with the hyperbolic metric.

With the same notations as above, letting $b \to a$ and $y \to x$ we can see that for any disjoint rectifiable curves $\gamma$, $\gamma'$ the integral

$$I(\gamma,\gamma') = \int_\gamma \int_{\gamma'} \frac{|dx|\ |da|}{|x-a|^2}$$

is a conformal invariant.

## References


[AVV1] Anderson, G. D., Vamanamurthy, M. K. and Vuorinen, M., Dimension-free quasiconformal distortion in n-space. Trans. Amer. Math. Soc. 297 (1986), 687–705.

[AVV2] Anderson, G. D., Vamanamurthy, M. K. and Vuorinen, M., Special functions of quasiconformal theory. Preprint (University of Helsinki, 1987).

[F1] Lelong-Ferrand, J., Invariants conformes globaux sur les variétés riemanniennes. J. Differential Geom. 8 (1973), 487–510.

[F2] Lelong-Ferrand, J., Construction de métriques pour lesquelles les transformations quasiconformes sont lipschitziennes. Ist. Nazionale di Alta Mat. Symposia Math. XVIII (1976), 407–420.

[F3] Ferrand, J., Invariant conformes dans l'espace de Möbius et caractérisations des applications quasiconformes. Rev. Roum. Math. Pures Appl. (to appear).

[G1] Gehring, F. W., Symmetrization of rings in space. Trans. Amer. Math. Soc. 101 (1961), 499–519.

[G2] Gehring, F. W., Rings and quasiconformal mappings in space. Trans. Amer. Math. Soc. 103 (1962), 353–393.

[GP] Gehring, F. W. and Palka, B. P., Quasiconformally homogeneous domains, J. Analyse Math. 30 (1976), 172–199.

[TV] Tukia, P. and Väisälä, J., Quasisymmetric embeddings of metric spaces. Ann. Acad. Sci. Fenn. Ser. A I Math. 5 (1980), 97–114.

[Vä 1] Väisälä, J., Lectures on n-dimensional quasiconformal mappings. Lecture Notes in Math., 229, Springer-Verlag, Berlin, 1971.

[Vä 2] Väisälä, J., Quasisymmetric embeddings in euclidean spaces. Trans. Amer. Math. Soc. 264 (1981), 191–204.

[Vä 3] Väisälä, J., Quasimöbius maps. J. Analyse Math. 44 (1984–1985), 218–234.

[Vu 1] Vuorinen, M., Conformal invariants and quasiregular mappings. J. Analyse Math. 45 (1985), 69–115.

[Vu 2] Vuorinen, M., On the distortion of n-dimensional quasiconformal mappings. Proc. Amer. Math. Soc. 96 (1986), 275–283.

[Vu 3] Vuorinen, M., On Teichmüller's modulus problem in $\mathbb{R}^n$. Preprint (University of Helsinki, 1987).

# ON A CLASS OF UNIVALENT FUNCTIONS [1]

**Jaroslav Fuka**
Matematický ústav ČSAV
Žitná 25, 11567 Praha 1, Czechoslovakia

**Introduction.** As usual, we shall denote by $\mathbf{C}$ the set of all complex numbers, $T_r = \{\zeta \in \mathbf{C} : |\zeta| = r\}$, $D_r = \{z \in \mathbf{C} : |z| < r\}$ (for $r = 1$ we omit the index), $S$ the class of functions $f$ one-to-one and holomorphic in $\mathbf{D}$, normalized by $f(0) = 0$, $f'(0) = 1$, $\partial B$ the boundary of a set $B$, diam $M$ the diameter of $M$. Given $\varepsilon > 0$, we put $H^1_\varepsilon(M) = \inf \Sigma_{n=1}^\infty \operatorname{diam} M_n$, where the infimum is taken over all sequences of sets $M_n \subset \mathbf{C}$ such that $\operatorname{diam} M \le \varepsilon$ and $M \subset \bigcup_{n=1}^\infty M_n$. The linear Hausdorff measure (= length) of $M$ is defined by $H^1(M) = \lim_{\varepsilon \to 0+} H^1_\varepsilon$. For a function $f$ defined on the interval $\langle a,b \rangle$ we shall denote by var $[f; \langle a,b \rangle]$ its total variation on $\langle a,b \rangle$.

For $z \in \mathbf{C}$ denote by $\pi_z : \zeta \to (\zeta - z)/(|\zeta - z|)$ the projection of $\mathbf{C} \setminus \{z\}$ onto $T$. For a compact set $K \subset \mathbf{C}$ we define for $\theta \in \langle 0, 2\pi \rangle$,

$$N^K_z(\theta) = \sum \chi_K(u), \quad u \in K \setminus \{z\}, \quad \pi_z(u) = e^{i\theta},$$

where the sum is taken over all $u \in \pi_z^{-1}(\theta)$ and $\chi_K$ denotes the characteristic function of $K$. Thus $N^K_z(\theta)$ is the number (finite or infinite) of all points lying in the intersection of $K$ with the open ray $\{z + te^{i\theta} : t > 0\}$. $N^K_z$ is Borel measurable (cf. [10], p. 217) and therefore we may adopt the following

**Definition.** Let $K \subset \mathbf{C}$ be a compact set. For each $z \in \mathbf{C}$ we define

$$v^K(z) = \int_0^{2\pi} N^K_z(\theta)\, d\theta.$$

This quantity is sometimes called the cyclic variation of $K$ at the point $z$. Further we

[1] This paper is in final form and no version of it will be submitted for publication elsewhere.

define

$$V(K) = \sup_{\zeta \in K} v^K(\zeta).$$

**Remark.** If $I$ is a segment, $v^I(z)$ is obviously the angle, under which $I$ is seen from the point $z$. By obversing more complicated examples we discover the geometric sense of the cyclic variation $v^K(z)$: it is "the variation of the angle under which $K$ is seen from the given point $z$, if we go continuously through all points of $K$". For the strict mathematical meaning of this sentence and also other important properties of the cyclic variation the reader may consult the paper [5], especially Proposition 1.7, p. 453.

The aim of the present paper is to introduce and to study a subclass $W \subset S$.

**Definition.** We say, that $f \in S$ belongs to the class $W$, if it is bounded and $V(\partial f(\mathbf{D})) < \infty$.

The introduction of this class was motivated by the reflections on the class of normalized univalent functions with bounded boundary rotation (beschränkte Randdrehung) introduced by Löwner [6] but systematically studied by Paatero [7], [8]. A domain with smooth (i.e. continuously differentiable) boundary is said to be of bounded boundary rotation, if the total variation of direction angle of the tangent to the boundary curve under a complete circuit is finite. For a general simply connected domain $D \subset \mathbf{C}$ the boundary rotation is defined in the following manner. We take an exhaustion $\{D_n\}$ of $D$, i.e. a sequence of domains $\{D_n\}$ such that $\bar{D}_n \subset D_{n+1} \subset \bar{D}_{n+1} \subset D$ and $\bigcup_{n=1}^{\infty} D_n = D$. Let $\alpha_n$ be the infimum of the boundary rotations of all smoothly bounded Jordan domains $G_n$ with $D_n \subset G_n \subset D$. Then $\alpha_n$ increases to a limit $\alpha(D)$, which is easily seen to be independent of the choice of the exhaustion $\{D_n\}$. This limit $\alpha(D)$ is called the boundary rotation of $D$. A function $f$ holomorphic and univalent in $\mathbf{D}$ is said to be of bounded boundary rotation, if $\alpha(f(\mathbf{D})) < \infty$. Paatero proved (cf. [7], pp. 13–19) a remarkable theorem that, if $f$ is of bounded boundary rotation $\alpha$, then $\alpha(f(\mathbf{D}_r)) \to \alpha$ for $r \to 1-$. This is in a certain sense an analogy to the classical theorem: $f \in S$ is convex (i.e. $f(\mathbf{D})$ is convex) if and only if $f(\mathbf{D}_r)$ is convex for every $r$, $0 < r < 1$. But there is an essential difference: the concept of convexity is defined in a unique manner for smooth domains as well as for general simply connected domains, while in the definition of the boundary rotation there is essential asymmetry between smooth and general simply connected domains. To remove this asymmetry we try to extend suitably the class of bounded univalent functions with bounded boundary rotation. The main results of this paper show that the class $W$ satisfies this requirement.

**Theorem 1.** *Let* $f \in S$ *be bounded. The following conditions are equivalent:*

(i) $f \in W$;

(ii) *there exists a constant* $M_f$ *so that*

$$V(f(T_r)) \le M_f < \infty \quad \text{for } 0 < r < 1.$$

**Theorem 2.** *The class of bounded normalized univalent functions with bounded boundary rotation is contained in* W.

The proof of Theorem 1 is given in section 1. The proof of the implication (i) ⇒ (ii) follows in general lines the proof of Theorem 3.2 in [2], but there are some technical simplifications. In Section 2 the proof of Theorem 2 is given. The key of this proof is the geometric observation made by Radon: for a Jordan domain D with rectifiable boundary and for an arbitrary point $z_0 \in D$ one has $v^{\partial D}(z_0) \le \alpha(D)$. Since Theorem 1 is already at our disposal, we may exploit the result of Radon in the analytic case only. For convenience of the reader we recapitulate the beautiful proof of Radon (cf. [9], pp. 1133–35) in Proposition 2.1. In Section 2 we give also a simple example of a bounded simply connected domain D such that $V(\partial D) < \infty$ but $\alpha(D) = \infty$. The Theorems 1 and 2 were announced together with the description of main ideas of their proof in [3] and [4].

## 1

**Lemma 1.1.** *Let* $G \subset \mathbf{C}$ *be a simply connected domain whose boundary contains at least two different points and* $H^1(\partial G) < \infty$. *Then the set of prime ends of* G *consists only of the prime ends of the first kind, that is the prime ends of* G *reduce to points.*

*Proof.* See [11], (11.3), p. 15.

**Proposition 1.1.** *Let* $f \in W$. *Then the function* $f(e^{it}) = \lim_{r \to 1-} f(re^{it})$, $t \in \langle 0, 2\pi \rangle$, *is absolutely continuous and the following holds:*

(a) $\frac{d}{dt} f(e^{it}) = ie^{it} \lim_{r \to 1-} f'(re^{it})$ *almost everywhere (a.e.) on* $\langle 0, 2\pi \rangle$;

(b) $f' \in H^1$.

*Proof.* By [5] Proposition 1.1, p. 448, we have $H^1(\partial f(\mathbf{D})) < \infty$, hence by Lemma 1.1 the set of prime ends of the simply connected bounded domain $f(\mathbf{D})$ consists only of the prime ends of the first kind. As a corollary we obtain, that $f(z)$ is continuous in $\bar{\mathbf{D}}$ and, if we denote the extension also by f, that $\partial f(\mathbf{D}) = f(\mathbf{T})$. The boundary length of the continuum $f(T)$ (in the sense of L. C. Young, see [12], p. 294) with respect to the domain $f(D)$ is not greater than $2H^1(f(T)) = 2H^1(\partial f(D))$ (by [12], Theorem 7.2, p. 294) and therefore it is finite. Hence (by [12], (7.3), p. 295) the length of the closed curve $f(e^{it})$, $0 \leq t \leq 2\pi$, is finite. Therefore the function $f(e^{it})$ is of bounded variation and by classical theorems (see [1], Theorem 3.10 and Theorem 3.11, p. 42) the assertions (a) and (b) follow.

**Proposition 1.2.** *Let* $f \in W$. *Let* $z_0 \in D$ *be a given point. Then the function* $\chi_{z_0}(t) := \arg(f(e^{it}) - f(z_0)) = \operatorname{Im} \log(f(e^{it}) - f(z_0))$ *is absolutely continuous on* $\langle 0,2\pi\rangle$, $\chi_{z_0}(2\pi) = \chi_{z_0}(0) + 2\pi$ *and*

$$\text{var}\,[\chi_{z_0};\ \langle 0,2\pi\rangle] \leq 2v^{f(T)}(f(z_0)). \tag{1}$$

*Proof.* From the assumption $f \in W$ it follows by Proposition 1.1 that the function

$$\varphi(z) = \frac{f(z) - f(z_0)}{z - z_0}, \quad \varphi(z_0) = f'(z_0)$$

is holomorphic in $D$ and continuous in $\bar{D}$. Clearly, $\varphi(z) \neq 0$ in $\bar{D}$, because $f$ is an open univalent mapping in $D$. Hence there exists a continuous branch of $\log(f(z) - f(z_0)/(z - z_0))$ on $\bar{D}$ and so the function $\chi_{z_0}(t)$ is defined on $\langle 0,2\pi\rangle$ as desired: $\chi_{z_0}(2\pi) = \chi_{z_0}(0) + 2\pi$. Its absolute continuity follows easily from the following facts: $f(e^{it})$ is absolutely continuous (Proposition 1.1); $|f(e^{it}) - f(z_0)| \geq \alpha > 0$ on $\langle 0,2\pi\rangle$; $\log(1 + u) = O(u)$ for $u \to 0$. So it remains to prove (1). Denote, for $\theta \in (-\infty,\infty)$, $A_{z_0}(\theta) = \{t \in \langle 0,2\pi\rangle : \chi_{z_0}(t) = \theta\}$ and $M_{z_0}$ the number of all points in $A_{z_0}(\theta)$ (possibly infinite). Because the function $t \to f(e^{it})$ realizes a one-to-one correspondence between the interval $\langle 0,2\pi\rangle$ and the prime ends of the simply connected domain $f(D)$, $t$ belongs to $A_{z_0}$ if and only if there is a prime end containing the point $w = f(e^{it})$ and $\arg(w - f(z_0)) = \theta$. So we may interpret $M_{z_0}(\theta)$ as the number of all prime ends of $f(D)$ containing some point $w \in \partial f(D)$ such that $\arg(w - f(z_0)) = \theta$. With this interpretation in mind we will estimate the integral $\int_{-\infty}^{\infty} M_{z_0}(\theta)\, d(\theta)$. With regard to $A_{z_0}(\theta+2j\pi)\cap A_{z_0}(\theta+2k\pi)=\emptyset$ for $j\neq k$ (because $\chi_{z_0}(t)$ is well defined) and because, in fact, $M_{z_0}(\theta)=0$ outside of a compact interval, we have

$$\int_{-\infty}^{\infty} M_{z_0}(\theta)\,d\theta=\sum_{k=-\infty}^{\infty}\int_{2k\pi}^{2(k+1)\pi} M_{z_0}(\theta)\,d\theta=\sum_{k=-\infty}^{\infty}\int_0^{2\pi} M_{z_0}(\theta+2k\pi)\,d\theta$$
$$=\int_0^{2\pi}\sum_{k=-\infty}^{\infty} M_{z_0}(\theta+2k\pi)\,d\theta.$$

The sum $\Sigma_{k=-\infty}^{\infty} M_{z_0}(\theta+2k\pi)$, $\theta\in\langle 0,2\pi\rangle$, gives the number of all prime ends of $f(D)$, which contain some point of $\partial f(D)$ lying on the ray $\{f(z_0+te^{i\theta}): t>0\}$. But, by [11], Theorem (20.3), p. 23, there are at most countably many points on $f(D)$, which are contained in at least three prime ends of $f(D)$. Therefore, with an execption of at most countable many $\theta\in\langle 0,2\pi\rangle$ we have

$$\sum_{k=-\infty}^{\infty} M_{z_0}(\theta+2k\pi)\le 2\,N_{f(z_0)}^{f(T)}(\theta),$$

consequently

$$\int_{-\infty}^{\infty} M_{z_0}(\theta)\,d\theta\le 2\int_0^{2\pi} N_{f(z_0)}^{f(T)}(\theta)\,d\theta.$$

On the other hand, $M_{z_0}(\theta)$ is the Banach indicatrix of the continuous function $\chi_{z_0}(t)$ and hence by the well known theorem of Banach

$$\int_{-\infty}^{\infty} M_{z_0}(\theta)\,d\theta=\mathrm{var}\,[\chi_{z_0};\langle 0,2\pi\rangle]$$

and the inequality (1) is proved.

**Theorem 1.1.** *Let* $f\in W$. *Then for every* $w_0\in f(D)$ *we have*

$$\int_0^{2\pi}\left|\mathrm{Im}\,\frac{ie^{it}f'(e^{it})}{f(e^{it})-w_0}\right|dt\le 2\,(V(\partial f(D))+\pi), \tag{2}$$

*where* $f'(e^{it}):=\lim_{r\to 1-} f'(re^{it})$ *a.e. on* $\langle 0,2\pi\rangle$.

*Proof.* By Proposition 1.1(b) $f' \in H^1$ and therefore $f'(e^{it})$ exists a.e. on $\langle 0,2\pi\rangle$. By the assumption $w_0 \in f(\mathbf{D})$, hence we may choose $z_0 \in \mathbf{D}$ so that $w_0 = f(z_0)$. By Proposition 1.2 the function $\chi_{z_0}(t)$ is absolutely continuous on $\langle 0,2\pi\rangle$, hence $\operatorname{var}\,[\chi_{z_0}; \langle 0,2\pi\rangle] = \int_0^{2\pi} |\chi'_{z_0}(t)|\,dt.$ By definition

$$\chi'_{z_0}(t) = \frac{d}{dt}(\operatorname{Im}\log(f(e^{it}) - f(z_0)) = \operatorname{Im}(\frac{d}{dt}\log(f(e^{it}) - f(z_0)) = \operatorname{Im}\frac{\frac{d}{dt}f(e^{it})}{f(e^{it}) - f(z_0)}.$$

By Proposition 1.1(a) $\frac{d}{dt}f(e^{it}) = ie^{it}f'(e^{it})$, hence

$$\int_0^{2\pi}\left|\operatorname{Im}\frac{ie^{it}f'(e^{it})}{f(e^{it}) - z_0}\right| dt = \operatorname{var}\,[\chi_{z_0}; \langle 0,2\pi\rangle].$$

By Proposition 1.2(1)

$$\int_0^{2\pi}\left|\operatorname{Im}\frac{ie^{it}f'(e^{it})}{f(e^{it}) - z_0}\right| dt \le 2\, v^{\partial f(\mathbf{D})}(f(z_0)).$$

By the "maximum principle" for $v^K(z)$ (see [5], Proposition 2.1, p.454) $v^{\partial f(\mathbf{D})}(f(z_0)) \le V(\partial f(\mathbf{D})) + \pi$ and Theorem 2.1 is proved.

Now we are in a position to prove the implication (i) ⇒ (ii) in the Theorem 1. We prove even more:

**Theorem 1.2.** *Let* $f \in W$. *Then for* $0 < r < 1$

$$V(f(T_r)) \le 2(V(\partial f(\mathbf{D})) + 3\pi). \tag{3}$$

*Proof.* By Proposition 1.1(b) $f' \in H^1$. Therefore the function

$$\chi_\zeta(z) := \frac{if'(z)}{f(z) - f(\zeta)} - \frac{iz}{z - \zeta},$$

which is evidently holomorphic in $\mathbf{D}$, is also in $H^1$ and so $\chi_\zeta(e^{it}) = \lim_{r\to 1-}\chi_\zeta(re^{it})$ exists for almost all $t \in \langle 0,2\pi\rangle$. By the classical theorem of F. Riesz (see [1], Theorem 2.6, p. 21), we have

$$\lim_{r\to 1-}\int_0^{2\pi} |\chi_\zeta(re^{it}) - \chi_\zeta(e^{it})|\, dt = 0,$$

hence

$$\lim_{r\to 1-}\int_0^{2\pi} |\mathrm{Im}\,\chi_\zeta(re^{it}) - \mathrm{Im}\,\chi_\zeta(e^{it})|\, dt = 0,$$

and

$$(4)\qquad \lim_{r\to 1-}\int_0^{2\pi} |\mathrm{Im}\,\chi_\zeta(re^{it})|\, dt = \int_0^{2\pi} |\mathrm{Im}\,\chi_\zeta(e^{it})|\, dt.$$

But since the function $|\mathrm{Im}\,\chi_\zeta(z)|$ is subharmonic, $\int_0^{2\pi} |\mathrm{Im}\,\chi_\zeta(re^{it})|\, dt$ is an increasing function of $r \in (0,1)$ and therefore from (4) we have

$$(5)\qquad \int_0^{2\pi} |\mathrm{Im}\,\chi_\zeta(re^{it})|\, dt \le \int_0^{2\pi} |\mathrm{Im}\,\chi_\zeta(e^{it})|\, dt \quad \text{for all } r,\ 0 < r < 1.$$

But

$$\int_0^{2\pi} |\mathrm{Im}\,\chi_\zeta(e^{it})|\, dt \le \int_0^{2\pi} \left|\mathrm{Im}\,\frac{ie^{it}\, f'(e^{it})}{f(e^{it}) - f(\zeta)}\right| dt + \int_0^{2\pi} \left|\mathrm{Im}\,\frac{ie^{it}}{e^{it} - \zeta}\right| dt.$$

By Theorem 1.1(2) the first integral is less than $2(V(\partial f(\mathbf{D})) + \pi)$. The second integral is equal to $v^T(\zeta)$ and it is clear, that $v^T(\zeta) = 2\pi$ for every $\zeta \in \mathbf{D}$. Hence

$$(6)\qquad \int_0^{2\pi} |\mathrm{Im}\,\chi_\zeta(re^{it})|\, dt \le 2(V(\partial f(\mathbf{D})) + \pi).$$

Finally, for every $r > |\zeta|$,

$$\int_0^{2\pi} \left|\mathrm{Im}\,\frac{ire^{it}\, f'(re^{it})}{f(re^{it}) - \zeta}\right| dt \le \int_0^{2\pi} \left|\mathrm{Im}\,\chi_\zeta(re^{it})\right| dt + \int_0^{2\pi} \left|\mathrm{Im}\,\frac{ire^{it}}{re^{it} - \zeta}\right| dt.$$

Because

$$\int_0^{2\pi} \left|\mathrm{Im}\,\frac{ire^{it}}{re^{it} - \zeta}\right| dt = 2\pi$$

for $|\zeta| < r$ and with regard to (6), we have proved the following assertion: for every fixed r, $0 < r < 1$, and for every $\zeta$, $|\zeta| < r$,

$$v^{f(T_r)}(f(\zeta)) = \int_0^{2\pi} \left| \operatorname{Im} \frac{ire^{it}\, f'(re^{it})}{f(re^{it}) - f(\zeta)} \right| dt \le 2(V(\partial f(D)) + 3\pi).$$

The inequality (5) follows now immediately from the upper semicontinuity of the function $\zeta \to v^{f(T_r)}(f(\zeta))$.

The key of the proof of the implication (ii) ⇒ (i) in Theorem 1 is the following

**Proposition 1.3.** *Let* $f \in S$ *be bounded and let* (ii) *hold. Then* $f' \in H^1$.

For better understanding the proof, we recall the geometric idea lying beyond the formal calculations: if we look at the curves $f(T_r)$, $r > r_0$, from three suitably chosen points of $D_{r_0}$, then from (ii) we obtain a uniform estimate for the length of such curves.

*Proof of Proposition* 1.3. Let $\zeta_1 \in \mathbf{C}$, $\zeta_2 \in \mathbf{C}$ be two different points and let $z \in \mathbf{D}$ be such a point, that $f(z) \neq f(\zeta_1)$, $f(z) \neq f(\zeta_2)$. Denote, for $j = 1, 2$,

$$\mu_j(z) = \operatorname{Im} \frac{iz\, f'(z)}{f(z) - \zeta_j}\,. \tag{7}$$

We may interpret (7) as the system of two linear equations for two unknowns $f'(z)$ and $\overline{f'(z)}$. By solving it we obtain

$$f'(z) = \frac{2[\mu_1(z)|f(z)-\zeta_1|^2(f(z)-\zeta_2) - \mu_2(z)|f(z)-\zeta_2|^2(f(z)-\zeta_1)]}{z[\overline{(f(z)-\zeta_1)}(f(z)-\zeta_2) - \overline{(f(z)-\zeta_2)}(f(z)-\zeta_1)]}\,. \tag{8}$$

Now let $f \in S$ be bounded, $|f(z)| \le L$ in $\mathbf{D}$. Clearly, there exists $r_0 > 0$ such that $D_{r_0} \subset f(\mathbf{D})$. Because $f^{-1}$ is continuous, there exists an $\alpha > 0$ so that $|f(z)| < \alpha$ implies $|z| < r_0$. Therefore

$$|f(z)| \ge \alpha > 0 \text{ for } |z| > r_0, \tag{9}$$

since $f$ is one-to-one. Choose in (8) $\zeta_1 = 0$, $\zeta_2 = \zeta$, $|\zeta| < r_0$, $z = re^{it}$, $r > r_0$. From (9) and (8) we obtain for $r > r_0$ the estimate

$$|f'(z)| \le \frac{4L^3}{r_0} \left( \left| \operatorname{Im} \frac{iz\, f'(z)}{f(z)} \right| + \left| \operatorname{Im} \frac{iz\, f'(z)}{f(z) - \zeta} \right| \right) \frac{1}{|\operatorname{Im} \bar{\zeta} f'(z)|}\,. \tag{10}$$

For fixed $r > r_0$ denote

$$E_r = \{t \in \langle 0,2\pi\rangle : |\mathrm{Im}\, f(re^{it})| \geq \tfrac{\alpha}{2}\},$$
$$F_r = \{t \in \langle 0,2\pi\rangle : |\mathrm{Im}\, f(re^{it})| < \tfrac{\alpha}{2}\}.$$

By (9) we have $|\mathrm{Re}\, f(re^{it})| > \alpha/2$ on $F_r$, i.e. $|\mathrm{Im}\, if(re^{it})| > \alpha/2$. Hence if we choose $\zeta = \rho$, $\zeta = i\rho$ for $t \in E_r$, $t \in F_r$ respectively, we obtain from (10)

$$\begin{aligned}\int_0^{2\pi} |f'(re^{it})|\, dt &= \int_{E_r} |f'(re^{it})|\, dt + \int_{F_r} |f'(re^{it})|\, dt \\ &\leq \frac{4L^3}{r_0\rho\frac{\alpha}{2}}\Bigg(\int_{E_r}\left|\mathrm{Im}\,\frac{iz\, f'(z)}{f(z)}\right| dt + \int_{E_r}\left|\mathrm{Im}\,\frac{iz\, f'(z)}{f(z)-\rho}\right| dt \\ &\quad + \int_{F_r}\left|\mathrm{Im}\,\frac{iz\, f'(z)}{f(z)}\right| dt + \int_{F_r}\left|\mathrm{Im}\,\frac{iz\, f'(z)}{f(z)-i\rho}\right| dt\Bigg) \\ &\leq \frac{8L^3}{r_0\rho\alpha}\left(v^{f(T_r)}(\rho) + v^{f(T_r)}(i\rho) + v^{f(T_r)}(0)\right).\end{aligned}$$

From this estimate we obtain, by the "maximum principle" for the quantity $v^K(z)$ (see [5], p. 454)

$$\int_0^{2\pi} |f'(re^{it})|\, dt \leq \frac{24L^3}{r_0\rho a}(M_f + \pi). \tag{12}$$

Since the function $|f'(z)|$ is subharmonic, (12) holds for all $r$, $0 \leq r < 1$, q.e.d.

From Proposition 1.3 it follows (see [1], Theorem 3.11, p. 42), that $f$ is continuous on $\bar{\mathbf{D}}$, the function $f(e^{it})$ is absolutely continuous on $\langle 0,2\pi\rangle$ and so the simply connected domain $f(\mathbf{D})$ has prime ends of the first kind only. As in the proof of Proposition 1.2 we show, that the function $\chi_{z_0}(t) = \arg(f(e^{it}) - f(z_0))$ is absolutely continuous and

$$\mathrm{var}\,[\chi_{z_0}; \langle 0,2\pi\rangle] = \int_{-\infty}^{\infty} M_{z_0}(\theta)\, d\theta = \int_0^{2\pi} \sum_{k=-\infty}^{\infty} M_{z_0}(\theta + 2k\pi)\, d\theta,$$

where $M_{z_0}(\theta)$ is the number of points $t \in \langle 0,2\pi\rangle$, for which $\chi_{z_0}(t) = \theta$. Because $\Sigma_{k=-\infty}^{\infty} M_{z_0}(\theta + 2k\pi)$ is the number of all prime ends of $f(\mathbf{D})$, which contain some point of $\partial f(\mathbf{D})$ lying on the ray $\{f(z_0) + te^{i\theta} : t > 0\}$, we clearly have

$$\sum_{k=-\infty}^{\infty} M_{z_0}(\theta + 2k\pi) \geq N_{f(z_0)}^{\partial f(\mathbf{D})}(\theta),$$

hence

$$\text{var}\,[\chi_{z_0}; \langle 0,2\pi\rangle] \geq \int_0^{2\pi} N_{f(z_0)}^{\partial f(\mathbf{D})}(\theta)\, d\theta = V^{\partial f(\mathbf{D})}(f(z_0)). \tag{13}$$

Consequently, with regard to the upper semicontinuity of the function $z \to v^{\partial f(\mathbf{D})}(f(z))$, the proof of the implication (ii) ⇒ (i) will be finished, if we show a uniform estimate for $\text{var}\,[\chi_{z_0}; \langle 0,2\pi\rangle]$. By Fatou's theorem, for almost every $t \in \langle 0,2\pi\rangle$ there exists $f'(e^{it}) := \lim_{r\to 1-} f'(re^{it})$ and so for such a $t$ and every $\zeta \in f(\mathbf{D})$, $\zeta = f(z_0)$, there exists (see [1], Theorem 11(4), p.42)

$$\lim_{r\to 1-} \frac{ire^{it} f'(re^{it})}{f(re^{it}) - f(z_0)} = \frac{ie^{it} f'(e^{it})}{f(e^{it}) - f(z_0)}.$$

By Proposition 1.3 and by theorem of F. Riesz it follows, that

$$\lim_{r\to 1-} \int_0^{2\pi} |f'(re^{it}) - f'(e^{it})|\, dt = 0,$$

from which we easily obtain (with regard to the continuity of $iz/(f(z) - f(z_0))$ for $1 \geq |z| \geq r > |z_0|$)

$$\lim_{r\to 1-} \int_0^{2\pi} \left| \frac{ire^{it} f'(re^{it})}{f(re^{it}) - f(z_0)} - \frac{ie^{it} f'(e^{it})}{f(e^{it}) - f(z_0)} \right| dt = 0,$$

hence also

$$\lim_{r\to 1-} \int_0^{2\pi} \left| \text{Im}\,\frac{ire^{it} f'(re^{it})}{f(re^{it}) - f(z_0)} - \text{Im}\,\frac{ie^{it} f'(e^{it})}{f(e^{it}) - f(z_0)} \right| dt = 0. \tag{14}$$

By the "maximum principle" (see [5], proposition 2.1, p. 454) we have $v^{f(T_r)}(f(z_0)) \leq V(f(T_r)) + \pi$ for $r > |z_0|$, hence

$$(15) \qquad \int_0^{2\pi} \left| \mathrm{Im} \frac{ire^{it} f'(re^{it})}{f(re^{it}) - f(z_0)} \right| dt \leq V(f(T_r)) + \pi \quad \text{for } r > |z_0|.$$

Combining (14), (15) and noting that

$$\mathrm{var}\, [\chi_{z_0}; \langle 0,2\pi \rangle] = \int_0^{2\pi} \left| \mathrm{Im} \frac{ie^{it} f'(e^{it})}{f(e^{it}) - f(z_0)} \right| dt,$$

we finally obtain $\mathrm{var}\, [\chi_{z_0}; \langle 0,2\pi \rangle] \leq M_f + \pi$, q.e.d.

## 2

**Proposition 2.1.** *Let* $f \in S$. *Then for every* $z_0 \in D_r$ *we have*

$$v^{f(T_r)}(f(z_0)) \leq \alpha(f(D_r)).$$

*Proof.* Since $\alpha(D)$ as well as $v^K(z)$ do not depend on translations, we may consider, for fixed $r$, the analytic curve $\lambda(\theta) = f(re^{i\theta}) - f(r_0e^{i\theta})$, $0 \leq \theta < 2\pi$ with $z_0 = r_0e^{i\theta_0}$, $\theta_0 \in \langle 0,2\pi)$, $r_0 < r$. Choose $r_1$, $r < r_1 < 1$ and consider the function

$$g(z) = \frac{iz\, f'(z)}{f(z) - f(z_0)}$$

holomorphic in the domain $D_{r_1} = D_{r_1} \setminus \{z : z = \rho e^{i\theta_0}, 0 \leq \rho \leq r_1\}$. Clearly, $g(z) \neq 0$ in $D_{r_1}$, hence any branch of $\log g(z)$ is holomorphic in $D_{r_1}$ and the curve $\psi(\theta) = \mathrm{Im} \log g(re^{i\theta})) = \mathrm{Im} \log \lambda'(\theta)/\lambda(\theta)$ is analytic on $\langle \theta_0,\theta_0+2\pi \rangle$. Therefore $\psi(\theta)$ is a constant curve and $\psi'(\theta) = 0$ identically in $\langle \theta_0,\theta_0+2\pi \rangle$, or $\psi'(\theta) = 0$ only in a finite number of points $\theta_i$, $i = 1,2,\ldots,n$, $\theta_0 \leq \theta_1 < \theta_2 < \ldots < \theta_n \leq \theta_0 + 2\pi$. Now, we claim: for every $i = 1,2,\ldots,n-1$ we have

$$\int_{\theta_i}^{\theta_{i+1}} \operatorname{sgn}(\sin \psi(\theta))\, \psi'(\theta)\, d\theta = 0.$$

If $\theta_1 \neq \theta_0$ (and hence $\theta_n \neq \theta_0 + 2\pi$), we have

$$\int_{\theta_n}^{\theta_0+2\pi} \operatorname{sgn}(\sin \psi(\theta))\, \psi'(\theta)\, d\theta + \int_{\theta_0}^{\theta_1} \operatorname{sgn}(\sin \psi(\theta))\, \psi'(\theta)\, d\theta = 0.$$

We shall prove the first equality, the proof of the second one is analogous. In view of $(\sin \psi(\theta))' = \cos \psi(\theta)\, \psi'(\theta)$ and $\psi'(\theta_i) = 0$, we have $\psi(\theta_i) = (2k_i + 1)\pi/2$. After the substitution $\psi(\theta) = u$ on $(\theta_i, \theta_{i+1})$ (which is legitimate, because $\psi'(\theta) > 0$ or $\psi'(\theta_0) < 0$ everywhere on $(\theta_i, \theta_{i+1})$) we obtain

$$\int_{\theta_i}^{\theta_{i+1}} \operatorname{sgn}(\sin \psi(\theta))\, \psi'(\theta)\, d\theta = \int_{(2k_i+1)\frac{\pi}{2}}^{(2k_{i+1}+1)\frac{\pi}{2}} \operatorname{sgn}(\sin u)\, du.$$

But the last integral is clearly zero with regard to

$$\int_{(2k-1)\frac{\pi}{2}}^{(2k+1)\frac{\pi}{2}} \operatorname{sgn}(\sin u)\, du = 0,$$

and the claim is proved. Now, we are ready to finish the proof of Proposition 2.1. We have to estimate the variation of the function $\arg \lambda(\theta)$ by the help of the variation of $\arg \lambda'(\theta)$, i.e. to prove

$$\int_{\theta_0}^{\theta_0+2\pi} \left|\frac{d \arg \lambda(\theta)}{d\theta}\right| d\theta \leq \int_{\theta_0}^{\theta_0+2\pi} \left|\frac{d \arg \lambda'(\theta)}{d\theta}\right| d\theta.$$

Elementary calculations give

$$\frac{d \arg \lambda(\theta)}{d\theta} = \operatorname{Im} \frac{\lambda'(\theta)}{\lambda(\theta)},$$

$$\frac{d \arg \lambda'(\theta)}{d\theta} = \operatorname{Im} \frac{\lambda''(\theta)}{\lambda'(\theta)},$$

$$\frac{d}{d\theta}\arg\frac{\lambda'(\theta)}{\lambda(\theta)} = \operatorname{Im}\frac{\lambda''(\theta)}{\lambda'(\theta)} - \operatorname{Im}\frac{\lambda'(\theta)}{\lambda(\theta)}.$$

Multiplying by $\operatorname{sgn}(\operatorname{Im}\lambda'(\theta)/\lambda(\theta))$ we obtain

$$\text{(16)}\qquad \operatorname{sgn}\left(\operatorname{Im}\frac{\lambda'(\theta)}{\lambda(\theta)}\right)\psi'(\theta) = \operatorname{sgn}\left(\operatorname{Im}\frac{\lambda'(\theta)}{\lambda(\theta)}\right)\operatorname{Im}\frac{\lambda''(\theta)}{\lambda'(\theta)} - \left|\operatorname{Im}\frac{\lambda'(\theta)}{\lambda(\theta)}\right|.$$

But

$$\operatorname{Im}\frac{\lambda'(\theta)}{\lambda(\theta)} = \left|\frac{\lambda'(\theta)}{\lambda(\theta)}\right| \sin\arg\frac{\lambda'(\theta)}{\lambda(\theta)}$$

and therefore

$$\operatorname{sgn}\operatorname{Im}\frac{\lambda'(\theta)}{\lambda(\theta)} = \operatorname{sgn}\sin\psi(\theta).$$

Combining with (16) we have

$$\text{(17)}\qquad \int_{\theta_0}^{\theta_0+2\pi} \operatorname{sgn}(\sin\psi(\theta))\,\psi'(\theta)\,d\theta = \int_{\theta_0}^{\theta_0+2\pi} \operatorname{sgn}(\sin\psi(\theta))\operatorname{Im}\frac{\lambda''(\theta)}{\lambda'(\theta)}\,d\theta - \int_{\theta_0}^{\theta_0+2\pi}\left|\operatorname{Im}\frac{\lambda'(\theta)}{\lambda(\theta)}\right| d\theta.$$

If $\psi'(\theta) = 0$ identically on $(\theta_0, \theta_0+2\pi)$, we have

$$\int_{\theta_0}^{\theta_0+2\pi} \operatorname{sgn}(\sin\psi(\theta))\,\psi'(\theta)\,d\theta = 0.$$

In the opposite case

$$\int_{\theta_0}^{\theta_0+2\pi} \operatorname{sgn}(\sin\psi(\theta))\,\psi'(\theta)\,d\theta = \int_{\theta_0}^{\theta_1} + \sum_{i=1}^{n-1}\int_{\theta_i}^{\theta_{i+1}} + \int_{\theta_n}^{\theta_0+2\pi} = 0$$

by the claim. In every case we obtain from (17)

$$\int_{\theta_0}^{\theta_0+2\pi}\left|\operatorname{Im}\frac{\lambda'(\theta)}{\lambda(\theta)}\right| d\theta = \int_{\theta_0}^{\theta_0+2\pi} \operatorname{sgn}(\sin\psi(\theta))\operatorname{Im}\frac{\lambda''(\theta)}{\lambda'(\theta)}\,d\theta \le \int_{\theta_0}^{\theta_0+2\pi}\left|\operatorname{Im}\frac{\lambda''(\theta)}{\lambda'(\theta)}\right| d\theta, \text{ q.e.d.}$$

**Example 2.1.** Denote G the domain in **C**, which arises from the open unit square with corners $z_1 = 0$, $z_2 = 1$, $z_3 = 1 + i$, $z_4 = i$ by omitting the points lying on the intervals $I_n = \{z = a_n + ita_n^2 : 0 \le t \le 1\}$, $a_n = 1/2^n$, $n = 1,2,\dots$ . We will show $V(\partial G) < \infty$ but $\alpha(G) = \infty$. Intuitively, it is clear, that for every compact $K \subset G$ and every $N > 0$ one may construct a smooth Jordan curve $\gamma \subset G\backslash K$ imitating the shape of N of the cuts $I_n$ so that $\alpha(\mathrm{Int}\,\gamma) > N$, and so $\alpha(G) = \infty$. Hence it suffices to prove $V(\partial G) < \infty$. From the definition of $v^K(z)$ it is not hard to show that in our case for every $z \in \mathbf{C}$

$$v^{\partial G}(z) = \sum_{i=1}^{4} \alpha_i(z) + \sum_{n=1}^{\infty} \varphi_n(z),$$

where $\alpha_i(z)$, i=1,2,3,4, is the angle under which the side $\langle z_i, z_{i+1}\rangle$ is seen from an arbitrary point z and $\varphi_n(z)$ is the angle under which $I_n$ is seen from the point z. Since every side $\langle z_i, z_{i+1}\rangle$ is seen from an arbitrary point $z \in \mathbf{C}$ under the angle not greater than $\pi$,

$$\sum_{i=1}^{4} \alpha_i(z) \le 4\pi.$$

Now let $z \in \partial G$. We distinguish four cases:

(a) $z \in \langle z_i, z_{i+1}\rangle$, $i = 2,3,4$. Then $\varphi_n(z) \le \varphi_n$, where $\varphi_n$ is the angle under which $I_n$ is seen from the point $\frac{1}{2} a_n^2 \cdot i$. Evidently $\tan \varphi_n/2 = (\frac{1}{2} a_n^2)/(\frac{1}{2} a_n) = a_n$, hence $a_n > \varphi_n/2$ and $\Sigma_{n=1}^{\infty} \varphi_n(z) \le 2\,\Sigma_{n=1}^{\infty} a_n = 2$.

(b) $z \in \langle a_{k+1}, a_k\rangle$, $k = 1,2,\dots$ . Then $\varphi_k(z) < \pi/2$, $\varphi_{k+1}(z) < \pi/2$, $\varphi_{k+p}(z) < \varphi_{k+p}$ for $p = 2,3,\dots$ and $\varphi_p(z) < \psi_{k,p}$ for $p = 1,2,\dots,k-1$, where $\psi_{k,p}$ is the angle under which $I_p$ is seen from the point $a_k + a_k^2 i$. But $\tan \psi_{k,p} < (a_p^2 - a_k^2)/(a_p - a_k)$ and so $\psi_{k,p} < a_k + a_p$. Hence

$$\sum_{n=1}^{\infty} \varphi_n(z) < \sum_{p=1}^{k-1} \varphi_p(z) + 2\cdot\frac{\pi}{2} + \sum_{p=k+2}^{\infty} \varphi_p(z)$$

$$< \frac{k-1}{2^k} + \sum_{p=1}^{k-1} \frac{1}{2^p} + \pi + \sum_{p=k+2}^{\infty} \frac{1}{2^p} < \pi + 2.$$

(c) $z \in \langle 1/2, 1\rangle$. Then $\varphi_1(z) < \pi/2$, $\varphi_n(z) < \varphi_n$ for $n = 2,3,\ldots$, hence $\Sigma_{n=1}^{\infty} \varphi_n(z) < \pi/2 + 1$.

(d) $z \in I_k$, $k = 1,2,\ldots$ . Then $\varphi_k(z) = 0$, $\varphi_{k+p}(z) \leq \varphi_{k+p}$ for $p = 1,2,\ldots$ and $\varphi_p(z) < \psi_{k,p}$ for $p = 1,2,\ldots,k-1$. Hence

$$\sum_{n=1}^{\infty} \varphi_n(z) < \frac{k-1}{2^k} + \sum_{p=1}^{k-1} \frac{1}{2^p} + \sum_{p=k+1}^{\infty} \frac{1}{2^p} < 2.$$

From (a)–(d) we have

$$v^{\partial G}(z) = \sum_{n=1}^{4} \alpha_n(z) + \sum_{n=1}^{\infty} \varphi_n(z) < 5\pi + 2.$$

hence $V(\partial G) < \infty$, q.e.d.

**Remark 2.1.** From the construction in Example 2.1 it is clear that it is possible to construct a Jordan domain $\tilde{G}$ with $\alpha(\tilde{G}) = \infty$ and $V(\partial G) < \infty$, if one removes from the unit square sufficiently thin rectangles instead of the intervals $I_n$. By smoothing the angles of $\partial\tilde{G}$ by small circles one might construct a smooth Jordan domain $H$ with $\alpha(H) = \infty$ and $V(\partial H) < \infty$.

Now we are in a position to supply the proof of Theorem 2. Let $f \in S$ be a given function of bounded boundary rotation $\alpha_f < \infty$. By the theorem of Paatero ([7], p. 17) we have $\alpha(f(D_r)) \leq \alpha_f$. By Proposition 2.1 and by the upper semicontinuity of $z \to v^K(z)$ we have $\alpha(f(D_r)) \geq V(f(T_r))$ for every $r$, $0 \leq r < 1$. Hence by Theorem 1, (ii) $\Rightarrow$ (i), we obtain $f \in W$. On the other hand, there exists the conformal representation $\varphi : D \to G$, where $G$ is the domain of Example 2.1. Clearly, normalizing the function $\varphi$ by a proper linear transformation, we obtain the function $f$ belonging to $W$, but not belonging to the Paatero class of functions with bounded boundary rotation.

## References


[1] P. L. Duren, Theory of $H^p$ spaces, Academic Press, 1970.

[2] J. Fuka, On the continuity of the Faber mapping, Ann. Polon. Math. **56** (1985), 91–103.

[3] J. Fuka, O pewnej podklasie klasy S, Materiały VII konferencji szkoleniowej z teorii zagadnien ekstremalnych, Uniwersytet Łódzki, 63–70.

[4] J. Fuka, O pewnej podklasie klasy S, II, Materiały VIII konferencji szkoleniowej

z teorii zagadnien ekstremalnych, Uniwersytet Łódzki (to appear).

[5] J. Fuka, J. Král, Analytic capacity and linear measure, Czechoslovak Math. J. **28(103)** (1978), 445–461.

[6] K. Löwner, Untersuchungen über die Verzerrung bei der konformen Abbildung des Einheitskreises $|z| < 1$, die durch Funktionen mit nicht verschwindender Ableitung geliefert werden, Ber. Verh. Sächs. Ges. Wiss. Leipzig **26** (1917), 93–100.

[7] V. Paatero, Über die konforme Abbildung von Gebieten deren Ränder von beschränkter Drehung sind, Ann. Acad. Sci. Fenn. Ser. A **33** (1931), No. 9.

[8] V. Paatero, Über Gebiete von beschränkter Randdrehung, Ann. Acad. Sci. Fenn. Ser. A **37** (1933), No. 9.

[9] J. Radon, Über die Randwertaufgaben beim logarithmischen Potential, Sitzungsber. Akad. Wiss. Wien **128** (1919), 1123–1167.

[10] T. Radó, P. V. Reichelderfer, Continuous transformations in analysis, Springer-Verlag, 1955.

[11] H. D. Ursell, L. C. Young, Remarks on the theory of prime ends, Mem. Amer. Math. Soc. **3** (1951), 1–29.

[12] L. C. Young, On area and length, Fund. Math. **35** (1948), 275–302.

# INTEGRAL MEANS OF $\alpha$-STRONGLY STARLIKE FUNCTIONS [1]

**Janusz Godula** and **Maria Nowak**
Institute of Mathematics, Marie Curie-Sklodowska University
20-031 Lublin, Poland

Let $S^*(\alpha)$, $0 < \alpha \le 1$, denote the class of all functions f, $f(z) = z + a_2 z^2 + \cdots$, that are analytic and univalent in the unit disc U and satisfy the condition

$$|\arg zf'(z)/f(z)| < \alpha \frac{\pi}{2}, \quad z \in U. \tag{1}$$

The functions of the class $S^*(\alpha)$ are called $\alpha$-strongly starlike, cf. [2], [3], [7]. In 1966 J. Stankiewicz [7] gave the geometric characterization of functions of $S^*(\alpha)$. Namely, if $f \in S^*(\alpha)$ and $w_0 \notin f(U)$, then $f(U)$ is contained in the domain bounded by two logarithmic spirals with foci at the origin and making with the radius vector the angle $(1-\alpha)\frac{\pi}{2}$ and joining the points $w_0$ and $-w_0 \exp(\pi \operatorname{tg} \alpha \frac{\pi}{2})$. The function

$$F(z) = z \exp \int_0^z (((1+t)/(1-t))^{\alpha} - 1)/t \, dt \tag{2}$$

belongs to the class $S^*(\alpha)$ and maps U onto the domain bounded by two logarithmic spirals joining the points $F(-1)$, $F(1)$. This function is extremal for several extremal problems for the class $S^*(\alpha)$.

In this paper we solve the integral means problem for $S^*(\alpha)$. Namely we have

**Theorem 1.** *Let $\phi$ be a convex nondecreasing function on $(-\infty, +\infty)$. If $f \in S^*(\alpha)$ and F is given by* (2), *then*

$$\int_{-\pi}^{\pi} \phi(\pm \log|f(re^{i\theta})|) \, d\theta \le \int_{-\pi}^{\pi} \phi(\pm \log|F(re^{i\theta})|) \, d\theta \tag{3}$$

*for all $r \in (0,1)$. If equality occurs for some r, $r \in (0,1)$ and some strictly convex $\phi$ then*

---

[1] This paper is in final form and no version of it will be submitted for publication elsewhere.

$f(z) = e^{-i\varphi}F(e^{i\varphi}z)$, $\varphi \in \mathbb{R}$.

Moreover, the integral means theorem extends to derivatives:

**Theorem 2.** *Let* $\phi$ *be a convex nondecreasing function on* $(-\infty, +\infty)$. *If* $f \in S^*(\alpha)$ *and* $F$ *is given by* (2), *then*

$$\int_{-\pi}^{\pi} \phi(\pm \log|f'(re^{i\theta})|)\, d\theta \le \int_{-\pi}^{\pi} \phi(\pm \log|F'(re^{i\theta})|)\, d\theta \tag{4}$$

*for all* $r \in (0,1)$.

The proofs depend on some lemmas. The following result may be of some independent interest. Let $g(x)$ be a real valued integrable function on $[-\pi,\pi]$. Define

$$g^*(\theta) = \sup_{|E|=2\theta} \int_E g(x)\, dx, \quad 0 \le \theta \le \pi.$$

**Lemma 1.** *Suppose* $(X,\mu)$ *is a measure space and* $f, g : [-\pi,\pi]\times X \to \mathbb{R}$ *are integrable with respect to the product of Lebesgue measure and* $\mu$. *Let* $F$, $G$ *be defined on* $[-\pi,\pi]$ *by*

$$F(\theta) = \int_X f(\theta,x)\, d\mu(x), \quad G(\theta) = \int_X g(\theta,x)\, d\mu(x).$$

*Suppose, for each* $x \in X$, $f(\cdot,x)$ *is symmetric on* $[-\pi,\pi]$ *and nonincreasing on* $[0,\pi]$ *and* $(g(\cdot,x))^* \le (f(\cdot,x))^*$. *Then* $G^* \le F^*$.

*Proof.* By the definition of the star-function

$$\begin{aligned} G^*(\theta) &= \sup_{|E|=2\theta} \int_E \Big(\int_X g(t,x)\, d\mu(x)\Big)\, dt \\ &\le \int_X \Big(\sup_{|E|=2\theta} \int_E g(t,x)\, dt\Big)\, d\mu(x) \\ &= \int_X g^*(\theta,x)\, d\mu(x) \\ &\le \int_X f^*(\theta,x)\, d\mu(x). \end{aligned}$$

Because $f(\cdot,x)$ is symmetric on $[-\pi,\pi]$ and nonincreasing on $[0,\pi]$, we have

$$\int_X f^*(\theta,x)\, d\mu(x) = \int_X \int_{-\theta}^{\theta} f(t,x)\, dt\, d\mu(x)$$

$$= \int_{-\theta}^{\theta} \int_X f(t,x)\, d\mu(x)\, dt$$

$$= \int_{-\theta}^{\theta} F(t)\, dt = F^*(\theta).$$

This proves Lemma 1.

**Lemma 2.** [5, p. 230]. *Let* u *and* v *be subharmonic in* U, *and suppose that* v *is subordinate to* u. *Then for each* r, $0 \le r < 1$,

$$v^*(re^{i\theta}) \le u^*(re^{i\theta}),\ 0 \le \theta \le \pi. \tag{5}$$

**Lemma 3.** [5, p. 230]. *Let* g *and* h *be real-valued functions integrable over* $[-\pi,\pi]$. *Then*

$$(g + h)^*(\theta) \le g^*(\theta) + h^*(\theta),\ 0 \le \theta \le \pi. \tag{6}$$

*Equality occurs if both* g, h *are symmetrically decreasing.*

*Proof of Theorem* 1. Let $f \in S^*(\alpha)$. First we will prove the inequality

$$\int_{-\pi}^{\pi} \phi(\log(|f(re^{i\theta})|/r))\, d\theta \le \int_{-\pi}^{\pi} \phi(\log(|F(re^{i\theta})|/r))\, d\theta. \tag{3'}$$

By Baernstein's results on the star-functions ([1], [5]) the inequality (3') is equivalent to the following

$$(\log(|f(re^{i\theta})|/r))^* \le (\log(|F(re^{i\theta})|/r))^*,\ 0 \le \theta \le \pi. \tag{7}$$

Notice that if $p(z) = (z\, f'(z))/f(z)$, $z \in U$, then

$$\log(f(z)/z) = \int_0^z (p(\xi) - 1)/\xi\, d\xi$$

and p is subordinate to the function $((1 + z)/(1 - z))^\alpha$. So $\mathrm{Re}(p(z) - 1)$ is subordinate to $\mathrm{Re}(((1 + z)/(1 - z))^\alpha - 1)$. By Lemma 2 we have

$$(\mathrm{Re}(p(re^{i\theta})-1))^* \le (\mathrm{Re}(((1+re^{i\theta})/(1-re^{i\theta}))^\alpha - 1))^*.$$

Therefore, in view of Lemma 1, it follows that

$$(\log(|f(re^{i\theta})|/r))^* \le (\log(|F(re^{i\theta})|/r))^*, \quad 0 \le \theta \le \pi.$$

Thus we have proved (7). Let observe that if $\phi(x)$ satisfies the assumption of our theorem then $\phi(x + \log r)$, for a fixed $r$, $r > 0$, does too. Then the inequality (3) in the case "+" follows from the inequality (3').

The proof that the inequality (3) holds in the case "−" follows, as the proof of Theorem 2 of [1].

Now let us consider the case of equality. If $\phi$ is as in our theorem then the inequality (5) is equivalent to

$$\int_{-\pi}^{\pi} \phi(v(re^{i\theta}))\, d\theta \le \int_{-\pi}^{\pi} \phi(u(re^{i\theta}))\, d\theta \tag{8}$$

and this inequality is considered in the proof of Lemma 1 in [5]. From this proof one can conclude that if the equality holds in (8) for some $r$, $r \in (0,1)$, and $\phi$ strictly increasing, then $v(z) = u(e^{it}z)$, $t \in \mathbb{R}$. So, if the equality holds in (3'), then

$$\log|f(z)/z| = \log|F(e^{it}z)/z| \quad \text{and} \quad f(z) = F(e^{it}z)e^{-it}.$$

This completes the proof.

*Proof of Theorem* 2. Let $f \in S^*(\alpha)$. As we noticed in the proof of Theorem 1,

$$z f'(z) = f(z)p(z)$$

where $p$ is subordinate to the function $((1+z)/(1-z))^\alpha$. Thus

$$\log|f'(z)| = \log|f(z)/z| + \log|p(z)|.$$

By Lemma 1 we have

$$(\log(|f(re^{i\theta})|/r))^* \le (\log(|F(re^{i\theta})|/r))^*, \quad 0 \le \theta \le \pi,$$

and by Lemma 2

$$(\log|p(re^{i\theta})|)^* \le (\log|((1+re^{i\theta})/(1-re^{i\theta}))^\alpha|)^*, \quad 0 \le \theta \le \pi.$$

Since the functions $\log|((1+re^{i\theta})/(1-re^{i\theta}))^\alpha|$ and $\log(|F(re^{i\theta})|/r)$ are both symmetrically decreasing, we have (by Lemma 3)

$$(\log|f'(re^{i\theta})|)^* \le (\log|F'(re^{i\theta})|)^*.$$

Thus, by Baernstein's results ([1],[5]) we obtain (4) for every convex nondecreasing function $\phi$.

Now we give some applications of Theorems 1 and 2. Choosing $\phi(x) = e^{px}$ we have

**Theorem 3.** *If* $f \in S^*(\alpha)$ *and* $r \in (0,1)$ *then for all* $p, p \in \mathbb{R}$,

$$\int_{-\pi}^{\pi} |f(re^{i\theta})|^p \, d\theta \le \int_{-\pi}^{\pi} |F(re^{i\theta})|^p \, d\theta$$

*with equality for some* $r, r \in (0,1)$, *if and only if* $f(z) = e^{-it}F(e^{it}z), t \in \mathbb{R}$.

**Theorem 4.** *For* $f$ *and* $r$ *as in Theorem 3 we have*

$$\int_{-\pi}^{\pi} |f'(re^{i\theta})|^p \, d\theta \le \int_{-\pi}^{\pi} |F'(re^{i\theta})|^p \, d\theta$$

*for all* $p, p > 0$.

**Corollary 1.** *If* $f(z) = z + a_2z^2 + \cdots \in S^*(\alpha)$ *and* $F(z) = z + A_2z^2 + \cdots$ *for* $|z| < 1$, *then*

$$\sum_{n=2}^{\infty} n|a_n|^2 \le \sum_{n=2}^{\infty} n|A_n|^2,$$

$$|a_n| \le \frac{1}{2\pi} \int_{-\pi}^{\pi} \exp \int_0^1 \mathrm{Re}(((1 + te^{i\theta})/(1 - te^{i\theta}))^\alpha - 1)/t \, dt \, d\theta.$$

**Corollary 2.** *If* $f \in S^*(\alpha)$ *and* $f^{-1}(w) = w + b_2w^2 + \cdots$ *in some neighbourhood of the origin, then*

$$|b_n| \le \frac{1}{2\pi n} \int_{-\pi}^{\pi} \exp \int_0^1 \mathrm{Re}(((1 + te^{i\theta})/(1 - te^{i\theta}))^\alpha - 1)/t \, dt \, d\theta.$$

These results follow from Theorem 3 in the same manner as in [6], [4].

## References

[1] A. Baernstein, Integral means, univalent functions and circular symmetrization, Acta Math. **133** (1974), 139–169.

[2] D. A. Brannan, J. Clunie, W. E. Kirwan, Coefficient estimates for a class of star-like functions, Canad. J. Math. **22** (1970), 476–485.

[3] D. A. Brannan, W. E. Kirwan, On some classes of bounded univalent functions, J. London Math. Soc. (2) 1 (1969), 431–443.

[4] J. Brown, Meromorphic univalent functions whose ranges contain a fixed disk, J. Analyse Math. **40** (1981), 155–165.

[5] P. Duren, Univalent Functions, Springer-Verlag, New York-Berlin-Heidelberg-Tokyo, 1983.

[6] W. E. Kirwan, G. Schober, Extremal problems for meromorphic univalent functions, J. Analyse Math. **30** (1976), 330–348.

[7] J. Stankiewicz, Quelques problèmes extremaux dans les classes des fonctions $\alpha$-angulairement étoilées, Ann. Univ. Mariae Curie-Skłodowska Sect. A **20** (1966), 59–75.

# ON p-ADIC NEVANLINNA THEORY [1]

**Hà Huy Khoái and My Vinh Quang**
Institute of Mathematics
P.O.Box 631, Bo Ho, 10000 Hanoi, Vietnam

## 1. Introduction

Classical Nevanlinna theory is so beautiful that one would naturally be interested in determining how such a theory would look in the p-adic case. There are two "fundamental theorems" which occupy a central place in Nevanlinna theory. In [2] we proved an analogue of the first theorem. However, in the general case we did not obtain the same type of theorem as in Nevanlinna theory. In this paper, using a minor modification of the definition of the characteristic function in [2] we prove p-adic analogues of the two "fundamental theorems" of Nevanlinna theory.

We first recall some facts from Nevanlinna theory. Let $f(z)$ be a meromorphic function in the complex plane $C$, and let $a \in C$ be a complex number. One asks the following question: How "large" is the set of points $z \in C$ at which $f(z)$ takes the value $a$ or values "close to $a$"? For every $a$ Nevanlinna constructed the following functions.

Let $n(f,a,r)$ denote the number of points $z \in C$ for which $f(z) = a$ and $|z| \leq r$, counting the multiplicity. We set:

$$N(f,a,r) = \int_0^r \frac{n(f,a,t) - n(f,a,0)}{t}\, dt + n(f,a,0) \log a,$$

$$m(f,a,r) = \frac{1}{2\pi} \int_0^{2\pi} \log^+ \frac{1}{|f(re^{i\varphi}) - a|}\, d\varphi,$$

where

$$\log^+ x = \begin{cases} \log x & \text{if } x > 1, \\ 0 & \text{if } x \leq 1. \end{cases}$$

We further set

[1] This paper is in final form and no version of it will be submitted for publication elsewhere.

$$T(f,a,r) = N(f,a,r) + m(f,a,r).$$

Nevanlinna's first fundamental theorem asserts that for every meromorphic function $f(z)$ there exists a function $T(f,r)$ such that for all $a \in C$

$$T(f,a,r) = T(f,r) + h(f,a,r),$$

where $h(f,a,r)$ is a bounded function of $r$. Since the function $T(f,r)$ does not depend on $a$, we can say that, roughly speaking, a meromorphic function takes every value $a$ the same number of times.

Nevanlinna's second fundamental theorem asserts that generally $m(f,a,r)$ is small compared with $T(f,r)$ and, consequently, $N(f,a,r)$ approximates $T(f,r)$. Namely, one defines the defect of $a$ as follows:

$$\delta(a,f) = \underline{\lim}_{r\to\infty} \frac{m(f,a,r)}{T(f,r)} = 1 - \overline{\lim}_{r\to\infty} \frac{N(f,a,r)}{T(f,r)}.$$

Then the set of defect values, i.e. of values $a$ such that $\delta(a) > 0$, is finite or countable, and

$$\sum \delta(a) \leq 2$$

where the sum extends over all defect values.

In §2 we shall construct an analogue of Nevanlinna's characteristic function $T(f,a,r)$. The construction uses in an essential way the notion of the Newton polygon of a p-adic analytic function. The Newton polygon gives expression to one of the most basic differences between p-adic analytic functions and complex analytic functions. Namely, the modulus of a p-adic analytic function depends only on the modulus of the argument, except at a discrete set of values of the modulus of the argument. The fact often makes it easier to prove p-adic analogues of classical results.

## 2. The function $T(\varphi,a,t)$

Let $p$ be a prime number, let $Q_p$ be the field of p-adic numbers, and let $C_p$ be the p-adic completion of the algebraic closure of $Q_p$. Let $D$ be the open unit disc in $C_p$: $D = \{z \in C_p ; |z| < 1\}$. The absolute value in $C_p$ is normalized so that $|p| = p^{-1}$. We further use the notion $v(z)$ for the additive valuation on $C_p$ which extends $ord_p$.

Let $f(z)$ be a p-adic analytic function on $D$ represented by a convergent power series:

$$f(z) = \sum_{n=0}^{\infty} a_n z^n.$$

For each $n$ we draw the graph $\Gamma_n$ which depicts $v(a_n z^n)$ as a function of $v(z)$. This graph is a straight line with slope $n$. Since we have

$$\lim_{n\to\infty} \{v(a_n) + nt\} = \infty$$

for all $t > 0$, it follows that for every $t$ there exists an $n$ for which $v(a_n) + nt$ is minimal. Let $v(f,t)$ denote the boundary of the intersection of all of the half-planes lying under the lines $\Gamma_n$. Then in any finite segment $[r,s]$, $0 < r < s < +\infty$ there are only finitely many $\Gamma_n$ which appear in $v(f,t)$. Thus, $v(f,t)$ is a polygon line. This line is what we call the Newton polygon of the function $f(z)$. The points $t > 0$ at which $v(f,t)$ has vertices are called the critical points of $f(z)$. A finite segment $[r,s]$ contains only finitely many critical points. It is clear that if $t$ is a critical point, then $v(a_n) + nt$ attains its minimum at least at two values of $n$. A basic property of the Newton polygon is that, if $t = v(z)$ is not a critical point, then $|f(z)| = p^{-v(f,t)}$.

The Newton polygon gives complete information about the number of zeros of the function $f(z)$. Namely, $f(z)$ has zeros when $v(z) = t_i$, where $t_0, t_1, t_2, \ldots$ is the sequence of critical points, and the number of zeros (counting multiplicity) for which $v(z) = t_i$ is equal to the difference $n_{i+1} - n_i$ between the slope of $v(f,t)$ at $t_i - 0$ and its slope at $t_i + 0$ (see [4]). It is easy to see that $n_i$ and $n_{i+1}$ are, respectively, the smallest and the largest values of $n$ at which $v(a_n) + nt$ attains its minimum. We shall call the monomials $a_{n_i} z^{n_i}$ the leading terms.

**Example.** Consider the function $\log(1 + z) = \Sigma_{n=1}^{\infty}(-1)^{n-1}z^n/n$. For every $t > 0$ we have

$$v((-1)^{n-1}/n) + nt \begin{cases} = nt - \log n/\log p & \text{if } n = p^k, \\ > nt - \log n/\log p & \text{if } n \neq p^k. \end{cases}$$

Thus, only the graphs $\Gamma_{p^k}$ $(k = 0,1,2,\ldots)$ appear in $v(f,t)$ and the function $\log(1 + z)$ has the following critical points: $t_k = 1/(p^k - p^{k-1})$ $(k = 1,2,\ldots)$. At each $t_k$,

$\log(1+z)$ has $p^k - p^{k-1}$ zeros, and we have $v(f,t) = -k + p/(p-1)$.

Now let $\varphi(z)$ be a meromorphic function on D. By definition, $\varphi(z) = f(z)/g(z)$, where $f(z)$ and $g(z)$ are analytic functions on D. We set:

$$v(\varphi,t) = v(f,t) - v(g,t).$$

It is clear that, if t is not a critical point for either $f(z)$ or $g(z)$, then $|\varphi(z)| = p^{-v(\varphi,t)}$. We call $t > 0$ a critical point for $\varphi(z)$ if it is a critical point of at least one of the functions $f(z)$, $g(z)$.

Let $a \in C_p$. We let $n(\varphi,a,t)$ denote the number of points $z \in D$ such that $v(z) = t$ and $\varphi(z) = a$, where every point is counted as many times as its multiplicity as a root of $\varphi(z) - a$. We set

$$T(\varphi,a,t) = \sum_{s>t} n(\varphi,a,s)(s-t) + v^+(\varphi - a,t)$$

where $v^+(\varphi - a,t) = \max\{v(\varphi - a,t),0\}$.

Note that if $\varphi$ is not the constant function a, then the sum on the right is a finite sum for every $t > 0$, because $n(\varphi,a,s) \neq 0$ only at the critical points s of the function $f(z) - ag(z)$, and there are only finitely many such critical points in the region $s > t > 0$.

We call $T(\varphi,a,t)$ the characteristic function for the meromorphic function $\varphi(z)$.

### 3. p-adic analogue of the first fundamental theorem

**Lemma 1.** (Jensen's formula). *Let* $\varphi(z)$ *be a meromorphic function on* D, $a_1,a_2,\ldots,a_N$ and $b_1,b_2,\ldots,b_P$, *respectively, be all the zeros and poles (counting multiplicity) of the function* $\varphi(z)$ *in* $\{|z| < r\}$, $r < R \leq 1$, *and let* $|\varphi(0)| \neq 0, \infty$. *Then we have:*

$$\log_p|\varphi(0)| + v(\varphi,t) = (N-P)t + \sum_{i=1}^{N} v(a_i) - \sum_{i=1}^{P} v(b_i)$$

*where* $t = \log_p r$.

*Proof.* First we prove Lemma 1 in the case of analytic functions. Let

$t_0 > t_1 > \dots > t_n > t$ be all the critical points of $v(\varphi,t)$ in $[t,+\infty)$. From the properties of the Newton polygon it follows that

$$v(\varphi,t) = v(\varphi,t_0) - v'(\varphi,t_0)(t_0 - t_1) - \cdots - v'(\varphi,t_n)(t_n - t)$$
$$= v(\varphi,t_0) - v'(\varphi,t_0)t_0 - \sum_{i=0}^{n-1} [v'(\varphi,t_{i+1}) - v'(\varphi,t_i)]\, t_{i+1} + v'(\varphi,t_n)t,$$

where $v'(\varphi,t)$ denotes the left derivative of $v(\varphi,t)$ at the point $t$. In view of the properties of the Newton polygon we obtain

$$-v(\varphi,t_0) + v(\varphi,t) = Nt + \sum_{i=1}^{N} v(a_i).$$

Consequently,

$$\log_p|\varphi(0)| + v(\varphi,t) = Nt + \sum_{i=1}^{N} v(a_i). \tag{1}$$

Now let $\varphi(z) = f(z)/g(z)$. Applying the formula (1) for the functions $f(z)$ and $g(z)$ we have

$$\log_p|\varphi(0)| + v(\varphi,t) = (N-P)t + \sum_{i=1}^{N} v(a_i) - \sum_{i=1}^{P} v(b_i).$$

**Remark 1.** If the function $\varphi(z)$ has a zero or a pole at $z = 0$ we consider the function $\psi(z) = r^k z^{-k}\varphi(z)$ for a suitable $k$.

**Theorem 1.** *Let* $\varphi(z)$ *be a meromorphic function in* D. *Then for every* $a \in C_p$ *we have*

$$T(\varphi,a,t) = T(\varphi,t) + h(\varphi,a,t),$$

*where* $h(\varphi,a,t)$ *is a bounded function.*

*Proof.* Let $a_1,a_2,\dots,a_N$ and $b_1,b_2,\dots,b_P$, respectively, be all the zeros and poles (counting multiplicity) of the function $\varphi(z)$ in $\{|z| < r\}$. We set

$$m(\varphi,t) = v^+(1/\varphi,t),$$
$$N(\varphi,t) = \sum_{i=1}^{N} \{v(b_i) - t\},$$

$$T(\varphi,t) = m(\varphi,t) + N(\varphi,t).$$

Then we have:

$$m(1/(\varphi - a),t) = v^+(\varphi - a,t),$$
$$N(1/(\varphi - a),t) = \sum_{s>t} n(\varphi,a,s)(s - t).$$

Consequently,

$$T(\varphi,a,t) = m\left(\frac{1}{\varphi - a}, t\right) + N\left(\frac{1}{\varphi - a}, t\right).$$

We first prove the following

**Lemma 2.** *Let* $\varphi$, $\varphi_i$ $(i = 1,2,\ldots,k)$ *be meromorphic functions on* D. *Then we have:*

1) $m(\sum_{i=1}^{k} \varphi_i, t) \leq \max_{1\leq i\leq k} m(\varphi_i, t)$,

2) $m(\prod_{i=1}^{k} \varphi_i, t) \leq \sum_{i=1}^{k} m(\varphi_i, t)$,

3) $N(\sum_{i=1}^{k} \varphi_i, t) \leq \sum_{i=1}^{k} N(\varphi_i, t)$,

4) $N(\prod_{i=1}^{k} \varphi_i, t) \leq \sum_{i=1}^{k} N(\varphi_i, t)$,

5) $T(\sum_{i=1}^{k} \varphi_i, t) \leq \sum_{i=1}^{k} T(\varphi_i, t)$,

6) $T(\prod_{i=1}^{k} \varphi_i, t) \leq \sum_{i=1}^{k} T(\varphi_i, t)$,

7) $T(\varphi,t)$ *is a decreasing function of* t,

8) $T(\varphi,t)$ *is a bounded function if and only if* $\varphi(z)$ *is a ratio of two bounded analytic functions.*

*Proof of Lemma.* 1) and 2) follow from the properties of the Newton polygon and the definition of the function $v^+$.

3) and 4) follow from the fact that if $\varphi(z)$ is a sum or a product of the functions $\varphi_i(z)$ then the order of a pole of $\varphi(z)$ at a point $z_0$ is not greater than the sum of the orders of the poles at $z_0$ of the functions $\varphi_i$.

5) and 6) are consequences of 1), 2), 3), 4).

7) is proved as follows. Suppose that $t > t'$. If $m(\varphi,t) = 0$ then $T(\varphi,t) = N(\varphi,t) \leq N(\varphi,t') \leq T(\varphi,t')$. When $m(\varphi,t) > 0$, we have $m(1/\varphi,t) = 0$ and then $T(1/\varphi,t) = N(1/\varphi,t) \leq N(1/\varphi,t') \leq T(1/\varphi,t')$. On the other hand, since $-v(\varphi,t) = -v^+(\varphi,t) + v^+(1/\varphi,t)$, it follows from Lemma 1 that

$$\log_p|\varphi(0)| + T(1/\varphi,t) = T(\varphi,t). \qquad (2)$$

In view of the equality (2) it holds that $T(\varphi,t) \le T(\varphi,t')$.

To prove 8), suppose that $\varphi(z) = f(z)/g(z)$, where $f(z)$ and $g(z)$ are two bounded analytic functions. From (2) it follows that

$$N(g,t) + m(g,t) = N(1/g,t) + m(1/g,t) + \log_p|g(0)|.$$

Then,

$$N(1/g,t) = m(g,t) - m(1/g,t) - \log_p|g(0)| = -v(g,t) - \log_p|g(0)|.$$

Thus $N(\varphi,t) = N(1/g,t)$ is bounded, and then $T(\varphi,t) = N(\varphi,t) + m(\varphi,t)$ is bounded.

Now let $T(\varphi,t)$ be bounded. Then $N(\varphi,t)$ is bounded, and since $T(1/\varphi,t)$ is bounded, $N(1/\varphi,t)$ is bounded. Suppose that $\varphi(z) = f(z)/g(z)$. It follows from (2) that

$$m(f,t) - m(1/f,t) = N(1/f,t) - N(f,t) + \log_p|f(0)|.$$

$-v(f,t) = N(1/f,t) - \log_p|f(0)|$ is bounded, because $N(1/f,t) = N(1/\varphi,t)$ is bounded. Thus, $f(z)$ is bounded. Similarly $g(z)$ is bounded.

We are now in a position to prove Theorem 1. We have

$$m\left(\frac{1}{\varphi - a}, t\right) + N\left(\frac{1}{\varphi - a}, t\right) = T\left(\frac{1}{\varphi - a}, t\right) = T(\varphi - a, t) - \log_p|\varphi(0) - a|.$$

Using Lemma 2 we obtain:

$$T(\varphi - a, t) \le T(\varphi,t) + \log_p^+ a,$$
$$T(\varphi,t) \le T(\varphi - a, t) + \log_p^+ a.$$

Since $T(\varphi,a,t) = T(\varphi - a, t)$, Theorem 1 is proved.

## 4. p-adic analogue of the second fundamental theorem

**Lemma 3.** (The basic inequality). *Let* $\varphi(z)$ *be a non-constant meromorphic function on* $\{|z| \le r\}$, $a_1, a_2, \ldots, a_q$ *be distinct numbers of* $C_p$, $|a_i - a_k| \ge \delta > 0$ *for* $i \ne k$, $t = -\log_p r$. *Then we have:*

$$m(\varphi,t) + \sum_{i=1}^{q} m\left(\frac{1}{\varphi - a_i}, t\right) \le 2T(\varphi,t) - N_1(t) + S(t),$$

*where* $N_1(t)$ *is a non-negative function given by the formula*

$$N_1(t) = N(1/\varphi',t) + 2N(\varphi,t) - N(\varphi',t)$$

*and*

$$S(t) = m(\varphi'/\varphi,t) + m\left(\sum_{i=1}^{q} \frac{\varphi'}{\varphi - a_i}, t\right) + q \log_p^+ 1/\delta - \log_p |\varphi'(0)|.$$

*Proof.* We consider the function

$$F(z) = \sum_{i=1}^{q} \frac{1}{\varphi(z) - a_i}.$$

If $|\varphi(z) - a_k| < \delta$ for some $k$, then for all $i \neq k$,

$$|\varphi(z) - a_i| = \max\{|\varphi_i(z) - a_k|, |a_i - a_k|\} \geq \delta$$

and then

$$\frac{1}{|\varphi(z) - a_k|} > \frac{1}{\delta}, \quad \frac{1}{|\varphi(z) - a_i|} \leq \frac{1}{\delta} \text{ for every } i \neq k.$$

Consequently,

$$|F(z)| = \frac{1}{|\varphi(z) - a_k|}.$$

Thus, in this case we obtain

$$v^+(F,t) = v^+\left(\frac{1}{\varphi - a_k}, t\right) \geq \sum_{i=1}^{q} v^+(\varphi - a_i, t) - q \log^+ 1/\delta.$$

In the case that $|\varphi(z) - a_i| \geq \delta$ for every $i$, this inequality is trivial, since the value on the right is negative. Thus, in the general case we have

$$m(F,t) \geq \sum_{i=1}^{q} m\left(\frac{1}{\varphi - a_i}, t\right) - q \log_p^+ 1/\delta. \tag{3}$$

On the other hand,

$$m(F,t) = m\left(\frac{1}{\varphi} \cdot \frac{\varphi}{\varphi'} \cdot \frac{\varphi'}{F}, t\right) \leq m\left(\frac{1}{\varphi}, t\right) + m\left(\frac{\varphi'}{\varphi}, t\right) + m\left(\frac{\varphi'}{F}, t\right).$$

In view of the formula (2) we have

$$m\left(\frac{\varphi}{\varphi'},t\right) = m\left(\frac{\varphi'}{\varphi},t\right) + N\left(\frac{\varphi'}{\varphi},t\right) - N\left(\frac{\varphi}{\varphi'},t\right) + \log_p\left|\frac{\varphi(0)}{\varphi'(0)}\right|,$$
$$m\left(\frac{1}{\varphi},t\right) = T(\varphi,t) - N\left(\frac{1}{\varphi},t\right) - \log_p|\varphi(0)|.$$

Thus,

$$m(F,t) \le T(\varphi,t) - N\left(\frac{1}{\varphi},t\right) + m\left(\frac{\varphi'}{\varphi},t\right) + N\left(\frac{\varphi'}{\varphi},t\right) - N\left(\frac{\varphi}{\varphi'},t\right) - \log_p|\varphi'(0)| + m\left(\frac{\varphi'}{\varphi},t\right).$$

Combining this inequality and (3) we obtain

$$\sum_{i=1}^{q} m\left(\frac{1}{\varphi - a_i}, t\right) + m(\varphi,t) \le T(\varphi,t) - N\left(\frac{1}{\varphi},t\right) + N\left(\frac{\varphi'}{\varphi},t\right) -$$
$$- N\left(\frac{\varphi}{\varphi'},t\right) + m\left(\frac{\varphi'}{\varphi},t\right) + T(\varphi,t) - N(\varphi,t) + m\left(\frac{\varphi'}{F},t\right) -$$
$$- \log_p|\varphi'(0)| + q\log_p^+ 1/\delta.$$

It follows from the formula (2) that

$$N\left(\frac{\varphi'}{\varphi},t\right) - N\left(\frac{\varphi}{\varphi'},t\right) = m\left(\frac{\varphi}{\varphi'},t\right) - m\left(\frac{\varphi'}{\varphi},t\right) + \log_p\frac{|\varphi'(0)|}{|\varphi(0)|}$$
$$= v(\varphi,t) - v(\varphi',t) + \log_p|\varphi'(0)| - \log_p|\varphi(0)|$$
$$= N\left(\frac{1}{\varphi},t\right) - N(\varphi,t) - N\left(\frac{1}{\varphi'},t\right) + N(\varphi',t).$$

Thus we have

$$m(\varphi,t) + \sum_{i=1}^{q} m\left(\frac{1}{\varphi - a_i}, t\right) \le 2T(\varphi,t) + N_1(t) + S(t),$$

where

$$S(t) = m\left(\frac{\varphi'}{\varphi},t\right) + m\left(\sum_{i=1}^{q}\frac{1}{\varphi - a_i}, t\right) + q\log_p^+ 1/\delta + \log_p|\varphi'(0)|,$$
$$N_1(t) = N\left(\frac{1}{\varphi'},t\right) + 2N(\varphi,t) - N(\varphi',t).$$

Since $2N(\varphi,t) \ge N(\varphi',t)$, $N_1(t)$ is non-negative.

**Lemma 4.** *Let* $\varphi(z)$ *be a meromorphic function on* D *and* $S(\varphi,t)$ *be the function defined in Lemma* 3. *Then* $S(\varphi,t)$ *is a bounded function.*

*Proof.* We first prove that $m(\varphi'/\varphi,t)$ is bounded. Suppose that $\varphi(z)$ is analytic on D, $\varphi(z) = \Sigma_{k=0}^{\infty} a_k z^k$. We set

$$\varphi_n(z) = \sum_{k=0}^{n} a_k z^k.$$

Then $\varphi_n'(z)/\varphi_n(z) \to \varphi'(z)/\varphi(z)$. For every $\alpha$, $0 < \alpha < 1$ and $t \le -\log_p \alpha$ we have:

$$\alpha\, v(\varphi',t) = \alpha \min_{1 \le K \le n} \{v(a_k k\, z^{k-1},t)\} \ge \min\{v(a_k z^k,t)\} = v(\varphi_n,t).$$

Hence, for every $t \le -\log_p \alpha$, $v(\varphi_n',t) - v(\varphi,t) \ge -v(\alpha)$. Consequently, $v(\varphi',t) - v(\varphi,t) \ge -v(\alpha)$. Thus, for $t \le -\log_p \alpha$,

$$m\left(\frac{\varphi'}{\varphi},t\right) \le \log_p^+ 1/\alpha.$$

Now let $\varphi = f/g$, where $f$ and $g$ are analytic functions. Then for every $t$ we have

$$m\left(\frac{\varphi'}{\varphi},t\right) = m\left(\frac{gf' - fg'}{g^2}.\frac{g}{f}\,,\,t\right) = m\left(\frac{f'}{f} - \frac{g'}{g}\,,\,t\right)$$

$$\le m\left(\frac{f'}{f},t\right) + m\left(\frac{g'}{g},t\right) \le 2\log_p^+ 1/\alpha.$$

We now consider the function

$$S(\varphi,t) = m\left(\frac{\varphi'}{\varphi},t\right) + m\left(\sum_{i=1}^{q} \frac{\varphi'}{\varphi - a_i}\,,\,t\right).$$

We set

$$\Phi(z) = \prod_{i=1}^{q} (\varphi(z) - a_i).$$

Then

$$\frac{\Phi'}{\Phi} = \sum_{i=1}^{q} \frac{\varphi'}{\varphi - a_i}\,,$$

$$S(\varphi,t) = m\left(\frac{\varphi'}{\varphi},t\right) + m\left(\frac{\Phi'}{\Phi},t\right) \le 2\log_p 1/\alpha + 2\log_p 1/\alpha = 4\log_p 1/\alpha$$

for every $t \le -\log_p \alpha$, i.e. $S(\varphi,t)$ is bounded.

We now turn to the second fundamental theorem. We set

$$N\left(\frac{1}{\varphi - a}\,,\,t\right) = \sum_{i=1}^{N} v(a_i) - Nt,$$

where $a_1, a_2, \ldots, a_N$ are the zeros of the function $\varphi(z) - a$ (counting multiplicity) in $\{v(z) \geq t\}$ and

$$\bar{N}\left(\frac{1}{\varphi - a}, t\right) = \sum_{k=1}^{M} v(a_{i_k}) - Mt,$$

where $a_{i_1}, \ldots, a_{i_M}$ are the distinct zeros of $\varphi(z) - a$ in $\{v(z) \geq t\}$,

$$\delta(a) = \delta(\varphi, a) = \underline{\lim}_{t \to 0} \frac{m\left(\frac{1}{\varphi - a}, t\right)}{T(\varphi, t)} = 1 - \overline{\lim}_{t \to 0} \frac{N\left(\frac{1}{\varphi - a}, t\right)}{T(\varphi, t)},$$

$$\Theta(\varphi, a) = 1 - \overline{\lim}_{t \to 0} \frac{\bar{N}\left(\frac{1}{\varphi - a}, t\right)}{T(\varphi, t)},$$

$$\theta(\varphi, a) = \underline{\lim}_{t \to 0} \frac{N\left(\frac{1}{\varphi - a}, t\right) - \bar{N}\left(\frac{1}{\varphi - a}, t\right)}{T(\varphi, a)}.$$

**Theorem 2.** *Let* $\varphi(z)$ *be a meromorphic function on* D. *Then the set of values* $a \in C_p$ *such that* $\Theta(\varphi, a) > 0$ *is finite or countable and we have:*

$$\sum_a (\delta(a) + \theta(a)) \leq \sum_a \Theta(a) \leq 2.$$

*Proof.* Given $\varepsilon > 0$, for $t$ sufficiently close to zero it holds

$$N\left(\frac{1}{\varphi - a}, t\right) - \bar{N}\left(\frac{1}{\varphi - a}, t\right) > (\theta(a) - \varepsilon)) T(\varphi, t),$$

$$N\left(\frac{1}{\varphi - a}, t\right) < (1 - \delta(a) + \varepsilon) T(\varphi, t).$$

Hence,

$$\bar{N}\left(\frac{1}{\varphi - a}, t\right) < (1 - \delta(a) - \theta(a) + 2\varepsilon) T(\varphi, t).$$

Thus,

$$\Theta(a) \geq \delta(a) + \theta(a).$$

Adding $N(\varphi, t) + \Sigma_{i=1}^{q} N(1/(\varphi - a_i), t)$ to both sides of the basic inequality we obtain

$$(q+1)T(\varphi,t) \le \sum_{i=1}^{q} N\left(\frac{1}{\varphi - a_i}, t\right) + N(\varphi,t) - N_1(t) +$$
$$+ 2T(\varphi,t) + 4 \log_p 1/\alpha - \log_p |\varphi'(0)|.$$

Since $N(\varphi',t) - N(\varphi,t) = \bar{N}(\varphi,t)$, we have

$$(q-1)T(\varphi,t) \le \sum_{i=1}^{q} N\left(\frac{1}{\varphi - a_i}, t\right) + \bar{N}(\varphi,t) - N(1/\varphi',t) +$$
$$+ 4 \log_p 1/\alpha - \log_p |\varphi'(0)|.$$

Since a root of multiplicity $k$ of the equation $\varphi(z) = a$ is a root of multiplicity $k$ - of the equation $\varphi'(z) = 0$, we obtain

$$\sum_{i=1}^{q} N\left(\frac{1}{\varphi - a_i}, t\right) - N(1/\varphi',t) \le \sum_{i=1}^{q} \bar{N}\left(\frac{1}{\varphi - a_i}, t\right).$$

Consequently, for every $t \le -\log_p \alpha$,

$$(q-1)T(\varphi,t) \le \sum_{i=1}^{q} \bar{N}\left(\frac{1}{\varphi - a_i}, t\right) + \bar{N}(\varphi,t) + 4 \log_p 1/\alpha - \log |\varphi'(0)|.$$

Therefore,

$$\sum_{i=1}^{q} \overline{\lim_{t\to 0}} \frac{\bar{N}\left(\frac{1}{\varphi - a_i}, t_i\right)}{T(\varphi,t)} + \overline{\lim_{t\to 0}} \frac{\bar{N}(\varphi,t)}{T(\varphi,t)} \ge q - 1,$$

$$\sum_{i=1}^{q} (1 - \theta(a_i)) + 1 - \theta(\infty) \ge q - 1,$$

$$\sum_{i=1}^{q} \theta(a_i) + \theta(\infty) \le 2.$$

This proves Theorem 2.

## References


[1] D. Barsky, Théorie de Nevanlinna p-adique d'après Hà Huy Khoái. Groupe d'étude d'analyse ultramétrique, Sécretariat de l'Institut Henri Poncaré, Mars, 1984.

[2] Hà Huy Khoái, On p-adic meromorphic functions. Duke Math. J. 50 (1983), 695–711.

[3] W. K. Hayman, Meromorphic functions. Oxford: Clarendon, 1964.

[4] Ju. I. Manin, p-adic automorphic functions. Current Problems in Mathematics 3 (1974) (in Russian).

[5] R. Nevanlinna, Le théorème de Picard-Borel et la théorie des fonctions méromorphes. Paris, 1929.

# AHLFORS' THEOREM ON ASYMPTOTIC VALUES AND SOME DEVELOPMENTS FROM IT [1]

**W. K. Hayman**
University of York, Department of Mathematics
Heslington, York, YO1 5DD, England

## 1. Introduction

We are here to honour Lars Valerian Ahlfors. At a recent meeting in Ann Arbor we heard eight lectures about his contributions to eight fields. However these lectures were necessarily a bit hurried. I would like to talk today of the contribution made just by a single paper of Lars Ahlfors, namely his thesis [1930]. I shall ignore the contribution his method made, although it enabled my teacher Mary Cartwright to solve an important problem on the growth of p-valent functions, which in turn started me off on my research. Then there is the whole of quasiconformal mapping theory and extremal lengths which developed from this paper and from parallel work of Grötzsch [1928], which was however less well known at the time. When a normal mathematician tackles a problem he looks around at what is to hand, and with luck and perseverance uses it to give a positive answer. In other words, he is only interested in the problem. When a genius does the same he invents a new technique which not only solves the problem in question but also many other problems and perhaps starts a new field of study. That is exactly what Lars Ahlfors did.

Nevertheless it may be interesting to look at the problem itself and the ramifications which have come from it.

Suppose that $f(z)$ is an entire function and that

$$f(z) - a \to 0 \text{ as } z \to \infty \text{ along a path } \Gamma. \tag{1}$$

Then we say that $f(z)$ has the asymptotic value $a$ along the path $\Gamma$. The conjecture of Denjoy [1907] was the following

**Conclusion.** *Suppose that* $f(z)$ *has* $N$ *distinct asymptotic values* $a_1$ *to* $a_N$. *Then the order of* $f(z)$ *is at least* $N/2$.

---

[1] This paper is in final form and no version of it will be submitted for publication elsewhere.

To show that the minimum value N/2 can occur consider

$$f(z) = \int_0^z \frac{\sin \xi^{N/2}}{\xi^{N/2}} d\xi .$$

It is clear that f has order N/2. In fact, if $|z| = r \geq 1$, then

$$|f'(z)| \leq e^{r^{N/2}}, \quad |f(z)| \leq r\, e^{r^{N/2}} .$$

Also as $z \to \infty$ along the positive axis,

$$f(z) \to \omega = \int_0^\infty \frac{\sin t^{N/2}}{t^{N/2}} dt ,$$

and $\omega > 0$. If we set $\xi = re^{i\theta}$, where $\theta = 2\pi\nu/N$, $\nu = 0,1,2,\ldots,N-1$, and let r tend to $\infty$, we see that $(\sin \xi^{N/2})/\xi^{N/2} = (\sin t^{N/2})/t^{N/2}$,

$$f(re^{i\theta}) = e^{i\theta} \int_0^r \frac{\sin t^{N/2}}{t^{N/2}} dt \to \omega e^{i\theta}.$$

Thus f does indeed have N distinct asymptotic values since $\omega \neq 0$.

Ahlfors' [1930] method of proof was the following. Suppose that $a_1,\ldots,a_N$ are distinct asymptotic values with corresponding paths $\Gamma_1,\ldots,\Gamma_N$. It is clear from (1) that the $\Gamma_i$ cannot intersect near $\infty$. Thus if $z_j = Re^{i\theta_j}$ are the last intersections of $\Gamma_j$ with the circle $|z| = R$, we may assume that the $z_j$ are distinct and that the $\Gamma_j$ do not meet each other in $|z| \geq R$. We replace the arcs $0,z_j$ by straight line segments and obtain N paths, going from 0 to $\infty$ and not meeting elsewhere, which we still denote by $\Gamma_j$. Further (1) holds with $a_j$, $\Gamma_j$ replacing a, $\Gamma$.

We may assume that the $\Gamma_j$ are in counterclockwise order and define $\Gamma_{N+1} = \Gamma_1$. Then, for $1 \leq j \leq N$, $\Gamma_j$, $\Gamma_{j+1}$ bound a Jordan domain $D_j$ in the closed plane. (If the $\Gamma_j$ are not Jordan arcs they will have loops which we can cut off.) We then map the half plane $H^+ : w = u + iv,\ u > 0$ onto $D_j$ by $z = \phi_j(w)$, so that $z = 0, \infty$ correspond to $w = 0, \infty$, and consider the corresponding functions $F_j(w) = f\{\phi_j(w)\}$ in $H^+$. Then $F_j(w) \to a_j, a_{j+1}$ as $w \to \infty$ along the negative and positive imaginary axis. Thus $F_j(w)$ cannot be bounded since otherwise

$$G_j(w) = (a_j - F_j)(F_j - a_{j+1}) \to 0 \tag{2}$$

along the imaginary axis and so uniformly in $H^+$ as $w \to \infty$. But going along an arc of $|w| = r$, we can see that there exists a point on which

$$|F_j - a_j| = |F_j - a_{j+1}| \geq \frac{1}{2} |a_{j+1} - a_j| = \delta, \quad |G_j(w)| \geq \delta^2.$$

This contradicts (2). Thus $G_j$ and so $F_j$ is unbounded and now a Phragmèn-Lindelöf argument shows that if

$$B_j(r) = \sup_{\substack{|w|=r \\ u>0}} |F_j(w)|$$

then $B_j(r) > cr$ for all large $r$, where $c$ is a positive constant.

If the $\Gamma_j$ are straight lines then

$$z = \phi_j(w) = cw^{\pi/\alpha_j},$$

where $\alpha_j$ is the angle between the lines. The minimum of $\alpha_j$ is at least $2\pi/N$. Hence in at least one of the angles $|z| > c|w|^{N/2}$. This was the case Denjoy [1907] considered and which motivated his conjecture. Ahlfors [1930] by a subtle study of the conformal mapping involved, showed that the conclusion holds in the general case for at least one of the domains $D_j$ depending on $r$.

I have spent so long on this set up, since it is important for generalisations and extensions.

## 2. Asymptotic functions

The first generalisation was supplied by Denjoy [1907] himself. Suppose that in (1) we replace the constants $a_j$ by small entire functions $a_j(z)$. In this case Denjoy conjectured that the conclusion still holds if the orders $\rho_a$ and $\rho_f$ of a and f are related by

$$\frac{1}{\rho_a} > 2 + \frac{1}{\rho_f}. \tag{3}$$

Denjoy [1956] proved this if the paths are straight lines. (Apparently the result was lost in his desk for fifty years.)

In the case $\rho_f = 1/2$, and $N = 1$, (3) becomes $\rho_a < 1/4$ and under this hypothesis, the conclusion has recently been proved by Fenton [1983]. Earlier Somorjai [1980] had obtained the conclusion when $\rho_a < 1/30$. Probably the right condition is $\rho_a < 1/2$. The result breaks down when $\rho_a = 1/2$, since in this case, we may choose

$$a(z) = \frac{\sin z^{1/2}}{z^{1/2}} .$$

Then $a(z) \to 0$ as $z \to 0$ along the positive real axis and, if $C$ is any constant, the function $Ca(z)$ has the same property. Thus all the functions $Ca(z)$ are asymptotic functions for $f(z) = a(z)$. On the other hand, if $a_1(z)$, $a_2(z)$ are distinct entire functions of order less than half (or order half minimal type) then, by a classical theorem of Wiman [1905], $a_1(z) - a_2(z)$ cannot tend to zero along a path $\Gamma$ and so (1) cannot hold for different $a_j$ and the same path. Thus there is hope that the result may be true.

If we try to use Ahlfors' argument we have two problems, (a) to show that paths $\Gamma_1$, $\Gamma_2$ corresponding to two different functions $a_1$, $a_2$ are finally nonintersecting and (b) that $f$ grows more quickly than $a_1$ or $a_2$ in the domain $D_1$. Fenton [1983] solved these problems using subharmonic techniques and the socalled $\cos(\pi\lambda)$ theorem applied to the functions $G_j(w)$ given by (2). This time we do not use conformal mapping, since it is hard to preserve the property that $a_1$, $a_2$ are entire under such a map. Instead we can use a technique of Tsuji [1959] going back to Carleman [1933] which works directly on the function $f(z)$ or the corresponding subharmonic function $u(z)$.

## 3. Tracts

To generalise Ahlfors' Theorem in a different direction we write

$$u(z) = \log|f(z)|$$

and suppose that $f$ has again $N$ distinct asymptotic values $a_j$. As we saw $u(z)$ is then bounded above on the curves $\Gamma_j$ and unbounded in between.

We may assume that $u(z) < K$ on the $\Gamma_j$, but that $u(z) > K$ somewhere in each of the domains $D_j$, where $K$ is a suitable constant. Thus if $C(K)$ is the set where $u \geq K$, then $C(K)$ has at least $N$ components for large $K$, one in each domain $D_j$. It is not difficult to see that the number $N(K)$ of components of $u > K$ is a nondecreasing function of $K$ for large $K$ and so tends to a limit $N_0$, which is either a positive integer or $+\infty$. We call $N_0$ the number of tracts of $u(z)$. This point of view was adopted by Heins [1959] who, using the Carleman convexity formula, proved directly that a function with $N_0$ tracts has at least order $N_0/2$ mean type.

The notion of tract extends to subharmonic functions in $R^m$ where $m \geq 2$, and one can ask for the lower bound $l(N,m)$ of the orders of such functions in $R^m$ having at least $N$ tracts. The result for $m = 2$ is due to Ahlfors [1930] and Heins [1959] and shows that $l(N,2) = N/2$. It follows from a theorem of Talpur [1967], that $l(N,3) \geq \frac{1}{2} j_0(N-1)^{1/2} - \frac{1}{2}$, where $j_0 = 2.4048\ldots$ . More general results were obtained by Friedland and Hayman [1976]. They showed that $l(2,m) = 1$ for all $m$. For $m > 2$ and $N > 2$ the exact value of $l(N,m)$ is not known, but the order of magnitude of $l(N,m)$ is $\log N$ if $N < 2^m$ and $mN^{1/(m-1)}$ if $N \geq 2^m$. As an example of the first situation consider the function $u = |x_1|\,|x_2| \cdots |x_p|$, where $p \leq m$. This has order $p$ and $2^p$ tracts. The example was suggested orally by T. Kövari.

The problem is that right circular cones do not stack nicely in $R^m$, $m > 2$. One might conjecture that $l(N,m) = N/2 = 3/2$, if $N = 3$. At any rate, Friedland and Hayman [1976] proved that $l(3,3) \geq l(3,4) \geq 1.41$, $l(4,3) \geq l(4,4) \geq 1.72$.

#### 4. Spiral asymptotic values

For the rest of this talk I would like to confine myself to the plane case. Suppose that one of the asymptotic paths $\Gamma = \Gamma_j$ is parametrized by $z = z(t)$, $0 \leq t < 1$. Following Ahlfors [1930] we define

$$\alpha = \overline{\lim_{t\to 1}} \frac{|\arg z|}{\log|z|}, \quad \beta = \underline{\lim_{t\to 1}} \frac{|\arg z|}{\log|z|} .$$

Let the domains $D_j$ be as above and suppose that $u(z)$ is a subharmonic function bounded above on the $\Gamma_j$ and unbounded on each $D_j$. After subtracting a constant we may assume that $u(z) \leq 0$ on $\Gamma_j$, $u(z) > 0$ somewhere in each $D_j$. It can be shown that the case of a function with $N$ tracts can always be reduced to the above case. We write

$$B_j(r,u) = \sup_{\substack{|z|=r \\ z\in D_j}} u(z), \quad B(r,u) = \sup_j B_j(r,u).$$

Then we define the order $\lambda_j$, and lower order $\mu_j$ of u in $D_j$ by

$$\lambda_j = \overline{\lim_{r\to\infty}} \frac{\log B_j(r)}{\log r}, \quad \mu_j = \underline{\lim_{r\to\infty}} \frac{\log B_j(r)}{\log r}.$$

The order $\lambda$ and lower order $\mu$ are similarly defined in terms of B(r). In his thesis Ahlfors stated the inequality

$$\lambda \geq \tfrac{1}{2} N(1 + \alpha^2). \tag{4}$$

There was a gap in the original proof, which has recently been elegantly filled by Jenkins [1987]. In the same way one can prove the inequality

$$\mu \geq \tfrac{1}{2} N(1 + \beta^2). \tag{5}$$

If we use the $\lambda_j$, we can obtain a sharper result, namely

$$\sum \frac{1}{\lambda_j} \leq \frac{2}{1+\alpha^2}, \tag{6}$$

and now the number of tracts can even be infinite. However $\Sigma \lambda_j^{-1}$ must converge, so that $\lambda_j \to \infty$. The inequality (6) is sharp. We can construct a subharmonic function with a sequence of tracts $T_j$, having order $\lambda_j$ in the j'th tract and rotation index $\alpha$, provided that $\alpha$ and the $\lambda_j$ satisfy (6). At least if the number of the tracts is finite one can even do this with a function of the form $u = \log f$, where f is entire with distinct asymptotic values. It is perhaps worth saying something about such constructions.

Given the paths $\Gamma_j$ and domains $D_j$ as above we return to the mapping functions $z = \phi_j(w)$ from the right half-plane H onto $D_j$ and their inverses

$$w = \phi_j^{-1}(z) = u_j + iv_j .$$

The function $u_j$ is harmonic in $D_j$, continuous in $\bar{D}_j$, positive in $D_j$ and zero on the boundary $\Gamma_j \cup \Gamma_{j+1}$. We obtain the required subharmonic function u, by defining

$$u = u_j \text{ in } \bar{D}_j .$$

Evidently $u \geq 0$, and u is continuous in the plane. In $D_j$ u is harmonic and on $\Gamma_j$ $u = 0$ so that the mean value inequality is trivially satisfied. A somewhat similar construction is available in $R^m$ if the $D_j$ are smooth enough, say right circular cones.

How can we obtain entire functions in the plane? The following method was first introduced by Kjellberg [1948], refined by Kennedy [1956] and perfected by Katifi [1966]. The function $u(z)$ constructed as above has a Riesz mass $\mu_j$ along the curves $\Gamma_j$. Let $n_j(t)$ be the total mass from the origin to the point $t$. We suppose that $t$ is arc length and define

$$U(z) = u(z) + \sum_{j=1}^{N-1} \int_0^\infty \log\left|1 - \frac{z}{z_j(t)}\right| d\{[n_j(t)] - n_j(t)\} .$$

It turns out that under fairly general hypotheses the integrals converge and now $U(z)$ has only positive integral Riesz mass, so that $U(z) = \log|f(z)|$ where $f(z)$ is entire. If $P_k(z)$ is the k'th partial sum of $f(z)$ and $k$ is large enough then

$$F(z) = \frac{f(z) - P_k(z)}{z^{k+1}} \to 0 \text{ as } z \to \infty \text{ along } \Gamma_j ,$$

while $F(z)$ is unbounded in $D_j$. Another trick is necessary to allow $F(z)$ to have different asymptotic values along the curves $\Gamma_j$. We need a suitable integral of a product of functions $F$ constructed as above [Hayman 1969].

## 5. Regularity theorems

It follows from Ahlfors' thesis that not only $\lambda \geq N/2$, but more precisely

$$\underline{\lim}_{r\to\infty} \frac{B(r)}{r^{N/2}} > 0$$

if $f$ has $N$ distinct asymptotic values and more generally if $u$ is a subharmonic function with $N$ tracts. Ahlfors asked what conclusions can be drawn if this minimal growth is attained, i.e. if

$$\underline{\lim}_{r\to\infty} \frac{B(r)}{r^{N/2}} < \infty .$$

There has been a great deal of progress in this direction also due largely to Kennedy [1955] who sharpened earlier results of Heins [1949]. It turns out that in this case the tracts must be almost straight and in fact (5) shows that $\beta = 0$. However a great deal more than this is true. It can be shown that there exists an absolutely continuous function $\theta(r)$, $1 < r < \infty$ such that

$$\int_1^\infty r\,\theta'(r)^2\,dr < \infty \tag{7}$$

and that

$$z = r\,e^{i\theta(r)+2\pi j/N+o(1)} \quad \text{as } z \to \infty \text{ on } \Gamma_j .$$

Kennedy [1955] proved that

$$\arg(z) = o(\log|z|)^{1/2} \quad \text{as } z \to \infty \text{ on } \Gamma_j .$$

A gap in the original argument was removed by Clunie [1958]. All the above results are sharp.

We also have very accurate information on the behaviour of $u(z)$ in the domains $D_j$. Let $r\theta_j(r)$ be the length of the intersection of $D_j$ with $|z| = r$. The above results show that

$$\theta_j(r) = \frac{2\pi}{N} + o(1).$$

However Kennedy [1955] proved more strongly that

$$\int_1^\infty \left(\theta_j(r) - \frac{2\pi}{N}\right)^2 \frac{dr}{r} < \infty$$

and that

$$\log B_j(r) = \pi \int_0^\infty \frac{dr}{r\theta_j(r)} + c_j + o(1) \quad \text{as } r \to \infty$$

where $c_j$ is a constant. He also proved that

$$\frac{\prod_{j=1}^{N} B_j(r)}{r^{N^2/2}} \to l \tag{8}$$

where $l$ is a positive constant. Heins [1948] had shown that the left hand side is bounded above and below. We can also prove that $u(z)$ behaves regularly in $D_j$ and

$$u(re^{i\theta}) = B_j(r)\,\{\sin N(\theta - \theta(r) - 2\pi j/N) + o(1)\}, \quad z \in D_j \tag{9}$$

outside a small set of $z$. For instance (9) holds uniformly on $|z| = r$ as $r \to \infty$ outside a set of finite logarithmic measure. For these results the original method of Ahlfors [1930] appears to be the ideal tool together with a regularity theorem of Eke [1967], which refines an earlier result of Kennedy [1955].

The above results are more or less sharp. Thus for instance (7) cannot be improved further. If (7) holds, the curve $z = re^{i\theta(r)}$ can serve as an asymptotic path in the above situation. The individual $B_j(r)$ do not have the same regularity as the product in (8). Here, as Kennedy [1956] showed, the most one can say is that

$$\log B_j(r) = \frac{N}{2} \log r + o(\log r)^{1/2} .$$

Heins [1949] had proved the corresponding result with $o(\log r)$ instead of $o(\log r)^{1/2}$.

We have seen that Ahlfors' thesis has given rise to a host of interesting extensions and consequences and, like its original author, the subject flourishes.

## References


[1] L. V. Ahlfors, Untersuchungen zur Theorie der konformen Abbildungen und der ganzen Funktionen, Acta Soc. Sci. Fenn. Nova Ser. 1, No. 9 (1930).

[2] T. Carleman, Sur une inégalité différentielle dans la théorie des fonctions analytiques, C. R. Acad. Sci. Paris 196 (1933), 995–997.

[3] J. Clunie, On a paper of Kennedy, J. London Math. Soc. 33 (1958), 118–120.

[4] A. Denjoy, Sur les fonctions entières de genre fini, C. R. Acad. Sci. Paris 145 (1907), 106–108.

[5] A. Denjoy, L'allure asymptotique des fonctions entières d'ordre fini, C. R. Acad. Sci. Paris 242 (1956), 213–218.

[6] B. G. Eke, Remarks on Ahlfors' distortion theorem, J. Analyse Math. 19 (1967), 97–134.

[7] P. C. Fenton, Entire functions having asymptotic functions, Bull. Austral. Math. Soc. 27 (1983), 321–328.

[8] S. Friedland and W. K. Hayman, Eigenvalue inequalities for the Dirichlet problem on spheres and the growth of subharmonic functions, Comment. Math. Helv. 51 (1976), 133–161.

[9] H. Grötzsch, Über einige Extremalprobleme der konformen Abbildung, Ber. sächs. Akad. Wiss. Leipzig, Math.-phys. Kl. 58 (1928), 367–376 and 497–502.

[10] W. K. Hayman, On integral functions with distinct asymptotic values, Proc. Cambridge Phil. Soc. 66 (1969), 301–315.

[11] M. Heins, On the Denjoy-Carleman-Ahlfors theorem, Ann. of Math. 49 (1948), 533–537.

[12] M. Heins, On a notion of convexity connected with a method of Carleman, J. Analyse Math. 7 (1959), 53–77.

[13] J. A. Jenkins, On Ahlfors' spiral generalisation of the Denjoy conjecture, Indiana Univ. Math. J. 36 (1987), 41–44.

[14] W. Al-Katifi, On the asymptotic values and paths of certain integral and meromorphic functions, Proc. London Math. Soc. (3) 16 (1966), 599–634.

[15] P. B. Kennedy, On a conjecture of Heins, Proc. London Math. Soc. (3) 5 (1955), 22–47.

[16] P. B. Kennedy, A class of integral functions bounded on certain curves, Proc. London Math. Soc. (3) 6 (1956), 518–547.

[17] B. Kjellberg, On certain integral and harmonic functions: a study in minimum modulus, Thesis (Uppsala, 1948).

[18] G. Somorjai, On asymptotic functions, J. London Math. Soc. (2) 21 (1980), 297–303.

[19] M. N. M. Talpur, On the sets, where a subharmonic funtion is large, Thesis (London, 1967), p. 155.

[20] M. Tsuji, Potential theory in modern function theory, Maruzen, Tokyo, 1959.

[21] A. Wiman, Sur une extension d'un théorème de M. Hadamard, Arkiv f. Mat., Astr. och Fys. 2, No. 14, 1905.

# REGULAR B-GROUPS AND REPEATED DEHN TWISTS [1]

**Dennis A. Hejhal** [2]
School of Mathematics, University of Minnesota
Minneapolis, MN 55455, U.S.A.

**OVERVIEW.** Let $\theta$ be a Dehn twist on $S_0$. Let $S_n$ be the marked surface [i.e. point of $T_g(S_0)$] corresponding to $\theta^n(S_0)$. Each $S_n$ can be represented via a quasi-Fuchsian group (ala the Bers embedding of $T_g$). Our primary aim in this paper is to understand what happens to the region of discontinuity of $G_n$ as $n \longrightarrow \infty$.

The final answer will involve a (certain) Kleinian group $\mathcal{G}$ which *strictly contains* the regular b-group $G_\infty$ obtained by passing to the *algebraic* limit of $G_n$.

The limit set of $G_\infty$ will be seen to act as a kind of "crinkled up mirror." A second regular b-group, $\mathcal{X}$, "dual" in some sense to $G_\infty$, will arise in this way.

§1. The connection between repeated Dehn twists and regular b-groups goes back a number of years, to at least [1, p. 33] and [16, p. 291 (para 2)]. Cf. theorem A in §7 for a precise statement.

Our primary aim in this paper is to *supplement* theorem A with a result *wherein* (the ideas of) Carathéodory convergence, extremal distance, Schwarzian derivatives, and linear fractional *re*normalization blend together in a rather attractive way.

To make things a bit more accessible: we begin with a quick review of certain preliminary facts, and then follow this up (in §8) with a detailed proof of theorem A.

Our basic references will be [12, 7, 19].

The main theorem (≡ theorem B) is developed in §9.

The idea for this result arose from [13], where 3-dimensional topological techniques are used to investigate several closely related matters. Our techniques will generally tend to *complement* those of [13, 16, 17].

The author wishes to thank Al Marden for directing his attention to [13].

---

[1] This paper is in final form and no version of it will be submitted for publication elsewhere.

[2] Supported in part by NSF Grant DMS 86-07958.

§2. The following notation will be used.

$\hat{\mathbb{C}}$ = the Riemann sphere, $\hat{\mathbb{R}}$ = the extended real line $\subseteq \hat{\mathbb{C}}$.

U = the upper half-plane, L = the lower half-plane.

QC = quasiconformal, RS = Riemann surface, LP = linearly polymorphic.

RHS = right-hand side, iff = if and only if, wrt = with respect to.

APT = accidental parabolic transformation, I = the identity.

$\pi$ always means projection (in the appropriate context).

Hôlomorphic means holomorphic except for a possible simple pole.

Lôxodromic means {loxodromic} $\cup$ {hyperbolic}.

$\xi_M$ = the attractive fixpoint of M, $\eta_M$ = the repulsive fixpoint of M.

A *pseudo-axis* for M (in, say, U) is an M-invariant Jordan arc which connects the points $\xi_M$ and $\eta_M$.

$\mathcal{A}_2(U,\Gamma)$ = the space of holomorphic quadratic differentials on $U/\Gamma$.

$\mathcal{A}_2(L,\Gamma)$ = the space of holomorphic quadratic differentials on $L/\Gamma$.

$[x_1,x_2,...]$ = the group generated by $x_1,x_2,...$ .

$\Omega(G)$ = the region of discontinuity of G.

$\Lambda(G) = \hat{\mathbb{C}} - \Omega(G)$ = the limit set of G, $G_A = \{T \in G : T(A) = A\}$.

$\rightrightarrows$ means uniform convergence, $\cong$ means conformally equivalent.

If $\gamma$ is a Jordan curve in $\mathbb{R}^2$, then $\hat{\mathbb{C}} - \gamma$ has two components. The one containing $\infty$ is called $\text{ext}(\gamma)$; the other is called $\text{int}(\gamma)$.

If R is a Jordan *region* on $\hat{\mathbb{C}}$, then ext(R) has the obvious meaning.

A topological deformation is a *self-homeomorphism* which is isotopic to the identity.

§3. Let $\Gamma$ be any strictly hyperbolic Fuchsian group with compact quotient (of genus $g \geq 2$). Choose any point $z_0 \in U$. Set $S_0 = U/\Gamma$ and $\bar{S}_0 = L/\Gamma$.

Supply $S_0$ with a canonical dissection centered at $\mathcal{O} = \pi(z_0)$. In this way: $S_0$ becomes a surface with Teichmüller marking. By lifting the dissection through $z_0$, we obtain a natural fundamental polygon for $U/\Gamma$ and a corresponding marking on $\Gamma$.

The Bers embedding of Teichmüller space $T_g(S_0)$ is familiar from [6, pp. 272–274], [7, pp. 589–592], [3, pp. 122–126]. In this paper, we'll use the normalization:

$$w^{\mu}(z) = \frac{1}{z - \bar{z}_0} + O[(z - \bar{z}_0)]. \tag{3.1}$$

For later use, it is important to recall the following fact. Given any

$S_1 \in T_g(S_0)$. Write $S_1 = U/\Gamma_1$ and let $f_t$ be any (orientation-preserving) diffeomorphism $S_0 \to S_1$ which depends *continuously* on an auxiliary parameter $t \in [0,1]$. There is no difficulty constructing a continuous family of lifts $\tilde{f}_t : U \to U$ which have exactly the same *monodromy homomorphism* (under $\Gamma$) as $t$ varies. Each such $\tilde{f}_t$ determines a Beltrami coefficient $\lambda_t$ on $U$ mod $\Gamma$. Let

$$\mu_t = \begin{Bmatrix} \lambda_t(z), & z \in U \\ 0 & , z \in L \end{Bmatrix}, \quad \hat{\mu}_t = \begin{Bmatrix} \lambda_t(z), & z \in U \\ \overline{\lambda_t(\bar{z})}, & z \in L \end{Bmatrix},$$

as usual. Form $w^{\mu_t}(z)$ and write $\{w^{\mu_t},z\} \equiv \varphi_t$ for $z \in L$. Then:

(i) each $\tilde{f}_t$ extends *uniquely* to a symmetric QC self-map of $\hat{\mathbb{C}}$ having complex dilatation $\hat{\mu}_t(z)$;

(ii) the resulting maps $\tilde{f}_t$ all *agree* along $\hat{\mathbb{R}}$;

(iii) the mappings $w^{\mu_t}(z)$ are conformal and independent of $t$ for $z \in L$;

(iv) the mappings $w^{\mu_t} \circ (\tilde{f}_t)^{-1}$ are conformal and independent of $t$ for $z \in U$;

(v) the quadratic differentials $\varphi_t \in A_2(L,\Gamma)$ are independent of $t$.

Cf. [3, pp. 123(bot)–124] regarding assertions (ii)–(iv).

It should also be noted that each $w^{\mu_t} \circ (\tilde{f}_t)^{-1}$ is globally QC on $\hat{\mathbb{C}}$ and LP with respect to $\Gamma_1$. It is therefore *natural* to take the Schwarzian derivative of $w^{\mu_t} \circ (\tilde{f}_t)^{-1}$ over $U$. (3)

§4. On the matter of b-groups, our basic references will be [7], [19]. The results in [19] are supplemented by [20, 21, 22].

In this paper: all Kleinian groups are understood to be finitely generated and torsion-free.

For background information concerning *regular* b-groups, we refer to (the review given in) chapter 1 of [12]. Cf. also [2].

---

3 Taking the Schwarzian over $U$ will turn out to be *quite* advantageous below. Heretofore, in the literature, primary emphasis has been placed on $\{w^{\mu_t},z\} \equiv \varphi_t$.

§5. For later use, we need to review some basic properties of the *lifts* of proper partitions of $S_0$.

To this end: let $\{\alpha_1,\dots,\alpha_N\}$ be any proper partition of $S_0$ by loops which avoid $\mathcal{O}$. Cf. [19, pp. 613, 614, 616]. (4) Denote the corresponding chunks of $S_0$ by $S_0^{(1)}, \dots, S_0^{(p)}$.

For simplicity: we *assume* (just as in [12]) that the $\alpha_j$ are derivable from their geodesic counterparts by a topological deformation of the underlying surface $S_0$.

Abbreviate $\alpha_1 \cup \dots \cup \alpha_N$ as $\alpha$ and consider the various components of $\pi^{-1}(\alpha)$ and $U - \pi^{-1}(\alpha)$. Let $\mathcal{E}$ be the set of all elements in $\Gamma$ which correspond to [integral] powers of $\alpha_1,\dots,\alpha_N$ on $S_0$.

The properties we require are as follows.

ⓐ Distinct components of $\pi^{-1}(\alpha)$ have disjoint closures in $U \cup \hat{\mathbb{R}}$.

ⓑ Let $\mathcal{A}_1$ and $\mathcal{A}_2$ be distinct components of $U - \pi^{-1}(\alpha)$. Then $\Gamma_{\mathcal{A}_1} \cap \Gamma_{\mathcal{A}_2}$ is nontrivial iff $\mathcal{A}_1$ and $\mathcal{A}_2$ abut. In that case: $\Gamma_{\mathcal{A}_1} \cap \Gamma_{\mathcal{A}_2}$ is cyclic with an obvious hyperbolic generator.

ⓒ In addition: card $[\hat{\mathbb{R}} \cap \partial\mathcal{A}_1 \cap \partial\mathcal{A}_2] = 2$ or $0$ depending on whether $\mathcal{A}_1$ & $\mathcal{A}_2$ abut or not.

ⓓ Elements of $\Gamma$ map components of $U - \pi^{-1}(\alpha)$ onto components of $U - \pi^{-1}(\alpha)$.

ⓔ Under deformations of $S_0$, the groups $\Gamma_{\mathcal{A}}$ and the sets $\partial\mathcal{A} \cap \hat{\mathbb{R}}$ are preserved.

ⓕ Given any $M$ in $\Gamma - \mathcal{E}$. The following statements are then equivalent:

($f_1$) $M \in \Gamma_{\mathcal{A}}$;

($f_2$) $\{\xi_M, \eta_M\} \subseteq \partial\mathcal{A} \cap \hat{\mathbb{R}}$;

($f_3$) $\xi_M \in \partial\mathcal{A} \cap \hat{\mathbb{R}}$;

($f_4$) $\eta_M \in \partial\mathcal{A} \cap \hat{\mathbb{R}}$;

($f_5$) there exists a pseudo-axis for $M$ contained entirely in $\mathcal{A}$.

ⓖ Let $\mathcal{A}$ be any component of $U - \pi^{-1}(\alpha)$. Let the corresponding chunk of $S_0$ be $S_0^{(k)}$. The region $\mathcal{A}$ can be regarded as a universal covering surface of (open RS) $S_0^{(k)}$. The group of cover transformations is simply $\Gamma_{\mathcal{A}}$. This shows, for instance, that

---

4 The $\alpha_j$ are *understood* to be oriented.

every group $\Gamma_{\mathcal{A}}$ is nonelementary.

[h] $\Gamma_{\mathcal{A}_1} \subseteq \Gamma_{\mathcal{A}_2}$ iff $\mathcal{A}_1 = \mathcal{A}_2$.

[i] Let $\mathcal{C}$ be any nonempty subset of $\mathcal{A}$. Then: $\Gamma_{\mathcal{C}} \subseteq \Gamma_{\mathcal{A}}$.

For the proofs, consult [12].

§6. Let G now denote *any* b-group with simply-connected invariant component $\Delta(G)$ satisfying $\Delta(G)/G \cong L/\Gamma$. Let f be any (suitably normalized) Riemann mapping function taking L onto $\Delta(G)$. Let $\chi_f$ be the corresponding monodromy homomorphism. Note that $\chi_f$ defines a natural isomorphism: $\Gamma \leftrightarrow G$. Let $\alpha_1,\dots,\alpha_N$ be the set of *structure loops* for G (realized on $S_0$) ala [19, 20, 21, 22]. The corresponding elements of G are APT's.

The quotient $\Omega(G)/G$ can be expressed as $\bar{S}_0 + T_1 + \cdots + T_m$ in accordance with the Ahlfors finiteness theorem [4]. The Maskit marking ([5]) will therefore correspond to some subset of $S_0^{(1)}, \dots, S_0^{(p)}$ having cardinality m. By relabelling things, we can arrange things so that $S_0^{(k)}$ corresponds to $T_k$ for $1 \le k \le m$. In line with [20, thm 6 (corrected)], there will now exist (orientation-preserving) homeomorphisms $\Psi_k$ mapping $S_0^{(k)}$ onto $T_k$ with obvious "blow-downs" along any structure loops.

Let $\Omega'$ be any component of $\Omega(G) - \Delta(G)$ corresponding to (say) $T_k$. We can then find a unique $\mathcal{A}$, corresponding to $S_0^{(k)}$, such that $\chi_f(\Gamma_{\mathcal{A}}) = G_{\Omega'}$. Cf. [19, thm 5 assertion 2].

In fact: by proper choice of the $\Psi_k$, we can arrange things so that there exists a lifting $\tilde{\Psi}_k : \mathcal{A} \to \Omega'$ satisfying

$$\tilde{\Psi}_k(M\xi) = \chi_f(M)[\tilde{\Psi}_k(\xi)] \quad \text{for } M \in \Gamma_{\mathcal{A}},\ \xi \in \mathcal{A}.$$

The functions $\tilde{\Psi}_k$ have blow-downs along the various components of $\pi^{-1}(\alpha)$.

The following properties round out the list that we started in §5.

[xa] Each $G_{\Omega'}$ is a quasi-Fuchsian group [of finite type] with invariant component $\Omega'$.

[xb] Let $\Omega'$ and $\Omega''$ be any two components of $\Omega(G) - \Delta(G)$. Then: $G_{\Omega'} = G_{\Omega''}$ iff $\Omega' = \Omega''$.

---

[5] on the complex $K(S_0,\alpha_1,\dots,\alpha_N)$

[xc] Assume that $\Omega' \neq \Omega''$. Write $G_{\Omega'} = \chi_f(\Gamma_{\mathcal{A}_1})$ and $G_{\Omega''} = \chi_f(\Gamma_{\mathcal{A}_2})$. Then $G_{\Omega'} \cap G_{\Omega''} = \{I\}$ unless $\mathcal{A}_1$ & $\mathcal{A}_2$ abut. In that case: $G_{\Omega'} \cap G_{\Omega''}$ is cyclic with an obvious APT as generator.

The groups $\chi_f(\Gamma_{\mathcal{A}})$ are known [for any $\mathcal{A}$] as factor subgroups of G. Cf. [22, p. 355].

§7. Given any $\{\alpha_1,...,\alpha_N\}$ as in §5. Let $\mathcal{R}_j$ be a narrow collar straddling $\alpha_j$. There is no loss of generality if we assume that the $\mathcal{R}_j$ are mutually disjoint & bounded away from $\mathcal{O}$. We can also require that the $\mathcal{R}_j$ have *analytic* boundaries.

Let $\mathcal{R}$ be an abbreviation for $\mathcal{R}_1 \cup ... \cup \mathcal{R}_N$.

The concept of a Dehn twist is familiar from [10, pp. 253, 360]. Cf. figure 1 (at the end of this paper).

Let $\theta_j$ be any sensible Dehn twist around $\alpha_j$ with "support" on $\mathcal{R}_j$. (Remember that $\alpha_j$ is oriented.) Each $\theta_j$ is a nice *self-map* of $S_0$. Since the $\mathcal{R}_j$ are disjoint, the transformations $\theta_1,...,\theta_N$ commute.

Let $\{\nu_n\}_{n=1}^{\infty}$ be *any* sequence in $\mathbb{Z}^N$ with the property that:

$$\lim_{n\to\infty} \nu_{nj} = +\infty \text{ for } 1 \leq j \leq N. \tag{7.1}$$

Here $\nu_n \equiv (\nu_{n1},...,\nu_{nN})$. Put

$$\Phi_n = \theta_1^{\nu_{n1}} \circ \theta_2^{\nu_{n2}} \circ \cdots \circ \theta_N^{\nu_{nN}}. \tag{7.2}$$

Note that each pair $[S_0, \Phi_n]$ defines a new point $S_n$ in $T_g(S_0)$. The Riemann surfaces $S_0$ & $S_n$ differ *only* in their choice of Teichmüller marking.

Let $A_n$ be *that* lift of $\Phi_n$ which is locally the identity around $z_0$. Define $\lambda_n$, $\mu_n$, $\hat{\mu}_n$, $\varphi_n$ by analogy with §3. The symmetric extension of $A_n$ can still be denoted be $A_n$.

Let $\chi_n$ be the monodromy homomorphism of $F_n \equiv w^{\mu_n}$. The groups $\chi_n(\Gamma)$ are (all) quasi-Fuchsian.

By passing to a subsequence, we can (*and do*) assume that

$$\varphi_n \longrightarrow \varphi_\infty \tag{7.3}$$

in $\mathcal{A}_2(L,\Gamma)$. Cf. [7, pp. 590–593]. Let $F_\infty(z)$ be the *normalized* LP function corresponding to $\varphi_\infty(z)$. Let $\chi_\infty$ be its monodromy homomorphism. One automatically knows that:

(i) $F_n(z) \rightrightarrows F_\infty(z)$ on L compacta;

(ii) $F_\infty$ is schlicht for $z \in L$;

(iii) $\lim_{n\to\infty} \chi_n(T) = \chi_\infty(T)$ for every $T \in \Gamma$ [6];

(iv) $\chi_\infty(\Gamma)$ is a b-group with a simply-connected invariant component $F_\infty(L)$.

**THEOREM A.** The group $\chi_\infty(\Gamma)$ is a *regular* b-group [in $\partial T_g(S_0)$] with structure loops corresponding to $\{\alpha_1,\ldots,\alpha_N\}$.

In accordance with §3, the groups $\chi_n(\Gamma)$, $\chi_\infty(\Gamma)$ and quadratic differentials $\varphi_n$, $\varphi_\infty$ are *not* affected by continuous deformations in $\{\theta_1,\ldots,\theta_N\}$.

§8. In this section, we *prove* theorem A. The method we use will serve as a convenient stepping stone for theorem B.

Let $G = \chi_\infty(\Gamma)$ and $\Delta(G) = F_\infty(L)$.

To prove theorem A, we are at liberty to pass to further subsequences as need be.

We propose to apply a *normal families* argument to $F_n(z)$ over $U - \pi^{-1}(\mathfrak{T})$. To this end: consider the various components of $U - \pi^{-1}(\mathfrak{T})$. Each such component is an obvious "retract" of a corresponding component of $U - \pi^{-1}(\alpha)$. The associated pairing will be denoted by $\mathcal{A}_r \leftrightarrow \mathcal{A}$; the "r" is short for "retract." Write $\pi(\mathcal{A}_r) = S_{0r}^{(k)}$ when $\pi(\mathcal{A}) = S_0^{(k)}$. Note that there is an immediate analog of §5(g), and that $\Gamma_{\mathcal{A}_r} \equiv \Gamma_{\mathcal{A}}$.

Standard distortion theorems (on L) coupled with the schlichtness of $F_n$ allow one to control $F_n(z)$ over U. Cf. [11, pp. 39, 42, 178 (lemma 2)] and (3.1). But: $F_n(z)$ is holomorphic on each $\mathcal{A}_r$. We can now apply the *usual* normal families argument. Since further subsequences are no problem, we are free to assume that:

$$F_n \rightrightarrows F_\infty(z) \text{ on } U - \pi^{-1}(\mathfrak{T}) \text{ compacta.} \tag{8.1}$$

Note that our *earlier* use of $F_\infty$ was restricted to L.

By enlarging $\mathfrak{T}_j$ (or adjusting $\theta_j$), we can arrange things so that the convergence in (8.1) continues to hold a little ways beyond the relative boundary of any given $\mathcal{A}_r$. Cf. [11, p. 13 (Satz 2)]. This remark will be employed (*without further ado*) in several places below.

---

[6] in the *standard* topology for $PSL(2,\mathbb{C})$ (compare [5, p. 78])

**FACT 1.** $F_\infty(z) \not\equiv$ constant on each $\mathcal{A}_r$.

Proof. If not, there would exist a common fixpoint for all the elements of $\chi_\infty(\Gamma_{\mathcal{A}})$. But: $\chi_\infty(\Gamma)$ is a torsion-free Kleinian group. This leads to an immediate contradiction using [14, p. 94 (2E)] and §5(g). Remember that $\chi_\infty$ is an isomorphism. □

By applying a Hurwitz-Rouché argument [to something like $F_n(z) - F_n(z_1)$], we conclude that $F_\infty$ is hôlomorphic + schlicht + LP on $L \cup [\cup \mathcal{A}_r]$. The monodromy homomorphism is just $\chi_\infty$.

Let $\omega_n$ be the monodromy homomorphism of $A_n(z)$. The correspondence $\omega_n$ defines an obvious *isomorphism* $\Gamma \longrightarrow \Gamma$.

A moment's thought shows that $A_n$ coincides with a unique element of $\Gamma$ (say, $R_{n\mathcal{A}}$) on each $\mathcal{A}_r$. Cf. §5(d)(g|| sentence 3). There is *no* difficulty giving the actual formula once the first few $\mathcal{A}_r$ have been handled. Simply use the following (easily demonstrated) facts:

$$R_{nT(\mathcal{A})} = \omega_n(T) R_{n\mathcal{A}} T^{-1} \quad \text{for } T \in \Gamma; \tag{8.2}$$

$$R_{n\mathcal{A}_2} = R_{n\mathcal{A}_1} M_j^{\varepsilon \nu_{nj}} \quad \text{when } \mathcal{A}_1 \;\&\; \mathcal{A}_2 \text{ abut along } \gamma. \tag{8.3}$$

Here $\gamma$ is *oriented* in accordance with (the obvious) $\alpha_j$; $M_j$ is the associated element of $\mathcal{E}$; and

$$\varepsilon = \begin{Bmatrix} +1, & \text{if } \mathcal{A}_1 \text{ lies to the left of } \gamma \\ -1, & \text{if } \mathcal{A}_1 \text{ lies to the right of } \gamma \end{Bmatrix}.$$

Equation (8.2) shows (incidentally) that:

$$\omega_n(T) = R_{n\mathcal{A}} \, T \, R_{n\mathcal{A}}^{-1} \quad \text{for } T \in \Gamma_{\mathcal{A}}. \tag{8.4}$$

**FACT 2.** Let $M_j$ be any hyperbolic element of $\Gamma$ corresponding to $\alpha_j$. Then: $\chi_\infty(M_j)$ is parabolic.

Proof. Suppose not. Then $\chi_\infty(M_j)$ must be lôxodromic. Let $\gamma$ be that component of $\pi^{-1}(\alpha_j)$ which corresponds to $M_j$. Let $\mathcal{C}$ be the corresponding lift of $\mathcal{Z}_j$ (straddling $\gamma$).

Let $\mathcal{A}^+$ be that $\mathcal{A}$ which abuts on the *left-hand* shore of $\gamma$. Select $R_n \in \Gamma$ so that $A_n(z) \equiv R_n z$ for $z \in \mathcal{A}_r^+$. Write $\hat{A}_n = R_n^{-1} \circ A_n$.

Within $\mathcal{C}$, there is *no* difficulty visualizing the action of $\hat{A}_n$. Cf. (7.2) & (8.3).

Needless to say: $\hat{A}_n(\mathcal{C}) = \mathcal{C}$.

Let $\beta$ be any simple arc (in $\mathcal{R}_j$) connecting the two components of $\partial\mathcal{R}_j$. By lifting $\mathcal{R}_j - \beta$, we obtain an obvious "quadrilateral" $z_1z_2z_3z_4$ inside $\mathcal{C}$. Cf. figure 2.

Note that $\mathcal{C}$ is invariant under $M_j$. One can now draw $F_n(\partial\mathcal{C})$ and $F_\infty(\partial\mathcal{C})$. The "endpoints" are located at the fixpoints of $\chi_n(M_j)$ & $\chi_\infty(M_j)$.

Since $\chi_n(M_j) \longrightarrow \chi_\infty(M_j)$ in PSL(2,ℂ), the multiplier of $\chi_n(M_j)$ must tend to that of $\chi_\infty(M_j)$. Similarly for the fixpoints. The crescents $F_n(\partial\mathcal{C})$ will therefore tend nicely to $F_\infty(\partial\mathcal{C})$ as $n \longrightarrow \infty$. [7] Cf. (8.1) and the group action under $M_j$. The hypothesis that $\chi_\infty(M_j)$ is *lôxodromic* plays a *crucial role* here. [8]

Let $\mathcal{D}_n$ be the region bounded by $F_n(\partial\mathcal{C})$. Similarly for $\mathcal{D}_\infty$. The sets $\mathcal{D}_n, \mathcal{D}_\infty$ are Jordan regions. One knows that $\mathcal{D}_n \longrightarrow \mathcal{D}_\infty$ nicely.

We now exploit (the concept of) the *conformal module*. Cf. [15, pp. 15, 16, 22, 23]. The notions of extremal distance and conformal module are mutually reciprocal.

Let $M[F_n(z_1),F_n(z_2) \parallel F_n(z_3),F_n(z_4)]$ be the obvious module wrt $\mathcal{D}_n$. By [15, p. 29]:

$$\lim_{n\to\infty} M[F_n(z_1),F_n(z_2) \parallel F_n(z_3),F_n(z_4)] = M[F_\infty(z_1),F_\infty(z_2) \parallel F_\infty(z_3),F_\infty(z_4)]. \quad (8.5)$$

Cf. (8.1) and remember that $\mathcal{D}_n \longrightarrow \mathcal{D}_\infty$. The RHS of (8.5) is positive (since $F_\infty$ is schlicht and $\mathcal{D}_\infty$ is nice).

Recall, however, that $F_n \circ (\hat{A}_n)^{-1}$ is conformal on U for each finite n. Cf. §3(iv). It follows that:

$$M[F_n(z_1),F_n(z_2) \parallel F_n(z_3),F_n(z_4)] \quad (8.6)$$
$$= \text{the module } M[\hat{A}_n(z_1),\hat{A}_n(z_2) \parallel \hat{A}_n(z_3),\hat{A}_n(z_4)] \text{ wrt } \mathcal{C}.$$

But: $\hat{A}_n(z_1) = z_1$, $\hat{A}_n(z_2) = z_2$ while $\hat{A}_n(z_3) \longrightarrow \xi_{M_j}$, $\hat{A}_n(z_4) \longrightarrow \xi_{M_j}$ as $n \uparrow \infty$. Cf. (7.1). The RHS of (8.6) will therefore approach 0. Cf. [15, pp. 26 (12), 25 (7)].

This *contradicts* our earlier remark about the RHS of (8.5). □

It follows that $\chi_\infty(T)$ is parabolic for every $T \in \mathcal{E} - \{I\}$.

At this point: we do *not* yet know the complete list of structure loops for G.

To continue: let $\mathcal{A}'$ and $\mathcal{A}''$ be any two abutting components of $U - \pi^{-1}(\alpha)$.

---

[7] $\partial\mathcal{C}$ is understood to be oriented.

[8] One should also recall that $F_\infty$ is hôlomorphic + schlicht on $L \cup [U - \pi^{-1}(\mathcal{R})]$.

Let the common side be $\gamma$ [corresponding to $M \in \mathcal{E}$].

By virtue of the group action, it is easily seen that:

$$F_\infty(\mathcal{A}_r') \subsetneq \Omega(G) - F_\infty(L), \; F_\infty(\mathcal{A}_r'') \subsetneq \Omega(G) - F_\infty(L).$$

Let $\Omega'$ be the component of $\Omega(G) - \Delta(G)$ which contains $F_\infty(\mathcal{A}_r')$. Similarly for $\Omega''$. By §6(xa): $\Omega', \Omega''$ are quasidisks. (9)

By applying §5(f), it is easily demonstrated that the fixpoints of any (non--trivial) element of $\chi_\infty(\Gamma_{\mathcal{A}'})$ are contained in $\partial\Omega' \cap \partial[F_\infty(\mathcal{A}_r')]$. Remember that $\Gamma_{\mathcal{A}'} \equiv \Gamma_{\mathcal{A}_r'}$. Similarly for $\chi_\infty(\Gamma_{\mathcal{A}''})$.

Since $F_\infty(L)$ is an invariant component of G, we also have:

$$\Lambda(G) = \partial[F_\infty(L)] = \partial[\Delta(G)] \text{ apriori.} \tag{8.7}$$

Cf. [14, p. 105 (line 14)].

**FACT 3.** $\Omega' \neq \Omega''$. In fact: $\partial\Omega' \cap \partial\Omega'' = \{\text{the fixpoint of } \chi_\infty(M)\}$.

Proof. Let P be the fixpoint of $\chi_\infty(M)$. Let $\gamma_M$ be the hyperbolic axis of M in L. The closure of Jordan arc $F_\infty(\gamma_M)$ is simply $F_\infty(\gamma_M) \cup \{P\}$. We denote this loop by $\beta_M$.

Let $C$ be the *crescent* situated between $\mathcal{A}_r'$ and $\mathcal{A}_r''$. Let $\gamma_1$ & $\gamma_2$ be the two components of $\partial C$. The Jordan arcs $F_\infty(\gamma_1)$ & $F_\infty(\gamma_2)$ can be completed [just like $F_\infty(\gamma_M)$] by adjoining $\{P\}$.

The resulting curves intersect *only* at P. It is also apparent that:

$$F_\infty(\gamma_M) \subsetneq \Delta(G), \; F_\infty(\gamma_1) \subsetneq \Omega', \; F_\infty(\gamma_2) \subsetneq \Omega''.$$

Since $\Omega'$ is a quasidisk, the interior of $F_\infty(\gamma_1)$ is contained in $\Omega'$. Similarly for $F_\infty(\gamma_2)$.

By relabelling things, there is no loss of generality in assuming that $C$ looks like figure 3. A moment's thought shows that $F_\infty(\mathcal{A}_r')$ must be situated *exterior* to $F_\infty(\gamma_1)$. [Otherwise there would be points of $\partial\Omega'$ inside $F_\infty(\gamma_1)$. Cf. the remark prior to (8.7).] It follows that $F_\infty(\gamma_1)$ is oriented counter-clockwise. Similarly for $F_\infty(\gamma_2)$.

Suppose for a moment that $F_\infty(\gamma_2)$ were *interior* to $F_\infty(\gamma_1)$. Since $F_\infty$ is schlicht, the region $F_\infty(\mathcal{A}_r'')$ would have to be situated between $F_\infty(\gamma_1)$ & $F_\infty(\gamma_2)$. This again leads to points of $\Lambda(G)$ inside $\Omega'$.

It follows that $F_\infty(\gamma_1)$ & $F_\infty(\gamma_2)$ are mutually exterior as in figure 4.

But, now, it is *apparent* [by $\chi_\infty(M)$ invariance] that $\beta_M$ must separate $F_\infty(\gamma_1)$

---

9 For present purposes, it's enough to know that $\Omega'$ & $\Omega''$ are simply-connected.

and $F_\infty(\gamma_2)$. Since $F_\infty(\gamma_M) \subsetneq \Delta(G)$, we conclude that $\partial\Omega' \cap \partial\Omega'' = \{P\}$. □

**FACT 4.** In the above context: $G_{\Omega'} = \chi_\infty(\Gamma_{\mathcal{A}'})$.

Proof. Since $F_\infty(\mathcal{A}'_\Gamma) \subsetneq \Omega'$, it is trivially checked that $\chi_\infty(\Gamma_{\mathcal{A}'}) \subsetneq G_{\Omega'}$.

Let $\{\alpha_1,...,\alpha_{N+q}\}$ be the complete list of structure loops for G. Cf. §6 and [19, section 5]. Let $\mathcal{B}$ be *the* component of $U - \pi^{-1}(\alpha_1 \cup ... \cup \alpha_{N+q})$ which satisfies $G_{\Omega'} = \chi_\infty(\Gamma_{\mathcal{B}})$. Let $\mathcal{B}_1$ be *any* component of $U - \pi^{-1}(\alpha_1 \cup ... \cup \alpha_{N+q})$ inside $\mathcal{A}'$. By §5(i):

$$\chi_\infty(\Gamma_{\mathcal{B}_1}) \subsetneq \chi_\infty(\Gamma_{\mathcal{A}'}) \subsetneq G_{\Omega'} = \chi_\infty(\Gamma_{\mathcal{B}}). \tag{8.8}$$

Accordingly: $\Gamma_{\mathcal{B}_1} \subsetneq \Gamma_{\mathcal{B}}$, whereupon $\mathcal{B}_1 = \mathcal{B}$. Cf. §5(h). The inclusion signs in (8.8) can *now* be replaced by equalities. □

By letting the original $\mathcal{A}'$ vary, we also see that $q = 0$. Hence $\{\alpha_1,...,\alpha_N\}$ is the *complete* list of structure loops for G.

**FACT 5.** Every component of $\Omega(G) - \Delta(G)$ contains [and is *induced* by] a unique region $F_\infty(\mathcal{A}'_\Gamma)$.

Proof. Let $\Delta$ be any component of $\Omega(G) - \Delta(G)$. By §6, we can write $G_\Delta = \chi_\infty(\Gamma_{\mathcal{A}'})$ for a uniquely determined "grid" component $\mathcal{A}'$. Let $\Omega' \supseteq F_\infty(\mathcal{A}'_\Gamma)$ as before. By FACT 4, we obtain:

$$G_\Delta = \chi_\infty(\Gamma_{\mathcal{A}'}) = G_{\Omega'}$$

whereupon $\Delta = \Omega'$ by §6(xb).

The uniqueness is proved similarly. One uses §5(h). □

In connection with fact 5, a moment's thought shows that:

$$\chi_\infty(T)\Omega_1 = \Omega_2 \text{ iff } T(\mathcal{A}_1) = \mathcal{A}_2. \tag{8.9}$$

Here $T \in \Gamma$. It is now *apparent* that G is a regular b-group. Cf. [2, p. 212 (line −8)], [19, theorem 5], or §6 paragraph 2. □□

**§9.** Consider the situation of theorem A. Let $\mathcal{S}$ be the subsequence of $\{n\}$ that we *implicitly* passed to in §8. Cf. (8.1). Let $\Omega_{\mathcal{A}}$ be that component of $\chi_\infty(\Gamma)$ which contains $F_\infty(\mathcal{A}_\Gamma)$.

Throughout §9, we *tacitly* assume that $n \in \mathcal{S}$.

Recall (too) that $\bar{S}_0 = L/\Gamma$. Let $\{\bar{\alpha}_1,...,\bar{\alpha}_N\}$ be the mirror image of $\{\alpha_1,...,\alpha_N\}$.

In line with §6: let $\Delta(\tilde{G})$ be the simply-connected invariant component for *any* b-group $\tilde{G}$.

Before turning to theorem B, it is useful to review the Carathéodory kernel theorem [11, pp. 46–50] and its proof. Note that slight changes in the basic normalization do not affect anything. One simply looks at $M_n[f_n(\zeta)]$ where the *linear* maps $M_n$ ultimately approach I. One is also free to make an auxiliary conformal mapping of the $\zeta$ variable at the very start.

In the case at hand: we immediately obtain

$$F_\infty(L) = \ker F_n(L) \quad (\text{wrt } \infty) \tag{9.1}$$

$$F_\infty(\mathcal{A}_\Gamma) = \ker F_n(\mathcal{A}_\Gamma) \quad (\text{wrt } F_\infty(\xi)) \tag{9.2}$$

for any $\xi \in \mathcal{A}_\Gamma$. Some *preliminary* information about $F_\infty(\mathcal{A}_\Gamma)$ was obtained during the proof of fact 3. To gain a *better* hold on the topology, note that fact 4 yields:

$$\frac{\Omega_\mathcal{A}}{G_{\Omega_\mathcal{A}}} = \frac{\Omega_\mathcal{A}}{\chi_\infty(\Gamma_\mathcal{A})} \supseteq \frac{F_\infty(\mathcal{A}_\Gamma)}{\chi_\infty(\Gamma_\mathcal{A})} \cong \frac{\mathcal{A}_\Gamma}{\Gamma_\mathcal{A}}. \tag{9.3}$$

Cf. $S_{0_\Gamma}^{(k)}$ in §8, and figure 5.

The next step would be to determine $\ker F_n(U)$ (wrt $F_\infty(\xi)$). A trivial application of [12, §1.3 fact VII] shows that the desired kernel is (in any event) a subset of $\Omega_\mathcal{A}$. This fact also follows from (9.1) & (8.7).

**THEOREM B** (for $n \in \mathcal{S}$). There exists a regular b-group $\mathcal{X} \in \partial T_g(\bar{S}_0)$ and a family of Möbius transformations $\{r_\mathcal{A}\}$ such that:

(i) $\mathcal{X}$ has structure loops $\{\bar{\alpha}_1,\dots,\bar{\alpha}_N\}$;

(ii) $\Delta(\mathcal{X})/\mathcal{X} \cong U/\Gamma$;

(iii) $\Delta(r_\mathcal{A}\mathcal{X}r_\mathcal{A}^{-1})$ is a *proper* subset of $\Omega_\mathcal{A}$ [containing $F_\infty(\mathcal{A}_\Gamma)$];

(iv) $\Delta(r_\mathcal{A}\mathcal{X}r_\mathcal{A}^{-1}) = \ker F_n(U)$ wrt $F_\infty(\xi)$ for any $\xi \in \mathcal{A}_\Gamma$;

(v) the *exterior* of $\partial\Omega_\mathcal{A}$ is a noninvariant component of $r_\mathcal{A}\mathcal{X}r_\mathcal{A}^{-1}$;

(vi) all other components of $r_\mathcal{A}\mathcal{X}r_\mathcal{A}^{-1}$ are situated *within* $\Omega_\mathcal{A}$;

(vii) {the nodes of G along $\partial\Omega_\mathcal{A}$} $\equiv$ {the nodes of $r_\mathcal{A}\mathcal{X}r_\mathcal{A}^{-1}$ along $\partial\Omega_\mathcal{A}$};

(viii) $[r_\mathcal{A}\mathcal{X}r_\mathcal{A}^{-1}]_{\text{ext}(\partial\Omega_\mathcal{A})} = G_{\Omega_\mathcal{A}}$.

The limit set $\Lambda(G)$ thus serves as a kind of "crinkled up" mirror. In this regard: it is useful to stare at [12, figure 12E] for a few minutes.

Assertions (iii) & (iv) show that $S_n$ approaches $\partial T_g(S_0)$ in a "tangential fashion." Cf. [16, p. 291]. In the case of *non*tangential convergence, one would have $\Omega_{\mathcal{A}} = \ker F_n(U)$ wrt $F_\infty(\xi)$. Cf. [12, theorem 1.17]. Of course: we'd *also* have $S_n \longrightarrow \partial\mathcal{M}_g$ as unmarked surfaces. [The latter is clearly violated by the Dehn twists in (7.2).]

Since $\partial[F_n(U)] = F_n(\hat{\mathbb{R}})$, equation (iv) also proves the *existence* of a massive "pinching action" inside the ovals of figure 5 (top) as $n \longrightarrow \infty$. This action is counter-balanced by (9.1).

*Cf.* §10 [F][H] for some additional assertions.

Proof of theorem B. In accordance with §§3 and 7, set $V_n \equiv F_n \circ A_n^{-1}$. For each finite $n$, $V_n$ is a QC self-mapping of $\hat{\mathbb{C}}$. Let its monodromy homomorphism be $\mathcal{V}_n$. It is immediately apparent that:

(a) $\chi_n = \mathcal{V}_n \circ \omega_n$ [cf. §8 (near fact 1) for $\omega_n$];

(b) $V_n$ is hôlomorphic + schlicht + LP on U.

Let

$$\{V_n, z\} = \psi_n(z) \text{ for } z \in U. \tag{9.4}$$

In an obvious way: $\psi_n \in T_g(\bar{S}_0) \cap \mathcal{A}_2(U,\Gamma)$.

Let $\mathcal{A}'$ be that component of $U - \pi^{-1}(\alpha)$ which contains $z_0$. Note that $V_n(z) \equiv F_n(z)$ on $\mathcal{A}'_r$. By applying (8.1) and the finite-dimensionality of $\mathcal{A}_2(U,\Gamma)$, we immediately see that $\{\psi_n\}_{n\in\mathcal{S}}$ converges. (10) Denote the limit function by $\psi_\infty$. We now find that:

(I) $V_n(z) \rightrightarrows V_\infty(z)$ on U compacta;

(IIa) $V_\infty$ is schlicht on U;

(IIb) $\{V_{\infty,z}\} = \psi_\infty(z)$.

Let $\mathcal{V}_\infty$ be the monodromy homomorphism of $V_\infty$. In accordance with [7, pp. 590–593]:

(III) $\lim_{n\to\infty} \mathcal{V}_n(T) = \mathcal{V}_\infty(T)$ for every $T \in \Gamma$;

(IV) $\mathcal{V}_\infty(\Gamma)$ is a b-group with simply-connected invariant component $V_\infty(U)$.

Incidentally: note that $V_n$ is uniformly bounded on U. Cf. the reference to [11] near (8.1) and observe that $V_n(L) \equiv F_n(L)$.

Let $\bar{\mathcal{R}}_j$, $\bar{\mathcal{R}}$, $\bar{\mathcal{A}}_r$, $\bar{\mathcal{A}}$ be the obvious counterparts of $\mathcal{R}_j$, $\mathcal{R}$, $\mathcal{A}_r$, $\mathcal{A}$. Recall that $A_n$ coincides with a unique transformation $R_{n\mathcal{A}} \in \Gamma$ on each $\mathcal{A}_r \cup \bar{\mathcal{A}}_r$. Cf. §8 after fact 1.

---

10 This fact can also be proved by combining (8.1) with Nehari's estimate & Vitali's theorem. Cf. [3, p. 126] and [11, p. 13 (Satz 2)].

By construction:

$$F_n(z) = V_n(R_{n\mathcal{A}}\,z) = \mathcal{V}_n(R_{n\mathcal{A}})\; V_n(z) \tag{9.5}$$

$$V_n(z) = \mathcal{V}_n(R_{n\mathcal{A}}^{-1})\; F_n(z) \tag{9.6}$$

for $z \in \mathcal{A}_r \cup \bar{\mathcal{A}}_r$. By taking $z$ in $\mathcal{A}_r$ and combining (I) with (8.1), we immediately establish that $\mathcal{V}_n(R_{n\mathcal{A}})$ converges in $PSL(2,\mathbb{C})$ for *every* choice of $\mathcal{A}$. Let

$$W_{\mathcal{A}} \equiv \lim_{n\to\infty} \mathcal{V}_n(R_{n\mathcal{A}}). \tag{9.7}$$

We can now take $z \in \bar{\mathcal{A}}_r$ in (9.6). It follows that

$$V_n(z) \rightrightarrows V_\infty(z) \quad \text{on} \quad L - \pi^{-1}(\bar{\mathfrak{X}}) \quad \text{compacta} \tag{9.8}$$

where $V_\infty(z) \equiv W_{\mathcal{A}}^{-1}[F_\infty(z)]$ on $\mathcal{A}_r$.

There is no difficulty extending (9.8) as in the remark following (8.1). (The essential thing is to remain on a compact subset of L.)

The identity $V_\infty(z) \equiv W_{\mathcal{A}}^{-1}[F_\infty(z)]$ also holds on $\bar{\mathcal{A}}_r$. In other words: $F_\infty(z) \equiv W_{\mathcal{A}}[V_\infty(z)]$. It is rather striking how *simple* the final limit in (8.1) has turned out to be... [11]

In any event: we can now repeat (with $V_n$) the reasoning that was used earlier in §8 with regard to $F_n$ and $\chi_n$. [L & U are switched and the normalization (at $z_0$) is a bit different, but these changes are inconsequential.]

Write:

$$\mathfrak{X} = \mathcal{V}_\infty(\Gamma) \;\text{ and }\; \Delta(\mathfrak{X}) = \mathcal{V}_\infty(U) \;\; [\text{in accordance with (IV)}]. \tag{9.9}$$

Assertions (i) & (ii) are (then) self-evident.

The *next* task is to verify (iii)–(viii) for $\mathcal{A} = \mathcal{A}'$. This is quite easy. In the first place:

$$V_n(U) = F_n(U), \;\; R_{n\mathcal{A}'} = I, \;\; V_n(z) \equiv F_n(z) \;\text{ on }\; \mathcal{A}'_r. \tag{9.10}$$

One can now apply the Carathéodory kernel theorem to $V_n(z)$ over U. This gives (iv) with $r_{\mathcal{A}'} = W_{\mathcal{A}'} = I$. (Let $\Omega' \equiv \Omega_{\mathcal{A}'}$ for notational simplicity.)

---

[11] Note too that $\mathcal{V}_\infty(T) \equiv W_{\mathcal{A}}^{-1}\, \chi_\infty(T)\, W_{\mathcal{A}}$ for $T \in \Gamma_{\mathcal{A}}$.

To go further: note that $\mathcal{V}_n \equiv \chi_n$ on $\Gamma_{\mathcal{A}'}$. Cf. (9.10) and recall that $\Gamma_{\mathcal{A}'_r} \equiv \Gamma_{\mathcal{A}'}$. Cf. also equation (a) near (9.4).

Let J' be that component of $\mathcal{X}$ which contains $V_\infty(\bar{\mathcal{A}}'_r) \equiv F_\infty(\bar{\mathcal{A}}'_r)$. In accordance with §8:

$$\mathcal{X}_{J'} = \mathcal{V}_\infty(\Gamma_{\bar{\mathcal{A}}'}) = \mathcal{V}_\infty(\Gamma_{\mathcal{A}'}) = \chi_\infty(\Gamma_{\mathcal{A}'}) = G_{\Omega'}. \tag{9.11}$$

But:

$$\partial\Omega' = [\text{closure of the set of fixpoints for } G_{\Omega'}],$$
$$\partial J' = [\text{closure of the set of fixpoints for } \mathcal{X}_{J'}].$$

Cf. §6(xa) and [14, p. 104]. Accordingly $\partial J' = \partial\Omega'$.

The region $V_\infty(\bar{\mathcal{A}}'_r) \equiv F_\infty(\bar{\mathcal{A}}'_r)$ contains $\infty$ by virtue of (3.1). It follows immediately that $J' = \text{ext}(\partial\Omega')$.

With this in hand: assertions (v), (vi), (viii), (iii) are trivial.

To obtain (vii), remember that the nodes of G along $\partial\Omega'$ are just the *fixpoints* of $\chi_\infty(M)$ with $M \in \Gamma_{\mathcal{A}'} \cap [\mathcal{E} - \{I\}]$. Similarly for $\mathcal{X}$ and $\partial J'$. But: $\chi_\infty \equiv \mathcal{V}_\infty$ on $\Gamma_{\mathcal{A}'}$.

It *remains* to treat the other choices of $\mathcal{A}$. To this end: select any component $\mathcal{B}$ of $U - \pi^{-1}(\alpha)$. We propose to *renormalize* $A_n$. This is done by writing

$$\hat{A}_n(z) \equiv R_{n\mathcal{B}}^{-1}[A_n(z)]. \tag{9.12}$$

Compare fact 2. By construction: $\hat{A}_n(z) \equiv z$ on $\mathcal{B}_r \cup \bar{\mathcal{B}}_r$. Set

$$\hat{V}_n \equiv F_n \circ (\hat{A}_n)^{-1}. \tag{9.13}$$

Note that $\hat{V}_n(z) = V_n[R_{n\mathcal{B}}\, z] = \mathcal{V}_n(R_{n\mathcal{B}})[V_n(z)]$ for $z \in \hat{\mathbb{C}}$.

We can now repeat the earlier arguments. One quickly finds that

$$\hat{A}_n(z) = \hat{R}_{n\mathcal{A}}\, z \text{ for } z \in \mathcal{A}_r \cup \bar{\mathcal{A}}_r,$$
$$\hat{R}_{n\mathcal{A}} = R_{n\mathcal{B}}^{-1} R_{n\mathcal{A}},$$
$$\hat{V}_n(z) = F_n(z) \text{ for } z \in \mathcal{B}_r,$$
$$\hat{\mathcal{V}}_n(K) \equiv \chi_n(K) \text{ for } K \in \Gamma_{\mathcal{B}},$$
$$\hat{\mathcal{V}}_n(T) = \mathcal{V}_n(R_{n\mathcal{B}})\, \mathcal{V}_n(T)\, \mathcal{V}_n(R_{n\mathcal{B}}^{-1}) \text{ for } T \in \Gamma,$$

$$\hat{W}_{\mathcal{A}} \equiv \lim_{n\to\infty} \hat{\mathcal{V}}_n(\hat{R}_{n\mathcal{A}}) = W_{\mathcal{A}} W_{\mathcal{B}}^{-1}, \tag{9.14}$$
$$\hat{V}_\infty(z) = W_{\mathcal{B}}[V_\infty(z)],$$
$$\hat{\mathcal{V}}_\infty(T) = W_{\mathcal{B}} \mathcal{V}_\infty(T)\, W_{\mathcal{B}}^{-1} \text{ for } T \in \Gamma,$$
$$\hat{\mathcal{X}} = W_{\mathcal{B}} \mathcal{X}\, W_{\mathcal{B}}^{-1}.$$

The Carathéodory kernel theorem immediately yields:

$$\Delta(\hat{\mathcal{X}}) = \hat{V}_\infty(U) = \ker F_n(U) \ \ (\text{wrt } F_\infty(\xi))$$

for $\xi \in \mathcal{B}_r$. Cf. $(\widehat{9.10})$. This proves (iv) with $r_{\mathcal{B}} = W_{\mathcal{B}}$.

Assertions (v), (vi), (viii), (iii), (vii) are then derived exactly as before. In the *analog* of (9.11), note that $\hat{V}_\infty(\bar{\mathcal{B}}_r) \equiv F_\infty(\bar{\mathcal{B}}_r)$ is a subset of $F_\infty(L) \equiv \Delta(G)$. As such: it cannot intersect $\Omega_{\mathcal{B}}$.

This completes the proof of theorem B. □ □

**§10.** Theorem B and its proof suggest a variety of further questions which deserve closer examination. Several of these will ([12]) be treated in a subsequent publication.

In this section: we use a series of 10 remarks to highlight some of the more immediate possibilities.

[A] Theorems A and B are both easily extended to the case of mappings having the form

$$\Phi_n^* \equiv \Phi_n \circ f, \tag{10.1}$$

where $f$ is a (given) QC transformation of $S_0^*$ *onto* $S_0$. It is understood here that $f$ preserves the Teichmüller marking.

The difficulties caused by this (very natural) change are mainly notational. The simplest approach is to make a systematic change-of-variable $z \equiv A(z^*)$, where $A$ is the (symmetrically extended) lift of $f$. The identity (10.1) immediately becomes $A_n^*(z^*) \equiv A_n(z)$. The normal families argument for $F_n^*(z^*)$ on $U - \pi^{-1}(\mathcal{X}^*)$ will *now* require QC mappings, but this is no problem. Cf. [15, pp. 30 (5.1), 31 (5.2), 68 (3.8) (3.9), 77 (Satz 5.3), 79 (Satz 5.4)]. Note that $V_n \equiv F_n^* \circ (A_n^*)^{-1}$ still "lives" on

[12] most likely

U mod $\Gamma$. For this reason: $\psi_n \in \mathcal{A}_2(U,\Gamma)$, while $\varphi_n^* \in \mathcal{A}_2(L,\Gamma^*)$.

[B] In [24], [25, lemma 1], Maskit obtains parabolicity by requiring that a certain family of closed loops on $S_n$ (the underlying RS) have extremal length $\ell_n \to 0$. The parabolics in theorem A arise because the "associated" extremal *distance* $\mathcal{D}_n$ tends to $\infty$. Smallness of $\ell_n$ implies largeness of $\mathcal{D}_n$, but not conversely.

There is an (apparent) allusion to this in [25, p. 663 (line 14)].

[C] It is possible to establish parabolicity in theorem A *without* any use of extremal distance whatsoever.

To illustrate the idea, suppose that $N = 1$. Let $\mathcal{A}'$ be the "grid" component which contains $z_0$. Let $\mathcal{A}''$ be one of its neighbors. After reaching (8.4), switch to §9 and review the arguments up to (9.7). Observe, however, that

$$R_{n\mathcal{A}''} = M_1^{\varepsilon\nu_{n1}}$$

by (8.3). We thus find [using $M_1 \in \Gamma_{\mathcal{A}''} \cap \Gamma_{\mathcal{A}'}$] that

$$W \equiv \lim_{n\to\infty} \chi_n(M_1^{\nu_{n1}}) \text{ exists in } PSL(2,\mathbb{C}) \text{ for } n \in \mathcal{S}.$$

But: $\lim_{n\to\infty} \chi_n(M_1) = \chi_\infty(M_1)$. It follows *immediately* that $\chi_\infty(M_1)$ must be parabolic.

By substituting back, we then obtain:

$$\lim_{n\to\infty} \lambda_n^{\nu_{n1}} = 1 \quad (n \in \mathcal{S}) \tag{10.2}$$

where $\lambda_n$ is the multiplier of $\chi_n(M_1)$. Note that $W$ commutes with $\chi_\infty(M_1)$. [(10.2) is just the assertion that $\mathrm{Tr}^2(W) = 4$].

The *key* ingredient in this argument is (in brief) Fact 1.

Equation (10.2) gives *new* meaning to the phrase "tangential convergence to $\partial T_g(S_0)$."

To treat $N \geq 2$, use the renormalization trick near (9.12).

[D] The ovals in figure 5 (top) are mapped *onto* one another under the action of $\chi_\infty(\Gamma_{\mathcal{A}})$. Cf. (9.3). The stabilizer of any particular oval is a cyclic group having an obvious APT as generator.

Let $\mathcal{A}$ correspond to $S_0^{(k)}$ and $T_k \equiv \Omega_{\mathcal{A}}/G_{\Omega_{\mathcal{A}}}$. Let P be any node of G along $\partial\Omega_{\mathcal{A}}$ with stabilizer $[\chi_\infty(M)]$. Point P determines a puncture $\mathcal{P}$ on $T_k$. Let $\mathcal{N}$

be that component of $r_{\mathcal{A}} \mathcal{X} r_{\mathcal{A}}^{-1}$ which is attached to $\mathrm{ext}(\partial\Omega_{\mathcal{A}})$ at P. Cf. thm B(v)(vii). By thm B(iii): $\mathcal{N}$ is a proper subset of the P-*oval*. [Cf. figure 6.]

Recall, however, that $\mathcal{N}$ & $\Omega_{\mathcal{A}}$ have identical horocycles at P (of sufficiently small radius). Cf. [4, p. 416 (lemma 1)]. It is now easily seen that $\mathcal{N}/[\chi_\infty(M)]$ is conformally embedded as a punctured quasidisk on $T_k$ (centered at $\mathcal{P}$).

By varying P, we conclude that the *projection* of $\Delta(r_{\mathcal{A}} \mathcal{X} r_{\mathcal{A}}^{-1})$ on $T_k$ is bounded away from the punctures and that it is topologically similar to $S_0^{(k)}$ (notwithstanding the *jaggedness* of its boundary).

[E] The transformations $W_{\mathcal{A}}$ are rather important. The simplest way to express $W_{\mathcal{A}}$ as a $\chi_n$ limit (as opposed to $\mathcal{V}_n$) is to use (9.14) and a succession of "transfer" matrices $W_{\mathcal{A}_2} W_{\mathcal{A}_1}^{-1}$ starting out with $\mathcal{A}_1 = \mathcal{A}'$. Compare (8.3) and remark C.

Incidentally: note that (8.2) yields

$$W_{T(\mathcal{A})} = \chi_\infty(T)\, W_{\mathcal{A}}\, \mathcal{V}_\infty(T^{-1}) \text{ for } T \in \Gamma. \tag{10.3}$$

This shows that $W_{T(\mathcal{A}_2)} W_{T(\mathcal{A}_1)}^{-1} = \chi_\infty(T)[W_{\mathcal{A}_2} W_{\mathcal{A}_1}^{-1}]\chi_\infty(T)^{-1}$, which agrees with (8.3) & (8.4).

[F] Let $\mathcal{N}_{\bar{\mathcal{A}}}$ be that component of $\mathcal{X}$ which contains $V_\infty(\bar{\mathcal{A}}_\Gamma)$. Cf. (the analog of) fact 5. [13] The following *facts* are easily demonstrated during the proof of theorem B.

(ix) $W_{\mathcal{A}}^{-1}[\mathrm{ext}(\Omega_{\mathcal{A}})] = \mathcal{N}_{\bar{\mathcal{A}}}$ [compare (v)];

(x) $W_{\mathcal{A}}[\mathrm{ext}(\mathcal{N}_{\bar{\mathcal{A}}})] = \Omega_{\mathcal{A}}$;

(xi) $G_{\Omega''} = W_{\mathcal{A}} \mathcal{X}_{\mathcal{N}''} W_{\mathcal{A}}^{-1}$ where $\Omega'' \equiv \Omega_{\mathcal{A}}$ and $\mathcal{N}'' \equiv \mathcal{N}_{\bar{\mathcal{A}}}$;

(xii) ker $V_n(L)$ wrt $V_\infty(\eta) = W_{\mathcal{A}}^{-1}[\Delta(G)]$ for every $\eta \in \bar{\mathcal{A}}_\Gamma$;

(xiii) $W_{\mathcal{A}}^{-1}[\Delta(G)]$ is a proper subset of $\mathcal{N}_{\bar{\mathcal{A}}}$.

Assertion (xii) is a kind of "dual" to (iv). Since $V_n(L) = F_n(L)$, it also serves to *augment* (9.1). [14]

To *prove* (xii), simply apply the CKT (over L) to

$$f_n(z) \equiv \mathcal{V}_n\,(R_{n\mathcal{A}}^{-1})\, F_n(z).$$

---

[13] And *remember* that $R_{n\mathcal{A}}$ & $W_{\mathcal{A}}$ both reduce to I when $\mathcal{A} = \mathcal{A}'$.

[14] Recall that $\mathcal{N}_{\bar{\mathcal{A}}} \subsetneq \Omega_{\mathcal{A}'}$ for every $\mathcal{A} \neq \mathcal{A}'$.

Since $\mathcal{V}_n \equiv \chi_n \circ \omega_n^{-1}$, one automatically knows that $V_n(L) = F_n(L) = f_n(L)$.

[G] Assertions (ix) & (x) readily show that

$$W_{\mathcal{A}}^k[\Delta(G) \cup \Delta(\mathcal{X})] \cap [\Delta(G) \cup \Delta(\mathcal{X})] = \emptyset \tag{10.4}$$

whenever $k \neq 0$ and $\mathcal{A} \neq \mathcal{A}'$. Cf. ([14]) and note (for instance) that

$$\Omega_{\mathcal{A}} = W_{\mathcal{A}}[\text{ext}(\mathcal{N}_{\bar{\mathcal{A}}})] \supseteq W_{\mathcal{A}}^2[\text{ext}(\mathcal{N}_{\bar{\mathcal{A}}})] \supseteq W_{\mathcal{A}}^3[\text{ext}(\mathcal{N}_{\bar{\mathcal{A}}})] \supseteq \dots . \tag{10.5}$$

As an immediate corollary, we see that the abelian group $[W, \chi_\infty(M_1)]$ in C is *both* discrete and of rank 2. Similarly for $W_{\mathcal{A}_2} W_{\mathcal{A}_1}^{-1}$ in E.

[H] ("A kaleidoscope of crinkled mirrors") Motivated by [13, theorem 4.9] one would like to know exactly which portions of $\hat{\mathbb{C}}$ are *eventually* contained in $F_n(L)$ – and which in $F_n(U)$. ([15])

The word "eventually" more-or-less means that we had better restrict ourselves to compacta. Even in this generality, the problem is probably hopeless. A more reasonable aim would be to shoot for something like a Carathéodory kernel; e.g. the set of all points having *neighborhoods* eventually contained in $F_n(L)$ (or $F_n(U)$).

Assertions (9.1), (9.2), (iv), (xii) represent *steps* in this direction.

The aforementioned theorem of Jørgensen/Marden gives a rather satisfactory solution to this problem *provided* one knows in advance that $\chi_n(\Gamma)$ tends geometrically ([16]) to a geometrically finite Kleinian group $\mathcal{G}$.

To avoid this difficulty, a more direct approach seems preferable [and, *indeed*, possible]. The basic idea is to combine our earlier techniques with a slight extension of the Klein-Maskit combination theorems [23].

The following outline shows how the case $N = 1$ should go.

There are two possibilities:

(a) $\alpha_1$ divides $S_0$;

(b) $\alpha_1$ does not divide $S_0$.

In either case, there is *no* difficulty applying [23, theorem I] to construct the group $G_0 \equiv [G, \mathcal{X}]$ amalgamated over $G_{\Omega'} = \mathcal{X}_{\mathcal{N}'}$. ([17])

---

15 We continue to assume that $n \in \mathcal{S}$.

16 Or: polyhedrally [13, proposition 3.10].

17 Compare: [20, p. 249] and [18, p. 456 (3b)].

Let $\mathcal{A}''$ be any neighbor of $\mathcal{A}'$. The inequivalent components of $G_0$ are then:

$$\Delta(G), \Delta(\mathcal{X}), \Omega'', \mathcal{N}'' \text{ in case (a)},$$
$$\Delta(G), \Delta(\mathcal{X}) \text{ in case (b)}.$$

$\mathcal{N}''$ is an abbreviation for $\mathcal{N}_{\mathcal{A}''}$. Similarly for $\Omega''$. In case (b), note that the transformations $W_{\mathcal{A}}$ *already* belong to $G_0$ by (10.3).

In case (a), one knows that $\mathcal{N}'' \subsetneq \text{int}(\partial\Omega')$. Cf. thm B(vi). It is now tempting to form $[G_0, W_{\mathcal{A}''}]$ in accordance with [23, theorem II]. Unfortunately: $\Omega''$ and $\mathcal{N}''$ are tangent at the fixpoint of $\chi_\infty(M_1) \equiv \mathcal{V}_\infty(M_1)$. Cf. fact 5, figure 6, and remark [D]. This violates hypothesis (k), so theorem II *cannot* be applied as it stands.

It is presumably possible (however) to *modify* theorem II to take account of limiting situations such as this. [Conclusion 5 will need some adjustment.]

In any event: by doing a bit of "image-chasing" with the *normal* forms of $[G_0, W_{\mathcal{A}''}]$ words {cf. [23], (xi), (10.5)}, one finds at least that $[G_0, W_{\mathcal{A}''}]$ is Kleinian with inequivalent components including $\Delta(G)$ and $\Delta(\mathcal{X})$. One naturally expects that this is the complete list. ([18]) ([19])

We now set:

$$\mathcal{G} = \left\{ \begin{array}{ll} [G_0, W_{\mathcal{A}''}] & \text{in case (a)} \\ G_0 & \text{in case (b)} \end{array} \right\}. \tag{10.6}$$

The group $\mathcal{G}$ will be Kleinian with 2 inequivalent components: $\Delta(G)$ & $\Delta(\mathcal{X})$. The limit set $\Lambda(\mathcal{G})$ will be nowhere dense. Cf. [14, p. 104].

**KEY LEMMA.** Let $T_n$ belong to $\Gamma$ and suppose that $\chi_n(T_n) \to Q$ in $PSL(2,\mathbb{C})$ over a certain *sub*sequence of $\mathcal{S}$. Then: $Q \in \mathcal{G}$.

Proof. Suppose not. By multiplying $Q$ by an appropriate element of $\mathcal{G}$, we can assume WLOG that $Q[F_\infty(L)]$ intersects $\Delta(G) \cup \Delta(\mathcal{X})$. But we can also apply the CKT to $F_n(T_n z)$ for $z \in L$. One immediately finds points of either $F_n(L)$ or $F_n(U)$ where they don't belong. ([20]) □

This result *proves* that $\chi_n(\Gamma)$ tends *geometrically* to $\mathcal{G}$. An additional moment's thought [using CKT for both $F_n$ & $V_n$] shows that the "L and U kernels"

---

[18] [Added in Proof.] The recent book of Maskit [27] contains *significantly* improved version of theorem II. Cf. page 161. This result shows that $[G_0, W_{\mathcal{A}''}]$ and $\Delta(G)$, $\Delta(\mathcal{X})$ behave exactly as expected.

[19] Compare: [23, thm II (conclusion 3)], [20, p. 249 (II3)], and [18, p. 456 (3a)].

[20] Cf. (9.1) and thm B(iv) for $\mathcal{A} = \mathcal{A}'$. Remember that $F_n$ is schlicht.

are simply the $\mathcal{G}$-orbits of $\Delta(G)$ and $\Delta(\mathfrak{X})$, respectively. (21)

With regard to our original question, we have *now* accounted for *all* points of $\hat{\mathbb{C}}$ except a closed set of measure 0 [namely $\Lambda(\mathcal{G})$].

The case $N \geq 2$ is similar.

[I] By virtue of (10.3): the group generated by $\chi_\infty(\Gamma)$ and the $W_{\mathcal{A}}$ is *identical* with the one generated by $\chi_\infty(\Gamma)$, $\mathcal{V}_\infty(\Gamma)$, $W_{\mathcal{A}}$. There is *thus* a certain resemblance here (when $N = 3g - 3$) with the construction used by Earle and Marden in [16, p. 288], [17, sec. 8].

In this regard: see also the generalization of [H] suggested by [A].

[J] It would be quite interesting to determine the precise relation between our techniques and those of [1, pp. 33–34], [16, p. 288], [17, sec. 8], [26]. Compare: [8], [9, pp. 143, 156, 160–162], [28, sec. 5].

## REFERENCES


[1] W. Abikoff, Degenerating families of Riemann surfaces, Ann. of Math. 105 (1977), 29–44.

[2] W. Abikoff, On boundaries of Teichmüller spaces and on Kleinian groups III, Acta Math. 134 (1975), 211–237.

[3] L. Ahlfors, Lectures on Quasiconformal Mappings, Van Nostrand, 1966.

[4] L. Ahlfors, Finitely generated Kleinian groups, Amer. J. Math. 86 (1964), 413–429 and 87 (1965), 759.

[5] A. Beardon, The Geometry of Discrete Groups, Springer-Verlag, 1983.

[6] L. Bers, Uniformization, moduli, and Kleinian groups, Bull. London Math. Soc. 4 (1972), 257–300.

[7] L. Bers, On boundaries of Teichmüller spaces and on Kleinian groups I, Ann. of Math. 91 (1970), 570–600.

[8] L. Bers, On iterates of hyperbolic transformations of Teichmüller space, Amer. J. Math. 105 (1983), 1–11.

[9] L. Bers, Finite dimensional Teichmüller spaces and generalizations, Bull. Amer. Math. Soc. (N.S.) 5 (1981), 131–172.

[10] M. Dehn, Papers on Group Theory and Topology (translated by J. Stillwell), Springer-Verlag, 1987. [See also: W. Lickorish, Proc. Cambr. Phil. Soc. 60 (1964), 769 and 62 (1966), 679.]

[11] G. M. Golusin, Geometrische Funktionentheorie, VEB Deutscher Verlag der Wiss., Berlin, 1957.

[12] D. A. Hejhal, Regular b-groups, degenerating Riemann surfaces, and spectral


21 Use [4, p. 414 (line 5)] to handle $\Lambda(\mathcal{G})$. One *finds* that neighborhoods of $\Lambda(\mathcal{G})$-points necessarily contain points of both $F_n(U)$ and $F_n(L)$ for large n.

theory, Chalmers Univ. Tech. Report 1987-19, 135 pp.

[13] T. Jørgensen and A. Marden, Algebraic and geometric convergence of Kleinian groups, MSRI (Berkeley) Preprint 10319-86, 39 pp.

[14] J. Lehner, Discontinuous Groups and Automorphic Functions, AMS Math. Surveys No. 8, 1964.

[15] O. Lehto and K. I. Virtanen, Quasikonforme Abbildungen, Springer-Verlag, 1965.

[16] A. Marden, Geometrically finite Kleinian groups and their deformation spaces, in Discrete Groups and Automorphic Functions (ed. by W. J. Harvey), Academic Press, 1977, pp. 259–293.

[17] A. Marden, Geometric complex coordinates for Teichmüller space, to appear in Proc. of the 1986 Conf. on Math. Aspects of String Theory (at UCSD), 14 pp.

[18] A. Marden, The geometry of finitely generated Kleinian groups, Ann. of Math. 99 (1974), 383–462.

[19] B. Maskit, On boundaries of Teichmüller spaces and on Kleinian groups II, Ann. of Math. 91 (1970), 607–639.

[20] B. Maskit, Decomposition of certain Kleinian groups, Acta Math. 130 (1973), 243–263.

[21] B. Maskit, On the classification of Kleinian groups II, Acta Math. 138 (1977), 17–42.

[22] B. Maskit, On the classification of function groups, in Discrete Groups and Automorphic Functions (ed. by W. J. Harvey), Academic Press, 1977, pp. 349–361.

[23] B. Maskit, On Klein's combination theorem III, *in* Annals of Math. Studies No. 66, 1971, pp. 297–316.

[24] B. Maskit, Comparison of hyperbolic and extremal lengths, Ann. Acad. Sci. Fenn. Ser. A I Math. 10 (1985), 381–386.

[25] B. Maskit, Parabolic elements in Kleinian groups, Ann. of Math. 117 (1983), 659–668.

[26] B. Maskit, Moduli of marked Riemann surfaces, Bull. Amer. Math. Soc. 80 (1974), 773–777.

[27] B. Maskit, Kleinian Groups, Springer-Verlag, 1988.

[28] W. Thurston, Three-dimensional manifolds, Kleinian groups, and hyperbolic geometry, Bull. Amer. Math. Soc. (N.S.) 6 (1982), 357–381.

## FIGURES

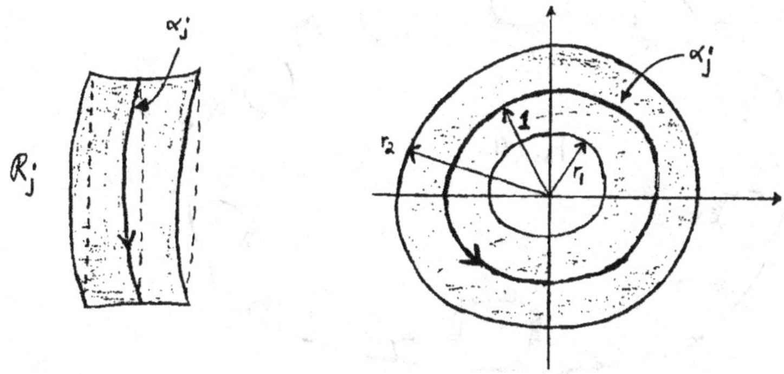


$$(r,\theta) \longrightarrow \left(r,\ \theta + \frac{r-r_1}{r_2-r_1}2\pi\right)$$

prototypical example of a Dehn twist

FIG. 1

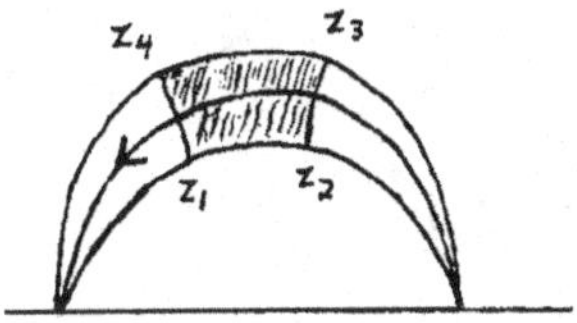


FIG. 2

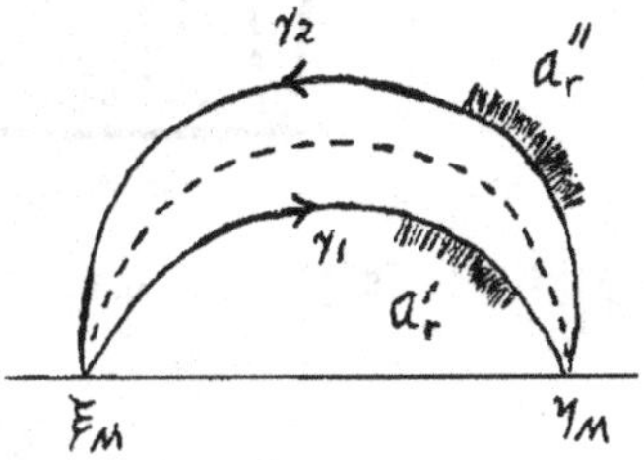


FIG. 3

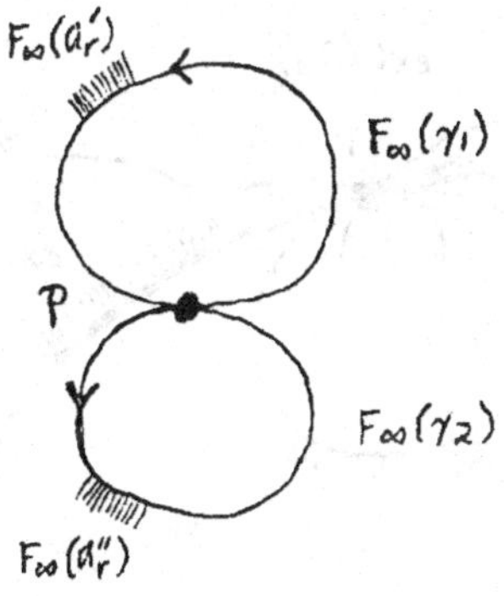


FIG. 4

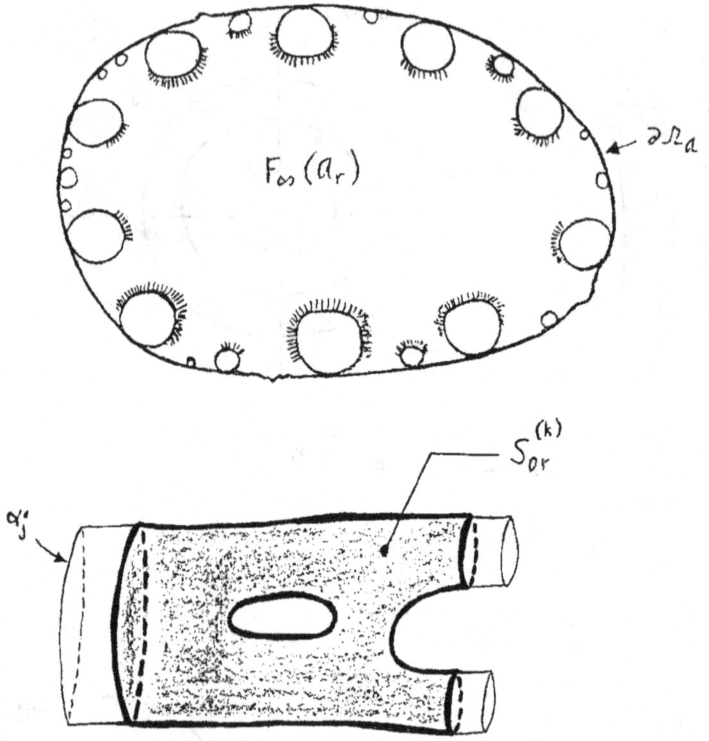


$S_{or}^{(k)}$ has universal covering surface $F_\infty(\mathcal{A}_r)$

FIG. 5

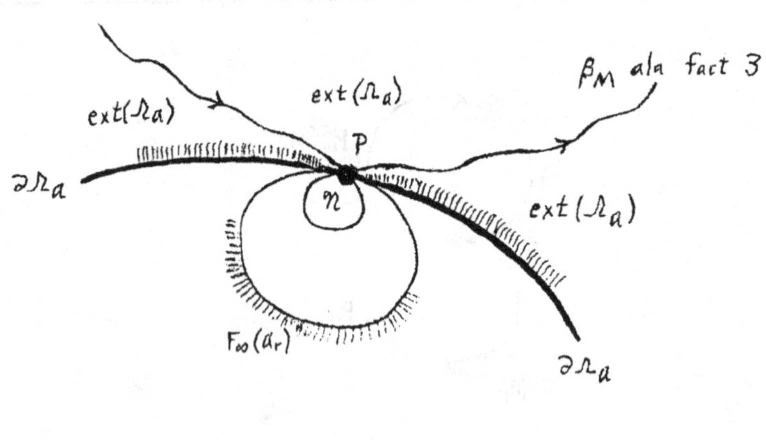


FIG. 6

# METRIC BOUNDARY CONDITIONS FOR PLANE DOMAINS [1]

**David A. Herron**
Department of Mathematical Sciences, University of Cincinnati
Cincinnati, Ohio 45221-0025, U.S.A.

## 1. Introduction

Let $(X,d)$ be a metric space and let a family of subdomains of $X$ with non-empty boundary be given. Suppose that for each domain $D$ in this family there exists a complete metric $\rho_D$ defined on $D$. Fix such a domain $D$ and fix a point $x_0$ in $D$. Often information describing the rate of growth of $\rho_D(x,x_0)$ as $d_D(x) = d(x,\partial D) \to 0$ can be used to ascertain geometric, topological, or function-theoretic properties of $\partial D$ or of $D$. We say that $D$ satisfies a *ρ–boundary condition* with *exponent* $\alpha$, *center* $x_0$ and *constant* $c$, denoted $D \in \rho\text{-BC}(\alpha,x_0,c)$, if $x_0 \in D$, $\alpha > 0$, $c > 0$ and

$$\rho_D(x,x_0) \le \alpha \log(c/d_D(x)) \quad \text{for all } x \in D. \tag{1.1}$$

We also use the notations $\rho$-BC, $\rho$-BC$(\alpha)$, $\rho$-BC$(\alpha,x_0)$; e.g., $\rho$-BC$(\alpha)$ is the class of all domains which satisfy (1.1) for some $x_0 \in D$, $c > 0$.

Becker and Pommerenke considered simply connected bounded domains in the complex plane $\mathbb{C}$ and *Poincaré's hyperbolic metric* h. They showed that the domains which satisfy a *hyperbolic boundary condition* are precisely the Hölder domains, i.e., images of the unit disk under Hölder continuous conformal mappings [BP]. Gehring and Martio studied the *quasihyperbolic boundary condition*, defined via the *quasihyperbolic metric* k, in connection with determining when quasiconformal mappings satisfy local Lipschitz conditions. They proved the proper inclusions $B \cap U \subset J \subset$ k-BC $\subset B$, where B, U and J are the classes of *bounded*, *uniform* and *John* domains respectively [GM, 2.18, 3.9, 3.11, 3.13].

In this article we study the classes j-BC and H-BC which are defined using metrics closely related to k. In Section 2 we consider the relative d metric j and investigate the j-boundary condition. In Section 3 we discuss the Harnack metric H

---

[1] This paper is in final form and no version of it will be submitted for publication elsewhere.

and the Harnack boundary condition.

**1.2. Remarks.** (a) Suppose $D \in \rho\text{-BC}(\alpha,x_0,c)$. Then:

(i) $d_D(x) \le c$ for all $x \in D$.

(ii) $D \in \rho\text{-BC}(\alpha,x_1,c^2/d_D(x_1))$ for all $x_1 \in D$.

(iii) If $E \subset \{x \in D : d_D(x) \ge \varepsilon > 0\}$, then $\rho_D(E) \le 2\alpha \log(c/\varepsilon)$.

(b) Observe that condition (1.1) is primarily a restriction on the growth of $\rho_D(x,x_0)$ as $d_D(x) \to 0$. First, it gives no information for $x$ near $x_0$. Second, if $D$ is any domain, then for all $E \subset D$ with $\rho_D(E) < +\infty$ and $d_D(x) \le \delta$ for all $x \in E$ we always have

$$\rho_D(x,x_0) \le \log(c/d_D(x)) \text{ for all } x, x_0 \in E,$$

where $c = \delta \exp(\rho_D(E))$. (c) Finally, we point out that (1.1) is related to the condition

$$\rho_D(x,y) \le \beta \log(c/\min\{d_D(x),d_D(y)\}) \text{ for all } x, y \in D.$$

It is easy to see that this condition implies that $d_D(x) \le c$ for all $x \in D$ and hence that $D \in \rho\text{-BC}(\beta,x_0,2c)$. On the other hand, since $ab \ge \min\{a^2,b^2\}$, it follows that $D \in \rho\text{-BC}(\alpha,x_0,c)$ implies the above condition with $\beta = 2\alpha$.

## 2. The class j-BC

Fix a subdomain $D$ of $X$ with nonempty boundary. The *relative* d *metric* in $D$ is

$$j_D = \log(1 + r_D),$$

where

$$r_D(x,y) = d(x,y)/\min\{d_D(x), d_D(y)\}$$

measures the "distance" between $x$ and $y$ *relative* to $D$.

In this section we give an obvious characterization for the class j-BC, obtaining the result $B = \text{j-BC}(1)$ as a corollary. Then we give examples to show that $B$ is a proper subclass of j-BC.

**2.1. Remark.** If $0 < \alpha < 1$, $\text{j-BC}(\alpha) = \emptyset$.

*Proof.* Suppose $D \in \text{j-BC}(\alpha,x_0,c)$. Choose any sequence $\{x_n\}$ in $D$ with $d_D(x_n) \to 0$. Then for all $n$ sufficiently large

$$\alpha \log(c/d_D(x)) \ge j_D(x_n,x_0) = \log(1 + d(x_n,x_0)/d_D(x_n)) \ge \log(d_D(x_0)/d_D(x_n)).$$

**2.2 Theorem.** *Fix* $\alpha \geq 1$. *The following are equivalent.*

(a) $D \in$ j-BC$(\alpha, x_0, a)$.

(b) $r_D(x,x_0) \leq b/d_D(x)^\alpha$ for all $x \in D$.

(c) $d(x,x_0) \leq c/d_D(x)^{\alpha-1}$ for all $x \in D$.

*Here the constants* a, b, c *depend only on each other and on* $\alpha$ *and* $d_D(x_0)$.

*Proof.* Since $j_D \geq \log r_D$, (b) follows from (a) with $b = a^\alpha$. Since $r_D(x,x_0) \geq d(x,x_0)/d_D(x)$ for all $x \in D$, (c) follows from (b) with $c = b$. We show that (c) implies (a).

Let $d = \min\{1, d_D(x_0)\}$, $E = \{x \in D : d_D(x) \geq d\}$. Now for all $x \in E$ we have

$$d_D(x) \leq d(x,x_0) + d_D(x_0) \leq c/d^{\alpha-1} + d_D(x_0) = \delta.$$

Hence if $x \in E$, then

$$j_D(x,x_0) \leq \log \beta \leq \log(\beta\delta/d_D(x))$$

where $\beta = 1 + c/d^\alpha$, while if $x \in D - E$, then

$$j_D(x,x_0) < \log((1 + d(x,x_0))/d_D(x)) < \log((1+c)/d_D(x)^\alpha).$$

Thus (a) holds with $a = \max\{\beta\delta, (1+c)^{1/\alpha}\}$.

**2.3. Corollary.** $B =$ j-BC(1).

In fact, if D is bounded, then $D \in$ j-BC$(1, x_0, \mathrm{dia}(D)^2/d_D(x_0))$ for each $x_0 \in D$, while if $D \in$ j-BC$(1, x_0, a)$, then $\mathrm{dia}(D) \leq 2a$.

**2.4. Remarks.** (a) A sufficient condition for membership in k-BC(1) is given in [HV, 2.12]. (b) As the examples below show, B is a proper subclass of j-BC$(\alpha)$ for all $\alpha > 1$.

**2.5. Examples.** (a) If $D = \{x \in \mathbb{R}^n : |x - x_0| < r\}$, then $j_D(x,x_0) = \log(r/d_D(x))$, so $D \in$ j-BC$(1, x_0, r)$.

(b) It is not hard to see that for each $\alpha > 1$ one can add an infinite spike to any ball and obtain a simply connected unbounded domain in j-BC$(\alpha)$; e.g., in $\mathbb{C}$ we have

$$\{z = x + iy : |z| < 1 \text{ or } |y| < x^{1/(1-\alpha)}, x \geq 1\} \in \text{j-BC}(\alpha, 0).$$

(c) The following example of a multiply connected unbounded plane domain in j-BC(2) is given partially to motivate an example in Section 3. The domain is given by

$D = \mathbb{C} - E$ where $E$ is a discrete set of points obtained as follows. First, divide the plane $\mathbb{C}$ into $2 \times 2$ squares whose edges are formed by the lines $|x| = 2n + 1$, $|y| = 2n + 1$ $(n = 0,1,2,\dots)$. Next, for $n = 1,2,3,\dots$ let $S_n$ denote the family of all such squares which lie inside the square $|x|, |y| \le 2n + 1$ but are not in $S_{n-1}$, where $S_0$ contains only the square $|x|, |y| \le 1$. Now put the centers of each of the squares in $S_1$ into $E$; so the eight points $\pm 2, \pm 2i, \pm(2 \pm 2i)$ belong to $E$. Then divide each of the squares in $S_2$ into four smaller congruent squares and put each of their centers into $E$. Continuing this, divide each of the squares in $S_n$ into $n^2$ smaller congruent squares and put their centers into $E$. Proceeding in this manner we obtain a discrete set $E$, and it is not difficult to see that $D = \mathbb{C} - E \in$ j-BC(2,0).

Finally, we point out that the class j-BC can sometimes be used as sort of a replacement for $\rho$-BC; e.g., we have k-BC $\cap$ U = j-BC $\cap$ U. The following is a related concept. We say that $D$ is *$\rho$–John* with *exponent* $\beta$ and *center* $x_0$, denoted $D \in \rho\text{-J}(\beta,x_0)$, if $x_0 \in D$, $\beta > 0$ and

$$\rho_D(x,x_0) \le \beta\, j_D(x,x_0) \quad \text{for all } x \in D.$$

**2.6. Theorem.** (a) $\text{j-BC}(\alpha,x_0,c) \cap \rho\text{-J}(\beta,x_0) \subset \rho\text{-BC}(\alpha\beta,x_0,c)$.

(b) *Suppose there exist* $0 < \varepsilon < 1$, $\delta > 0$ *such that*

$$\rho_D(x,x_0) \le \delta\, j_D(x,x_0) \quad \textit{when } d(x,x_0) \le \varepsilon\, d_D(x_0).$$

*Then*

$$\rho\text{-BC}(\alpha,x_0,c) \subset \rho\text{-J}(\beta,x_0)$$

*where*

$$\beta = \max\{\delta,\ \alpha\,(1 + \log(2c/d_D(x_0))/\log(1+\varepsilon))\}.$$

*Proof* (of (b)). Let $D \in \rho\text{-BC}(\alpha,x_0,c)$. Fix $x \in D$. Assume $d(x,x_0) \ge \varepsilon\, d_D(x_0)$. Then

$$j_D(x,x_0) \ \ge\ \log(1 + d(x,x_0)/d_D(x_0)) \ \ge\ \log(1 + \varepsilon).$$

Next,either $d(x,x_0) \ge d_D(x_0)/2$ or $d(x,x_0) \le d_D(x_0)/2$ and $d_D(x) \ge d_D(x_0)/2$. Thus

$$\min\{d_D(x),d_D(x_0)\} + d(x,x_0) \ \ge\ d_D(x_0)/2,$$

so

$$1 + r_D(x,x_0) \ \ge\ d_D(x_0)/2d_D(x)$$

and hence

$$\rho_D(x,x_0) \le \alpha \log\left(\frac{c}{d_D(x)}\,\frac{d_D(x_0)/2}{d_D(x_0)/2}\right) \le \alpha \log\left((1+r_D(x,x_0))\,\frac{2c}{d_D(x_0)}\right)$$
$$= \alpha\, j_D(x,x_0) + b \le (\alpha + b/\log(1+\varepsilon))\, j_D(x,x_0)$$

where $b = \alpha \log(2c/d_D(x_0))$.

**2.7. Remark.** The hypothesis of Theorem 2.6 (b) is satisfied provided

$$\partial\rho_D(x_0) = \limsup_{x \to x_0} \frac{\rho_D(x,x_0)}{d(x,x_0)} < +\infty .$$

**2.8. Corollary.** *If* $\partial\rho < +\infty$, *then* $\rho$-BC $\cap$ B $=$ $\rho$-J $\cap$ B.

## 3. The class H-BC

The *Harnack metric*, introduced by Köhn [K], is defined for $x, y \in D \subset \mathbb{R}^n$ by

$$H_D(x,y) = \sup_u |\log(u(x)/u(y))| ,$$

where the supremum is taken over all functions $u$ positive and harmonic in $D$; see also [H] and the references mentioned there. We point out that $D \in$ H-BC$(\alpha,x_0,c)$ if and only if the inequality

$$d_D(x)^\alpha/b \le u(x)/u(x_0) \le b/d_D(x)^\alpha \quad \text{for all } x \in D$$

holds for all $u$ positive and harmonic in $D$, where $b = c^\alpha$. Thus whether or not a domain $D$ satisfies a Harnack boundary condition is closely related to the rate of growth of positive harmonic functions in $D$.

In this section we present an example of an unbounded plane domain in H-BC. This shows that the class k-BC is a proper subclass of H-BC.

**3.1. Properties of $H_D$.** (a) *For simply connected proper subdomains* $D$ *of* $\mathbb{C}$

$$1/2\, k_D \le H_D = h_D \le 2\, k_D .$$

(b) *For general proper subdomains* $D$ *of* $\mathbb{C}$

$$H_D \le h_D \le 2\, k_D .$$

(c) *For general proper subdomains* D *of* $\mathbb{R}^n$

$$H_D \leq n\, k_D .$$

(d) *If* $D = G - E$ *where* $G \subset \mathbb{R}^n$ *is a domain with a Green's function* $g(x,y)$ *and* $E \subset G$ *is a relatively closed polar set, then for all* $x, y \in D$

$$H_D(x,y) = \max \{H_G(x,y), \sup_{z \in E} |\log g(x,z)/g(y,z)|\}.$$

(e) $H_D(x,y) \geq (n-2)|\log d_D(x)/d_D(y)|$ *for all* $x, y \in D$.

*Proofs.* The equality in (a) and the inequality in (c) [H, Theorem 2] both follow from Harnack's inequality. The right-hand inequalities in both (a) and (b) follow from the monotonicity of the hyperbolic metric. The left-hand inequality in (a) follows from Koebe's one-quarter theorem while that in (b) holds because the Harnack metric is distance decreasing with respect to holomorphic mappings. The formula in (d) [H, Theorem 3] follows from the Riesz representation formula. The inequality in (e) holds because $u(x) = 1/|x|^{n-2}$ is harmonic in $\mathbb{R}^n - \{0\}$ [L, 5.4].

**3.2. Corollaries.** (a) *For simply connected proper subdomains of* $\mathbb{C}$

$$k\text{-BC}(\alpha/2) \subset H\text{-BC}(\alpha) = h\text{-BC}(\alpha) \subset k\text{-BC}(2\alpha).$$

*In fact,* $H\text{-BC}(\alpha) = \emptyset$ *if* $0 < \alpha < 1$.

(b) *For general proper subdomains of* $\mathbb{C}$

$$k\text{-BC}(\alpha/2) \subset H\text{-BC}(\alpha) \subset h\text{-BC}(\alpha).$$

(c) *For general proper subdomains of* $\mathbb{R}^n$

$$k\text{-BC}(\alpha) \subset H\text{-BC}(n\alpha).$$

*Also,* $H\text{-BC}(\alpha) = \emptyset$ *if* $0 < \alpha < n-2$.

**3.3. Theorem.** *For general plane domains* $H\text{-BC} \neq k\text{-BC}$.

*Proof.* As every domain which satisfies a quasihyperbolic boundary condition is bounded [GM, 3.9], it suffices to exhibit an unbounded domain which satisfies a Harnack boundary condition. Indeed, we prove that $D = H - F \in H\text{-BC}(5,i,5)$ where $H = \{z : \text{Im}(z) > 0\}$, $F = \{z + i : z \in E\}$ and $E$ is the set described in Example 2.5(c).

Using 3.1 (a), (d) we obtain

$$H_D(z,i) = \max\{h(z,i)\ ,\ \sup_{\zeta\in F} |\log g(z,\zeta)/g(i,\zeta)|\}$$

for all $z \in D$, where $h$ and $g$ are the hyperbolic metric and Green's function in $H$. We prove

$$h(z,i) \le 3\log(9/d_D(z)) \quad \text{for all } z \in D \tag{3.4}$$

and

$$\frac{d_D(z)^3}{1800/\log 5} \le q(\zeta) \le \frac{1536/\log 5}{d_D(z)^5} \quad \text{for all } \zeta \in F\ ,\ z \in D, \tag{3.5}$$

where $q(\zeta) = g(z,\zeta)/g(i,\zeta)$. The desired conclusion then follows.

We need the following elementary facts:

$$t^2 \ge 1 + t \quad \text{for all } t \ge (1+\sqrt{5})/2; \tag{3.6}$$

$$6t^2 \ge a(a+t) \quad \text{for all } t \ge a/2 > 0; \tag{3.7}$$

$$\log(1+ab)\,t \le b\log(1+at) \quad \text{for all } 0 \le t \le b; \tag{3.8}$$

$$g(z,\zeta) = \log\left|\frac{z-\bar\zeta}{z-\zeta}\right| = \frac{1}{2}\log\left(1 + \frac{4y\eta}{|z-\zeta|^2}\right); \tag{3.9}$$

$$\min_{|\zeta-z|\le y/2} g(z,\zeta) = \log 3 = \max_{|\zeta-z|\ge y} g(z,\zeta). \tag{3.10}$$

Here, and below, we write $z = x + iy$, $\zeta = \xi + i\eta$.

Next, it is not difficult to show that

$$d_D(z) \le 6\min\{1,\ y,\ 1/y,\ 1/|z-i|\} \quad \text{for all } z \in D. \tag{3.11}$$

In fact, $d_D(z) \le \sqrt{2}$ for all $z \in D$.

From [G, 4.4, p. 35]

$$h(z_1,z_2) \le \log(1 + |z_1-z_2|/y_1)(1 + |z_1-z_2|/y_2),$$

so

$$h(z,i) \le \log(1 + |z-i|/y)(1 + |z-i|)$$

and (3.4) now follows from (3.11).

Next, we claim that

$$d_D(z)/5 \le \left|\frac{\zeta-i}{\zeta-z}\right| \le 8/d_D(z)^2 \quad \text{for all } \zeta \in F,\ z \in D. \tag{3.12}$$

Indeed, from (3.11) we get

$$\left|\frac{\zeta-i}{\zeta-z}\right| \le 1+\left|\frac{z-i}{\zeta-z}\right| \le 1+6/d_D(z)^2 \le 8/d_D(z)^2,$$

and since $|\zeta-i| \ge 2$, again (3.11) yields

$$\left|\frac{\zeta-i}{\zeta-z}\right| \ge 1/(1+\left|\frac{z-i}{\zeta-i}\right|) \ge 1/(1+3/d_D(z)) \ge d_D(z)/5.$$

To prove the upper bound in (3.5) we combine (3.6), (3.8), (3.9), (3.11), (3.12) and $|\zeta-i| \ge 2$ to obtain

$$q(\zeta) \le \frac{4y\eta/|z-\zeta|^2}{\log 5 \;\; \eta/|i-\zeta|^2} = \frac{4y}{\log 5}\left|\frac{\zeta-i}{\zeta-z}\right|^2 \le \frac{1536/\log 5}{d_D(z)^5}.$$

To prove the lower bound we first consider the case $|\zeta-z| \le y/2$. Then by (3.10) we have $g(z,\zeta) \ge \log 3 \ge g(i,\zeta)$, whence $q(\zeta) \ge 1$. Finally, when $|\zeta-z| \ge y/2$ we combine (3.7), (3.8), (3.9), (3.11) and (3.12) to get

$$q(\zeta) \ge \frac{(\log 5/3)(y\eta/|z-\zeta|^2)}{4\eta/|i-\zeta|^2} = \frac{y\log 5}{12}\left|\frac{\zeta-i}{\zeta-z}\right|^2 \ge \frac{d_D(z)^3}{1800/\log 5}.$$

## References


[BP] Becker, J. and Ch. Pommerenke, Hölder continuity of conformal mappings and non-quasiconformal Jordan curves, Comment. Math. Helv. **57** (1982), 221–225.

[G] Gehring, F. W., Characteristic properties of quasidisks, Les Presses de l'Université de Montréal, Montreal, 1982.

[GM] Gehring, F. W. and O. Martio, Lipschitz classes and quasiconformal mappings, Ann. Acad. Sci. Fenn. Ser. A I Math. **10** (1985), 203–219.

[H] Herron, D. A., The Harnack and other conformally invariant metrics, Kodai Math. J. **10** (1987), 9–19.

[HV] Herron, D. A. and M. Vuorinen, Positive harmonic functions in uniform and admissible domains, to appear in Analysis.

[K] Köhn, J., Die Harnacksche Metrik in der Theorie der harmonischen Funktionen, Math. Z. **91** (1966), 50–64.

[L] Leutwiler, H., On a distance invariant under Möbius transformations in $\mathbb{R}^n$, Ann. Acad. Sci. Fenn. Ser. A I Math. **12** (1987), 3–17.

# COMPACTIFICATIONS OF HARMONIC SPACES AND HUNT PROCESSES [1]

Teruo Ikegami
Department of Mathematics, Osaka City University
Osaka, 558, Japan

## Introduction

In the classical case of Green spaces, J. L. Doob [4] investigated a probabilistic potential theory under the notions of conditional Browinian motions and the Martin boundary. His results formed a probabilistic counterpart of the Naïm theory [12]. Recently, the author defined compactifications of Martin type [9] to develop the Naïm theory in an axiomatic framework. It is well-known that for a $\mathscr{P}$-harmonic space X in the sense of Constantinescu-Cornea [2] there corresponds a Hunt process $\mathscr{X}$ with state space X such that the family of excessive functions of $\mathscr{X}$ coincides with the set of non-negative hyperharmonic functions on X – a very accessible exposition of this important fact may be found in the book of Bliedtner-Hansen [1]. In the present paper, we treat this Hunt process under the assumption that the constant functions are harmonic.

After a brief introduction of notations, we give a probabilistic characterization of a metrizable and resolutive compactification $X^*$ of X, that is by means of the Hunt process $\mathscr{X}$ (Theorem 2.3). The following section is devoted to a special compactification of Martin type. It turns out that if we consider a Hunt process ${}^z\mathscr{X}$ corresponding to the quotient harmonic space by the kernel function $k_z$, ${}^z\mathscr{X}$ plays the role of a conditional Brownian motion in the classical case. We obtain that ${}^zP^x$-almost all paths converge to the same point z (Theorem 3.2), and from this fact we can derive a probabilistic interpretation of Naïm's result. In the final section, as an application, we consider a harmonic morphism $\varphi$ between harmonic spaces. A potential theoretic consideration of the boundary behavior of $\varphi$ was given for analytic mappings of Riemann surfaces in [8] and for an axiomatic setting concerning compactifications of Martin type in [10]. The

---

[1] This paper is in final form and no version of it will be submitted for publication elsewhere.

study of this problem in this article is a generalization of Doob's result [5] into the context of axiomatic potential theory.

## 1. Preliminaries

Let $X$ be a $\mathcal{P}$-harmonic space in the sense of Constantinescu-Cornea. We assume that $X$ has a countable base and the constant functions are *harmonic.* Let $\mathcal{U}$ be the hyperharmonic sheaf which defines the harmonic structure of $X$ and ${}^*\mathcal{H}^+ = \{v \in \mathcal{U}(X)\ ;\ v \geq 0\}$. Then, there is a Hunt process of function space type with state space $X$,

$$\mathcal{X} = (\Omega, \mathcal{F}, \mathcal{F}_t, Y_t, \theta_t, P^x)$$

such that the set of excessive functions coincides with ${}^*\mathcal{H}^+$ (in this case, we say that $\mathcal{X}$ *links* with $(X, {}^*\mathcal{H}^+)$) and the path $t \to Y_t(\omega)$ is continuous for $P^x$-almost every $\omega$ and for every $x \in X$ (this is called *almost surely* and abbreviated to a.s.) on $[0,\zeta)$, where $\zeta$ is the life time of $\omega$ ([1], Chapter IV, Chapter VI §3).

For every real $v \in {}^*\mathcal{H}^+$,

$$Z_t = \begin{cases} v[Y_t] & t < \zeta, \\ 0 & t \geq \zeta, \end{cases}$$

is an $(\mathcal{F}_t)$-supermartingale on $(\Omega, \mathcal{F}, P^x)$ for each $x \in X$ ([1], Chapter IV §7). By a supermartingale inequality [3], if $v \in {}^*\mathcal{H}^+$ and $v(x) < \infty$ for some $x \in X$, then there exists $\lim_{t\uparrow\zeta} v(Y_t)$ $P^x$-a.e.

A *resolutive compactification* $X^*$ of $X$ is defined to be a compact space such that $X$ is a dense open subset of $X^*$ and for every real continuous function $f$ on $\Delta = X^* \backslash X$, there corresponds a harmonic function $H_f$ on $X$ by the method of Perron-Wiener-Brelot.

Let $f$ be a strictly positive continuous function on $X$ and ${}^f\mathcal{U}(U) = \{u/f\ ;\ u \in \mathcal{U}(U)\}$. ${}^f\mathcal{U}$ is a hyperharmonic sheaf and the harmonic space with this hyperharmonic sheaf is a $\mathcal{P}$-harmonic space. A metrizable compactification is f-*resolutive* if it is resolutive with respect to the harmonic sheaf $\mathcal{H}_{{}^f\mathcal{U}}$.

## 2. A characterization of resolutive compactifications

In the sequel, let $X^*$ be a metrizable compactification of $X$ and $\{X_n\}$ be a compact exhaustion of $X$, i.e., $\{X_n\}$ is a covering of $X$ by relatively compact open sets such that $\bar{X}_n \subset X_{n+1}$. We define, for each $X_n$, the stopping time $T_n$ (the first hitting time of $X\backslash X_n$),

$$T_n(\omega) := \inf\{t > 0 ; Y_t(\omega) \in X\backslash X_n\}$$

under the convention $\inf \emptyset = \infty$. Then $T_n < \zeta$ for all $n$ a.s.

**Lemma 2.1.** $\lim_{n\to\infty} T_n(\omega) = \zeta(\omega)$ *a.s.*

*Proof.* Fix $x \in X$ and let $x \in X_n$. From $P^x[Y_{T_n} \in X\backslash X_n] = \varepsilon_x^{X\backslash X_n}(X\backslash X_n)$ ([1] Chapter VI 3.16) and $\varepsilon_x^{X\backslash X_n}(1) = \hat{R}_1^{X\backslash X_n}(x) = 1$, since $1$ is harmonic by assumption, we have $Y_{T_n} \in X\backslash X_n$ $P^x$-a.e. The fact that paths are continuous a.s. on $[0,\zeta)$ means $Y_{T_n} \in \partial X_n$ (the boundary of $X_n$), which implies $T_n \le \zeta$ $P^x$-a.e., i.e., $T_n(\omega) \le \zeta(\omega)$ for every $\omega \in \Omega_1$, where $P^x(\Omega\backslash\Omega_1) = 0$. Obviously $\{T_n\}$ is increasing. We assert that $\lim_{n\to\infty} T_n(\omega) = \sup_n T_n(\omega) = \zeta(\omega)$ for every $\omega \in \Omega_1$. Otherwise we would have $\sup_n T_n(\omega) < t_0 < \zeta(\omega)$ for some $\omega \in \Omega_1$, but this is impossible, since $K = \{Y_t(\omega) ; t \in [0,t_0]\}$ is a compact subset of $X$ and therefore $K \subset X_n$ for sufficiently large $n$.

**Lemma 2.2.** *For a potential* $p$ *of* $X$ *and* $x \in X$ *such that* $p(x) < \infty$, $\lim_{t\uparrow\zeta} p(Y_t) = 0$ $P^x$-*a.e.*

*Proof.* We have remarked in the previous section that for $P^x$-a.e. $\omega$, $p(Y_t(\omega))$ has the limit $\varphi(\omega)$ as $t \to \zeta(\omega)$. Then, by the above lemma and Fatou's lemma,

$$E^x[\varphi] = \int \varphi(\omega)\, P^x(d\omega) \le \liminf_{n\to\infty} E^x[p(Y_{T_n})] = \liminf_{n\to\infty} \hat{R}_p^{X\backslash X_n}(x) = 0.$$

**Theorem 2.3.** *Let* $X^*$ *be a metrizable compactification of* $X$. *Then the following statements are equivalent:*

(1) $X^*$ *is resolutive,*

(2) $\lim_{t\uparrow\zeta} Y_t(\omega) \in \Delta$ *a.s.*

*Proof.* (1) $\Rightarrow$ (2): Let $f$ be a finite non-negative continuous function on $X^*$ and $u = H_{f|\Delta}$ ($f|\Delta$ denotes the restriction of $f$ on $\Delta$). Then there is a potential $p$

on $X$ such that $|q| \leq p$, where $q = f - u$ ([7], Proposition 1.4.5). We may suppose that $p$ is bounded and we have, by Lemma 2.2, $\lim_{t\uparrow\zeta} q(Y_t) = 0$ a.s.. Thus, the existence of almost sure limits $\lim_{t\uparrow\zeta} u(Y_t)$ implies $\lim_{t\uparrow\zeta} f(Y_t) = \lim_{t\uparrow\zeta} u(Y_t)$ a.s. for each $f$. Now, let us consider a countable family $\{f_n\}$ of non-negative functions which are continuous and total on $X^*$. The above consideration yields easily the existence of almost sure limits $\lim_{t\uparrow\zeta} Y_t$ and $\lim_{t\uparrow\zeta} Y_t \in \Delta$ since, by Lemma 2.1, $\lim_{t\uparrow\zeta} Y_t = \lim_{n\to\infty} Y_{T_n}$ and $Y_{T_n} \in \partial X_n$.

(2) ⇒ (1): It is sufficient to prove that every finite continuous function $f$ on $X^*$ is harmonizable ([7], Lemma 2.2.1). This is also equivalent to the following:

(*) for every exhaustion $\{X_n\}$ of $X$ the limit $\lim_{n\to\infty} H_f^{X_n}$ exists

([11] Theorem 2.2.21; the proof in [11] is still valid with minor modifications in our axiomatic setting). Let $\{X_n\}$ be an exhaustion of $X$ and $T_n$ be the first hitting time of $X\backslash X_n$. Fix $x \in X$ and let $x \in X_n$. Then

$$\tilde{f}(\omega) := \lim_{n\to\infty} f[Y_{T_n}(\omega)]$$

exists $P^x$-a.e. since $\tilde{f}(\omega) = f[\lim_{t\to\zeta} Y_t(\omega)]$. The assertion (*) is an immediate consequence of Lebesgue's bounded convergence theorem:

$$E^x[\tilde{f}] = \lim_{n\to\infty} E^x[f(Y_{T_n})] = \lim_{n\to\infty} H_f^{X_n}(x).$$

The last equality is a consequence of the approximation theorem ([1], I,1.2).

**Remark.** If $X^*$ is metrizable and resolutive, then

$$E^x[f(\lim_{t\uparrow\zeta} Y_t)] = H_f(x)$$

for every finite continuous function $f$ on $\Delta$.

In the following we shall assume that $X^*$ is a metrizable and resolutive compactification of $X$. Then $H_f(x) = \int f(y)\, \lambda_x(dy)$ for every finite continuous function $f$ on $\Delta$. The Radon measure $\lambda_x$ on $\Delta$ is called a *harmonic measure* of $x$. We define the *harmonic boundary* $\Gamma = \overline{\cup_{x\in X}(\operatorname{supp} \lambda_x)}$. $\Gamma$ is characterized as $\Gamma = \{z \in \Delta\,;$ $\liminf_z p = 0$ for every potential $p$ on $X\}$. For a fixed $x \in X$ we denote by $Y^*$ the $P^x$-a.e. limit of $Y_t$, i.e.,

$$Y^*(\omega) := \lim_{t \uparrow \zeta} Y_t(\omega) \quad P^x\text{- a.e.}$$

**Corollary 2.4.** $E^x[f(Y^*)] = H_f(x)$ *for every* $x \in X$ *and every bounded Borel function* $f$ *on* $\Delta$.

This is an immediate consequence of Theorem 2.3 and the monotone class theorem. A special case of Corollary 2.4 is

**Corollary 2.5.** $P^x[\{\omega\,;\, Y^*(\omega) \in A\}] = H_{1_A}(x)$ *for every* $x \in X$ *and every Borel subset* $A$ *of* $\Delta$, *where* $1_A$ *is the characteristic function of* $A$ *on* $\Delta$.

Corollary 2.5 implies

**Corollary 2.6.** *For every* $x \in X$, $Y^*(\omega) \in \Gamma$ *for* $P^x$*-a.a.* $\omega$.

We conclude this section by proving the following result.

**Proposition 2.7.** *Suppose that* $f$ *and* $|f|$ *are resolutive. Then for every* $x \in X$,

$$\lim_{t \uparrow \zeta} H_f(Y_t) = f(Y^*) \quad P^x\text{-}a.e.$$

*Proof.* Fix $x \in X$. We may suppose $f \geq 0$. By the resolutivity of $f$ we have two decreasing sequences of lower bounded hyperharmonic functions $\{-u_k\}$, $\{v_k\}$ such that

(1) $\limsup u_k \leq f \leq \liminf v_k$ on $\Delta$ for every $k$,

(2) $0 \leq v_k(x) - u_k(x) \leq 1/k$ for every $k$.

We remark that the functions $v_k$ and $u_k$ are non-negative. By a remark in §1, we can define $P^x$-almost everywhere limits

$$\tilde{\varphi}_k := \lim_{t \uparrow \zeta} v_k(Y_t)$$

and

$$\tilde{\psi}_k := \lim_{t \uparrow \zeta} u_k(Y_t).$$

Since $\tilde{\varphi}_k = \lim_{n \to \infty} v_k(Y_{T_n})$, by Fatou's lemma and a property of supermartingales, $E^x[\tilde{\varphi}_k] \leq v_k(x)$, $E^x[\tilde{\psi}_k] \geq u_k(x)$. Putting $\eta = \lim_{t \uparrow \zeta} H_f(Y_t)$ we have $\tilde{\psi}_k \leq \eta \leq \tilde{\varphi}_k$, $\{-\tilde{\psi}_k\}$ and $\{\tilde{\varphi}_k\}$ are decreasing. Then

$$\tilde{\varphi}_k(\omega) = \lim_{t \uparrow \zeta} v_k[Y_t(\omega)] \geq \liminf_{y \to Y^*(\omega)} v_k(y) \geq f(Y^*(\omega)) \quad P^x\text{-a.e.}$$

and similarly,

$$\tilde{\psi}_k(\omega) \leq f(Y^*(\omega)) \ \ P^x\text{-a.e.}$$

Thus, we have

$$\lim_{k\to\infty} E^x[\tilde{\varphi}_k] = \lim_{k\to\infty} E^x[\tilde{\psi}_k] = E^x[\eta],$$

which means that $\tilde{\varphi}_k \downarrow \eta$, $\tilde{\psi}_k \uparrow \eta$ $P^x$-a.e., i.e., $\eta = f(Y^*)$ $P^x$-a.e.

### 3. Compactifications of Martin type

The compactifications of Martin type were introduced by the author [9] to develop Naïm's theory in an axiomatic framework. We recall the definition: a compactification $X^*$ of $X$ is said to be *of Martin type* if

1) $X^*$ is metrizable and resolutive; the harmonic measures are denoted by $\lambda_x$;

2) there exists a finite continuous function $k(x,z)$ $(=k^x(z) = k_z(x))$ defined on $X \times \Delta$ such that $k_z$ is non-negative and harmonic for every $z \in \Delta$;

3) there exist a Radon measure $\mu$ on $\Delta$ and a boundary set $\Delta_1$ satisfying

(i) $\Delta_1 \subset \{z \in \Delta\,;\, k_z$ is minimal harmonic$\}$, $\mu(\Delta \backslash \Delta_1) = 0$,

(ii) $\mu(T) = 0$ if $\lambda_x(T) = 0$ for every $x \in X$;

4) for every bounded harmonic function $u$ on $X$ there is a resolutive function $f$ on $\Delta$ such that

$$u(x) = H_f(x) = \int k(x,z) f(z)\, \mu(dz) \quad \text{for every } x \in X.$$

For $z \in \Delta_1$, we defined the filter $\mathcal{G}_z = \{E\,;\, X\backslash E$ is thin at $z$, i.e., $R_{k_z}^{X\backslash E} \neq k_z\}$ and proved that $\mathcal{G}_z$ is finer than the trace filter of neighborhoods of $z$ in $X^*$ for $\mu$-a.a. $z$ ([9], Theorem 3.3). Thus we may suppose that this is the case for every $z \in \Delta_1$.

**Lemma 3.1.** *Let* $z \in \Delta_1$ *and suppose that* $k_z$ *is strictly positive. Then* $X^*$ *is* $k_z$*-resolutive and the* $k_z$*-harmonic measure is* ${}^{k_z}\lambda_x = \varepsilon_z$ *for every* $x \in X$.

*Proof.* The $k_z$-resolutivity of $X^*$ is derived in the same way as in [11], Theorem 3.2.23. Denoting by ${}^{k_z}H_f$ the $k_z$-Dirichlet solution, we have ${}^{k_z}H_1 = 1$, since $1$ is $k_z$-harmonic. This implies $k_z \cdot {}^{k_z}H_f \leq (\sup f) k_z$ for every finite continuous function $f \geq 0$ on $\Delta$ and further, since $k_z \cdot {}^{k_z}H_f$ is non-negative and harmonic and $k_z$ is minimal, ${}^{k_z}H_f$ is constant, i.e., the $k_z$-harmonic boundary ${}^{k_z}\Gamma$ is a singleton. For $z' \in \Delta \setminus \{z\}$, take a neighborhood $U(z')$ of $z'$ with $z \notin \overline{U(z')}$. Since $(\hat{R}_{k_z}^{U(z')\cap X})/k_z$ is a $k_z$-potential and $\liminf_{x \to z'} (\hat{R}_{k_z}^{U(z')\cap X}(x))/k_z(x) = 1$ we have $z' \notin {}^{k_z}\Gamma$, which means ${}^{k_z}\Gamma = \{z\}$ and ${}^{k_z}\lambda_x = c\varepsilon_z$ for $c > 0$. Furthermore, $1 = {}^{k_z}H_1(x) = \int 1 \, d^{k_z}\lambda_x = c$.

We have thus a Hunt process ${}^z\mathscr{X} = ({}^z\Omega, {}^z\mathscr{F}, {}^z\mathscr{F}_t, {}^zY_t, {}^z\theta_t, {}^zP^x)$ which links with $(X, {}^{k_z *}\mathscr{H}^+)$ where ${}^{k_z *}\mathscr{H}^+ = \{u/k_z \,;\, u \in {}^*\mathscr{H}^+\}$. If we consider a Hunt process of function space type, we may assume ${}^z\Omega = \Omega$, ${}^zY_t = Y_t$ and ${}^z\theta_t = \theta_t$. We remark that the transition functions are ${}^zP_t(x,B) = \int_B k_z(y)/k_z(x) \, P_t(x,dy)$ for every Borel subset $B$ of $X$ and $x \in X$.

From Corollary 2.6 we obtain

**Theorem 3.2.** *Let* $z \in \Delta_1$ *and* $k_z > 0$. *Then,* ${}^zP^x$*-almost all paths of* ${}^z\mathscr{X}$ *converge to the point* $z$ *for every* $x \in X$.

In $X^*$ every superharmonic function $s \geq 0$ on $X$ is decomposed into

$$s = H_f + h^s + p,$$

where $p$ is a potential, $f$ is a resolutive function on $\Delta$ and $h^s$ is the singular part of G.H.M. s. Then, we have

$$\text{f-lim}\, s = f \quad \mu\text{-a.e. on } \Delta,$$

where f-lim means the limit along the filter $\mathscr{G}_z$ ([9], Theorem 5.5).

A counterpart of this result in the probabilistic consideration is the following:

**Theorem 3.3.** *For a superharmonic function* $s \geq 0$ *on* $X$ *we have*

$$\lim_{t \uparrow \zeta} s(Y_t) = f(Y^*) \quad P^x\text{-}a.e. \text{ for every } x \in X.$$

This is a consequence of Proposition 2.7 and the proof of Lemma 2.2.

Theorem 3.3 is a version of the counterpart which seems rather superficial. To get more crucial information we follow Doob [6], Part 2, Chapter X.

In the sequel, we always suppose that $k_z > 0$ for every $z \in \Delta_1$.

**Lemma 3.4.** *Let* $x \in X$ *and let* $T$ *be an* $\mathcal{F}_t$*-stopping time such that* $T$ *is a* ${}^z\mathcal{F}_t$*-stopping time for every* $z \in \Delta_1$*, and*

$$P^x[\{T \geq \zeta\}] = 0 \text{ and } {}^zP^x[\{T \geq \zeta\}] = 0 \text{ for every } z \in \Delta_1.$$

*Then,* ${}^zP^x[\Lambda] = \int_\Lambda k_z(Y_T)/k_z(x)\, dP^x$ *for every* $\Lambda \in \mathcal{F}_T \cap [\cap_{z\in\Delta_1} {}^z\mathcal{F}_T]$.

This is proved quite in the same way as in [6], pp. 672, 676.

Now, fix a compact exhaustion $\{X_n\}$ of $X$ and let $T_n$ be the first hitting time of $X \setminus X_n$. Then we have

**Proposition 3.5.** *Let* $x \in X$ *and let* $g$ *be a Borel function on* $X$, $\varepsilon > 0$ *and*

$$\Lambda_\varepsilon = \bigcap_n \bigcup_{k \geq n} \{|g(Y_{T_k}) - g(Y_{T_n})| \geq \varepsilon\}.$$

*Then*

$$P^x[\Lambda_\varepsilon] = \int {}^zP^x[\Lambda_\varepsilon] k_z(x)\, \mu(dz).$$

*Proof.* Applying the above lemma to

$$\Lambda_{n,m,\varepsilon} := \bigcup_{k=n+1}^{m} \{|g(Y_{T_k}) - g(Y_{T_k})| \geq \varepsilon\} \quad (n < m)$$

we have

$${}^zP^x[\Lambda_{n,m,\varepsilon}] = \int_{\Lambda_{n,m,\varepsilon}} k_z(Y_{T_m})/k_z(x)\, dP^x.$$

Therefore, since $\int k_z(\cdot)\, \mu(dz) = H_1(\cdot) = 1$,

$$\int {}^zP^x[\Lambda_{n,m,\varepsilon}] k_z(x)\, \mu(dz) = \int_{\Lambda_{n,m,\varepsilon}} [\int k_z(Y_{T_m})\, \mu(dz)]\, dP^x = P^x[\Lambda_{n,m,\varepsilon}].$$

Letting $m \to \infty$ for a fixed $n$, and then letting $n \to \infty$, the proof is complete.

**Corollary 3.6.** *The following assertions are equivalent:*

(a) $\lim_{n\to\infty} g(Y_{T_n})$ *exists* $P^x$*-a.e.,*

(b) $\lim_{n\to\infty} g(Y_{T_n})$ *exists* ${}^zP^x$*-a.e. for* $\mu$*-a.a.* $z \in \Delta$.

*Proof.* Using the notation of Proposition 3.5, (a) $\Leftrightarrow P^x[\Lambda_\varepsilon] = 0$ for every $\varepsilon > 0$ $\Leftrightarrow {}^zP^x[\Lambda_\varepsilon] = 0$ for $\mu$-a.a. $z$ and every $\varepsilon > 0$ (by Proposition 3.5) $\Leftrightarrow$ (b).

Now, we have the following conclusion which gives a reasonable certification that $\hat{g}$ is the fine boundary function of $g$.

**Theorem 3.7.** *Let* $g$ *be a Borel function on* $X$ *and* $\hat{g}$ *be a* $\mu$*-measurable function on* $\Delta$. *Then* $\lim_{n\to\infty} g(Y_{T_n}) = \hat{g}(Y^*)$ $P^x$*-a.e. for every* $x \in X$ *if and only if* $\lim_{n\to\infty} g(Y_{T_n}) = \hat{g}(z)$ ${}^zP^x$*-a.e. for* $\mu$*-a.a.* $z$ *for every* $x \in X$.

Applying the theorem to a superharmonic function $s \geq 0$ on $X$, $s = H_f + h^s + p$, we have

$$\left.\begin{array}{l} \lim\limits_{n\to\infty} h^s(Y_{T_n}) = \lim\limits_{n\to\infty} p(Y_{T_n}) = 0 \\ \lim\limits_{n\to\infty} s(Y_{T_n}) = f(z) \end{array}\right\} {}^zP^x\text{-a.e. for } \mu\text{-a.a. } z \text{ for every } x \in X.$$

## 4. Applications

Let $\varphi$ be a harmonic morphism from $X$ into a second harmonic space $X'$ [13]. We assume that $X$ and $X'$ are $\mathcal{P}$-harmonic spaces in the sense of Constantinescu-Cornea and the constant functions are harmonic on the both spaces. Furthermore, we assumed that $X^*$ is a compactification of Martin type such that $k_z > 0$ for every $z \in \Delta_1$ and $X'^*$ is a metrizable and resolutive compactification. In order to discuss the boundary behavior of $\varphi$, we define for each $z \in \Delta_1$ the fine cluster set

$$\hat{\varphi}(z) = \cap \{\overline{\varphi(E)}\,;\, E \in \mathcal{G}_z\}.$$

Among various informations, we shall underline the following prominent theorems: Fatou's theorem, Riesz theorem and a theorem on boundary characterization of Bl-mappings. These were proved by the author for analytic mappings of Riemann surfaces [8] and for harmonic morphisms in a slightly more general setting [10]. Here we develop the theorems in probabilistic language. This yields an axiomatic version of the work of Doob [5].

We use the notations from the previous sections.

**Theorem 4.1** (Fatou). $\lim_{t\uparrow\zeta} \varphi(Y_t(\omega))$ *exists for* $P^x$*-almost every* $\omega$ *for every* $x \in X$.

*Proof.* Let $f'$ be a finite continuous function on $X'^*$. Then $f' \circ \varphi$ is a

Wiener function on X and $f' \circ \varphi = H_{\hat{g}} + q$, where $\hat{g}$ is a resolutive function on $\Delta$ and q is a Wiener potential [9], [13]. Therefore $\lim_{t\uparrow\zeta} f'[\varphi(Y_t)]$ exist almost surely, which proves the theorem since $X'^*$ is separable.

To obtain a more crucial interpretation of Fatou's theorem, let $\{X_n\}$ be a compact exhaustion of X and $T_n$ be the first hitting time of $\mathscr{X}$ on $X\backslash X_n$. We have then

**Theorem 4.1'.** $\lim_{n\to\infty} \varphi(Y_{T_n}) = \hat{\varphi}(z)$ ${}^zP^x$*-a.e. for* $\mu$*-a.a.* z *for every* $x \in X$.

*Proof.* Fix $x \in X$ and a finite continuous function $f'$ on $X'^*$. Since, by Theorem 4.1, there exists $\lim_{t\uparrow\zeta} f'[\varphi(Y_t)]$ $P^x$-a.e., we have the existence of the limit $\lim_{n\to\infty} f'[\varphi(Y_{T_n})]$ $P^x$-a.e., which implies, by Corollary 3.6, that

$$\lim_{n\to\infty} f'[\varphi(Y_{T_n})] \text{ exists } {}^zP^x\text{-a.e. for } \mu\text{-a.a. } z \in \Delta.$$

Considering as above a countable family which is uniformly dense in the space of finite continuous functions on $X'^*$ we have

$$\lim_{n\to\infty} \varphi(Y_{T_n}(\omega)) \text{ exists for } {}^zP^x\text{-a.a. } \omega \text{ for } \mu\text{-a.a. } z.$$

Since $Y_{T_n} \to z$ ${}^zP^x$-a.e. we may denote this limit by $\tilde{\varphi}(z)$. With the representation $f' \circ \varphi = H_{\hat{g}} + q$ in the proof of Theorem 4.1,

$$f'[\tilde{\varphi}(z)] = \lim_{n\to\infty} f'[\varphi(Y_{T_n})] = \hat{g}(z) \quad {}^zP^x\text{-a.e. for } \mu\text{-a.a. } z.$$

On the other hand, by the consideration of the Wiener compactification of X [10], we have

$$\hat{g}(z) = \text{f-}\lim_z f' \circ \varphi = f'[\hat{\varphi}(z)] \quad \mu\text{-a.e.}$$

Thus, $f'[\tilde{\varphi}(z)] = f'[\hat{\varphi}(z)]$ $\mu$-a.e. for every finite continuous function $f'$ on $X'^*$, which implies $\tilde{\varphi}(z) = \hat{\varphi}(z)$ $\mu$-a.e.

As a corollary to this theorem we prove the following theorem of Riesz.

A set $A'$ of $X'^*$ is said to be *polar in* $X'^*$, if for each $a_0' \in X'$ there is

$u' \in {}^*\mathscr{H}^+(X')$ such that $u'(a_0') < \infty$ and $\lim_{X' \ni a' \to x'} u'(a') = \infty$ for every $x' \in A'$. Since $X'$ is $\mathscr{P}$-harmonic we may take $u'$ to be superharmonic on $X'$ ([13] Theorem 6.11).

**Theorem 4.2** (Riesz). *Let $\varphi$ be locally polarly non-constant, i.e., for every open subset $U$ of $X$, $\varphi(U)$ is a non-polar set containing at least two points. If for a $\mu$-measurable set $A \subset \Delta_1$,*

$$A' = \bigcup_{z \in A} \hat{\varphi}(z)$$

*is polar in $X'^*$, then $\mu(A) = 0$.*

*Proof.* Let $x \in X$, $s'$ be a non-negative superharmonic function on $X'$ such that $s'[\varphi(x)] < \infty$ and $\lim_{a' \to x'} s'(a') = \infty$ for every $x' \in A'$ and $s = s' \circ \varphi$. Then, since $s \in {}^*\mathscr{H}^+(X)$ and $s(x) < \infty$, the limit

$$\lim_{n \to \infty} s(Y_{T_n}) \text{ exists } {}^zP^x\text{-a.e. for } \mu\text{-a.a. } z.$$

This limit is finite, since

$$E^x[\lim_{n \to \infty} s(Y_{T_n})] \le \liminf_n E^x[s(Y_{T_n})] \le \liminf_n \hat{R}_n^{X \setminus X_n}(x) \le s(x) < \infty.$$

Suppose for a moment that $\mu(A) > 0$. Then there is $z \in A$ such that

$$\lim_{n \to \infty} s(Y_{T_n}) < \infty \quad {}^zP^x\text{-a.e.}$$

This contradicts with

$$\lim_{n \to \infty} s(Y_{T_n}) = \lim_{n \to \infty} s'[\varphi(Y_{T_n})] \ge \liminf_{\substack{a' \to \hat{\varphi}(z) \\ a' \in X'}} s'(a') = \infty.$$

Finally, we consider a mapping $\varphi$ of type Bl, i.e., every $x' \in X'$ has an open neighborhood $U'$ such that $\varphi^{-1}(U') = \emptyset$ or $H_1^{\varphi^{-1}(U')} = 1$. It has been proved that $\varphi$ is of type Bl if and only if $\hat{\varphi}(z) \cap X' = \emptyset$ $\mu$-a.e. [8], [10].

**Theorem 4.3.** *A harmonic morphism $\varphi : X \to X'$ is of type* Bl *if and only if*

$$(*) \qquad \lim_{t \uparrow \zeta} f'[\varphi(Y_t)] = 0 \text{ almost surely}$$

*for every finite continuous function* f' *of* X' *with compact support.*

*Proof.* Fix $x \in X$. Suppose $\varphi$ is of type Bl and $f' \in \mathfrak{C}_K(X')$, i.e., f' is a finite continuous function of X' with compact support. By Theorem 4.1, $\lim_{t\uparrow\zeta} f'[\varphi(Y_t)]$ exists $P^x$-a.e. and this limit is $\lim_{n\to\infty} f'[\varphi(Y_{T_n})]$. Thus,

$$\lim_{n\to\infty} f'[\varphi(Y_{T_n})] \text{ exists } {}^zP^x\text{-a.e. for } \mu\text{-a.a. } z.$$

On the other hand, we have in Theorem 4.1'

$$\lim_{n\to\infty} \varphi(Y_{T_n}) = \hat{\varphi}(z) \ {}^zP^x\text{-a.e. for } \mu\text{-a.a. } z.$$

Since $\hat{\varphi}(z) \in X'^*\backslash X'$ for $\mu$-a.a. z, the above consideration implies

$$\lim_{n\to\infty} f'[\varphi(Y_{T_n})] = 0 \ P^x\text{-a.e.}$$

and finally, since $T_n \uparrow \zeta$,

$$\lim_{t\uparrow\zeta} f'[\varphi(Y_t)] = 0 \ P^x\text{-a.e.}$$

Conversely, suppose that (*) is fulfilled. Let $f' \in \mathfrak{C}_K(X')$. We extend f' to be zero on $X'^*\backslash X'$. In view of Theorem 4.1', we have $f'[\hat{\varphi}(z)] = 0$ $\mu$-a.e.

Now, letting $\{X'_n\}$ be a compact exhaustion of X' and defining $f'_n \in \mathfrak{C}_K(X'^*)$ to be $f'_n = 1$ on $\bar{X}'_n$ and $f'_n = 0$ on $X\backslash X'_{n+1}$, it is derived from the above consideration that $\mu(N) = 0$, where

$$N = \bigcup_n \{z \in \Delta_1 \,;\, f'_n[\hat{\varphi}(z)] > 0\}.$$

Then, $\hat{\varphi}(z) \cap X' = \emptyset$ for every $z \in \Delta_1\backslash N$, which completes the proof.

## References

[1] J. Bliedtner - W. Hansen, Potential theory – An analytic and probabilistic approach to balayage, Universitext, Springer-Verlag, 1986.

[2] C. Constantinescu - A. Cornea, Potential theory on harmonic spaces, Grundl. der math. Wiss., 158, Springer-Verlag, 1972.

[3] C. Dellacherie - P. A. Meyer, Probabilities and potential B, North-Holland Math. Studies, 72, North-Holland, 1982.

[4] J. L. Doob, Conditional Brownian motion and the boundary limits of harmonic functions, Bull. Soc. Math. France 85 (1957), 431–458.

[5] J. L. Doob, Conformally invariant cluster value theory, Illinois J. Math. 5 (1961), 521–549.

[6] J. L. Doob, Classical potential theory and its probabilistic counterpart, Grundl. der math. Wiss., 262, Springer-Verlag, 1984.

[7] J. Hyvönen, On resolutive compactifications of harmonic spaces, Ann. Acad. Sci. Fenn. Ser A I Math. Dissertationes 8 (1976).

[8] T. Ikegami, The boundary behavior of analytic mappings of Riemann surfaces, Complex Analysis (Joensuu, 1978), 161–166, Lecture Notes in Math., 747, Springer-Verlag, 1979.

[9] T. Ikegami, Compactifications of Martin type of harmonic spaces, Osaka J. Math. 23 (1986), 653–680.

[10] T. Ikegami, On the boundary behavior of harmonic morphisms at the boundary of compactifications of Martin type, to appear in Osaka J. Math.

[11] C. Meghea, Compactification des espaces harmoniques, Lecture Notes in Math., 222, Springer-Verlag, 1971.

[12] L. Naïm, Sur le rôle de la frontière de R. S. Martin dans la théorie du potentiel, Ann. Inst. Fourier 7 (1957), 183–281.

[13] K. Oja, On cluster sets of harmonic morphisms between harmonic spaces, Ann. Acad. Sci. Fenn. Ser. A I Math. Dissertationes 24 (1979).

# CONVEX FUNCTIONS AND THE NEHARI UNIVALENCE CRITERION [1]

**Wolfram Koepf**
Fachbereich Mathematik der FU Berlin
Arnimallee 3, 1000 Berlin 33

Nehari [4] showed that a convex function, i.e. a function which maps the unit disk $\mathbb{D}$ univalently onto a convex domain, satisfies the Nehari univalence criterion, i.e.

$$(1-|z|^2)^2|S_f(z)| \le 2, \; z \in \mathbb{D},$$

where $S_f := \left(\frac{f''}{f'}\right)' - \frac{1}{2}\left(\frac{f''}{f'}\right)^2$ denotes the Schwarzian derivative, and that this result is sharp, as the function $f(z) = \frac{1}{2}\log\frac{1+z}{1-z}$ shows. The method of the proof does not give all sharp functions. We shall show that the sharp function is essentially unique using another approach implying Nehari's result.

This shows furthermore that all convex domains except of parallel strip domains are Jordan domains in $\hat{\mathbb{C}}$, using a result of Gehring and Pommerenke [3].

Moreover we give a geometrical description of convex domains whose corresponding convex functions satisfy the stronger relation

$$\sup_{z\in\mathbb{D}} (1-|z|^2)^2|S_f(z)| < 2. \tag{1}$$

This result generalizes Nehari's [4] that bounded convex functions satisfy (1).

Without loss of generality we may assume that $f(z) = z + a_2z^2 + a_3z^3 + \cdots$ is normalized. The following result is well-known.

**Theorem 1** (see e.g. [6]). *If* $f(z) = z + a_2z^2 + a_3z^3 + \cdots$ *is convex, then*

$$|a_3 - a_2^2| \le \frac{1-|a_2|^2}{3}.$$

As a consequence we have

---

[1] This paper is in final form and no version of it will be submitted for publication elsewhere.

**Corollary 1.** *If* $f(z) = z + a_2z^2 + a_3z^3 + \cdots$ *is convex, then*

$$|a_3 - a_2^2| \le \frac{1}{3}$$

*with equality if and only if*

$$f(z) = \frac{1}{2x} \log \frac{1 + xz}{1 - xz}, \quad x \in \partial\mathbb{D}. \tag{2}$$

*Proof.* The inequality obviously follows from the theorem. Equality can only occur if $a_2 = 0$ and $|a_3| = 1/3$. Now we have to show that in this case $f$ is of the form (2). If $f(z) = z + a_3z^3 + \cdots$ is convex, then $1 + z\frac{f''(z)}{f'(z)} = 1 + 6a_3z^2 + \cdots$ has positive real part (see e.g. [5], Theorem 2.7). So the second coefficient of this function is bounded by 2 (see e.g. [5], Corollary 2.3), and it follows that $|a_3| \le 1/3$. Equality holds if and only if

$$1 + z\frac{f''(z)}{f'(z)} = \frac{1 + x^2z^2}{1 - x^2z^2}, \quad x \in \partial\mathbb{D},$$

(see e.g. [5], Corollary 2.3) , which is equivalent to (2). □

Hence we get

**Theorem 2.** *If* $f(z) = z + a_2z^2 + a_3z^3 + \cdots$ *is convex, then*

$$(1 - |a|^2)^2 |S_f(a)| \le 2, \tag{3}$$

*for all* $a \in \mathbb{D}$, *with equality if and only if* $f(\mathbb{D})$ *is a parallel strip domain, i.e.*

$$f(z) = \frac{1}{x + y} \log \frac{1 + xz}{1 - yz}, \quad x, y \in \partial\mathbb{D}, \; y \ne -x.$$

*Proof.* The theorem is true for $a = 0$, as Corollary 1 shows. Let now $a \in \mathbb{D}\setminus\{0\}$. Then the function $g(z) = z + b_2z^2 + b_3z^3 + \cdots$, defined by

$$g(z) = \frac{f\left(\frac{z + a}{1 + \bar{a}z}\right) - f(a)}{(1 - |a|^2)\, f'(a)}, \tag{4}$$

is also convex. On the other hand it is easily seen that

$$(1 - |a|^2)^2\, S_f(a) = S_g(0) = 6(b_3 - b_2^2),$$

so that inequality (3) follows. If equality holds, then $g$ has the form (2), and because of (4) $f$ must have a similar range, which gives the result. □

As a consequence one has

**Corollary 2.** *If $f$ is convex, and if $f(\mathbb{D})$ is no parallel strip domain, then $f(\mathbb{D})$ is a Jordan domain in $\hat{\mathbb{C}}$.*

*Proof.* This follows from [3], Theorem 1. □

Finally we shall give a complete geometrical description for convex domains whose corresponding convex functions satisfy the stronger relation (1).

**Theorem 3.** *If $f(z) = z + a_2z^2 + a_3z^3 + \cdots$ is convex, then the following conditions are equivalent:*

(a) $\sup_{z\in\mathbb{D}} (1 - |z|^2)^2 |S_f(z)| = 2$;

(b) *there is a sequence of domains $G_k$ which are similar to $f(\mathbb{D})$, such that the Carathéodory kernel of $(G_k)$ (see e.g. [5], p. 28) is a parallel strip;*

(c) $f(\mathbb{D})$ *is a parallel strip or is unbounded such that $\partial f(\mathbb{D})$ has an angle $\alpha = 0$ at $\infty$;*

(d) $\partial f(\mathbb{D})$ *is not a quasicircle in $\hat{\mathbb{C}}$.*

*Proof.* Suppose that $f(\mathbb{D})$ is a parallel strip. Then all conditions are true, (a) by Theorem 2 and (b) by choosing the constant sequence. In the sequel let $f(z) = z + a_2z^2 + a_3z^3 + \cdots$ be convex and $f(\mathbb{D})$ not a parallel strip. Then by Corollary 2 it is a Jordan domain in $\hat{\mathbb{C}}$.

(a) ⇔ (b). Suppose, condition (a) holds. If the supremum is attained at an interior point $a \in \mathbb{D}$, then Theorem 2 implies that $f(\mathbb{D})$ is a parallel strip, which we assumed not to be the case. So there is a sequence $(z_k)$ of numbers $z_k \in \mathbb{D}$ with

$$(1 - |z_k|^2)^2 |S_f(z_k)| \to 2$$

as $k \to \infty$. We define functions $g_k$ by

$$g_k(z) := \frac{f\left(\dfrac{z + z_k}{1 + \bar{z}_k z}\right) - f(z_k)}{(1 - |z_k|^2)\ f'(z_k)} = z + b_{2k}z^2 + b_{3k}z^3 + \cdots,$$

and get

$$(1 - |z_k|^2)^2 |S_f(z_k)| = 6|b_{3k} - b_{2k}^2| \to 2,$$

so that it follows with aid of Corollary 1 that the sequence $(g_k)$ or a subsequence converges locally uniformly to some function of the form $\frac{1}{2x}\log\frac{1+xz}{1-xz}$, $x \in \partial\mathbb{D}$. By definition all functions $g_k$ have ranges which are similar to $f(\mathbb{D})$ implying (b). Similarly one sees that (b) implies (a), too.

(a), (b) ⇒ (c). We have to show that

(i) $f(\mathbb{D})$ is unbounded,

(ii) $\partial f(\mathbb{D})$ has an angle $\alpha$ at $\infty$,

(iii) $\alpha = 0$.

Step (i) is Nehari's result [4] that bounded convex functions satisfy (1), (ii) is a property of convex domains, and (iii) will be deduced from (b).

Because $f(\mathbb{D})$ is convex, the complement is the union of halfplanes $\mathbb{C}\setminus f(\mathbb{D}) = \bigcup_{t\in T} H_t$. Consider now the images of the corresponding lines $\partial H_t$ on the Riemann sphere. They represent a family of circles on the sphere, having the north pole as a common point. So there exist two extremal directions, which correspond to the asymptotic directions of $\partial f(\mathbb{D})$ and give the semitangents of $\partial f(\mathbb{D})$ at $\infty$, and (ii) is verified.

Suppose now, $\alpha > 0$. We consider the images of the asymptotic lines on the Riemann sphere. These circles intersect at the north pole and at some finite point under the same angle $\alpha$. So the same is true for the asymptotic lines theirselves, because the stereographic projection is angle-preserving. For an arbitrary similar region, i.e. the image of $f(\mathbb{D})$ under the conformal mapping $az + b$, this angle remains invariant, so that each similar domain has the same fixed angle $\alpha$ at $\infty$.

Now, let $(G_k)$ be a given sequence of domains similar to $f(\mathbb{D})$, which converges to a parallel strip. With aid of a suitable Möbius transformation we transform $\infty$ into a finite point $w_0$. Of course this transformation does not preserve convexity, but it preserves univalence and the angle $\alpha$. So we get a new sequence of regions $(\tilde{G}_k)$, which have the point $w_0$ as a common boundary point, having there the fixed angle $\alpha$, and which converge to a region, having $w_0$ as a (doubly) boundary point with a zero angle there. This easily gives a contradiction (see e.g. [5], p. 31, problem 3), and so $\alpha = 0$.

(c) ⇒ (d). As a well-known consequence of Ahlfors' intrinsic characterization [1] quasicircles don't have zero angles.

(d) ⇒ (a). This is equivalent to the result of Ahlfors and Weill [2], who showed that (1) implies that $f(\mathbb{D})$ is a quasicircle in $\hat{\mathbb{C}}$. □

I want to thank Professor J. Becker for discussing the object with me.

## References

[1] Ahlfors, L. V., Quasiconformal reflections. Acta Math. **109** (1963), 291–301.

[2] Ahlfors, L. V. and Weill, G., A uniqueness theorem for Beltrami equations. Proc. Amer. Math. Soc. **13** (1962), 975–978.

[3] Gehring, F. W. and Pommerenke, Ch., On the Nehari univalence criterion and quasicircles. Comment. Math. Helv. **59** (1984), 226–242.

[4] Nehari, Z., A property of convex conformal maps. J. Analyse Math. **30** (1976), 390–393.

[5] Pommerenke, Ch., Univalent functions. Göttingen: Vandenhoeck & Ruprecht, 1975.

[6] Trimble, S. Y., A coefficient inequality for convex univalent functions. Proc. Amer. Math. Soc. **48** (1975), 266–267.

# A GREEN FORMULA WITH MULTIPLICITIES [1]

**Tapani Kuusalo**
Department of Mathematics, University of Jyväskylä
SF-40100 Jyväskylä, Finland

E. Artin introduced in [2] the idea of basing the combinatorial topology needed in the study of complex line integrals wholly on the skilful use of the winding number of a closed curve with respect to a point. But it was the famous textbook [1] of L. Ahlfors that made this approach available to a wider audience. We propose to give here an extension of the homological integration theory of E. Artin and L. Ahlfors in the plane to domains in $R^n$.

**1. Introduction.** The *Green formula* is usually given as

$$\int_{\partial U} \alpha = \int_{U} d\alpha\,, \quad \alpha \in \Lambda_c^{n-1}\bar{U}, \tag{1}$$

where the smooth (n–1)-form $\alpha$ has a compact support in the closure of the smoothly bordered open set $U \subset R^n$, or equivalently for compactly supported vector fields $Z = (Z_1,\dots,Z_n)$

$$\int_{\partial U} Z\cdot e \, dm_{n-1} = \int_{U} \operatorname{div} Z \, dm_n\,, \tag{1'}$$

where $e$ is the outward unit normal of $\partial U$. The identity (1') is reduced to (1) by taking as $\alpha$ the interior product

$$\alpha = Z \lrcorner\, \omega = \sum_{i=1}^{n} (-1)^{i-1} Z_i \, dx_1 \wedge \dots \wedge d\hat{x}_i \wedge \dots \wedge dx_n \in \Lambda_c^{n-1}\bar{U}$$

of the vector field $Z$ with the volume form $\omega$ of $R^n$,

[1] This paper is in final form and no version of it will be submitted for publication elsewhere.

$$\omega = dx_1 \wedge \ldots \wedge dx_n.$$

We propose now the following generalization: For all smooth (n–1)-forms $\alpha \in \Lambda^{n-1}R^n$ and for all smooth singular (n–1)-cycles c we have

$$\int_c \alpha = \int_{R^n} n(c,y)\, d\alpha(y), \tag{2}$$

where the integral valued *Kronecker index*

$$n(c,a) = \frac{1}{\Omega_{n-1}} \int_c \kappa(x-a),$$

$$\kappa(y) = |y|^{-n}\, y \neg \omega = \frac{\sum_{i=1}^{n} (-1)^{i-1}\, y_i\, dx_1 \wedge \ldots \wedge d\hat{x}_i \wedge \ldots \wedge dx_n}{|y|^n} \tag{3}$$

is defined and is locally constant in the complement of the support supp(c) of the cycle c with integral coefficients ($\Omega_{n-1}$ is the (n–1)-dimensional surface area of the unit sphere $S^{n-1}$ of $R^n$).

**2. An integral representation formula.** The Lie derivative $L_Z\alpha$ of a k-form $\alpha \in \Lambda^k R^n$ with respect to a vector field Z can be expressed using exterior derivatives as follows,

$$L_Z\alpha = Z \neg d\alpha + d(Z \neg \alpha).$$

Thus we have for any compactly supported form $\alpha$ and for any unit vector e the identity

$$\alpha(x) = -\int_0^\infty e \neg d\alpha(x+te)\, dt - d\int_0^\infty e \neg \alpha(x+te)\, dt.$$

Averaging over the unit sphere $S^{n-1}$ gives then the following *integral representation formula*

$$\alpha(x) = \frac{1}{\Omega_{n-1}} \int_{R^n} \frac{(x-y) \neg d\alpha(y)}{|x-y|^n}\, dm_n(y) + d\left( \frac{1}{\Omega_{n-1}} \int_{R^n} \frac{(x-y) \neg \alpha(y)}{|x-y|^n}\, dm_n(y) \right) \tag{4}$$

for every k-form $\alpha \in \Lambda_c^k R^n$. Especially, the exterior derivative of an (n–1)-form $\alpha$ is a multiple of the volume form $\omega$,

$$d\alpha(x) = b(x)\omega,$$

so that any form $\alpha \in \Lambda_c^{n-1} R^n$ is cohomologuous to the form

$$K(d\alpha) = \frac{1}{\Omega_{n-1}} \int_{R^n} \kappa(x-y) b(y)\, dm_n(y) \in \Lambda^{n-1} R^n, \tag{5}$$

where $\kappa$ is the *Kronecker form* defined above in (3). Thus we have for all compactly supported (n–1)-forms $\alpha \in \Lambda_c^{n-1} R^n$ and for all (n–1)-cycles c

$$\int_c \alpha = \int_c K(d\alpha). \tag{6}$$

**3. The Green formula with multiplicities.** Let c be a smooth singular (n–1)-cycle in an open subset U of $R^n$. If the Kronecker index $n(c,x)$ vanishes for all points x outside the set U, we say following Artin and Ahlfors that c is *homologuous to zero in* U, $c \sim 0$ in U.

**Theorem.** *If a smooth singular* (n–1)-*cycle* c *is homologuous to zero in an open subset* U *of* $R^n$, *we have for all* (n–1)-*forms* $\alpha \in \Lambda^{n-1} U$ *the* Green formula with multiplicities,

$$\int_c \alpha = \int_U n(c,y)\, d\alpha(y). \tag{7}$$

*Proof.* As $c \sim 0$ in U, the *extended support of* c,

$$E = \operatorname{supp}(c) \cup \{x \in R^n \setminus \operatorname{supp}(c) : n(c,x) \neq 0\}$$

is a compact subset of U. Thus we can find a compactly supported form $\tilde{\alpha} \in \Lambda_c^{n-1} U$ which agrees with $\alpha$ in an open neighbourhood of E. If the exterior derivatives of $\alpha$ and $\tilde{\alpha}$ are denoted

$$d\alpha = b\omega, \quad d\tilde{\alpha} = \tilde{b}\omega,$$

then b and $\tilde{b}$ agree on E, so that we have by the Fubini theorem

$$\int_c \alpha = \int_c \tilde{\alpha} = \int_c \frac{1}{\Omega_{n-1}} \int_{y \in R^n} \kappa(x-y)\tilde{b}(y)\, dm_n(y)$$
$$= \int_{R^n} \left( \frac{1}{\Omega_{n-1}} \int_c \kappa(x-y) \right) \tilde{b}(y)\, dm_n(y) = \int_{R^n} n(c,y)\tilde{b}(y)\, dm_n(y)$$
$$= \int_U n(c,y) b(y)\, dm_n(y) = \int_U n(c,y)\, da(y),$$

and we are done.

Incidentally, it follows from the above theorem via the de Rham cohomology theory that the homology class of a smooth singular (n–1)-cycle c of U vanishes in $H_{n-1}(U)$ if and only if c is homologuous to zero in the open set $U \subset R^n$ according to the definition of Artin and Ahlfors.

**4. The planar case.** If we are considering only the (1,0)-differentials

$$\alpha = f\, dz \in \Lambda_c^{10} \mathbb{C}$$

in the complex plane $\mathbb{C}$, the representation formula (4) is reduced to the well-known $\bar{\partial}$-formula

$$f(z) = \frac{1}{\pi} \int_{\mathbb{C}} \frac{\partial f / \partial \bar{z}(\zeta)}{z - \zeta}\, dm_2(\zeta),$$

and the Green formula (7) is replaced by a *Pompeiu formula with multiplicities*

$$\int_c f\, dz = \int_U n(c,z)\, \partial f/\partial \bar{z}(z)\, d\bar{z} \wedge dz = 2i \int_U n(c,z)\, \partial f/\partial \bar{z}(z)\, dm_2(z) \qquad (8)$$

for all smooth complex-valued functions f in the open set $U \subset \mathbb{C}$.

## References


[1] Ahlfors, L. V., Complex analysis. McGraw-Hill Book Company, Inc., New York-Toronto-London, 1953.

[2] Artin, E., On the theory of complex functions. Notre Dame Math. Lectures 4, 55–70, 1944.

# A RIEMANNIAN METRIC INVARIANT UNDER MÖBIUS TRANSFORMATIONS IN $\mathbb{R}^n$ [1]

Heinz Leutwiler
Universität Erlangen-Nürnberg, Mathematisches Institut
Bismarckstrasse 1 1/2, D-8520 Erlangen, F. R. Germany

## 1. Introduction

A conformally invariant Riemannian metric is introduced as follows: Let $\Omega$ be a domain in $\mathbb{R}^n$ $(n \geq 3)$ and $G_\Omega$ its Green function. For $x \in \Omega$ we set

$$\kappa_\Omega(x) = \lim_{y \to x} (G(x,y) - G_\Omega(x,y))^{\frac{1}{n-2}}, \tag{1.1}$$

where $G = G_{\mathbb{R}^n}$ designates Green's function with respect to $\mathbb{R}^n$, i.e.

$$G(x,y) = \frac{1}{\|x-y\|^{n-2}}. \tag{1.2}$$

As usual, $\|x\| = (\Sigma x_i^2)^{1/2}$ denotes the euclidean norm of $x = (x_1,\dots,x_n) \in \mathbb{R}^n$.

Note that by definition of the Green function $G_\Omega$ the limit in (1.1) always exists. Introducing for fixed $x \in \Omega$ the greatest harmonic minorant $m_\Omega^x$ of the superharmonic function $G(\cdot,x)$, restricted to $\Omega$, we may in fact write

$$\kappa_\Omega(x) = (m_\Omega^x(x))^{\frac{1}{n-2}} \quad (x \in \Omega). \tag{1.3}$$

With $\kappa_\Omega$ at our disposal we now introduce a Riemannian metric on $\Omega$, setting

$$ds = \kappa_\Omega(x)\|dx\|. \tag{1.4}$$

---

[1] This paper is in final form and no version of it will be submitted for publication elsewhere.

In case of the unit ball $B_n$, resp. the upper half space $\mathbb{R}^n_+$, it agrees – up to the factor 1/2 – with the Poincaré (hyperbolic) metric.

In §2 we shall verify that the quotient $G_\Omega/G$ of the Green functions $G_\Omega$ and $G$ is invariant under Möbius transformations. This result will be used in §3 to show that the metric (1.4) is conformally invariant.

In §4 we compare our metric with the quasi-hyperbolic metric, introduced by Gehring and Palka in [3]. It will be shown that the two metrics are equivalent, provided the domain satisfies what we shall call a "uniform cone condition".

In §5 we investigate the sectional curvature of the metric (1.4), as well as some related ones, in case of the quarter space, respectively the half ball.

We are indebted to Professor Robert Burckel and Dr. Pawel Kröger for valuable discussions.

## 2. The invariance of $G_\Omega/G$

The conformal invariance of the metric (1.4) is based on the following behaviour of the quotient $G_\Omega/G$ of the Green functions $G_\Omega$ resp. $G$.

**Theorem 2.1.** *Let* $\gamma : \Omega_1 \to \Omega_2$ *be a Möbius transformation, mapping a domain* $\Omega_1 \subset \mathbb{R}^n$ *bijectively onto a domain* $\Omega_2 \in \mathbb{R}^n$. *Then*

$$\frac{G_{\Omega_1}(x,y)}{G(x,y)} = \frac{G_{\Omega_2}(\gamma x, \gamma y)}{G(\gamma x, \gamma y)}, \tag{2.1}$$

*for all* $x, y \in \Omega_1$, $x \neq y$.

The proof is based on the following observation. Let $\gamma'(x)$ denote the Jacobian matrix of the Möbius transformation $\gamma$ at the point $x \in \mathbb{R}^n$ and $|\gamma'(x)|$ the "linear change of scale" at $x$, i.e. the unique positive real number with the property that $\gamma'(x)/|\gamma'(x)|$ is an orthogonal matrix. Then we have

**Lemma 2.2.** *Under the hypothesis of Theorem* 2.1, *every harmonic, resp. superharmonic, function* $u_2$ *on* $\Omega_2$ *is transformed into a harmonic, resp. superharmonic, function* $u_1$ *on* $\Omega_1$, *provided we set*

$$u_1(x) = |\gamma'(x)|^{\frac{n-2}{2}} u_2(\gamma x), \tag{2.2}$$

*for all* $x \in \Omega_1$.

A proof of this result has been given in [6] (Lemma 2.1) in case the function $u_2$ is harmonic. Using the well known characterization of superharmonic functions as those "lying above the harmonic ones" (see e.g. [4], p. 60, Second Definition) it is clear that the statement of Lemma 2.2 also holds for this class of functions.

*Proof of Theorem* 2.1. Let us introduce the function

$$\varphi(x) = |\gamma'(x)|^{\frac{n-2}{2}}, \tag{2.3}$$

defined on $\mathbb{R}^n \setminus \{\gamma^{-1}(\infty)\}$. Fixing $y \in \Omega_2$ we conclude from Lemma 2.2, applied to $u_2 = G_{\Omega_2}(\cdot, \gamma y)$, that the function

$$u_1(x) = \varphi(x) G_{\Omega_2}(\gamma x, \gamma y) \tag{2.4}$$

is superharmonic on $\Omega_1$ and harmonic on $\Omega_1 \setminus \{y\}$. Since $u_2$ is in fact a potential, the same is true for $u_1$. Consequently $u_1$ must be a multiple of the Green function $G_{\Omega_1}(\cdot, y)$ with pole at $y$, i.e. there is a positive constant $c = c(y)$, depending on $y$, such that

$$G_{\Omega_1}(x,y) = c(y)\varphi(x) G_{\Omega_2}(\gamma x, \gamma y), \tag{2.5}$$

for all $x \in \Omega_1$. Interchanging the role of $x$ and $y$ and taking into account the symmetry of the Green functions we conclude that there is also a constant $d = d(x)$ such that

$$G_{\Omega_1}(x,y) = d(x)\varphi(y) G_{\Omega_2}(\gamma x, \gamma y), \tag{2.6}$$

for all $y \in \Omega_1$. Equating (2.5) and (2.6) yields

$$\frac{c(y)}{\varphi(y)} = \frac{d(x)}{\varphi(x)}$$

for arbitrary $x, y \in \Omega_1$. Consequently both sides are constant; in particular there is a constant $c > 0$ such that $c(y) = c\varphi(y)$, for all $y \in \Omega_1$. Inserted in (2.5) this yields

$$G_{\Omega_1}(x,y) = c\varphi(x)\varphi(y) G_{\Omega_2}(\gamma x, \gamma y), \tag{2.7}$$

for all $x, y \in \Omega_1$.

Choosing temporarily $\Omega_1 = \mathbb{R}^n \setminus \{\gamma^{-1}(\infty)\}$ and $\Omega_2 = \mathbb{R}^n \setminus \{\gamma(\infty)\}$, we infer from

(2.7), and the fact that in this case $G_{\Omega_1} = G_{\Omega_2} = G$, that

$$\frac{1}{\|x-y\|^{n-2}} = c\varphi(x)\varphi(y)\,\frac{1}{\|\gamma x-\gamma y\|^{n-2}} \tag{2.8}$$

holds for all $x, y \in \mathbb{R}^n\setminus\{\gamma^{-1}(\infty)\}$, $x \neq y$, and some constant $c > 0$. Actually $c = 1$ so that (2.8) agrees with formula (15) on page 19 in Ahlfors Lecture notes [1], but we shall not make use of this fact here. Inserted in (2.7) the relation (2.8) yields the identity

$$\frac{G_{\Omega_1}(x,y)}{G(x,y)} = a\,\frac{G_{\Omega_2}(\gamma x,\gamma y)}{G(\gamma x,\gamma y)}, \tag{2.9}$$

valid for all $x, y \in \Omega_1$, $x \neq y$, and some constant $a \in \mathbb{R}$.

From $\lim_{y\to x} G_{\Omega_i}(x,y)/G(x,y) = 1$ $(i = 1, 2)$, we then conclude that $a = 1$, finishing the proof of Theorem 2.1.

In case of the upper half space $\mathbb{R}^n_+ = \{(x_1,\dots,x_n) \in \mathbb{R}^n : x_n > 0\}$ Theorem 2.1 implies that the quotient $G_{\mathbb{R}^n_+}/G$ must be a function of $\delta$, the hyperbolic distance on $\mathbb{R}^n_+$. And indeed we have

$$\frac{G_{\mathbb{R}^n_+}(x,y)}{G(x,y)} = 1 - \left(\frac{e^{\delta(x,y)}-1}{e^{\delta(x,y)}+1}\right)^{n-2}$$

for all $x, y \in \mathbb{R}^n_+$, $x \neq y$. A similar formula of course holds in case of the unit ball $B_n$.

**Remark.** It might be worthwhile to point out that the argument used to prove Theorem 2.1 also works if we replace the domains $\mathbb{R}^n\setminus\{\gamma^{-1}(\infty)\}$ and $\mathbb{R}^n\setminus\{\gamma(\infty)\}$ by arbitrary Brelot spaces $X_1$ and $X_2$, having the proportionality property and admitting symmetric Green functions $G_1$ and $G_2$. The Möbius transformation $\gamma$ obviously has to be replaced by a bijective $\varphi$-morphism (in the sense of Fuglede [2]) in order to conclude that for any two domains $\Omega_i \subset X_i$ $(i = 1, 2)$ with $\gamma(\Omega_1) = \Omega_2$ we have the following equality: $G_{\Omega_1}(x,y)/G_1(x,y) = G_{\Omega_2}(\gamma x,\gamma y)/G_2(\gamma x,\gamma y)$, provided we normalize the Green functions accordingly.

### 3. The conformal invariance of the metric

The Riemannian metric, introduced in (1.4) and (1.1), resp. (1.3), obviously has the following property: Let $\Omega_1, \Omega_2$ be arbitrary domains in $\mathbb{R}^n$ $(n \geq 3)$. Then

$$\Omega_1 \subset \Omega_2 \Rightarrow \kappa_{\Omega_1} \geq \kappa_{\Omega_2} \text{ on } \Omega_1. \tag{3.1}$$

In addition we have

**Theorem 3.1.** *The Riemannian metric* $ds = \kappa_\Omega(x)\|dx\|$, *defined by* (1.1), *is conformally invariant, i.e. for each Möbius transformation* $\gamma$ *mapping a domain* $\Omega_1 \subset \mathbb{R}^n$ *bijectively onto a domain* $\Omega_2$ *of* $\mathbb{R}^n$,

$$\kappa_{\Omega_1}(x) = \kappa_{\Omega_2}(\gamma x) \cdot |\gamma'(x)|, \tag{3.2}$$

*for* $x \in \Omega_1$.

*Proof.* From (2.8) and the fact that $c = 1$ (as mentioned) there follows that

$$\lim_{y\to x} \frac{G(x,y)}{G(\gamma x, \gamma y)} = \lim_{y\to x} \varphi(x)\varphi(y) = |\gamma'(x)|^{n-2},$$

for all $x \in \mathbb{R}^n \backslash \{\gamma^{-1}(\infty)\}$. Hence, by Theorem 2.1,

$$\begin{aligned} \kappa_{\Omega_1}^{n-2}(x) &= \lim_{y\to x} (G(x,y) - G_{\Omega_1}(x,y)) \\ &= \lim_{x\to y} \frac{G(x,y)}{G(\gamma x, \gamma y)} \cdot \lim_{y\to x} (G(\gamma x, \gamma y) - G_{\Omega_2}(\gamma x, \gamma y)) \\ &= |\gamma'(x)|^{n-2} \cdot \kappa_{\Omega_2}^{n-2}(\gamma x), \end{aligned}$$

for all $x \in \Omega_1$, and hence (3.2) follows.

On $\mathbb{R}^n_+$, the upper half space, we have

$$\kappa_{\mathbb{R}^n_+}(x) = \frac{1}{2x_n}, \tag{3.3}$$

$x = (x_1,\dots,x_n) \in \mathbb{R}^n_+$, and hence for $B(0,R)$ the ball of radius $R$ and midpoint 0, the metric (1.4) is given by

$$\kappa_{B(0,R)}(x) = \frac{R}{R^2 - \|x\|^2} \quad (\|x\| < R). \tag{3.4}$$

On $\mathbb{R}^n \backslash \overline{B}(0,R)$ we have

$$\kappa_{\mathbb{R}^n \backslash \overline{B}(0,R)}(x) = \frac{R}{\|x\|^2 - R^2} \quad (\|x\| > R). \tag{3.5}$$

Finally let us mention the connection of $\kappa_\Omega(x)$ from (1.3) with the so-called *harmonic radius* $R_\Omega$, studied by J. Hersch in [5]. By definition

$$R_\Omega = (\inf_{x \in \Omega} \kappa_\Omega(x))^{-1}. \tag{3.6}$$

Hence, by (3.4), $R_\Omega$ agrees with R in case of the ball $B(x_0,R)$. Since $R_\Omega$ is explicitly known in some other cases (see [5]) it yields a lower bound for the metric (1.4).

## 4. The quasi-hyperbolic metric

In 1976 F. Gehring and B. Palka introduced the following metric: Let $\Omega$ be a proper subdomain of $\mathbb{R}^n$ and set

$$k_\Omega(x) = \frac{1}{\operatorname{dist}(x,\partial\Omega)} \quad (x \in \Omega), \tag{4.1}$$

where $\operatorname{dist}(x,\partial\Omega)$ denotes the euclidean distance from x to $\partial\Omega$. Then

$$ds = k_\Omega(x)\|dx\| \tag{4.2}$$

defines a complete metric, called the *quasi-hyperbolic metric* of $\Omega$. On $\mathbb{R}^n_+$ it agrees with the Poincaré metric, but on $B_n$ it does not. Hence this metric cannot be conformally invariant. Nevertheless it still behaves reasonable under Möbius transformations (see [3], Corollary 2.5).

It will be of interest to know when the metrics (1.4) and (4.2) are equivalent. An estimate from above is readily at hand, since

**Lemma 4.1.** *For any proper subdomain* $\Omega \subset \mathbb{R}^n$ $(n \geq 3)$,

$$\kappa_\Omega(x) \leq k_\Omega(x), \tag{4.3}$$

*for all* $x \in \Omega$.

*Proof.* Let $x \in \Omega$ and $r = \operatorname{dist}(x,\partial\Omega)$. By definition the ball $B = B(x,r)$ will be contained in $\Omega$. Thus, by (3.1), $\kappa_B(x) \ge \kappa_\Omega(x)$ and (4.3) follows from the fact that $k_B(x) = \kappa_B(x)$.

Obviously, (4.3) is sharp and so is the following estimate

**Lemma 4.2.** *For any convex, proper subdomain* $\Omega$ *of* $\mathbb{R}^n$,

$$\frac{1}{2} k_\Omega(x) \le \kappa_\Omega(x) \le k_\Omega(x), \tag{4.4}$$

*for all* $x \in \Omega$.

*Proof.* Given $x \in \Omega$, there exists $y \in \partial\Omega$ such that $\|x-y\| = \operatorname{dist}(x,\partial\Omega)$. Let $H$ denote a supporting hyperplane of $\Omega$ at $y$. Since $H$ is also a supporting plane for the ball $B(x,r)$ with radius $r = \|x-y\|$ it is clear that $\operatorname{dist}(x,H) = r$. Applying (3.1) to $\Omega$ and the half space generated by $H$ we conclude from (3.3) that

$$\kappa_\Omega(x) \ge \frac{1}{2\operatorname{dist}(x,H)} = \frac{1}{2} k_\Omega(x).$$

In the non-convex case the metrics (1.4) and (4.2) will not be equivalent in general, as the case of the punctured ball shows. Removing a closed polar set from $\Omega$ does not change the metric (1.4), showing that in general it will not be complete, in contrast to the quasi-hyperbolic one. However we have

**Lemma 4.3.** *Let* $\Omega$ *be a domain in* $\mathbb{R}^n$, *satisfying the condition*

$$b = \sup_{x\in\Omega} \operatorname{dist}(x,\partial\Omega) < \infty. \tag{4.5}$$

*Assuming that there is a positive number* $a$ *with the property that at each boundary point* $y \in \partial\Omega$ *there is a ball* $B$ *of radius* $a$ *with* $B \cap \Omega = \emptyset$ *and* $y \in \partial B$, *we have*

$$\frac{a}{2a+b} k_\Omega(x) \le \kappa_\Omega(x) \le k_\Omega(x) \tag{4.6}$$

*for all* $x \in \Omega$.

For bounded domains, more generally domains with the property (4.5), (4.6) implies (4.4), letting $a$ tend to $\infty$. In contrast to Lemma 4.2, however, Lemma 4.3 cannot hold for arbitrary unbounded domains. Indeed, on the complement of the ball $B_n$ the metric (1.4) is not equivalent to the quasi-hyperbolic one, although the "ball condition" is trivially satisfied.

*Proof of Lemma* 4.3. Given $x \in \Omega$ there is a point $y \in \partial\Omega$ such that $r = \|x-y\| = \operatorname{dist}(x,\partial\Omega)$. By hypothesis there is a ball $B$ of radius $a$, contained in the complement of $\Omega$, such that $y \in \partial B$. Hence, by (3.1) and (3.5),

$$\kappa_\Omega(x) \geq \frac{a}{(a+r)^2 - a^2} = \frac{a}{2a+r}\cdot\frac{1}{r} \geq \frac{a}{2a+b}\, k_\Omega(x),$$

proving Lemma 4.3.

It has been noticed by Pawel Kröger that the equivalence of the metrics (1.4) and (4.2) is already at hand, if we replace the "uniform ball condition" in Lemma 4.2 by the following one:

**Definition.** A domain $\Omega \in \mathbb{R}^n$ is said to satisfy a *uniform cone condition* if

1) at each boundary point $y \in \partial\Omega$ there is a truncated, solid cone $S$, with vertex at $y$, contained in the complement of $\Omega$.
2) The truncated cones are all congruent, i.e. differ only by an euclidean motion.

We then have:

**Theorem 4.4.** *Let $\Omega \subset \mathbb{R}^n$ be a domain fulfilling property* (4.5). *Assuming that in addition $\Omega$ satisfies a uniform cone condition, the metrics* (1.4) *and* (4.2) *are equivalent.*

Note that the equivalence of the two metrics implies in particular that – under the hypothesis of Theorem 4.4 – the metric (1.4) is *complete.*

The proof of Theorem 4.4 is most conveniently based on the following somewhat technical lemma.

**Lemma 4.5.** *Let $\Omega$ be a domain in $\mathbb{R}^n$, having the following property: To each $x \in \Omega$ there is a ball $B = B(z,R)$, contained in the complement of $\Omega$, such that with respect to some numbers $\alpha, \beta > 0$ $(\alpha > \beta)$, depending only on $\Omega$,*

$$\|x - z\| \leq \alpha\cdot \operatorname{dist}(x,\partial\Omega) \ \ \textit{and}\ \ R \geq \beta\cdot\operatorname{dist}(x,\partial\Omega). \tag{4.7}$$

*Under this hypothesis the metrics* (1.4) *and* (4.2) *are equivalent.*

*Proof.* Let $x \in \Omega$ and $r = \operatorname{dist}(x,\partial\Omega)$. By (1.3), (3.5) and (4.3)

$$k_\Omega(x) \geq \kappa_\Omega(x) \geq \kappa_{\mathbb{R}^n\setminus\overline{B}(z,R)}(x) = \frac{R}{\|x-z\|^2 - R^2}$$

$$\geq \frac{\beta r}{\alpha^2 r^2 - \beta^2 r^2} = \frac{\beta}{\alpha^2 - \beta^2}\, k_\Omega(x),$$

for all $x \in \Omega$.

*Proof of Theorem* 4.4. Let $\varphi$ be the half-angle and $h$ the height of the congruent, truncated cones. Given $x \in \Omega$, let $y \in \partial\Omega$ be a point with the property that $\operatorname{dist}(x,\partial\Omega) = \|x - y\| =: r$.

*Case* 1: Let $r \leq \frac{h}{2}$. Then choose $z$ to be the unique point on the axis of revolution with the property that $\|z-y\| = r$, and take $R = r \sin \varphi$.

*Case* 2: $r > \frac{h}{2}$. Let $z$ be the unique point on the axis with $\|z-y\| = \frac{h}{2}$ and set $R = \frac{h}{2} \sin \varphi$.

Then the hypotheses of Lemma 4.5 are satisfied, provided we put $\alpha = 2$ and $\beta = \sin \varphi \cdot \min(1, h/2b)$, where $b$ is defined by (4.5). Thus Theorem 4.4 is established.

Note that in the hypothesis of Theorem 4.4 the condition (4.5) may be dropped if we do not allow to truncate the cones.

**5. The quarter space**

Besides the unit ball $B_n$, resp. the upper half space $\mathbb{R}^n_+$, explicit formulas for the Green function are also known in case of the half ball resp. the quarter space $Q_n = \{(x_1,...,x_n) \in \mathbb{R}^n : x_1 > 0 \text{ and } x_2 > 0\}$. Since $Q_n$ is conformally equivalent to the half ball in $\mathbb{R}^n$, it suffices to concentrate in what follows on the quarter space.

On $Q_n$ $(n \geq 3)$ the metric (1.4) is given by

$$\kappa_{Q_n}(x) = \frac{1}{2}\left[\frac{1}{x_1^{n-2}} + \frac{1}{x_2^{n-2}} - \frac{1}{(x_1^2 + x_2^2)^{(n-2)/2}}\right]^{\frac{1}{n-2}}. \tag{5.1}$$

It is easy to see that there is a positive, real analytic function $f$ on $[2,\infty[$ with the property that $\kappa_{Q_n}$ can be written in the form

$$\kappa(x) = f\left(\frac{x_1}{x_2} + \frac{x_2}{x_1}\right)\frac{1}{x_1 + x_2}, \tag{5.2}$$

for all $x \in Q_n$.

On the other hand, given any positive function $f : [2,\infty[ \to \mathbb{R}$ of class $C^2$, the Riemann metric

$$ds = \kappa(x)\|dx\|, \tag{5.3}$$

with $\kappa$ defined by (5.2), is invariant under the group $\Gamma_n$ of Möbius transformations, mapping the quarter space $Q_n$ bijectively onto itself. Indeed, it can be shown (see e.g. [8]) that the orbits of the group $\Gamma_n$ on $Q_n$ are precisely those given by the equation

$$\frac{x_1}{x_2} + \frac{x_2}{x_1} = \text{const.} \tag{5.4}$$

What we are interested in is the curvature of all metrics of the form (5.3), resp. (5.2). We first remark

**Lemma 5.1.** *For any function* $f : [2,\infty[ \to \mathbb{R}^+$ *of class* $C^2$, *the function*

$$u(x_1,x_2) = \log f\left(\frac{x_1}{x_2} + \frac{x_2}{x_1}\right) - \log(x_1 + x_2),$$

*defined on* $\{(x_1,x_2) \in \mathbb{R}^2 : x_1 > 0, x_2 > 0\}$ *is a solution of the following Monge-Ampère equation*

$$(u_{11} - u_1^2)(u_{22} - u_2^2) = (u_{12} - u_1u_2)^2, \tag{5.5}$$

*where* $u_i = \partial u/\partial x_i$, $u_{ij} = \partial^2 u/\partial x_i \partial x_j$ $(i, j = 1, 2)$.

In case of (5.3) the sectional curvature $K(W)$ of the 2-plane $W$, generated by the linearly independent vectors $\xi = (\xi^i)$ and $\eta = (\eta^i)$ $(i, j = 1,\dots,n)$, is given by the formula

$$K(W) = \frac{R_{hijk}\xi^h\xi^j\eta^i\eta^k}{\kappa^4(\delta_{hj}\delta_{ik} - \delta_{hk}\delta_{ij})\xi^h\xi^j\eta^i\eta^k}, \tag{5.6}$$

where $R_{hijk}$ denotes the curvature tensor, provided we use the summation convention.

Setting

$$u = \log \kappa, \tag{5.7}$$

with $\kappa$ defined by (5.2), it is easy to verify that, on account of Lemma 5.1, the sectional curvature can be written in the following form:

$$K(W) = -\frac{\|A\xi - B\eta\|^2}{\|\xi\|^2\|\eta\|^2 - (\xi,\eta)^2} \cdot \frac{1}{\kappa^2(u_{11} - u_1^2)} - \frac{\|\nabla u\|^2}{\kappa^2} \tag{5.8}$$

where

$$A = (u_{11} - u_1^2)\eta^1 + (u_{12} - u_1u_2)\eta^2,$$
$$B = (u_{11} - u_1^2)\xi^1 + (u_{12} - u_1u_2)\xi^2.$$

We thereby assumed that $u_{11} - u_1^2 \neq 0$. Locally this can always be achieved, provided there are no points $x$ with $(u_{11} - u_1^2)(x) = (u_{22} - u_2^2)(x) = 0$. In the latter case, the first term on the right hand side of (5.8) vanishes.

Assuming now that the 2-plane $W$ is tangent to the orbit (5.4) (i.e., if $x_2 = cx_1$, then $\xi^2 = c\xi^1$ and $\eta^2 = c\eta^1$), it is readily seen that $A = B = 0$. There results

**Theorem 5.2.** *Assuming that the 2-plane* $W$ *is tangent to the orbit* (5.4), *the sectional curvature* $K(W)$ *is given by the formula*

$$K(W) = -\frac{\|\nabla\kappa\|^2}{\kappa^4} \tag{5.9}$$

*provided that the metric* (5.3) *is of the form* (5.2), *where* $f$ *denotes an arbitrary* $C^2$*-function on* $[2,\infty[$. *It is hence constant along each orbit.*

We remark that the scalar curvature is also constant along each orbit.

Finally let us look at the Laplace-Beltrami equation, associated to the metric (5.2), (5.3) on $Q_n$:

$$\Delta h + (n-2)\left\{\left[\left(\frac{1}{x_2} - \frac{x_2}{x_1^2}\right)g - \frac{1}{x_1 + x_2}\right]\frac{\partial h}{\partial x_1} + \left[\left(\frac{1}{x_1} - \frac{x_1}{x_2^2}\right)g - \frac{1}{x_1 + x_2}\right]\frac{\partial h}{\partial x_2}\right\} = 0, \tag{5.10}$$

where $g = f'/f$ and $\Delta$ denotes the usual Laplace operator. What we want is to determine the set of all those positive solutions which satisfy the Laplace-Beltrami equation (5.10) with respect to *every* $C^2$-function $g$ on $[2,\infty[$. A solution of this sort is the function $h(x) = (x_1^2 + x_2^2)^{(n-2)/2}$. Applying the inversion $J : x \to x/\|x\|^2$ to $Q_n$ and $h$, we obtain a further solution:

$$h(x) = \left(\frac{x_1^2 + x_2^2}{\|x\|^4}\right)^{\frac{n-2}{2}}.$$

Translating this kernel yields the solutions

$$h(x) = c(x_1^2 + x_2^2)^{\frac{n-2}{2}} + \tag{5.11}$$

$$\int_{\mathbb{R}^{n-2}} \frac{(x_1^2 + x_2^2)^{\frac{n-2}{2}}}{\left[x_1^2 + x_2^2 + (x_3 - \xi_3)^2 + \cdots + (x_n - \xi_n)^2\right]^{n-2}} d\mu(\xi_3,\ldots,\xi_n),$$

generated by some Borel measure $\mu$ on $\mathbb{R}^{n-2}$ and some constant $c \geq 0$. Our goal is to show that there are no other solutions of the desired kind, i.e.

**Theorem 5.3.** *Each positive* $C^2$*-function* $h$ *on* $Q_n$ *which satisfies, for every* $C^2$*-function* $g$, *the Laplace-Beltrami equation* (5.10) *admits the representation* (5.11).

For the proof, just note that every solution of (5.10) which is independent of $g$ must be of the form

$$h(x) = v(x_1^2 + x_2^2, x_3, \ldots, x_n)$$

where $v = v(y)$ $(y \in \mathbb{R}^{n-1},\ y_1 > 0)$ satisfies the equation

$$4\, y_1 v_{11} + v_{22} + \cdots + v_{n-1,n-1} - 2(n-4)v_1 = 0. \tag{5.12}$$

Setting $u(x_1,\ldots,x_{n-1}) = v(4x_1, x_2, \ldots, x_{n-1})$, equation (5.12) transforms into

$$x_1 u_{11} + u_{22} + \cdots + u_{n-1,n-1} - \tfrac{1}{2}(n-4)u_1 = 0. \tag{5.13}$$

Hence the proof of Theorem 5.3 is reduced to a representation theorem for the positive solutions of this equation on the open set $\{x \in \mathbb{R}^{n-1} : x_1 > 0\}$.

Rather than for (5.13) we shall solve the problem for the following slightly more general equation of Grušhin-type:

$$x_n^\alpha \left(\frac{\partial^2 u}{\partial x_1^2} + \cdots + \frac{\partial^2 u}{\partial x_{n-1}^2}\right) + \frac{\partial^2 u}{\partial x_n^2} + \frac{\beta}{x_n}\frac{\partial u}{\partial x_n} + \frac{\gamma}{x_n^2} u = 0 \tag{5.14}$$

$(\alpha > -2,\ 4\gamma \leq (\beta - 1)^2)$, considered on $\mathbb{R}_+^n = \{x \in \mathbb{R}^n : x_n > 0\}$.

In case $\beta = 0$, a representation for the positive solutions has already been given in [7, Theorem 6.1]. Using the same idea, we write $u$ in the form

$$u(x) = w(x_1, \ldots, x_{n-1}, \tfrac{1}{\delta} x_n^\delta), \text{ where } \delta = \frac{\alpha + 2}{2},$$

and check that the function $w$ satisfies the (generalized) Weinstein equation (1.1) treated in [7]. Then Theorem 2.2 of [7], essentially due to B. Brelot - M. Brelot and A. Huber, yields the following result.

**Theorem 5.4.** *Every positive solution* $u$ *on* $\mathbb{R}^n_+$ *of the equation* (5.14) *admits the unique representation*

$$u(x) = c(x_n)^{\frac{1}{2}(1-\beta+q)} + \int_{\mathbb{R}^{n-1}} \frac{(x_n)^{\frac{1}{2}(1-\beta+q)}}{\left[(\alpha+2)^2\|x'-\xi\|^2 + 4x_n^{\alpha+2}\right]^{\frac{n-1}{2}+\frac{q}{\alpha+2}}}\, d\mu(\xi)$$

*where* $x = (x',x_n) \in \mathbb{R}^n_+$, $q = \sqrt{(\beta-1)^2 - 4\gamma}$, $c$ *is nonnegative constant and* $\mu$ *a positive measure on* $\mathbb{R}^{n-1}$.

**Added in proof.** Meantime J. Ferrand found a conformally invariant version of the quasi-hyperbolic metric (see her article in this volume). In addition R. S. Kulkarni and U. Pinkall in their forthcoming paper "A canonical metric for Möbius structures" introduced yet another invariant Riemannian metric. It is of class $C^{1,1}$, so that a.e. the curvature tensor makes sense. Finally let us mention that in [6] a non-Riemannian, conformally invariant distance had been studied.

## References


[1] Ahlfors, L. V., Möbius transformations in several dimensions. Ordway Lectures in Mathematics, University of Minnesota, 1981.

[2] Fuglede, B., Harmonic morphisms between Riemannian manifolds. Ann. Inst. Fourier 28:2 (1978), 107–144.

[3] Gehring, F. W. and Palka, B. P., Quasiconformally homogeneous domains. J. Analyse Math. 30 (1976), 172–199.

[4] Helms, L. L., Introduction to potential theory. Wiley-Interscience, New York, 1969.

[5] Hersch, J., On the torsion function, Green's function and conformal radius: an isoperimetric inequality of Pólya and Szegö, some extensions and applications. J. Analyse Math. 36 (1979), 102–117.

[6] Leutwiler, H., On a distance invariant under Möbius transformations in $\mathbb{R}^n$. Ann. Acad. Sci. Fenn. Ser. A I Math. 12 (1987), 3–17.

[7] Leutwiler, H., Best constants in the Harnack inequality for the Weinstein equation. Aequationes Math. 34 (1987), 304–315.

[8] Sluka, B., Über Möbius Transformationen und eine invariante Metrik im $\mathbb{R}^n$. Diplomarbeit, Universität Erlangen-Nürnberg, 1987.

# DEFORMATION OF ONE-DIMENSIONAL QUASISYMMETRIC EMBEDDINGS

JOUNI LUUKKAINEN
Department of Mathematics, University of Helsinki
Hallituskatu 15, 00100 Helsinki, Finland

## 1. Introduction

1.1. Let $\mathcal{H}$ be the set of all homeomorphisms $\eta: [0,\infty) \to [0,\infty)$ with $\eta(1) \geq 1$. Consider an embedding $f: (X,d) \to (Y,d)$ of metric spaces. As in [TV], we say that $f$ is *$\eta$-quasisymmetric* ($\eta$–QS) for $\eta \in \mathcal{H}$ if $d(a,x) \leq td(b,x)$ implies $d(f(a),f(x)) \leq \eta(t)d(f(b),f(x))$ whenever $a,b,x \in X$ and $t \geq 0$. If this implication is only supposed to hold for $t = 1$, with $\eta(1)$ replaced by a constant $K \geq 1$, we say that $f$ is *$K$-quasisymmetric* ($K$–QS; in [TV] weakly $K$–QS). If each point $x \in X$ has a neighbourhood $U_x \subset X$ such that $f|U_x$ is $\eta_x$–QS for some $\eta_x \in \mathcal{H}$, we say that $f$ is *locally quasisymmetric* (LQS). If $Y = \mathbb{R}$ is the real line and $X \subset \mathbb{R}$ is an interval, $K$-quasisymmetry implies $\eta$-quasisymmetry for $\eta \in \mathcal{H}$ depending only on $K$ by [TV, 2.16] and reduces to $K$-quasisymmetry in the usual sense of [K], where it is assumed that $a - x = x - b$ in the above inequality.

The main purpose of this paper is to verify the fact needed in [L] that in $\mathbb{R}$ quasisymmetry is preserved in two extension operators for embeddings close to the identity map, one of which is related to homeomorphisms commuting with a homothety, the other to the hyperbolic metric of $(-1,1)$. In [L] we studied generalizations of these operators to the Euclidean space $\mathbb{R}^n$, applying for each $n \geq 1$ the fact mentioned above, and used them to proving a result on deformation of embeddings in $\mathbb{R}^n$. The following theorem is a simple corollary of the case $n = 1$ of this result [L, 5.7]. We denote any inclusion map by id.

1.2. THEOREM. *There is $\delta > 0$ with the following property: If $h: [-2,2] \to \mathbb{R}$ is an embedding with $|h(x) - x| < \delta$ for each $x \in [-2,2]$, there is a homeomorphism $h^*: \mathbb{R} \to \mathbb{R}$ depending continuously on $h$ in the topologies of uniform convergence and satisfying the following conditions:*

(1) *$h = \mathrm{id}$ implies $h^* = \mathrm{id}$.*

(2) *$h^*(x) = h(x)$ if $|x| \leq 1$.*

(3) *$h^*(x) = x$ if $|x| \geq 2$.*

This paper is in final form and no version of it will be submitted for publication elsewhere.

(4) *If* $K \geq 1$ *and* $h$ *is* $K$–QS, *then* $h^*$ *is* $K^*$–QS *with* $K^* \geq 1$ *depending only on* $K$.

## 2. Periodicity and quasisymmetry

2.1. In this section we prove Theorem 2.8 needed in [L] (see 2.7). We make use of two lemmas on homeomorphisms of $\mathbb{R}$ which commute with a homothety, i.e., are periodic.

2.2. LEMMA. *Let* $a = c_0 < c_1 < \cdots < c_p = b$ $(p \geq 2)$ *be a subdivision of an interval* $[a, b]$, *let* $m \in \mathbb{N}$ *satisfy* $m \geq (b-a)/(c_{j+1} - c_j)$ *for* $1 \leq j < p$, *let* $K \geq 1$, *and let* $f\colon [a, b] \to \mathbb{R}$ *be an embedding such that* $f|[c_{j-1}, c_{j+1}]$ *is* $K$–QS *for* $1 \leq j < p$. *Then* $f$ *is* $K^m$–QS.

*Proof.* We may assume that $f$ is increasing. Let $x \in \mathbb{R}$ and $t > 0$ with $[x-t, x+t] \subset [a, b]$. Let $x_i = x + it/m$ for $i \in \mathbb{Z}$, $|i| \leq m$. Then for each $i \in \mathbb{Z}$, $|i| < m$, there is $j \in \mathbb{Z}$, $1 \leq j < p$, such that $[x_{i-1}, x_{i+1}] \subset [c_{j-1}, c_{j+1}]$. In fact, if $1 \leq j < p$ and $c_{j-1} \leq x_{i-1} \leq c_j$, then $x_{i+1} - x_{i-1} = 2t/m \leq (b-a)/m \leq c_{j+1} - c_j$, whence $x_{i+1} \leq c_{j+1}$, and if $x_{i-1} > c_{p-1}$, then we can choose $j = p-1$. Let $d_i = f(x_i) - f(x_{i-1})$ for $-m < i \leq m$. Then $1/K \leq d_i/d_{i-1} \leq K$ for $-m+2 \leq i \leq m$. Hence,

$$f(x+t) - f(x) = \sum_{i=1}^{m} d_i \leq K \sum_{i=1}^{m} d_{i-1} \leq \cdots \leq K^m \sum_{i=1}^{m} d_{i-m} = K^m\big(f(x) - f(x-t)\big),$$

and, similarly, $f(x+t) - f(x) \geq \big(f(x) - f(x-t)\big)/K^m$. It follows that $f$ is $K^m$–QS. □

2.3. TERMINOLOGY. If $X$ denotes $\mathbb{R}$ or $\mathbb{R}_+ = [0, \infty)$ or $\mathbb{R}_- = (-\infty, 0]$, if $\varrho > 1$, and if $f\colon X \to X$ is a homeomorphism such that $f(\varrho x) = \varrho f(x)$ for each $x \in X$, we say that $f$ is *$\varrho$-periodic*. Note that $f(0) = 0$.

2.4. LEMMA. *Let* $\varrho > 1$, *let* $K \geq 1$, *and let* $f\colon \mathbb{R}_+ \to \mathbb{R}_+$ *be a* $\varrho$*-periodic homeomorphism such that* $f$ *is* $K$–QS *on* $[a/\varrho, \varrho a]$ *for some* $a > 0$. *Then* $f$ *is* $K'$–QS *with* $K' \geq 1$ *depending only on* $\varrho$ *and* $K$.

*Proof.* Obviously, $f|[a, \varrho^2 a]$ is $K$–QS. Hence, 2.2 implies that $f|[a/\varrho, \varrho^2 a]$ is $K_1$–QS with $K_1 \geq 1$ depending only on $\varrho$ and $K$. Since for each $x > 0$ there is $k \in \mathbb{Z}$ with $[x/\varrho, \varrho x] \subset [\varrho^{k-1}a, \varrho^{k+2}a]$ and since $f(\varrho^k x) = \varrho^k f(x)$ for all $x \in \mathbb{R}_+$ and $k \in \mathbb{Z}$, it follows that $f|[x/\varrho, \varrho x]$ is $K_1$–QS for each $x > 0$. Choose $p = p(\varrho) \in \mathbb{N}$ with $\varrho^p \geq 2$.

Now consider $Q = \big(f(x+t) - f(x)\big)/\big(f(x) - f(x-t)\big)$ with $0 < t \leq x$. Set $s = x - x/\varrho > 0$. Then $x - s = x/\varrho$ and $x + s \leq \varrho x$. Thus, $1/K_1 \leq Q \leq K_1$ if $t \leq s$.

Suppose $t > s$. Then

$$Q \geq \frac{f(x+s)-f(x)}{f(x)-0} \geq \frac{1}{K_1}\frac{f(x)-f(x-s)}{f(x)} = \frac{1-1/\varrho}{K_1}.$$

On the other hand, since $x+t \leq 2x \leq \varrho^p x$, we have

$$Q \leq \frac{f(\varrho^p x)-f(x)}{f(x)-f(x/\varrho)} = \frac{\varrho^p-1}{1-1/\varrho}.$$

Thus, $f$ is $K'$-QS, $K' = (1-1/\varrho)^{-1}\max(K_1, \varrho^p - 1)$. □

2.5. LEMMA. *Let $\varrho > 1$, $K \geq 1$, and $M \geq 1$, and let $f\colon \mathsf{R} \to \mathsf{R}$ be an increasing $\varrho$-periodic homeomorphism such that $f|\mathsf{R}_+$ and $f|\mathsf{R}_-$ are $K$–QS and such that*

$$1/M \leq f(c)/|f(-c)| \leq M$$

*for some $c > 0$. Then $f$ is $K'$-QS with $K' \geq 1$ depending only on $\varrho$, $K$, and $M$.*

*Proof.* If $x > 0$, choose $k \in \mathsf{Z}$ with $\varrho^k c \leq x \leq \varrho^{k+1}c$. Then $\varrho^k f(c) \leq f(x) \leq \varrho^{k+1}f(c)$ and $\varrho^{k+1}f(-c) \leq f(-x) \leq \varrho^k f(-c)$, whence $1/\varrho M \leq f(x)/|f(-x)| \leq \varrho M$. Thus, [K, Theorem 3] implies that $f$ is $K'$-QS with $K' = \varrho M(1+K+K^2)$. □

2.6. FACT ([K, p. 245], [TV, 2.2]). If $f\colon J \to \mathsf{R}$ and $g\colon J' \to J$ are $K$–QS embeddings of intervals in $\mathsf{R}$, then the inverse $f^{-1}\colon fJ \to J$ and the composition $fg$ are $K'$–QS with $K'$ depending only on $K$.

2.7. As in [L], if $U$ is an open subset of a metric space $X$, we let $E(U;X)$ denote the set of all open embeddings of $U$ into $X$ equipped with the topology of uniform convergence on compact sets. The following theorem 2.8 is a slight strengthening of a part of the case $n = 1$ of [L, 2.13], a quantitative canonical Schoenflies extension theorem for $\mathsf{R}^n$. The proof of [L, 2.13] will become complete as we prove 2.8 using exactly the same construction for the map $\varphi$ as in the proof of [L, 2.13] in [L] for $n = 1$.

2.8. THEOREM. *Let $X = \mathsf{R}$ or $\mathsf{R}_+$, let $A = ((-1,1) \setminus [-\frac{1}{3}, \frac{1}{3}]) \cap X$, and let $B = (-1,1) \cap X$. Then there exist a neighbourhood $\mathcal{P}$ of id: $A \to X$ in $E(A;X)$ and a continuous map $\varphi\colon \mathcal{P} \to E(B;X)$ such that the following conditions hold for each $h \in \mathcal{P}$:*

(1) *$h = \text{id}$ implies $\varphi(h) = \text{id}$.*

(2) *$\varphi(h) = h$ on $B \setminus (-\frac{2}{3}, \frac{2}{3})$.*

(3) *$\varphi(h)$ is increasing and $\varphi(h)(0) = 0$.*

(4) *If $Y \in \{\mathsf{R}, \mathsf{R}_+, \mathsf{R}_-\}$, if $Y \subset X$, and if (a) $h|A \cap Y$ is LQS or (b) $K \geq 1$ and $h$ is $K$–QS on each component of $A \cap Y$, then $\varphi(h)|B \cap Y$ is, respectively, (a) LQS or (b) $K^*$-QS with $K^* \geq 1$ depending only on $K$.*

*Proof.* As in [L], it suffices to prove the theorem with (1) omitted. In fact, since $\varphi(\mathrm{id})B = B$ and since $\varphi(\mathrm{id})$ is $K_0$–QS for some $K_0 \geq 1$, replacing $\varphi$ by $\varphi(\cdot)\varphi(\mathrm{id})^{-1}$ yields then the full theorem.

Choose a so small neighbourhood $\mathcal{P}_0$ of id: $A \to X$ in $E(A;X)$ that $h$ is increasing, $h(-\frac{1}{2}) < 0$ if $X = \mathrm{R}$, and $h(\frac{1}{2}) > 0$ for each $h \in \mathcal{P}_0$. Let the numbers $\frac{1}{2} < s < t < u < c = \frac{2}{3} < r < 1$ and $\varrho = \frac{4}{3}$ be as in [L]. Note that $c/\varrho = \frac{1}{2}$. Let $\mathcal{P}$ be a neighbourhood of id: $A \to X$ in $\mathcal{P}_0$ so small that we can construct a continuous map $\varphi\colon \mathcal{P} \to E(B;X)$ as in [L] such that if $h \in \mathcal{P}$, then $\hat{h} = \varphi(h)$ satisfies the conditions (2) and (3) of the theorem and the formulas (2.17), (2.18), (2.20), and (2.21) of [L], where $\bar{h} = \hat{h}|B \setminus (-\frac{1}{2}, \frac{1}{2})$. If $h \in \mathcal{P}$, then $\hat{h}(\varrho x) = \varrho \hat{h}(x)$ for each $x \in [-\frac{1}{2}, \frac{1}{2}] \cap X$ by [L, (2.17)], which implies that there is a unique $\varrho$-periodic homeomorphism $h^*\colon X \to X$ extending $\hat{h}|[-c,c] \cap X$. We also note that if $h \in \mathcal{P}$, then, as $h \in \mathcal{P}_0$, the construction of $\hat{h}$ in [L] shows that $\hat{h}|[0,r]$ depends only on $h|[\frac{1}{2}, r]$.

Now let $Y$, $h$, and $K$ be as in (4b). Suppose first $Y = \mathrm{R}_+$. Then [L, (2.21), (2.18), and (2.20)] and 2.6 imply that there is $K_1 \geq 1$ depending only on $K$ such that $\hat{h}|[u/\varrho, t]$, $\hat{h}|[s,c]$, and $\hat{h}|[u,1)$ are $K_1$–QS. By 2.2, $\hat{h}|[u/\varrho, c]$ is $K_2$–QS with $K_2 \geq 1$ depending only on $K_1$. Then $\hat{h}|[c/\varrho^2, \frac{1}{2}]$, too, is $K_2$–QS. By 2.2, $\hat{h}|[c/\varrho^2, c]$ is $K_3$–QS with $K_3 \geq 1$ depending only on $K_2$. Hence, 2.4 implies that $h^*|\mathrm{R}_+$ is $K_4$–QS with $K_4 \geq 1$ depending only on $K_3$. It follows that $\hat{h}|[0,c]$ is $K_4$–QS. Then 2.2 yields that $\hat{h}|[0,1)$ is $K_5$–QS with $K_5 \geq 1$ depending only on $K_4$ and $K_1$.

Let $X = \mathrm{R}$ in (4b). The cases $Y = \mathrm{R}_+$ and $Y = \mathrm{R}_-$ being symmetric, suppose that $Y = \mathrm{R}$. Then $h^*|\mathrm{R}_+$ and $h^*|\mathrm{R}_-$ are $K_4$–QS. We may assume that $\mathcal{P}$ is so small that $\frac{1}{2} \leq h(c)/|h(-c)| \leq 2$. Then 2.5 implies that $h^*$ is $K_6$–QS with $K_6 \geq 1$ depending only on $K_4$. Thus, $\hat{h}|[-c,c]$ is $K_6$–QS. By 2.2, $\hat{h}$ is $K_7$–QS with $K_7 \geq 1$ depending only on $K_6$ and $K_1$.

Finally, let $Y$ and $h$ be as in (4a). Then $h$ is $K$–QS for some $K \geq 1$ on each component of $([-r, -\frac{1}{2}] \cup [\frac{1}{2}, r]) \cap Y$. Thus, by the proof of (4b), $\hat{h}|(-r,r) \cap Y$ is LQS. Hence, $\hat{h}|B \cap Y$ is LQS. □

## 3. Hyperbolic metric and quasisymmetry

3.1. The hyperbolic metric $\sigma$ on $B^1 = (-1,1)$ is given by

$$\sigma(x,y) = \int_x^y \frac{2\,dt}{1-t^2} = \log \frac{(1-x)(1+y)}{(1+x)(1-y)} \quad \text{if } -1 < x \leq y < 1.$$

Let $\overline{B}_\sigma(x,r) = \{y \in B^1 \mid \sigma(x,y) \leq r\}$ for $x \in B^1$, $r > 0$. Let $M_\varepsilon = \frac{1}{4}\varepsilon e^{-2\varepsilon} > 0$ for $\varepsilon > 0$. The purpose of this section is to prove the following theorem needed in the proof of [L, 4.7].

3.2. THEOREM. *Let $\varepsilon > 0$, let $K \geq 1$, and let $f\colon B^1 \to B^1$ be an increasing homeomorphism such that*

(1) $f|\colon (\overline{B}_\sigma(x,\varepsilon),\sigma) \to (B^1,\sigma)$ *is $K$–QS for each $x \in B^1$,*

(2) $\sigma(f(x),x) \leq M_\varepsilon$ *for each $x \in B^1$.*

*Then the extension of $f$ by id to a homeomorphism $f^*\colon \mathbb{R} \to \mathbb{R}$ is $K^*$–QS with $K^* \geq 1$ depending only on $\varepsilon$ and $K$.*

3.3. LEMMA. *Let $x \in B^1$ and $r > 0$. Then*

$$\frac{1}{e^r} \leq \frac{1-x^2}{2}\,\frac{\sigma(a,b)}{|a-b|} \leq e^r \quad \text{if } a,b \in \overline{B}_\sigma(x,r),\ a \neq b, \tag{3.4}$$

$$\frac{1}{e^{2r}} \leq \frac{\sigma(a,c)}{\sigma(b,c)}\,\frac{|b-c|}{|a-c|} \leq e^{2r} \quad \text{if } a,b,c \in \overline{B}_\sigma(x,r),\ a \neq c \neq b. \tag{3.5}$$

*Proof.* (3.4): Since $\sigma(-y,-z) = \sigma(y,z)$ for all $y,z \in B^1$, we may assume that $x \geq 0$. Write $\overline{B}_\sigma(x,r) = [x_1,x_2]$. Suppose $x_1 \leq a < b \leq x_2$. Since $|x_1| \leq x_2$, we get

$$\frac{\sigma(a,b)}{|a-b|} = \frac{1}{b-a}\int_a^b \frac{2\,dt}{1-t^2} \leq \frac{1}{b-a}\int_a^b \frac{2\,dt}{1-x_2^2} = \frac{2}{1-x_2^2} = \frac{2e^r(1+x)}{(1-x)(1+x_2)^2} \leq \frac{2e^r}{1-x^2}.$$

On the other hand, if $x_1 \geq 0$, then

$$\frac{\sigma(a,b)}{|a-b|} \geq \frac{1}{b-a}\int_a^b \frac{2\,dt}{1-x_1^2} = \frac{2}{1-x_1^2} = \frac{2(1+x)}{e^r(1-x)(1+x_1)^2} \geq \frac{2}{e^r(1-x^2)},$$

whereas if $x_1 \leq 0$, then $\sigma(0,x) \leq r$, whence

$$\frac{\sigma(a,b)}{|a-b|} \geq \frac{1}{b-a}\int_a^b 2\,dt = 2 \geq \frac{2}{(1+x)^2} \geq \frac{2}{e^r(1-x^2)}.$$

(3.5): This follows from (3.4). □

3.6. COROLLARY. $|x-y| \leq (1-|x|)e^{\sigma(x,y)}\sigma(x,y)$ *for all $x,y \in B^1$.*

*Proof.* This follows from (3.4). □

3.7. LEMMA. *If $\varepsilon > 0$, if $f\colon B^1 \to B^1$ is a function satisfying 3.2(2), and if $x,y \in B^1$ with $\sigma(x,y) \geq \varepsilon$, then $|x-y|/2 \leq |f(x)-f(y)| \leq 2|x-y|$.*

*Proof.* It suffices to show that $|f(x)-x| \leq \frac{1}{4}|x-y|$. Choose a point $z$ between $x$ and $y$ with $\sigma(x,z) = \varepsilon$. Then, since $\sigma(f(x),x) \leq M_\varepsilon \leq \varepsilon$, it follows from (3.5) that

$$\frac{|f(x)-x|}{|x-y|} \leq \frac{|f(x)-x|}{|x-z|} \leq e^{2\varepsilon}\,\frac{\sigma(f(x),x)}{\sigma(x,z)} \leq \frac{e^{2\varepsilon}M_\varepsilon}{\varepsilon} = \tfrac{1}{4}. \ \square$$

3.8. LEMMA. *Let $\varepsilon$, $K$, and $f$ be as in 3.2. Then $f$ is $K'$–QS in Euclidean metrics with $K' \geq 1$ depending only on $\varepsilon$ and $K$.*

*Proof.* Consider $B_x = \overline{B}_\sigma(x,\varepsilon)$ for $x \in B^1$. It is easy to see that in the terminology of [TV, 2.7], $(B^1,\sigma)$ and, thus, $(B_x,\sigma)$ are $k$–HTB with $k(\alpha) = 2\alpha+1$ and $(B_x,\sigma)$ is 1–BT. Hence, by 3.2(1) and [TV, 2.9 and 2.15], there is $\eta_K \in \mathcal{H}$ depending only on $K$ such that $f|\colon (B_x,\sigma) \to (B^1,\sigma)$ is $\eta_K$–QS. Let $r = \varepsilon + M_\varepsilon$. Then $fB_x \subset B'_x = \overline{B}_\sigma(x,r)$ by 3.2(2). Let $d$ denote the Euclidean metric of $B^1$. Then (3.5) implies that id: $(B'_x,d) \to (B'_x,\sigma)$ and id: $(B'_x,\sigma) \to (B'_x,d)$ are $\eta_\varepsilon$–QS, where $\eta_\varepsilon(t) = e^{2r}t$ for $t \geq 0$. It follows that $f|\colon (B_x,d) \to (B^1,d)$ is $\theta$–QS, where $\theta = \eta_\varepsilon\eta_K\eta_\varepsilon$.

Now consider three distinct points $a,b,x \in B^1$ with $|a-x| \leq |b-x|$. Let $Q = |f(a)-f(x)|/|f(b)-f(x)|$. We must find $K' \geq 1$ depending only on $\varepsilon$ and $K$ such that $Q \leq K'$. We divide the estimation into four separate cases.

*Case 1:* $\sigma(a,x) \leq \varepsilon$ and $\sigma(b,x) \leq \varepsilon$. Now $Q \leq \theta(1)$.

*Case 2:* $\sigma(a,x) > \varepsilon \geq \sigma(b,x)$. Choose a point $c$ between $a$ and $x$ such that $\sigma(c,x) = \varepsilon$. Then

$$\frac{|f(a)-f(x)|}{|f(c)-f(x)|} \leq 4\frac{|a-x|}{|c-x|} \leq 4\frac{|b-x|}{|c-x|} \leq 4\eta_\varepsilon\left(\frac{\sigma(b,x)}{\sigma(c,x)}\right) \leq 4\eta_\varepsilon(1)$$

by 3.7, and

$$\frac{|f(c)-f(x)|}{|f(b)-f(x)|} \leq \theta\left(\frac{|c-x|}{|b-x|}\right) \leq \theta(1)\,.$$

Multiplying these inequalities yields $Q \leq 4\eta_\varepsilon(1)\theta(1)$.

*Case 3:* $\sigma(a,x) \leq \varepsilon < \sigma(b,x)$. Choose a point $c$ between $b$ and $x$ such that $\sigma(c,x) = \varepsilon$. Then

$$Q \leq \frac{|f(a)-f(x)|}{|f(c)-f(x)|} \leq \eta_\varepsilon\left(\eta_K\left(\frac{\sigma(a,x)}{\sigma(c,x)}\right)\right) \leq \eta_\varepsilon(\eta_K(1))\,.$$

*Case 4:* $\sigma(a,x) > \varepsilon$ and $\sigma(b,x) > \varepsilon$. Now $Q \leq 4$ by 3.7. □

3.9. LEMMA. *Let $K \geq 1$, let $\alpha \in (0,1)$, and let $f\colon B^1 \to B^1$ be an increasing homeomorphism such that*

(1) *$f$ is $K$–QS in Euclidean metrics,*

(2) *$|f(x)-x| \leq \alpha(1-|x|)$ for each $x \in B^1$.*

*Then the extension of $f$ by* id *to a homeomorphism $f^*\colon \mathbb{R} \to \mathbb{R}$ is $K^\dagger$–QS with $K^\dagger \geq 1$ depending only on $K$ and $\alpha$.*

*Proof.* Set

$$Q(x,t) = \frac{f^*(x+t)-f^*(x)}{f^*(x)-f^*(x-t)} \quad \text{for } x \in \mathbb{R},\ t > 0.$$

We show that $f^*|(-1,\infty)$ and $f^*|(-\infty,1)$ are $K_*$–QS with $K_* \geq 1$ depending only on $K$ and $\alpha$. Since $1/K \leq Q(0,t) \leq K$ if $0 < t < 1$ and $Q(0,t) = 1$ if $t \geq 1$, it then follows from [K, Theorem 3] that $f^*$ is $K^\dagger$–QS with $K^\dagger = K(1+K_*+K_*^2)$.

We only consider $f^*|(-1,\infty)$. We modify the proof of [K, Theorem 4]. Estimate first $Q(1,t)$ for $0<t<2$. We have

$$Q(1,t)=\frac{(1+t)-1}{1-f(1-t)}=\frac{t}{t+((1-t)-f(1-t))}.$$

Here $|(1-t)-f(1-t)|\leq\alpha(1-|1-t|)\leq\alpha t$. Hence,

$$1-\alpha\leq\frac{1}{1+\alpha}\leq Q(1,t)\leq\frac{1}{1-\alpha}.$$

Thus, by [K, Theorem 3], $f^*|(-1,3)$ is $K_1$-QS, where $K_1=(1+K+K^2)/(1-\alpha)$.

Now estimate $Q(x,t)$ for all $x>-1$ and $t>0$ with $x-t>-1$. If $x-t\geq 1$, then $Q(x,t)=1$. If $x-t<1$ and $x\geq 3$, then, as $x-1<t<x+1$, we get that

$$\tfrac{1}{2}\leq\frac{x-1}{x+1}\leq Q(x,t)=\frac{t}{x-f(x-t)}\leq\frac{x+1}{x-1}\leq 2.$$

Thus, $\frac{1}{2}\leq Q(3,t)\leq 2$ if $t<4$. Hence, by [K, Theorem 3], $f^*|(-1,7)$ is $K_2$-QS, where $K_2=2(1+K_1+K_1^2)\geq 2$. Therefore, $1/K_2\leq Q(x,t)\leq K_2$ if $x<3$ and, thus, always, i.e., $f^*|(-1,\infty)$ is $K_2$-QS. □

3.10. PROOF OF 3.2. By 3.8 and 3.6, we can apply 3.9 with the constant $K'$ of 3.8 substituted for $K$ and with $\alpha=M_\varepsilon e^{M_\varepsilon}<1$. Theorem 3.2 follows with $K^*=(K')^\dagger$. □

## References


[K] J. A. Kelingos, *Boundary correspondence under quasiconformal mappings.* Michigan Math. J. 13 (1966), 235–249.

[L] J. Luukkainen, *Respectful deformation of bi-Lipschitz and quasisymmetric embeddings.* Ann. Acad. Sci. Fenn. Ser. A I Math. 13 (to appear).

[TV] P. Tukia and J. Väisälä, *Quasisymmetric embeddings of metric spaces.* Ann. Acad. Sci. Fenn. Ser. A I Math. 5 (1980), 97–114.

# ON THE EXPONENT OF CONVERGENCE FOR INFINITELY GENERATED FUCHSIAN GROUPS [1]

**Thomas A. Metzger**
Department of Mathematics and Statistics, University of Pittsburgh
Pittsburgh, PA 15260, U.S.A.

## §1. Introduction and Preliminaries

Let $\Delta$ be the unit disk in the complex plane and suppose that $\Gamma$ is a Fuchsian group acting on $\Delta$. In general, we will be considering the case that $\Gamma$ is infinitely generated, thus if one looks at the quotient Riemann surface $\Delta/\Gamma$, one gets an arbitrary open surface and not the classical case of a compact or compact bordered Riemann surface. The exponent of convergence, $\delta(\Gamma) = \delta$, of the group $\Gamma$ is defined to bc

$$\delta = \inf \{t > 0 : \Sigma_{\gamma \in \Gamma} |\gamma'(0)|^t < \infty\}.$$

It is well known that $\delta \leq 1$. If $\delta = 1$ and $\Sigma_{\gamma \in \Gamma} |\gamma'(0)| < \infty$ we say $\Gamma$ is of convergence type (C.T.) otherwise we say $\Gamma$ is of divergence type (D.T.). If we denote by L the limit set of $\Gamma$, then for a finitely generated Fuchsian group we have: $\Gamma$ is of D.T. if and only if $m(L) = 2\pi$. The case of C.T. is more interesting in that we have, $\Gamma$ is of C.T. implies $\delta(\Gamma) < 1$ and in fact, one has $\delta(\Gamma) = \Lambda(L)$ (Here, $\Lambda$ is the Hausdorff dimension of L). These results are due to various authors. One should look at [B1], [Pa1], [Pa2] and [S1] for many interesting results and methods. In our case much less is known unless $\Gamma$ is a subgroup of a finitely generated group. In this note we shall attempt to make the first steps toward a better understanding of the infinitely generated case.

## §2. The Action of $\Gamma$ on $\partial\Delta$

Without loss of generality assume that $\Gamma$ does not fix the origin and let $\Omega$ be

---

[1] This paper is in final form and no version of it will be submitted for publication elsewhere.

the Ford fundamental region for $\Gamma$ centered at 0. Define the set $e = \partial\Omega \cap \partial\Delta$ and let $E = \cup\gamma e$. Clearly E is measurable and $m(E) > 0$ if and only if $m(e) > 0$. Moreover, since every point of e lies outside every isometric circle or lies on the boundary of some isometric circle we see that $\gamma e \cap e$ is at most countable and thus it is of measure zero. It follows immediately that one has

$$m(E) = \int_E |d\zeta| = \int_{\cup\gamma e} |d\zeta| = \sum_{\gamma\in\Gamma} \int_e |\gamma'(\zeta)|\,|d\zeta| = \int_e \sum_{\gamma\in\Gamma} |\gamma'(\zeta)|\,|d\zeta| \tag{1}$$

and we have shown

**Lemma 1.** *For a.e.* $\zeta$ *in* E *we have*

$$\sum_{\gamma\in\Gamma} |\gamma'(\zeta)| < \infty.$$

The points on $\partial\Delta - E$ have been studied previously by various authors, for example see [N1]. This set is where the group $\Gamma$ acts ergodically on L, in the sense that there is no measurable fundamental set for $\partial\Delta - E$. For our purposes it is best to see this by proving

**Lemma 2.** *If* $\zeta \in \partial\Delta - E$ *then* $\Sigma_{\gamma\in\Gamma} |\gamma'(\zeta)| = \infty$.

In order to prove this we let $c(\zeta, r)$ be the horocycle at $\zeta \in \partial\Delta$ given by

$$c(\zeta, r) = \{z \in \Delta : |z - (1-r)\zeta| = r\}$$

and let $D(\zeta, r)$ be the interior of this horocycle. P. Nicholls [N1] has shown

**Proposition A.** *If* $\zeta \in \partial\Delta - E$ *then for any* $r > 0$ *we have*

$$D(\zeta, r) \cap \Gamma 0 \neq \emptyset.$$

Here, as usual, $\Gamma z_0 = \{\gamma z_0 : \gamma \in \Gamma\}$ denotes the orbit of $z_0$ under $\Gamma$.

*Proof* (Lemma 2). If $\gamma \in \Gamma$ then

$$\gamma z = e^{i\alpha} \frac{z - w}{1 - z\bar{w}}$$

where $w = \gamma^{-1}(0)$. By Proposition A it suffices to show that the fact $M_s = \{\gamma : \gamma^{-1}(0) \in D(\zeta, s)\}$ is an infinite set implies that the series diverges at $\zeta$.

To see this, we fix $s > 0$ and let $w = te^{ix}$ and $\rho = e^{i\varphi}$. The fact that $w \in D(\rho, s)$ implies that $|te^{ix} - (1-s)e^{i\varphi}| < s^2$, i.e., we have

$$1 - 2s + t^2 - 2t\cos(\varphi - x) + 2ts\cos(\varphi - x) < 0. \tag{2}$$

Thus (2) implies that if $w \in D(\zeta, s)$ then

$$|1 - \zeta\bar{w}|^2 < 2s(1 - t\cos(\varphi - x)).$$

But (2) also implies

$$1 - t\cos(\varphi - x) \le 1 + \frac{2s - 1 - t^2}{2(1-s)} = \frac{1 - t^2}{s(1-s)}.$$

This and the definition of $M_s$ implies

$$\sum_{\gamma\in\Gamma} |\gamma'(\zeta)| \ge \sum_{\gamma\in M_s} \frac{1 - |w|^2}{|1 - \zeta\bar{w}|^2} \ge \sum_{\gamma\in M_s} \frac{1 - t^2}{2s(1 - t\cos(\varphi - x))}$$
$$\ge \frac{1-s}{s} \sum_{\gamma\in M} 1 = \infty.$$

As we shall see this very elementary and naive computation and counting argument will be very useful for us. Also it is worth noting that Lemmas 1 and 2 and the remarks preceeding Lemma 1 yield

**Proposition 3.** L *has a measurable fundamental set if and only if* $m(E) = 2\pi$.

This follows immediately from the remarks above since if L has a measurable fundamental set $\lambda$, then by (1) above

$$m(L) = \int_\lambda \sum |\gamma'(\zeta)|\,|d\zeta|$$

and so $\Sigma|\gamma'(\zeta)| < \infty$ a.e. on $\lambda$. We note that in general one cannot take e (or $e \cap L$) for this $\lambda$, the reason for this is that $\gamma e \cap e \neq \emptyset$. However, since $\gamma e \cap e$ is at most countable it works perfectly well for the integration theory and is "close enough" to a fundamental set for our purposes. In particular, Proposition 3 above says that if $\Gamma$ is of D.T. then there is no measurable fundamental set for the action of $\Gamma$ on $\partial\Delta$, i.e., the action of $\Gamma$ on $\partial\Delta$ is ergodic.

**Remark 1.** In [Pol] Ch. Pommerenke basically proved all of the results above. We have reproved these results, in order to set the stage for the tangential approximation which is done in §4.

**Remark 2.** In [K1] Kruškal asserted that L never has a measurable fundamental set if $m(L) > 0$. Proposition 3 above clarifies this situation once we note that there exist groups $\Gamma$ with $m(L) = m(E) = 2\pi$. This group $\Gamma$ is infinitely generated Fuchsian group of the first kind, thus Kruškal's assertion is only true in the finitely generated case.

## §3. The Absolute Poincaré Series

Following Pommerenke [Po1] we shall say that a group $\Gamma$ is fully accessible if $m(E) = 2\pi$, as is well known such $\Gamma$ are of C.T. Many interesting examples of fully accessible groups have been instructed. Perhaps the most interesting example is due to S. J. Patterson [Pa4] who showed that there exists an infinitely generated Fuchsian group of the first kind, with $m(L) = 2\pi$, such that $\Sigma_{\gamma\in\Gamma}|\gamma'(0)|^p < \infty$ for all $p > 0$. The fact that this group is fully accessible follows from

**Proposition 4.** *If* $\Sigma_{\gamma\in\Gamma}|\gamma'(0)|^p < \infty$ *for some* $p < 1/2$ *then* $\Sigma|\gamma'(\zeta)| < \infty$ *a.e. on* $\partial\Delta$.

The proof of this follows from some known work in Hardy space theory. For completeness we recall

$$H^p(\Delta) = \{f : f \text{ is holomorphic on } \Delta \text{ and } \sup_{r<1} \int_0^{2\pi} |f(re^{i\theta})|^p \, d\theta < \infty\}.$$

Now let $g(z) = \Pi_{\gamma\in\Gamma}\gamma z$ be the Green's function of the group $\Gamma$ and we have

**Proposition B.** *Let* $g(z)$ *be as above, then*

(i) *if* $0 < p < 1/2$, $\Sigma|\gamma'(0)|^p < \infty$ *if and only if* $g' \in H^p(\Delta)$;

(ii) *if* $\alpha \in (0,1)$, $g' \in H^\alpha(\Delta)$ *if and only if* $\Sigma_{\gamma\in\Gamma}|\gamma'(\zeta)|$ *belongs to* $L^\alpha(\partial\Delta)$.

Thus we see that if $\Sigma|\gamma'(0)|^p < \infty$ for some $p < 1/2$ then $g' \in H^p$ and so $\Sigma|\gamma'(\zeta)| \in L^p$. This implies that $\Gamma$ is a fully accessible group.

**Remark.** It should be noted that Patterson proved Proposition 4 by an entirely different method and used it to look at the spectral theory of the Laplace Beltrami operator for these groups.

At this point we have begun to realize that the manner in which $\Gamma 0$, the orbit of the origin, approaches a $\zeta \in L$ (whether or not $\zeta \in E$) has an effect on $\Sigma_{\gamma\in\Gamma}|\gamma'(\zeta)|$ and thus on $\delta(\Gamma)$. (See Lemma 2 and Proposition 4). In particular we saw that if too many elements of $\Gamma 0$ lie in a horocycle at $\zeta$ then $\Sigma_{\gamma\in\Gamma}|\gamma'(\zeta)|$ diverges. But, when $\delta(\Gamma)$ is small, $\Sigma_{\gamma\in\Gamma}|\gamma'(\zeta)|$ converges. Thus we need to consider tangential approach, i.e., outside of horocycles. This type of approach will be the thrust of the next section.

## §4. Tangential Approach

Our method for denoting this tangential approach will be based on the idea of looking at $\zeta$ as the origin and considering the curves $y = |x|^p$, $p = 2,3,4,\ldots$ Here the $y$ value will be the argument and $x$ will be distance to the boundary. Accordingly we define for all integers $p \geq 2$

$$\Lambda_p(\zeta) = \{z \in \Delta : |z| \geq 1/2 \text{ and } (1-|z|)^{1/(p-1)} \leq |\arg\zeta - \arg z| \leq (1-|z|)^{1/p}\}.$$

We note that for $p = 2$ one gets the points outside a Stolz angle and inside a horocycle at $\zeta$. Therefore Proposition A above can be reformulated as

**Proposition A'.** *If* $\zeta \in \partial\Delta - E$ *then* $|M_p(\zeta, \Gamma)| = \infty$.

Here we have used the notation $M_p(\zeta, \Gamma) = \Gamma 0 \cap \Lambda_p(\zeta)$ and $|M_p(\zeta, \Gamma)|$ equals the cardinality of this set.

The usefulness of the $\Lambda_p(\zeta)$ comes from reasonably elementary computations and a counting according to Lemma 2. Let

$$\gamma z = e^{i\alpha} \frac{z-w}{1-z\bar{w}}$$

with $w = \gamma^{-1}(0)$, set $\zeta = e^{i\varphi}$ and $w = re^{i\theta}$ and it is elementary that

$$|\gamma'(\zeta)| = \frac{1-r^2}{1+r^2-2r\cos(\theta-\varphi)}.$$

Since $\cos t$ is a decreasing function of $t$ (for $t$ near zero) the fact the $w = \gamma^{-1}(0)$ lies in $\Lambda_p(\zeta)$ will imply that

$$1+r^2-2r\cos(1-r)^{1/p} < |1-\zeta\bar{w}|^2 < 1+r^2-2r\cos(1-r)^{1/(p-1)}. \tag{3}$$

Upon using the fact that

$$1-t^2/2 < \cos t < 1-t^2/2+t^4/24$$

and setting $t = (1-r)^\alpha$ we get

$$\delta(1-r)^{2\alpha} \leq 1+r^2-2r\left(1-\frac{(1-r)^{2\alpha}}{2}\right) \tag{4}$$

and

$$1 + r^2 - 2r\left(1 - \frac{(1-r)^{2\alpha}}{2} + \frac{(1-r)^{4\alpha}}{24}\right) \le 2(1-r)^{2\alpha}. \tag{5}$$

Thus, if we use the symbol $a \equiv b$ to mean that there exist positive constants m and M such that

$$ma \le b \le Ma,$$

it follows immediately that (3), (4) and (5) above yield

**Lemma 5.** *If* $|\arg\zeta - \arg w| \equiv (1-r)^{\alpha}$ *then* $|1 - \zeta\bar{w}|^2 \equiv (1-r)^{2\alpha}$.

It follows immediately that we have

**Lemma 6.** *If* $\gamma^{-1}(0) \in \Lambda_p(\zeta)$ *then*

$$\delta|\gamma'(0)|^{1-2/(p-1)} \le |\gamma'(\zeta)| \le 2|\gamma'(0)|^{1-2/p}.$$

It is now easy to see how to relate this tangential approach to the exponent of convergence, in fact, we have

**Theorem 7.** *If* $\Sigma_{\gamma\in\Gamma}|\gamma'(\zeta)| = \infty$ *and* $\Sigma_{M_q(\zeta,\Gamma)}|\gamma'(\zeta)| < \infty$ *for* $q \le k$ *then* $\delta(\Gamma) \ge 1 - 2/(k+1)$.

*Proof.*

$$\infty = \sum_{\Gamma}|\gamma'(\zeta)| \ge \sum_{q=1}^{\infty}\sum_{M_q(\zeta,\Gamma)}|\gamma'(\zeta)| + S_2.$$

Here $S_2$ is the sum of those $\gamma$ in $\Gamma$ for which $\gamma^{-1}(0) \notin \Lambda_q(\zeta)$ but for the $\gamma$, we have $|1 - \zeta\,\overline{\gamma(0)}| \ge \delta$ and so this second term is finite. Now

$$\sum_{q=1}^{\infty}\Big(\sum_{M_q}|\gamma'(\zeta)|\Big) = \Big(\sum_{q=1}^{k} + \sum_{k+1}^{\infty}\Big)\Big(\sum_{M_q}|\gamma'(\zeta)|\Big)$$

$$\le M + \Big(\sum_{k+1}^{\infty}\sum_{M_q}|\gamma'(\zeta)|\Big)$$

$$\le M + \sum_{k+1}^{\infty}\sum_{M_q}|\gamma'(0)|^{1-2/q}.$$

Now for all $q \ge k+1$ we have $|\gamma'(0)|^{1-2/q} \le |\gamma'(0)|^{1-2/(k+1)}$ and so putting everything together we get

$$\infty \leq \sum_{K+1}^{\infty} \sum_{M_q} |\gamma'(0)|^{1-2/(k+1)} \leq \sum_{\gamma\in\Gamma} |\gamma'(0)|^{1-2/(k+1)}$$

and the proof is complete.

We see from this that if $\delta(\Gamma) < 1 - 2/(k+1)$ then at each $\zeta$ for which $\Sigma_{M_q(\zeta,\Gamma)}|\gamma'(\zeta)| < \infty$ for $q \leq k$ we have $\Sigma_{\gamma\in\Gamma}|\gamma'(\zeta)| < \infty$.

**Corollary 8.** *If* $|M_q(\zeta,\Gamma)| < \infty$ *for* $q \leq k$ *and* $\Sigma_{\gamma\in\Gamma}|\gamma'(\zeta)| = \infty$ *then* $\delta(\Gamma) \geq 1 - 2/(k+1)$.

For an upper bound on $\delta(\Gamma)$ we have

**Theorem 9.** *If* $\Sigma_{k+1}^{\infty}\Sigma_{M_q}|\gamma'(\zeta)| < \infty$ *and* $\Sigma_{k+1}^{\infty}\Sigma_{M_q}|\gamma'(0)|^{1-2/(k+1)} < \infty$ *for* $\zeta \in \{\zeta_1,\ldots,\zeta_N\}$ *such that* $|\arg\zeta_i - \arg\zeta_{i+1}| < \pi/4$, *then* $\delta(\Gamma) \leq 1 - 2/(k+1)$.

*Proof.*

$$\begin{aligned}\sum |\gamma'(0)|^{1-2/(k+1)} &\leq \sum_{i=1}^{N}\sum_{q=0}^{\infty}\sum_{M_q(\zeta_i)} |\gamma'(0)|^{1-2/(k+1)} \\ &= \Big(\sum_{i=1}^{N}\sum_{q=1}^{k}\sum_{M_q} + \sum_{i=1}^{N}\sum_{k+1}^{\infty}\sum_{M_q(\zeta_i)}\Big)|\gamma'(0)|^{1-2/(k+1)} \\ &= S_1 + S_2.\end{aligned}$$

Now $S_2$ is finite by our hypothesis. To see $S_1$ is finite we note that for $\gamma^{-1}(0)$ in $M_q(\zeta,\Gamma)$ we have $|\gamma'(0)|^{1-2/(k+1)} \leq |\gamma'(0)|^{1-2/q}$ and so

$$S_1 \leq \sum_{i=1}^{N}\sum_{q=1}^{k}\sum_{M_q(\zeta_i)} |\gamma'(0)|^{1-2/q} \leq \sum_{i=1}^{N}\sum_{q=1}^{k}\sum_{M_q(\zeta_i)} |\gamma'(\zeta_i)| < \infty,$$

and the proof is complete.

It should be noted that merely hypothezing that $\Sigma_{\gamma\in\Gamma}|\gamma'(\zeta)| < \infty$ at a single point of $\partial\Delta$ does not in general seem to be sufficient to conclude $\delta(\Gamma) < 1$. The reason for this is that $\Gamma 0$ can approach one point very tangentially but another point very non-tangentially so one needs a set $\{\zeta_1,\ldots,\zeta_n\}$ above.

## §5. Concluding Remarks

We first note a result analogous to Lemma 2 holds for Kleinian groups. The reason for this is that Proposition A, suitably modified, is still valid in this case. It is also easy to see that Lemma 1 follows immediately since it only involves switching sum and integral over a measurable set. To see this it is somewhat easier to consider $\Gamma$ acting on $H^3 = \{(z,t) : z \in \mathbb{C}, t > 0\}$. If $\Lambda$ denotes the limit set of $\Gamma$, we shall assume that $\Gamma$ is of the second kind, i.e., $\Gamma$ is discontinuous at some point of $\mathbb{C}$ and by conjugation we can assume that $\infty$ is an ordinary point of $\Gamma$. Then, if $\lambda$ is a measurable fundamental set for $\Lambda$, we have in analogy to (1)

$$\infty > m(\Lambda) = \int_{\lambda} \sum |\gamma'(z)|^2 \, dxdy. \tag{6}$$

Let $r(\gamma)$ and $c(\gamma)$ denote the radius and center of the isometric sphere for $\gamma$ in $\Gamma$. It follows immediately that

$$|\gamma'(z)|^2 = \frac{r^4(\gamma)}{|z - c(\gamma)|^2}.$$

Thus using [N1] we see that if $z$ is not in $E = \partial\Lambda \cap \partial H^3 = \partial\Lambda \cap \hat{\mathbb{C}}$ then for infinitely many $\gamma$ in $\Gamma$, $|z - c(\gamma)| = O(r(\gamma))$ and so for $z$ not in $E$ we have $\Sigma_{\gamma\in\Gamma} |\gamma'(z)|^2 = \infty$. Once this start has been achieved one can define solid $\Lambda_p(\zeta)$ for $\zeta$ in $\mathbb{C}$ and get results analogous to those above.

It should also be noted that one need not only to consider intergers in defining $\Lambda_p(\zeta)$. Clearly one need only chosen a sequence $\alpha_n \to 0$ and define $\Lambda_{\alpha_n}(\zeta)$ and everything goes through exactly as above. For specific groups this may have advantages but letting $\alpha_n = 1/n$ (as we have done) certainly seems to give the main flavor of the tangential approximation.

Finally we note that for approach within a Stolz angle or within a horocycle there is a large body of results on transitivity and ergodic theory of flows, etc. It would be interesting to know if such a theory can be worked out for this type of tangential approach (see for example [BM] and [N1]).

## References


[AC] P. Ahern and D. Clark, "On inner functions with $H^p$ derivative", Michigan Math. J. 21 (1974), 115–127.

[B1] A. F. Beardon, "Inequalities for certain Fuchsian groups", Acta Math. 127 (1971), 221–250.

[BM] A. F. Beardon and B. Maskit, "Limit points of Kleinian groups and finite sided fundamental polyhedra", Acta Math. 132 (1973), 1–12.

[K1] S. L. Kruškal, "On a property of limit sets of Kleinian groups", Soviet Math. Dokl. 16 (1975), 1497–1499.

[N1] P. J. Nicholls, "Transitive horocycles for Fuchsian groups", Duke Math. J. 42 (1975), 307–312.

[Pa1] S. J. Patterson, "The exponent of convergence of Poincaré series", Monatsh. Math. 82 (1976), 297–316.

[Pa2] S. J. Patterson, "The limit set of a Fuchsian group", Acta Math. 136 (1976), 241–273.

[Pa3] S. J. Patterson, "Spectral Theory and Fuchsian groups", Math. Proc. Cambridge Philos. Soc. 81 (1977), 59–76.

[Pa4] S. J. Patterson, "Some examples of Fuchsian groups", Proc. London Math. Soc. 39 (1979), 276–298.

[Po1] Ch. Pommerenke, "On the Greens function of Fuchsian groups", Ann. Acad. Sci. Fenn. Ser. A I Math. 2 (1976), 409–447.

[Pr1] D. Protas, "Blaschke products with derivatives in $B^p$ and $H^p$", Michigan Math. J. 20 (1973), 393–396.

[S1] D. Sullivan, "Related aspects of positivity in Riemannian geometry", J. Diff. Geom. 25 (1987), 327–351.

# THE RELATION BETWEEN THE BOUNDED POINT DERIVATION AND THE FINE DERIVATION [1]

**Nguyen Xuan-Loc**
Institute of Computer Science and Cybernetics
Lieu Giai, Ba dinh, Hanoi, Vietnam

Let $\mathbb{C}$ be the complex plane, $K \subset \mathbb{C}$ be a compact set with empty interior and $R(K)$ be the function algebra whose elements can be approximated uniformly on $K$ by functions holomorphic in the neighborhoods of $K$. It is proved by Hallstrom (see [4]) that $R(K)$ has a bounded point derivation of an arbitrary order, i.e., for any integer $p = 0, 1, \ldots$ and any non-peak point $x$ (of order $p$) of $R(K)$ there is a positive constant $M_p(x)$ such that

$$|\partial^p f(x)| \le M_p(x) |f|_K , \tag{1}$$

where $f$ is any function holomorphic in a neighborhood of $K$. Moreover if the fine interior $K'$ of $K$ is not empty then it is also known that the function algebra of finely holomorphic functions in $K'$ contains the restriction of $R(K)$ on the same set (see for instance [7], [3]) and one may ask the following: *Is there a relation between the bounded point derivation and the fine derivation of a given finely holomorphic function in* $K'$? Several authors have exploited the method of analytic capacity used by Hallstrom in order to study either the analyticity properties of finely holomorphic functions (see [1] and [7]) or the differentiability properties of finely harmonic functions (see [8]). The aim of this note is twofold:

a) We show by elementary arguments a convergence theorem for finely hypoharmonic functions (Proposition 1) which permits us to sharpen the pointwise estimate (1) in the sense that $M_p(x)$ (as a function of $x$) can be chosen to be finely locally bounded from above.

b) As an application the above mentioned result can be used to give an elementary proof of the identification of the bounded point derivation and the fine derivation of a given finely holomorphic function. In fact, we can answer a question raised by us in [6a] and also recover some deep results of Fuglede in [3].

---

[1] This paper is in final form and no version of it will be submitted for publication elsewhere.

## A convergence theorem and the bounded point derivation

Let $(E,\mathscr{H})$ be a strong harmonic space equipped with the domination principle (D). Let us denote by $\rho_x^A(\cdot)$ the swept-out measure (balayage) of the Dirac mass at x on a universally measurable set $A \subset E$. For any measurable, numerical function f defined in E such that $\int_E \inf(-f,0)\, \rho_x^A < +\infty$, let us define

$$R_f^A(x) := \int^* f(y)\, \rho_x^A(dy). \tag{2}$$

Note that the upper integral on the right hand side of (2) is always well-defined and equal either to $\int f(y)\, \rho_x^A(dy)$ or to $+\infty$ according as f is $\rho_x^A$-integrable or not.

The following result (see [5a], Proposition 5) will be used in the proof of Proposition 1: If A is finely closed then $R_f^A$, as a function of x, is well-defined and finely u.s.c. in the finely open set

$$U := (E\backslash A) \cap [\, x \mid \int \inf(-f,0)\, \rho_x^A < +\infty]. \tag{3}$$

**Proposition 1.** *Let* $(v_n)$ *be a sequence of non-negative finely hypoharmonic functions in a finely open set* $U \subset E$. *Define*

$$u_n := \sum_{i=1}^{n} v_i\,.$$

*Then the following statements are equivalent:*

a) *the function* $u := \sup_n u_n$ *is* $< +\infty$ *in* U,

b) *the function* $u := \sup_n u_n$ *is finely locally bounded from above in* U,

c) *the function* $u := \sup_n u_n$ *is nearly finely hypoharmonic in* U.

*Proof.* Let $V \subset U$ be a compact subset with non-empty fine interior V'. Since $(u_n)$ increases to u we have:

$$\liminf_n \int^* u_n\, \rho_z^{V^c} = \sup_n \int^* u_n\, \rho_z^{V^c} = \int^* u\, \rho_z^{V^c}, \quad \text{for all } z \in V'.$$ [2]

On the other hand, since $u_n$ $(n = 1,2,\ldots)$ is finely hypoharmonic, we also have:

$$u(z) = \liminf_n u_n(x) = \liminf_n \int^* u_n\, \rho_z^{V^c} \le \int^* u\, \rho_z^{V^c}, \quad \text{for all } z \in V'. \tag{4}$$

---

[2] In this paper $X^c$ always stands for the complement of $X \subset E$.

a) ⇒ b). Let x be a given point in U, $\gamma$ be a base of compact fine neighborhoods of x in U. For simplicity, denote by $f^V(x)$ the upper integral $R_f^{V^c}(x)$ of f as defined by (2). We claim that the following a priori estimate (5) is valid for any sequence $(V_i)$ of elements in $\gamma$:

$$\inf_{V\in\gamma}\Big(\sum_{i=1}^{\infty} v_i^V(x)\Big) \le \sum_{i=1}^{\infty} v_i^{V_i}(x). \tag{5}$$

If the right hand side of (5) is equal to $+\infty$ then there is nothing to prove. Let us suppose that this sum is $< +\infty$. For any integer $n = 1,2,\ldots$ define $\tilde{V}_n := V_1 \cap V_2 \cap \cdots \cap V_n$. Then

$$\inf_{V\in\gamma}\Big(\sum_{i=1}^{n} v_i^V(x)\Big) \le \sum_{i=1}^{n} v_i^{\tilde{V}_n}(x), \text{ for } n = 1,2,\ldots,$$

hence

$$\sup_n \inf_{V\in\gamma}\Big(\sum_{i=1}^{n} v_i^V(x)\Big) \le \sup_n \Big(\sum_{i=1}^{n} v_i^{\tilde{V}_i}(x)\Big) \le \sum_{i=1}^{\infty} v_i^{\tilde{V}_i}(x) \le \sum_{i=1}^{\infty} v_i^{V_i}(x) < +\infty.$$

On the other hand, if we fix $V \in \gamma$ and let $\varepsilon > 0$, then there is an integer $n_0$ (depending on V and $\varepsilon$) such that

$$\sum_{i=1}^{n} v_i^V(x) \ge \sum_{i=1}^{\infty} v_i^V(x) - \varepsilon, \text{ for all } n \ge n_0.$$

Hence

$$\sup_n \inf_{V\in\gamma}\Big(\sum_{i=1}^{n} v_i^V(x)\Big) \ge \inf_{V\in\gamma}\Big(\sum_{i=1}^{n_0} v_i^V(x)\Big) \ge \inf_{V\in\gamma}\Big(\sum_{i=1}^{\infty} v_i^V(x)\Big) - \varepsilon.$$

It follows that

$$\sup_n \inf_{V\in\gamma}\Big(\sum_{i=1}^{n} v_i^V(x)\Big) = \inf_{V\in\gamma}\Big(\sum_{i=1}^{\infty} v_i^V(x)\Big) \le \sum_{i=1}^{\infty} v_i^V(x) < +\infty,$$

i.e., (5) is proved.

Suppose now that $u < +\infty$ in U. Since $v_i$ $(i = 1,2,\ldots)$ is finely hypoharmonic it follows that (see [2], Lemma 11.4):

$$v_i(x) = \inf_{V\in\gamma} v_i^V(x) < +\infty, \text{ for all } x \in U \text{ and } i = 1,2,\ldots .$$

Thus for each $\varepsilon > 0$ there corresponds to each $i = 1,2,\ldots$ an element $V_i \in \gamma$ ($i = 1,2,\ldots$) such that

$$v_i^{V_i}(x) < v_i(x) + 2^{-i}\varepsilon.$$

It follows from (5) that

$$\begin{aligned}\inf_{V\in\gamma} u^V(x) &\le \inf_{V\in\gamma}\Big(\sum_{i=1}^{\infty} v_i^V(x)\Big) \le \sum_{i=1}^{\infty} v_i^{V_i}(x)\\ &\le \varepsilon + \sum_{i=1}^{\infty} v_i(x) = u(x) + \varepsilon < +\infty.\end{aligned} \tag{6}$$

Now take a $\tilde{V} \in \gamma$ such that $u^{\tilde{V}}(x) < u(x) + \varepsilon < +\infty$. Since the function $y \to u^{\tilde{V}}(y)$ is finely u.s.c. in the fine interior of $\tilde{V}$ (see (3)) there are a $W \in \gamma$ and a constant $M > 0$ such that

$$u(y) \le u^V(y) \le M, \text{ for all } y \in W,$$

and b) is proved.

b) ⇒ c). Suppose that b) is satisfied. Then it follows from the inequality (4) that $u = \sup_n u_n$ is in fact nearly finely hypoharmonic in $U$ (see Definition 11.1 in [2a]).

c) ⇒ a). Since nearly finely hypoharmonic functions are finely locally bounded from above by the very definition, the desired implication is trivial. □

**Corollary 2** (Convergence theorem). a) *Let* $(u_n)$ *be a sequence of finely hyperharmonic functions in a finely open set* $U \subset E$ *such that for any* $i = 1,2,\ldots$ *the difference* $u_{i+1} - u_i$ *is non-positive and finely hypoharmonic. Then the pointwise infimum*

$$u = \inf_n u_n \quad (\textit{within } F(U) := [-\infty,+\infty]^U)$$

*is nearly finely hyperharmonic in* $U$ *provided that* $u > -\infty$ *in* $U$.

b) *Let* $(u_n)$ *be a specifically decreasing sequence of finely hyperharmonic functions in a finely open set* $U \subset E$. *Then the pointwise infimum* $u = \inf_n u_n$ *is nearly finely hyperharmonic in* U *provided that* $u > -\infty$ *in* U.

*Proof.* a) Put $\tilde{u}_i = -u_i$ for $i = 1,2,\dots$ Then $(\tilde{u}_n)$ is a sequence of functions such that

$$\tilde{u}_n + u_1 = \sum_{i=1}^{n} v_i ,$$

where $v_i = u_i - u_{i+1}$ is non-negative and finely hypoharmonic by hypotheses, hence $\sup_n(u_1 - u_n) = u_1 - \inf_n u_n$ is nearly hypoharmonic by Proposition 1. Since $u_1$ is finely hyperharmonic by hypotheses,

$$u = \inf_n u_n = u_1 + \sup_n(u_1 - u_n)$$

is nearly finely hyperharmonic in U.

b) The proof of this part is similar to that of Proposition 1. □

**Theorem 3.** a) *Let* $K \subset \mathbb{C}$ *be compact and* $B_n(z) = K \cap A_n(z)$, *where*

$$A_n(z) := \{\xi \mid 2^{-(n+1)} \le |\xi - z| \le 2^{-n}\}.$$

*If* $\nu_n(d\xi)$ *is the equilibrium measure in* $B_n$, *then*

$$-\frac{n\ \log 2}{\log C_\ell(B_n)} \le \int_{B_n} \log|\xi - z|^{-1}\ \nu_n(d\xi) \le -\frac{(n+1)\log 2}{\log C_\ell(B_n)}, \tag{7}$$

*where* $C_\ell(B_n)$ *denotes the logarithmic capacity of* $B_n(z)$. *Furthermore, the equilibrium potential*

$$p_n(z) = \int_{B_n(z)} \log|\xi - z|^{-1}\ \nu_n(d\xi) \tag{8}$$

*is finely harmonic in the fine interior of* $\mathbb{C}\backslash B_n$.

b) *Let* $K \subset \mathbb{C}$ *be a compact set with non-empty fine interior* K' *and* $p_n$ $(n = 0,1,\dots)$ *be the equilibrium potential in*

$$B_n(z) := (\mathbb{C}\backslash K) \cap A_n(z).$$

*Then for any given integer* $p = 0,1,\dots$ *and any* z *in* K',

$$\sum_{n=0}^{\infty} 2^{(p+1)n}\, C_\ell(B_n(z)) \le \sum_{n=0}^{\infty} 2^{(p+1)n} \exp\left(-\frac{n\log 2}{p_n(z)}\right) < +\infty. \tag{9}$$

*Furthermore, the function*

$$C_p(z) := \sum_{n=0}^{\infty} 2^{(p+1)n} \exp\left(-\frac{n\log 2}{p_n(z)}\right), \tag{10}$$

*is nearly finely subharmonic in* $K'$.

c) *Let* $f$ *be a function holomorphic in a neighborhood of* $K$ *and* $p \in \mathbb{N}_0$ *be a given integer. Then there is a finely subharmonic function* $M_p(z)$ *in* $K'$ *such that*

$$|\partial^p f(z)| \le \frac{C}{\pi}(1 + C_p(z))|f|_K \le M_p(z)|f|_K, \text{ for all } z \in K', \tag{11}$$

*where* $C$ *is a universal constant and* $|\cdot|_K$ *denotes the supremum norm of* $f$ *in* $K$.

*Proof.* a) Let $K \subset B_n$ be compact and let $\nu_K(d\xi)$ be the equilibrium measure of $K$. Then

$$(\log 2^n)\nu_K(K) \le \int_K \log|\xi - z|^{-1}\, \nu_K(d\xi) \le (\log 2^{n+1})\nu_K(K).$$

Now, $\nu_K(K)$ being identified with the Wiener capacity $W(K)$ of $K$ and $W(K)$ with $-(\log C_\ell(K))^{-1}$ by the very definition, it turns out that

$$-n(\log 2)(\log C_\ell(K))^{-1} \le p_K(z) \le -(n+1)(\log 2)(\log C_\ell(K))^{-1}, \tag{12}$$

where $p_K$ denotes the equilibrium potential generated by $\nu_K$.

Now by a standard potential theoretic argument it is possible to construct an increasing sequence of compact sets $(K_m)$ in $B_n$ such that their associated equilibrium measures $(\nu_K)$ converge vaguely to $\nu_n$ and their equilibrium potentials $p_K$ increase pointwise to $p_n$. Hence (7) is obtained by taking the limits of both sides of (12) as $m$ tends to $+\infty$. Let us suppose without loss of generality that $\emptyset \ne K' \subset K \subset \omega$, where $\omega$ is a Green domain in $\mathbb{C}$. Since the equilibrium potential $p_n(z)$ admits the following probabilistic interpretation:

$$p_n(z) = P^z(T_{B_n(z)} < \tilde{\varphi}),$$

where $\tilde{\varphi}$ denotes the exit-time for $\omega$, the same arguments as in the proof of [5b], Lemma 2.2 show that $p_n(\cdot)$ is invariant under balayage on compact fine neighborhoods in K'. Hence, by the very definition, $p_n$ is finely harmonic in K'.

b) Let z belong to the fine interior K' of K. Since $\mathbb{C}\backslash K$ is thin at z, it follows from the Wiener criterium that

$$\sum_{n=0}^{\infty} -n(\log 2)(\log C_{\ell}(B_n(z)))^{-1} \leq +\infty, \tag{13}$$

hence by (7),

$$p_n(z) \leq \sum_{n=0}^{\infty} -(n+1)(\log 2)(\log C_{\ell}(B_n(z)))^{-1} \leq +\infty.$$

Since the series $\Sigma_n p_n(z)$ is convergent, $\lim_n p_n(z)$ is zero, and it follows that

$$\lim_n \left(2^{(p+1)n} \exp(-n \log 2/p_n(z))\right)^{1/n} = 0.$$

Thus the series on the right hand side of (9) converges for any given integer $p = 0,1,\ldots$ and any point z in K' and (9) is an immediate consequence of (7).

Now we claim that the function $u_n(z) := \exp(-n \log 2/p_n(z))$ $(n = 0,1,\ldots)$ is finely subharmonic in K'. Indeed, the $u_n$ are finely continuous there since the $p_n$ are. On the other hand, if V is any compact fine neighborhood of z in K', then

$$\begin{aligned} u_n(z) &= \exp\left\{n \log 2/ \int -p_n(\xi)\, \rho_z^V(d\xi)\right\} \\ &\leq \int \exp(n \log 2/-p_n(\xi))\, \rho_z^V(d\xi) = \int u_n(\xi)\, \rho_z^V(d\xi), \end{aligned}$$

where $\rho_z^V(d\xi)$ denotes the swept-out mass of $\delta_z$ on $\mathbb{C}\backslash V$.

The first equality holds since $p_n$ $(n = 0,1,\ldots)$ is finely harmonic in K' and the second inequality follows from the convexity of the function $\exp(n \log 2/y)$ (the Jensen inequality).

By taking into account the finiteness of the right hand side of (9), the function $C_p(z)$ defined by (10) turns out to be nearly finely subharmonic in K' by the Proposition 1.

c) Suppose that K is contained in the open disk $D(z,2^M)$ for an integer M. Let U be an open neighborhood of K such that f is holomorphic in a neighborhood of $\bar{U}$, $|f|_{\bar{U}} \leq 2|f|_K$ and $U \subset D(z,2^M)$. Let $\tilde{f}$ be a continuous extension of $(f|U)$ into

$D(z,2^M)$ such that $|\tilde{f}| \le 2|f|_K$ and $\tilde{f}(\xi) = 0$ for $|\xi - z| = 2^M$. Then, by the Cauchy theorem,

$$\partial^p f(z) = \frac{1}{2\pi i}\int_{\Gamma_z^N} \frac{f(\xi)\, d\xi}{(\xi - z)^{p+1}}$$

$$= \sum_{n=-M}^{N-1} \frac{1}{2\pi i}\left(\int_{\Gamma_z^{n+1}} \frac{\tilde{f}(\xi)\, d\xi}{(\xi - z)^{p+1}} - \int_{\Gamma_z^{n}} \frac{f(\xi)\, d\xi}{(\xi - z)^{p+1}}\right) + \frac{1}{2\pi i}\int_{\Gamma_z^{M}} \frac{f(\xi)\, d\xi}{(\xi - z)^{p+1}},$$

where $N$ is chosen such that $D(z,2^{-N}) \subset U$ and $\Gamma_z^k$ denotes the boundary of the disk $D(z,2^{-k})$. By the Melnikov estimate (see [9], §6),

$$|\partial^p f(z)| \le \frac{1}{2\pi}\sum_{n=-M}^{N-1}\left|\int_{\Gamma_z^{n+1}} \frac{\tilde{f}(\xi)\, d\xi}{(\xi - z)^{p+1}} - \int_{\Gamma_z^{n}} \frac{f(\xi)\, d\xi}{(\xi - z)^{p+1}}\right|$$

$$= \frac{2C}{2\pi}\left(\sum_{n=-M}^{N-1} 2^{(p+1)n}\gamma(B_n(z))\right)|f|_K$$

$$= \frac{C}{\pi}\left(\sum_{n=1}^{M}\frac{1}{2^n} + \sum_{n=0}^{N-1} 2^{(p+1)n}\gamma(B_n(z))\right)|f|_K,$$

where $C$ is a universal constant independent of $f$, $n$ and $z$ and $\gamma$ denotes the analytic capacity.

On the other hand, since the analytic capacity of a subset of $\mathbb{C}$ is always dominated by its logarithmic capacity, it follows from the estimate (9) that

$$|\partial^p f(z)| \le \frac{C}{\pi}\left(1 + \sum_{n=0}^{\infty} 2^{(p+1)n} C_\ell(B_n(z))\right)|f|_K \le \frac{C}{\pi}(1 + C_p(z))|f|_K.$$

By the part b), the function $C_p$ is nearly finely subharmonic in $K'$. It suffices to define $M_p(z)$ as the fine u.s.c. regularization of $(C/\pi)(1 + C_p)$ to obtain the desired estimate (11). □

**Remarks.** 1) As we have noted in the introduction, our estimate (11) is sharper than that of Hallstrom in the sense that the constant $M_p$ can be chosen to be finely locally bounded in $K'$, since our function $M_p(z)$ in (11) is finely subharmonic there.

2) Another estimate is given by Debiard and Gaveau in [1], Lemma 5, where the constant $M_p$ could be chosen to be l.s.c. in $z$. However, their constant $C_p(z,f)$

depends essentially on the function f itself!

## Applications

Lemma 4 below gives us an estimate for the rest term of the Taylor expansions of functions belonging to the algebra R(K) and is used to characterize the point derivative of such a function in the fine interior of K as its fine derivative. Recall that the complex derivative $\frac{1}{2}(\frac{d}{dx} - i\frac{d}{dy})$ is always denoted by $\partial$.

**Lemma 4.** *Let* $K \subset \mathbb{C}$ *be a compact set with non-empty fine interior* $K'$, $p \in \mathbb{N}_0$ *be given and* $z_0$ *be a point in* $K'$. *Then there is a (compact) fine neighborhood* $V_0$ *of* $z_0$ *in* $K'$ *such that*

$$|\partial^p f(\gamma(t')) - \partial^p f(\gamma(t)) - (\gamma(t') - \gamma(t))\partial^{p+1} f(\gamma(t))| \leq \int_t^{t'} |d\gamma|(s_1)\Big(\int_t^{s_1} M_{p+2}(\gamma(s))|d\gamma|(s)\Big)|f|_K, \tag{14}$$

$$\Big|f(\gamma(t')) - \sum_{r=0}^{p} \frac{\partial^r f(\gamma(t))}{r!}(\gamma(t') - \gamma(t))^r\Big| \leq \int_t^{t'} |d\gamma|(s_p) \int_t^{s_p} |d\gamma|(s_{p-1}) \cdots \Big(\int_t^{s_1} M_{p+1}(\gamma(s))|d\gamma|(s)\Big)|f|_K \tag{15}$$

*for* $\alpha \leq t \leq t' \leq \beta$, *where* f *is any function holomorphic in the neighborhood of* K *and* $\gamma : [\alpha,\beta] \to V_0$ *any piecewise* $C^1$*-curve in* $V_0$ *and* $M_r(z)$ $(r = p+1, p+2)$ *the finely subharmonic function in* $K'$ *defined by* (11).

*Proof.* a) Since f is holomorphic in the neighborhood of K and since the graph of $\gamma$ is contained in $K'$ we have:

$$\lim_{s' \to s} \frac{f(\gamma(s')) - f(\gamma(s))}{\gamma(s') - \gamma(s)} \cdot \frac{\gamma(s') - \gamma(s)}{s' - s} = \partial f(\gamma(s)) \cdot \gamma'(s).$$

Hence

$$f(\gamma(t')) - f(\gamma(t)) = \int_t^{t'} \partial f(\gamma(s)) \cdot d\gamma(s), \quad \alpha < t \leq t' \leq \beta. \tag{16}$$

Now, since $\partial^i f$ $(i = 1, \ldots, p+1)$ are also holomorphic in the neighborhoods of K, the same arguments applied iterately to the integrand of (16) give rise to the following Taylor expansion of $f(\gamma(t))$ :

$$f(\gamma(t')) - f(\gamma(t)) = \sum_{i=1}^{p} \frac{\partial^i f(\gamma(t))}{i!} (\gamma(t') - \gamma(t))^i + \cdots$$
$$+ \int_t^{t'} d\gamma(s_p) \int_t^{s_p} d\gamma(s_{p-1}) \cdots \int_t^{t_1} \partial^{p+1} f(\gamma(s))\, d\gamma(s),\ t' \le \beta.$$

It follows from (11) that

$$\left| f(\gamma(t')) - \sum_{i=0}^{p} \frac{\partial^i f(\gamma(t))}{i!} (\gamma(t') - \gamma(t))^i \right|$$
$$\le \int_t^{t'} |d\gamma|(s_p) \cdots \int_t^{s_1} |\partial^{p+1} f(\gamma(s))|\ |d\gamma|(s)$$
$$\le \int_t^{t'} |d\gamma|(s_p) \int_t^{s_p} |d\gamma|(s_{p-1}) \cdots \int_t^{s_1} M_{p+1}(\gamma(s))\ |d\gamma|(s).$$

b) Similarly,

$$|\partial^p f(\gamma(t')) - \partial^p f(\gamma(t)) - \partial^{p+1} f(\gamma(t))(\gamma(t') - \gamma(t))|$$
$$\le \int_t^{t'} |d\gamma|(s_1) \int_t^{s_1} |\partial^{p+2} f(\gamma(s))|\,|d\gamma|(s)$$
$$\le \int_t^{t'} |d\gamma|(s_1) \int_t^{s_1} M_{p+2}(\gamma(s))|d\gamma|(\gamma(s)). \ \square$$

**Theorem 5.** *Let* $U \subset \mathbb{C}$ *be a finely open set and* $f : U \to \mathbb{C}$ *be finely holomorphic in* U. *Then at any point* z *in* U *there is a finely open neighborhood* $W_z$ *of* z *with compact closure* $\bar{W}$ *contained in* U *and a sequence of functions* $(f_n)$ *holomorphic in the neighborhoods of* $\bar{W}$ *such that:*

a) $(f_n)$ *converges uniformly to* f *in* $\bar{W}$ *and furthermore their derivatives* $(\partial^p f_n)$ *converge pointwise in* W *to a finely holomorphic function* $\partial^p f$ $(p = 1,2,\ldots)$. *The function* $\partial^p f$ $(p = 1,2,\ldots)$ *is called the* p-*th complex derivative of* f *in* U.

b) *there is a constant* $N_p(z)$ (*resp.* $\tilde{N}_p(z)$) *depending on* p *and* z *only such that*

$$|\partial^p f(z'') - \partial^p f(z') - \partial^{p+1} f(z')(z''-z')| \leq N_p(z)|z''-z'|, \ \textit{for all}\ z', z'' \in W_z \quad (17)$$

(*resp.*

$$\left|f(z'') - \sum_{i=0}^{p} \frac{\partial^i f(z')}{i!}(z''-z')^i\right| \leq \tilde{N}_p(z')|z''-z'|^{p+1}, \ \textit{for all}\ z'', z' \in W_z). \quad (18)$$

*Proof.* Let z be in U, by the very definition of a finely holomorphic function (see [2b]) there is a compact fine neighborhood $V_z$ of z in U such that the restriction of f to $V_z$ is in $R(V_z)$, i.e., f can be approximated uniformly on $V_z$ by the restriction to $V_z$ of a sequence $(f_n)$ of holomorphic functions in open neighborhoods of $V_z$.

a) *Existence of the point derivative* $\partial^p f$ (p = 1,2,...) *of* f *in* U. Fix a natural number p, and take $K = V_z$ and $f = f_n - f_m$ in the estimate (11) of the part c) of Theorem 3. Then

$$|\partial^p(f_n - f_m)(z)| \leq M_p(z)|f_n - f_m|_{V_z}, \ \text{for all}\ z \in V_z'.$$

Since $(f_n)$ is a Cauchy sequence with respect to the norm $|\cdot|_{V_z}$ so is $(\partial^p f_n(z))$ in $\mathbb{C}$, hence the p-th point derivative of f is given by

$$\partial^p f(z) = \lim_n \partial^p f_n(z), \ \text{for all}\ z \in W. \quad (19)$$

Note that the right hand side of (19) is independent of the sequence $(f_n)$.

b) *Identification of* $\partial^p f$ (p = 1,2,...) *with the fine derivative.* Since the fine topology in $\mathbb{C}$ is locally connected and $M_p(z)$, as a function of z, is finely subharmonic in U, there is no loss of generality to assume on the one hand that $V_z$ is a fine domain and on the other hand that $M_p(z') \leq M_p(z) + \varepsilon$ for any z' in $V_z$. Now by the polygonally arcwise connectedness property of a fine domain (see [7]) there is a fine neighborhood W of z in $V_z$ such that two points z', z'' in W can be joined by a finite polygonal line $\gamma: [\alpha,\beta] \to V_z$ in $V_z$ (not necessarily in W) and of length less than $\alpha|z''-z'|$, $\alpha > 1$ being given.

It suffices now to apply the estimate (14) (resp. (15)) to such a curve and to each function $f_n$ (n = 1,2,...) in order to obtain

$$|\partial^p f_n(z'') - \partial^p f_n(z') - \partial^{p+1} f_n(z')(z''-z')| \le \frac{(M_{p+2}(z)+\varepsilon)}{2} |f_n|_{V_z} |z''-z'|^2 \alpha^2$$

(resp.

$$|\partial^p f_n(z'') - \sum_{i=0}^{p} \frac{\partial^i f_n(z')}{i!}(z''-z')^i| \le \frac{(M_{p+1}(z)+\varepsilon)}{(p+1)!} |f_n|_{V_z} |z''-z'|^{p+1} \alpha^{p+1}).$$

Let n tend to $+\infty$ in both sides of the above inequalities, by taking into account (19), we get (17) and (18) where,

$$N_p(z) = \frac{1}{2}\alpha^2(M_{p+2}(z)+\varepsilon)|f|_{V_z}$$

(resp.

$$\tilde{N}_p(z) = \frac{\alpha^{p+1}}{(p+1)!}(M_{p+1}(z)+\varepsilon)|f|_{V_z}). \quad \square$$

**Remark.** The estimate (18) is closely related to a theorem of Fuglede ([3], Theorem 11), which was obtained there by rather sophisticated technics such as the theory of BLD functions and the Cauchy-Pompeiu formula.

**References**


[1] A. Debiard and B. Gaveau, Potentiel fin et algèbres de fonctions analytiques. II, J. Funct. Anal. 17 (1974), 296–310.

[2a] B. Fuglede, Finely harmonic functions, Lecture Notes in Math., 289, Springer-Verlag, 1972.

[2b] B. Fuglede, Fine topology and finely holomorphic functions, 18th Scandinavian Congress of Mathematicians (Aarhus, 1980), 22–38, Birkhäuser, 1981.

[3] B. Fuglede, Sur les fonctions finement holomorphes, Ann. Inst. Fourier (Grenoble) 31:4 (1981), 57–88.

[4] A. Hallstrom, On bounded point derivations and analytic capacity, J. Funct. Anal. 4 (1969), 153–165.

[5a] Nguyen Xuan-Loc, Fine boundary minimum principle and dual processes, Z. Wahrsch. Verw. Gebiete 27 (1973), 233–256.

[5b] Nguyen Xuan-Loc and T. Watanabe, A characterization of fine domains for a certain class of Markov processes with applications to Brelot harmonic spaces, Z. Wahrsch. Verw. Gebiete 21 (1972), 167–178.

[6a] Nguyen Xuan-Loc, Sur la théorie des fonctions finement holomorphes, Bull. Sc. Math. 102 (1978), 337–364.

[6b] Nguyen Xuan-Loc, Stochastic differentiation formula and convergence theorem

for finely hypoharmonic functions (to appear).

[7] T. Lyons, Finely holomorphic functions, J. Funct. Anal. 37 (1980), 1–8.

[8] A. M. Davie and B. Øksendal, Analytic capacity and differentiability properties of finely harmonic functions, Acta Math. 149 (1982), 127–152.

[9] L. Zalcman, Analytic capacity and rational approximation, Lecture Notes in Math., 50, Springer-Verlag, 1968.

# MODULI OF CONTINUITY AND A HARDY-LITTLEWOOD THEOREM

CRAIG A. NOLDER and DANIEL M. OBERLIN
Florida State University
Tallahassee, FL 32306-3027, U.S.A.

1.1. INTRODUCTION. We denote by $\mathbf{D}$ the unit disk of the complex plane $\mathbf{C}$ and by $\partial\mathbf{D}$ its boundary. A differentiable increasing function $\lambda(t) : [0,\infty) \to [0,\infty)$ with $\lambda(0) = 0$ is called a majorant if $\lambda'(t)$ is decreasing. Although our results hold for more general majorants, we use this definition for simplicity. When $f : \mathbf{D} \to \mathbf{C}$ and $\lambda(t)$ is a majorant we write $f \in \mathrm{Lip}_\lambda(\mathbf{D})$ if there exists a constant $M < \infty$ such that

$$|f(z_1) - f(z_2)| \leq M\lambda(|z_1 - z_2|)$$

for all $z_1, z_2 \in \mathbf{D}$. We denote the infimum of all such $M$ by $\|f\|_\lambda$. When $\lambda(t) = t^\alpha$, $0 < \alpha \leq 1$, the classes $\mathrm{Lip}_\lambda(\mathbf{D})$ are the usual Lipschitz classes. Theorem 1.2 is due to Hardy and Littlewood [5]. See also [2] p. 74.

1.2. THEOREM. *If $f$ is analytic in $\mathbf{D}$, if $\lambda(t) = t^\alpha, 0 < \alpha \leq 1$, and if $f \in \mathrm{Lip}_\lambda(\mathbf{D})$, then there is a constant $C$ such that*

$$|f'(z)| \leq C\lambda'(1 - |z|)$$

*for all $z \in \mathbf{D}$. The value of $C$ depends only on $\alpha$ and $\|f\|_\lambda$.*

Versions of Theorem 1.2 and its converse are valid in more general domains than $\mathbf{D}$ and for quasiconformal and quasiregular mappings in space. See [4], [7], [1] and [9].

The first result of this paper, Theorem 1.3, characterizes the majorants for which a result such as Theorem 1.2 holds.

1.3. THEOREM. *Suppose that $\lambda(t)$ is a majorant. The following statements are equivalent:*

(1) *If $f$ is analytic in $\mathbf{D}$ and if $f \in \mathrm{Lip}_\lambda(\mathbf{D})$, then there is a constant $C$, depending only on $\lambda$ and $\|f\|_\lambda$, such that*

$$|f'(z)| \leq C\lambda'(1 - |z|)$$

This paper is in final form and no version of it will be submitted for publication elsewhere.
Both authors are partially supported by grants from the National Science Foundation.

*for all* $z \in \mathbf{D}$;

(2)
$$\limsup_{t\to 0+} \frac{\lambda(t)}{t\lambda'(t)} < \infty.$$

Our second result concerns the following theorem, which is closely related to Theorem 1.2 (see [3] and [2]).

1.4. THEOREM. *Suppose that* $\lambda(t) = t^\alpha$, $0 < \alpha \le 1$. *If* $f = u + iv$ *is analytic in* $\mathbf{D}$ *and if* $u \in \mathrm{Lip}_\lambda(\mathbf{D})$, *then* $f \in \mathrm{Lip}_\lambda(\mathbf{D})$. *Furthermore* $\|f\|_\lambda \le C\|u\|_\lambda$ *where* $C$ *is a constant which depends only on* $\alpha$.

A version of Theorem 1.4 also holds for quasiconformal and quasiregular mappings in space [8]. In section 3 we prove the following result.

1.5. THEOREM. *Suppose that* $\lambda(t)$ *is a majorant. Consider the following statements.*

(1) *If* $f = u + iv$ *is analytic in* $\mathbf{D}$ *and if* $u \in \mathrm{Lip}_\lambda(\mathbf{D})$, *then* $f \in \mathrm{Lip}_\lambda(\mathbf{D})$.

(2) *There exist constants* $\delta_0$ *and* $C$ *such that*

$$\int_0^\delta \frac{\lambda(t)}{t}dt \le C\lambda(\delta)$$

*for all* $0 < \delta \le \delta_0$.

*Statement* (2) *implies* (1). *If we further assume that*

$$\liminf_{t\to 0+} \frac{\lambda(t)}{t\lambda'(t)} > 1,$$

*then* (1) *implies* (2).

1.6. REMARK. Statement (2) of Theorem 1.3 implies statement (2) of Theorem 1.5. If $\lim_{t\to 0+} \lambda(t)/t\lambda'(t)$ exists, either finite or infinite, then statement (2) of Theorem 1.3 is equivalent to statement (2) of Theorem 1.5 (and therefore also equivalent to statement (1) of Theorem 1.5).

2.1. PROOF OF THEOREM 1.3. We first show that (2) implies (1). There exist positive constants $t_0$ and $C_0$ such that

$$\frac{\lambda(t)}{t} \le C_0\lambda'(t) \tag{2.1}$$

for $t \in (0, t_0]$. Hence (2.1) holds for $t \in (0,1]$ with $C_0$ replaced by $C_1 = \max(C_0, \frac{\lambda(t_0)}{t_0\lambda'(1)})$. Fix $z \in \mathbf{D}$ and let $0 < R < 1 - |z|$. Assuming that $f \in \mathrm{Lip}_\lambda(\mathbf{D})$ and using (2.1) we have

$$\begin{aligned}|f'(z)| &= \left|\frac{1}{2\pi R}\int_0^{2\pi} (f(z+Re^{i\theta}) - f(z))e^{-i\theta}d\theta\right| \\ &\le \|f\|_\lambda \frac{\lambda(R)}{R} \le C_1\|f\|_\lambda \lambda'(R).\end{aligned}$$

Now (1) follows by letting $R$ tend to $1-|z|$.

We next assume that (2) fails and exhibit a power series in $\mathrm{Lip}_\lambda(\mathbf{D})$ for which (1) fails. There exists a rapidly decreasing sequence $\{t_j\}_{j=1}^{\infty}$ such that

$$\frac{\lambda(t_j)}{t_j\lambda'(t_j)} \geq 4^j,\ j=1,2,\dots . \tag{2.2}$$

Next define

$$f(z)=\sum_{j=1}^{\infty} a_j z^{n_j}$$

where $a_j=\lambda(t_j)/2^j$ and $n_j$ is the greatest integer in $1/t_j$. Since

$$|z_1^k - z_2^k| \leq \min(2, k|z_1-z_2|)$$

for $k=0,1,2,\dots$ and $z_1,z_2\in\mathbf{D}$, we have

$$\sup_{z_1,z_2\in\mathbf{D}} \frac{|z_1^k-z_2^k|}{\lambda(|z_1-z_2|)} \leq \max\left\{\sup_{0<t<2/k}\frac{kt}{\lambda(t)},\ \sup_{2/k<t\leq 2}\frac{2}{\lambda(t)}\right\} < \frac{2}{\lambda(1/k)}.$$

Hence

$$\begin{aligned}\sup_{z_1,z_2\in\mathbf{D}} \frac{|f(z_1)-f(z_2)|}{\lambda(|z_1-z_2|)} &\leq \sum_{j=1}^{\infty} a_j \sup_{z_1,z_2\in\mathbf{D}}\frac{|z_1^{n_j}-z_2^{n_j}|}{\lambda(|z_1-z_2|)}\\ &\leq \sum_{j=1}^{\infty} 2^{1-j}\frac{\lambda(t_j)}{\lambda(1/n_j)} \leq 2.\end{aligned}$$

However

$$\begin{aligned}\sup_{z\in\mathbf{D}}\frac{|f'(z)|}{\lambda'(1-|z|)} &\geq \sup_{0<r<1}\frac{\sum_{j=1}^{\infty} a_j n_j r^{n_j-1}}{\lambda'(1-r)}\\ &\geq \sup_j \sup_{0<r<1}\frac{a_j n_j r^{n_j-1}}{\lambda'(1-r)}\\ &\geq \sup_j \frac{a_j n_j (1-1/n_j)^{n_j-1}}{\lambda'(1/n_j)}\\ &\geq \frac{1}{e}\sup_j \frac{\lambda(t_j)}{2^{j+1}t_j\lambda'(t_j)} = \infty,\end{aligned}$$

using (2.2).

3.1. Proof of Theorem 1.5. First we show that (2) implies (1). Suppose that $z\in\mathbf{D}$ and $0<R<1-|z|$. With $h=|h|e^{i\phi}$ and $|h|<R$ we obtain

$$\begin{aligned}u(z+h)-u(z) &= \frac{1}{2\pi}\int_0^{2\pi}\left(\frac{R^2-|h|^2}{|Re^{i\theta}-h|^2}-1\right)(u(z+Re^{i\theta})-u(z))d\theta\\ &= \frac{|h|}{\pi}\int_0^{2\pi}\frac{R\cos(\theta-\phi)-|h|}{|Re^{i\theta}-h|^2}(u(z+Re^{i\theta})-u(z))d\theta.\end{aligned}$$

Hence, with $u \in \mathrm{Lip}_\lambda(\mathbf{D})$,

$$\frac{|u(z+h)-u(z)|}{|h|} \le \frac{\lambda(R)}{\pi}\|u\|_\lambda \int_0^{2\pi} \frac{R|\cos(\theta-\phi)|+|h|}{(R-|h|)^2}\,d\theta$$

and so, letting $R$ tend to $1-|z|$, we have

$$|\nabla u(z)| \le \frac{4\|u\|_\lambda}{\pi}\frac{\lambda(1-|z|)}{1-|z|}. \tag{3.1}$$

Fix $z_1, z_2 \in \mathbf{D}$ and let $\gamma$ be the unique circular segment, between $z_1$ and $z_2$ and contained in $\mathbf{D}$, of the circle passing through $z_1$ and $z_2$ which is orthogonal to the boundary of $\mathbf{D}$ (i.e. the hyperbolic geodesic). We have

$$\ell(\gamma) = \text{ Euclidean length of } \gamma \le \frac{\pi}{2}|z_1 - z_2| \tag{3.2}$$

and

$$\min(s, \ell(\gamma)-s) \le \frac{\pi}{2}(1-|z'|) \tag{3.3}$$

for all $z' \in \gamma$. Here $s$ is the arclength along $\gamma$ from $z_1$ to $z'$. Using (3.1) and (3.3) we obtain

$$\begin{aligned} |v(z_1)-v(z_2)| &\le \int_\gamma |\nabla v|ds = \int_\gamma |\nabla u|ds \\ &\le \frac{4\|u\|_\lambda}{\pi}\int_\gamma \frac{\lambda(1-|\gamma(s)|)}{1-|\gamma(s)|}ds \le C_0\int_0^{\ell(\gamma)/2}\frac{\lambda(s)}{s}ds \end{aligned} \tag{3.4}$$

where $C_0 = 4(1+2/\pi)\|u\|_\lambda$. From (2) of Theorem 1.5 it follows that

$$\int_0^\delta \frac{\lambda(t)}{t}dt \le C_1\lambda(\delta) \tag{3.5}$$

for $0 < \delta \le \pi/2$ where $C_1 = \max(C, \int_0^{\pi/2}\frac{\lambda(t)}{t}dt/\lambda(\delta_0))$. Using (3.5) and (3.2), (3.4) becomes

$$|v(z_1)-v(z_2)| \le C_0C_1\lambda(\frac{\ell(\gamma)}{2}) \le 2C_0C_1\lambda(|z_1-z_2|)$$

and so $v \in \mathrm{Lip}_\lambda(\mathbf{D})$.

To prove that (1) implies (2) we use the following result.

3.2. LEMMA. *Suppose that $\lambda(t)$ is a majorant. The following statements are equivalent.*

(1) *There exists a constant $C_0$ such that*

$$t\int_t^\pi \frac{\lambda(s)}{s^2}ds \le C_0\lambda(t)$$

*for all $0 \le t \le \pi$.*

(2) *If $u \in \mathrm{Lip}_\lambda(\partial\mathbf{D})$, then its harmonic extension*

$$u(re^{i\theta}) = \frac{1}{2\pi}\int_0^{2\pi} u(e^{it})\frac{1-r^2}{1+r^2-2r\cos(\theta-t)}dt$$

*is in* $\mathrm{Lip}_\lambda(\mathbf{D})$.

PROOF. We first show that (1) implies (2). Fix $\theta$ and assume that $r \geq \frac{1}{2}$. We have

$$\begin{aligned}|u(e^{i\theta}) - u(re^{i\theta})| &\leq \frac{1}{2\pi}\int_{\theta-\pi}^{\theta+\pi} |u(e^{i\theta}) - u(e^{it})|\frac{1-r^2}{(1-r)^2+4r\sin^2(\frac{\theta-t}{2})}dt \\ &\leq \frac{\pi}{2}\int_{\theta-\pi}^{\theta+\pi} |u(e^{i\theta}) - u(e^{it})|\frac{1-r}{(1-r)^2+(\theta-t)^2}dt \\ &\leq \frac{\pi}{2}\frac{1}{1-r}\int_{|\theta-t|\leq 1-r} |u(e^{i\theta}) - u(e^{it})|dt \\ &\quad + \frac{\pi}{2}(1-r)\int_{1-r\leq|\theta-t|\leq\pi} \frac{|u(e^{i\theta}) - u(e^{it})|}{(\theta-t)^2}dt \\ &\leq \pi\|u\|_\lambda\{\lambda(1-r) + (1-r)\int_{1-r}^{\pi}\frac{\lambda(t)}{t^2}dt\}.\end{aligned} \tag{3.6}$$

Here we have assumed that $u \in \mathrm{Lip}_\lambda(\partial\mathbf{D})$ with constant $\|u\|_\lambda$. Hence, using (1), (3.6) gives the following estimate:

$$\begin{aligned}|u(re^{i\theta}) - u(e^{i\phi})| &\leq |u(re^{i\theta}) - u(e^{i\theta})| + |u(e^{i\theta}) - u(e^{i\phi})| \\ &\leq C_1\{\lambda(1-r) + \lambda(|e^{i\theta} - e^{i\phi}|)\} \leq 3C_1\lambda(|re^{i\theta} - e^{i\phi}|).\end{aligned} \tag{3.7}$$

Here $C_1 = \|u\|_\lambda(\pi(1+C_0)+1)$. On the other hand, if $r \leq \frac{1}{2}$, then

$$\frac{|u(re^{i\theta}) - u(e^{i\phi})|}{\lambda(|re^{i\theta} - e^{i\phi}|)} \leq \frac{2\sup_{z\in\mathbf{D}}|u(z)|}{\lambda(\frac{1}{2})}. \tag{3.8}$$

To complete the proof we apply the following result, (3.9), which follows from the maximum principle for harmonic functions and is proved in [6]. For each $\delta > 0$,

$$\begin{aligned}&\sup\{|u(z_1) - u(z_2)| : |z_1 - z_2| \leq \delta, |z_1| \leq 1, |z_2| \leq 1\} \\ &\quad = \sup\{|u(z_1) - u(z_2)| : |z_1 - z_2| \leq \delta, |z_1| \leq 1, |z_2| = 1\}.\end{aligned} \tag{3.9}$$

Combining (3.7), (3.8), and (3.9) we see that $u \in \mathrm{Lip}_\lambda(\mathbf{D})$.

Next we show that (2) implies (1). Define $u$ on $\partial\mathbf{D}$ as follows:

$$u(e^{it}) = \begin{cases}\lambda(t) & \text{if } 0 \leq t \leq \pi, \\ \lambda(-t) & \text{if } -\pi \leq t \leq 0.\end{cases}$$

Then $u \in \mathrm{Lip}_\lambda(\partial\mathbf{D})$ and so we assume that $u(re^{i\theta}) \in \mathrm{Lip}_\lambda(\mathbf{D})$. We obtain for $\frac{1}{2} \le r < 1$,

$$\begin{aligned}(1-r)\int_{1-r}^{\pi} \frac{\lambda(s)}{s^2}ds &\le 2(1-r)\int_{1-r}^{\pi} \frac{\lambda(s)}{(1-r)^2+s^2}ds \\ &\le 8\int_{1-r}^{\pi} \frac{(1-r^2)\lambda(s)ds}{(1-r)^2+4r\sin^2(s/2)} \\ &\le 16\pi u(r) = 16\pi|u(r)-u(1)| \le 16\pi\|u\|_\lambda \lambda(1-r).\end{aligned}$$

Hence (1) holds for $0 \le t \le \frac{1}{2}$. If $\frac{1}{2} \le t \le \pi$, then

$$\begin{aligned}t\int_t^{\pi} \frac{\lambda(s)}{s^2}ds &\le 2t\frac{1}{2}\int_{\frac{1}{2}}^{\pi} \frac{\lambda(s)}{s^2}ds \\ &\le 32\pi^2\lambda(\frac{1}{2}) \le 32\pi^2\lambda(t).\end{aligned}$$

We now continue with the proof of Theorem 1.5, showing that (1) implies (2) under the assumption that

$$\liminf_{t\to 0+} \frac{\lambda(t)}{t\lambda'(t)} > 1. \tag{3.10}$$

We define the boundary values of $u$, with $m = -2\lambda(\frac{\pi}{2})/\pi$, as follows:

$$u(e^{it}) = \begin{cases} \lambda(t) & \text{if } 0 < t \le \frac{\pi}{2} \\ m(t-\pi) & \text{if } \frac{\pi}{2} \le t \le \pi \\ 0 & \text{otherwise.} \end{cases}$$

Next set $\beta(t) = \lambda(t) - t\lambda'(t)$. By (3.10) there exist constants $t_0 > 0$ and $\epsilon > 0$ such that

$$\lambda(t) \ge (1+\epsilon)t\lambda'(t)$$

for all $0 < t \le t_0$. Hence

$$\lambda(t) \le (1+\epsilon)\beta(t)/\epsilon \quad \text{for } 0 < t \le t_0$$

and we have

$$\lambda(t) \le C_2\beta(t) \quad \text{for } 0 < t \le \pi$$

where $C_2 = (1+\epsilon)\lambda(\pi)/\epsilon\beta(t_0)$. Next

$$\int_t^{\pi} \frac{\lambda(s)}{s^2}ds \le C_2\int_t^{\pi} \frac{\beta(s)}{s^2}ds = C_2\int_t^{\pi} -\frac{d}{ds}\frac{\lambda(s)}{s}ds = C_2\{\frac{\lambda(t)}{t} - \frac{\lambda(\pi)}{\pi}\},$$

which gives

$$t\int_t^{\pi} \frac{\lambda(s)}{s^2}ds \le C_3\lambda(t).$$

By Lemma 3.2 the harmonic extension of $u, u(re^{i\theta})$, is in $\mathrm{Lip}_\lambda(\mathbf{D})$. If we assume that (1) holds, it then follows that $v \in \mathrm{Lip}_\lambda(\mathbf{D})$. For $\delta > 0$,

$$v(e^{-i\delta}) = \frac{1}{2\pi}\int_0^\pi \cot(\frac{-\delta - s}{2})u(e^{is})ds.$$

It follows from the Monotone Convergence Theorem that

$$v(1) = \lim_{\delta\to 0+} v(e^{-i\delta}) = \frac{1}{2\pi}\int_0^\pi \cot(\frac{-s}{2})u(e^{is})ds.$$

Hence for $0 < \delta \le \frac{\pi}{2}$

$$\begin{aligned} v(e^{-i\delta}) - v(1) &= \frac{1}{2\pi}\int_0^\pi [\cot(\frac{s}{2}) - \cot(\frac{s+\delta}{2})]u(e^{is})ds \\ &= \frac{1}{2\pi}\int_0^{\pi+\delta} \cot(\frac{s}{2})[u(e^{is}) - u(e^{i(s-\delta)})]ds. \end{aligned}$$

We estimate as follows

$$\begin{aligned} 2\pi|v(e^{-i\delta}) - v(1)| &\ge \int_0^\delta \cot(\frac{s}{2})u(e^{is})ds \\ &\quad - |\int_{\frac{\pi}{2}}^{\frac{\pi}{2}+\delta} \cot(\frac{s}{2})[u(e^{is}) - u(e^{i(s-\delta)})]ds| \\ &\quad - |\int_{\frac{\pi}{2}+\delta}^{\pi+\delta} \cot(\frac{s}{2})[u(e^{is}) - u(e^{i(s-\delta)})]ds| \\ &\ge \int_0^\delta \cot(\frac{s}{2})u(e^{is})ds - 2\lambda(\frac{\pi}{2})\cot(\frac{\pi}{4})\delta - \lambda(\frac{\pi}{2})\cot(\frac{\pi}{4}+\frac{\delta}{2})\delta \\ &> \int_0^\delta \frac{\lambda(s)}{s}ds - C_3\delta \end{aligned} \tag{3.11}$$

where $C_3 = 3\lambda(\frac{\pi}{2})$. Next since $\lambda(t)$ is subadditive there exist constants $C_4$ and $\delta_1$ such that $\delta \le C_4\lambda(\delta)$ for $0 < \delta \le \delta_1$. Hence for any $\delta$ with $0 < \delta \le \delta_1$ either

$$\int_0^\delta \frac{\lambda(s)}{s}ds < 2C_3\delta \le 2C_3C_4\lambda(\delta)$$

or

$$\int_0^\delta \frac{\lambda(s)}{s}ds \ge 2C_3\delta. \tag{3.12}$$

If (3.12) holds, then combined with (3.11) we obtain

$$\frac{1}{4\pi}\int_0^\delta \frac{\lambda(s)}{s}ds \le |v(e^{-i\delta}) - v(1)| \le \|v\|_\lambda \lambda(|e^{i\delta} - 1|) \le \|v\|_\lambda \lambda(\delta).$$

Hence statement (2) of Theorem 1.5 holds with $C = \max(2C_3C_4, 4\pi\|v\|_\lambda)$ and $\delta_0 = \delta_1$.

## References

[1] K. Astala and F. W. Gehring, Quasiconformal analogues of theorems of Koebe and Hardy and Littlewood, Michigan Math. J. 32 (1985), 99–108.

[2] P. L. Duren, Theory of $H^p$ spaces, Academic Press, Orlando, Florida, 1970.

[3] F. W. Gehring and O. Martio, Quasidisks and the Hardy-Littlewood property, Complex Variables Theory Appl. 2 (1983), 67–78.

[4] ———, Lipschitz classes and quasiconformal mappings, Ann. Acad. Sci. Fenn. Ser. A I Math. 10 (1985), 203–219.

[5] G. H. Hardy and J. E. Littlewood, Some properties of fractional integrals II, Math. Z. 34 (1932), 403–439.

[6] A. Hinkkanen, Modulus of continuity of harmonic functions, preprint.

[7] V. Lappalainen, $\mathrm{Lip}_\lambda$-extension domains, Ann. Acad. Sci. Fenn. Ser. A I Math. Dissertationes 56 (1985).

[8] C. A. Nolder, Conjugate functions and moduli of continuity, Illinois J. Math. 31, no. 4 (1987).

[9] ———, A quasiregular analogue of a theorem of Hardy and Littlewood, preprint.

# A SHORT PROOF OF A THEOREM OF AHLFORS [1]

**Osmo Pekonen** [2]
University of Jyväskylä, Department of Mathematics
Seminaarinkatu 15, SF-40100 Jyväskylä, Finland

## 1. Introduction

In [1] Ahlfors proved that the Weil-Petersson metric of the Teichmüller space is Kähler. A new proof was given by Fischer and Tromba [5] in a purely Riemannian setting of Teichmüller theory [3]. We shall provide yet another proof that slightly shortens the argument of Fischer and Tromba.

We begin with a brief review of the Fischer-Tromba approach to Teichmüller theory [3–5].

Let $M$ be a compact connected oriented 2-dimensional manifold without boundary.

Let $C^\infty(T_1^1(M))$ be the space of smooth tensor fields of type (1,1) on $M$. An *almost complex structure* $\mathcal{J}$ on $M$ is an element $\mathcal{J} \in C^\infty(T_1^1(M))$ such that $\mathcal{J}^2 = -\mathcal{I}$; i.e., at each $x \in M$, $\mathcal{J}(x) \circ \mathcal{J}(x) = -\mathcal{I}(x)$, where $\mathcal{I}(x) : T_xM \to T_xM$ is the identity map.

Moreover, it is well known that $\mathcal{J}$ induces a natural orientation on $M$ and the space of complex structures on $M$, denoted by $\mathcal{A}$, can be identified with the space of such $\mathcal{J}$ that the induced orientation is that of $M$.

Let $\mathcal{D}$ be the group of $C^\infty$ orientation preserving diffeomorphisms of $M$ and $\mathcal{D}_0$ its connected component of the identity. As a group $\mathcal{D}$ (and hence $\mathcal{D}_0$) then acts on $\mathcal{A}$ via pull-back; i.e. if $f \in \mathcal{D}$, then the action sends $\mathcal{J} \mapsto f^*\mathcal{J}$, where

$$f^*\mathcal{J} = f_*^{-1} \circ \mathcal{J}_f \circ f_*$$

and the lower star denotes the tangent map.

---

[1] This paper is in final form and no version of it will be submitted for publication elsewhere.

[2] This research was supported by the Oskar Huttunen Foundation.

The spaces $\mathcal{R} = \mathcal{A}/\mathcal{D}$ and $\mathcal{T} = \mathcal{A}/\mathcal{D}_0$ are nothing but the classical *Riemann moduli space* and the *Teichmüller space*, respectively. They differ from each other by the action of the *modular group* $\Gamma = \mathcal{D}/\mathcal{D}_0$, which is known to be discrete. Clearly $\mathcal{T}/\Gamma = \mathcal{R}$.

Let us suppose henceforward that *the genus of* M *is greater than one.* Under this assumption we may recapitulate some of the main theorems of Fischer and Tromba in the following way:

**Theorem 1.** *In the* ILH *category due to Omori* [8] *$\mathcal{A}$ is a smooth infinite dimensional manifold and $\mathcal{D}$ and $\mathcal{D}_0$ are infinite dimensional Lie groups. The tangent space of $\mathcal{A}$ at $\mathcal{J} \in \mathcal{A}$ is characterized by*

$$T_{\mathcal{J}}A = \{\dot{\mathcal{J}} \in C^{\infty}(T_1^1(M)) \mid \dot{\mathcal{J}} \circ \mathcal{J} + \mathcal{J} \circ \dot{\mathcal{J}} = 0\}.$$

**Theorem 2.** *The Teichmüller space $\mathcal{T} = \mathcal{A}/\mathcal{D}_0$ carries the sructure of a smooth finite dimensional manifold which, moreover, is a cell of dimension* $6 \cdot (genus) - 6$. *The canonical fibration $\pi : \mathcal{A} \to \mathcal{T}$ carries the structure of a smooth* ILH *principal fibre bundle with structure group $\mathcal{D}_0$.* (See also [2].)

**Theorem 3.** *There is a natural $\mathcal{D}_0$-invariant, Nijenhuis torsionless almost complex structure on $\mathcal{A}$, as follows. For each $\mathcal{J} \in \mathcal{A}$ define $\Phi \in C^{\infty}(T_1^1(\mathcal{A}))$ by*

$$\Phi_{\mathcal{J}} : T_{\mathcal{J}}\mathcal{A} \to T_{\mathcal{J}}\mathcal{A}, \quad \Phi_{\mathcal{J}}(\dot{\mathcal{J}}) = \mathcal{J} \circ \dot{\mathcal{J}}.$$

*Then $\Phi$ induces an integrable almost complex structure $\Phi_{\mathcal{T}}$ on the finite dimensional Teichmüller space $\mathcal{T}$ thus providing it with a natural structure of a complex manifold.*

**Theorem 4.** *There is a natural $\mathcal{D}_0$-invariant,* $L^2$ *Riemannian metric $\langle\!\langle \cdot,\cdot \rangle\!\rangle$ on $\mathcal{A}$, as follows. For $\mathcal{J} \in \mathcal{A}$ and $X, Y \in T_{\mathcal{J}}\mathcal{A}$ let*

$$\langle\!\langle X,Y \rangle\!\rangle_{\mathcal{J}} = \int_M \operatorname{tr}(X \circ Y)\, \nu_{g(\mathcal{J})}$$

*where* tr *means the trace of a* (1,1)*-tensor and $\nu_{g(\mathcal{J})}$ denotes the volume element of the Poincaré metric $g(\mathcal{J})$ canonically associated to $\mathcal{J}$.* (Thus $g = g(\mathcal{J})$ is the Riemannian metric with constant scalar curvature equal to minus one uniquely determined by the condition

$$-g^{-1}(\nu_g) = \mathcal{J}$$

as explained in [3].)

*Furthermore,* $\langle\!\langle\cdot,\cdot\rangle\!\rangle$ *projects to the Teichmüller space* $\mathcal{T}$ *to a Riemannian metric* $\langle\cdot,\cdot\rangle$ *which is, in fact, the classical Weil–Petersson metric. The Teichmüller space is a Hermitian manifold, i.e.*

$$\langle\Phi_{\mathcal{T}}X,\Phi_{\mathcal{T}}Y\rangle = \langle X,Y\rangle$$

*for all smooth vector fields* X, Y *on* $\mathcal{T}$.

**Theorem 5.** *The Teichmüller space is a Kähler manifold.*

Fischer and Tromba [5] proceed to prove the theorem 5 by showing directly that the Kähler 2-form

$$\omega = \langle\Phi_{\mathcal{T}}(\cdot),\cdot\rangle$$

on the Teichmüller space is closed.

However, it turns out to be simpler, in this particular case, to apply an equivalent characterization of a Kähler manifold [7] and show that the almost complex structure $\Phi_{\mathcal{T}}$ on $\mathcal{T}$ is parallel with respect to the Levi-Civita connection of the Weil-Petersson metric. In other words, our proof consists of showing directly, that

$$\nabla_{\mathcal{T}}\Phi_{\mathcal{T}} = 0 \tag{1.1}$$

where $\nabla_{\mathcal{T}}$ is the covariant derivative associated to the Weil-Petersson metric.

## 2. A decomposition theorem

In this section we review some more information about the canonical fibration $\pi : \mathcal{A} \to \mathcal{T}$.

Let $\mathcal{X}(M)$, $\mathcal{X}(\mathcal{A})$, $\mathcal{X}(\mathcal{T})$ denote the space of smooth vector fields on M, $\mathcal{A}$, $\mathcal{T}$, respectively.

The tangent space $T_{\mathcal{J}}\mathcal{A}$ of $\mathcal{A}$ at any $\mathcal{J} \in \mathcal{A}$ has a canonical subspace consisting of vectors tangent to the orbit through $\mathcal{J}$ of the action of the group $\mathcal{D}_0$. Obviously, this so-called *vertical* subspace is just the image of the Lie derivative map

$$\alpha_{\mathcal{J}} : \mathcal{X}(M) \to C^\infty(T^1_1(M))$$

which sends $\xi \to \frac{1}{2} L_\xi \mathcal{J}$.

Furthermore, the group $\mathcal{D}_0$ acts as an isometry on $\mathcal{A}$ with respect to the Riemannian metric $\langle\!\langle\cdot,\cdot\rangle\!\rangle$. The symbol of $\alpha_{\mathcal{J}}$ being injective, we have according to a general theorem in global analysis [9] the following decomposition theorem [10]:

**Theorem 6.** *The tangent space* $T_{\mathcal{J}}\mathcal{A}$ of $\mathcal{A}$ at any $\mathcal{J} \in \mathcal{A}$ *splits into an orthogonal sum*

$$T_{\mathcal{J}}\mathcal{A} = \operatorname{Ker} \alpha^*_{\mathcal{J}} \oplus \operatorname{Im} \alpha_{\mathcal{J}}$$

*with respect to the Riemannian metric* $\langle\langle \cdot, \cdot \rangle\rangle$. *Here*

$$\alpha^*_{\mathcal{J}} : T_{\mathcal{J}}\mathcal{A} \to \mathcal{X}(M)$$

*is the formal adjoint of* $\alpha_{\mathcal{J}}$ *with respect to the Riemannian metric on* $\mathcal{X}(M)$ *given by the Poincaré metric of* $\mathcal{J}$.

Geometrically, the kernel of $\alpha^*_{\mathcal{J}}$, or the *horizontal* subspace of $T_{\mathcal{J}}\mathcal{A}$, thus gives a local *slice* of vectors transversal to the orbit through $\mathcal{J}$. More explicitly, one finds by an integration by parts the following expression for $\alpha^*_{\mathcal{J}}$ in local coordinates [5]. For $X \in T_{\mathcal{J}}\mathcal{A}$

$$(\alpha^*_{\mathcal{J}}(X))^i = -(\bar{\nabla}_j X \circ \mathcal{J})^{ij}$$

where $\bar{\nabla}$ means the covariant derivative of the Poincaré metric of $\mathcal{J}$ on M and the raising of indices is also with respect to the Poincaré metric. Thus the horizontal tangent vectors at $\mathcal{J} \in \mathcal{A}$ are the divergence free smooth (1,1) tensor fields that anticommute with $\mathcal{J}$.

Any smooth vector field $X \in \mathcal{X}(\mathcal{A})$ can be decomposed into the sum of its horizontal and vertical parts. On the other hand, according to the general theory of principal fibre bundles, any smooth vector field $X_{\mathcal{T}} \in \mathcal{X}(\mathcal{T})$ on the Teichmüller space admits a unique horizontal lift; i.e. there is a unique purely horizontal smooth vector field $X \in \mathcal{X}(\mathcal{A})$ that projects down to $X_{\mathcal{T}}$.

### 3. The natural algebraic covariant derivative on $\mathcal{A}$

We next review a nice characterization of the covariant derivative of the Weil-Petersson metric used by Tromba [10] to compute the curvature of the Teichmüller space.

A natural symmetric covariant derivative $\nabla$ on $\mathcal{A}$ is algebraically constructed starting from the usual Fréchet derivative D (see e.g. [6]), which is symmetrized by subtracting a suitable term as follows. For $X, Y \in \mathcal{X}(\mathcal{A})$ define at any $\mathcal{J} \in \mathcal{A}$

$$\nabla_Y X = DX(Y) - \frac{1}{2}\mathcal{J}(XY + YX). \tag{3.1}$$

Then $\nabla$ satisfies indeed the algebraic properties of a covariant derivative. Moreover, one has

**Theorem 7.** *The covariant derivative* $\nabla$ *defined by* (3.1) *projects to the covariant derivative* $\nabla_{\mathcal{T}}$ *of the Weil–Petersson metric on the Teichmüller space* $\mathcal{T}$. *More precisely, for any* $X_{\mathcal{T}}, Y_{\mathcal{T}} \in \mathcal{X}(\mathcal{T})$ *let* $X, Y \in \mathcal{X}(\mathcal{A})$ *be their horizontal lifts. Then the horizontal lift of* $(\nabla_{\mathcal{T}})_{Y_{\mathcal{T}}} X_{\mathcal{T}}$ *agrees with the horizontal component of* $\nabla_Y X$.

## 4. A simple new proof of Theorem 5

In virtue of our preliminary discussion, showing that the Teichmüller space is Kählerian or, equivalently, establishing the formula (1.1) boils down to proving the

**Lemma.** $\nabla\Phi = 0$.

*Proof.* By a direct computation for $\mathcal{J} \in \mathcal{A}$, $X, Y \in T_{\mathcal{J}}\mathcal{A}$ arbitrary, one finds:

$$\begin{aligned}
(\nabla_Y \Phi_{\mathcal{J}})(X) &= \nabla_Y(\Phi_{\mathcal{J}} X) - \Phi_{\mathcal{J}}(\nabla_Y X) \\
&= \nabla_Y(\mathcal{J}X) - \mathcal{J}(\nabla_Y X) \\
&= D(\mathcal{J}X)(Y) - \frac{1}{2}\mathcal{J}(\mathcal{J}XY + Y\mathcal{J}X) - \mathcal{J}DX(Y) + \frac{1}{2}\mathcal{J}^2(XY + YX) \\
&= D(\mathcal{J}X)(Y) - \frac{1}{2}(-XY + YX) - \mathcal{J}DX(Y) - \frac{1}{2}(XY + YX) \\
&= D(\mathcal{J}X)(Y) - \mathcal{J}DX(Y) - YX \\
&= 0. \quad \square
\end{aligned}$$

The lemma together with the theorem 7 implies the theorem 5.

As a matter of fact, our lemma also snatches a proof for the fact that the Nijenhuis tensor of the almost complex structure $\Phi_{\mathcal{T}}$ vanishes [7], which was the main computation of [4]. Thus the integrability part of the proof of the theorem 3 is also contained in our lemma.

## References

[1] L. V. Ahlfors, Some remarks on Teichmüller's space of Riemann surfaces, Ann. of Math. (2) 74 (1961), 171–191.

[2] C. Earle and J. Eells, A fibre bundle description of Teichmüller space, J. Differential Geom. 3 (1969), 19–43.

[3] A. E. Fischer and A. J. Tromba, On a purely "Riemannian" proof of the structure and dimension of the unramified moduli space of a compact Riemann surface, Math. Ann. 267 (1984), 311–345.

[4] A. E. Fischer and A. J. Tromba, Almost complex principal fibre bundles and the complex structure on Teichmüller space, J. Reine Angew. Math. 352 (1984), 151–160.

[5] A. E. Fischer and A. J. Tromba, On the Weil-Petersson metric on Teichmüller space, Trans. Amer. Math. Soc. 284 (1984), 319–335.

[6] R. S. Hamilton, The inverse function theorem of Nash and Moser, Bull. Amer. Math. Soc. 7 (1982), 65–222.

[7] S. Kobayashi and K. Nomizu, Foundations of differential geometry, Vol. II, Interscience Publishers, New York, 1969.

[8] H. Omori, On the group of diffeomorphisms on a compact manifold, Proc. Sympos. Pure Math., Vol. 15, Amer. Math. Soc., Providence, R.I. (1970), 167–183.

[9] R. S. Palais, Seminar on the Atiyah-Singer index theorem, Princeton University Press, Princeton, N.J., 1965.

[10] A. J. Tromba, On a natural algebraic affine connection on the space of almost complex structures and the curvature of Teichmüller space with respect to its Weil-Petersson metric, Manuscripta Math. 56 (1986), 475–497.

# REMARKS ON THE PICARD THEOREM

H. RENGGLI
Department of Mathematics, Kent State University
Kent, Ohio 44242, U.S.A.

**0.** We have two purposes. The first is to give new simple proofs of some known facts. The second is to present a generalization that has so far been overlooked.

We begin by showing that to a parabolic cover transformation (of the universal covering surface) of a hyperbolic Riemann surface $R$ corresponds a puncture of $R$.

Next we prove that a holomorphic mapping of a punctured disc into a hyperbolic Riemann surface $R$ has either the puncture as a removable singularity or the puncture corresponds to a puncture of $R$. In the latter case if the puncture is patched in $R$ – i.e., added to $R$ – then the puncture of the disc is once more a removable singularity.

Some corollaries then follow. Finally using transfinite ordinals we show that any closed countable subset $P$ of a Riemann surface $R$ is removable for a holomorphic mapping $f : R\backslash P \to M$ – where $M$ is a hyperbolic Riemann surface – as long as in each step $M$ stays hyperbolic even after it has been patched.

**1.** We start by defining a puncture and by explaining the procedure of patching a puncture.

DEFINITION. Take a chart of a Riemann surface $R$ such that its domain $U$ is homeomorphic to the punctured disc $D' = \{z : 0 < |z| < 1\}$. If for a sequence $\{z_n\}$ $(n = 1, 2, \ldots)$, $z_n \to 0$, the corresponding sequence $\{p_n\}$, $p_n \in U$, diverges on $R$, then we say that $\{p_n\}$ converges to a puncture $p$ of $R$.

Next we add a point $p$ to $R$ and the disc $D = \{z : |z| < 1\}$ to the conformal atlas of $R$, where 0 corresponds to $p$. This procedure will be quoted as patching a puncture of $R$. The thus extended Riemann surface is denoted by $R \cup \{p\}$. Then the sequence $\{p_n\}$ converges on $R \cup \{p\}$ to $p$.

Using those notions we state the following

LEMMA. *Let $R$ be a hyperbolic Riemann surface. Suppose that the group $G$ of cover transformations contains a parabolic element $g$. Then there is a puncture $p$ of*

---

This paper is in final form and no version of it will be submitted for publication elsewhere.

*R that corresponds to g. Moreover a Dirichlet region is given where the puncture is explicitly patched.*

PROOF. Since $R$ is hyperbolic its universal covering surface can be mapped conformally onto the upper half plane $H = \{z = x + iy \ : \ y > 0\}$. Take $\infty$ as the fixed point of $g$. By the discontinuity of $G$ there must be a primitive element $z + h, h > 0$, such that every parabolic element of $G$ with $\infty$ as the fixed point is a multiple $z + mh, m$ integer, $m \neq 0$. Choose $z_0$ with $y_0 \geq h/2$ and construct with $z_0$ as center the Dirichlet region $\Delta_0$. Recall that it consists of all points that have smaller hyperbolic distance from $z_0$ than from any other point equivalent to $z_0$ under $G$. Choose a hyperbolic disc $F$ with center at $z_0$ that lies entirely in $\Delta_0$.

We claim that the orbit of $z_0$, i.e., all points $z_i$ equivalent to $z_0$ under $G$, satisfies the inequality $y_i \leq d$ for some $d$. Because otherwise $F$ cannot fit for $y_i \to \infty$ into some vertical open strip of width $h$ that contains possibly nonequivalent points. Once more by the discontinuity of $G$ there exist in the orbit of $z_0$ two elements $z_1$ and $z_2$ with maximal imaginary part, such that $x_2 > x_1$ and $x_2 - x_1$ is minimal. Denote by $g^*$ the element $g^* \in G$ with $g^*(z_1) = z_2$. If $g^*$ is parabolic with $\infty$ as the fixed point then we have necessarily $g^* = z + h$.

Otherwise $g^*$ is hyperbolic or is parabolic with some finite fixed point. We shall show that this is impossible. Obviously $x_2 - x_1 \leq h/2$. The Dirichlet regions $\Delta_1$ and $\Delta_2$ of $z_1$ and $z_2$ must have some vertical ray that extends to $\infty$ as their common boundary. It is part of the hyperbolic perpendicular bisector $\ell$ of the line through $z_1$ and $z_2$, and its points have smaller hyperbolic distance from $z_1$ and $z_2$ than from any other point of the orbit of $z_0$. Observe that $\ell$ is part of a level curve of $g^*$ and that distinct level curves cannot intersect except at a parabolic fixed point. Since $g^*(\Delta_1) = \Delta_2$ and $g^*(\ell)$ does not go through $\infty$, $g^*(\ell)$ must be a Euclidean half circle with its center on the real axis and a diameter of at most $h/2$. This is so because $\Delta_2$ is convex and must therefore lie in a vertical strip of width at most $h/2$. That strip must also contain $g^*(\ell)$. Observe that $y_1 = y_2 \geq h/2$ and that $x_2 - x_1 \leq h/2$, while the imaginary part of each point of $g^*(\ell)$ is at most $h/4$. Therefore the hyperbolic distance from $z_1$ to $\ell$ is less than that measured on the horizontal line through $z_1$ where we have $\int \frac{ds}{y_1} \leq \frac{h}{4}\frac{2}{h} = 1/2$. But the hyperbolic distance from $z_2 = g^*(z_1)$ to $g^*(\ell)$ is at least that measured on a vertical line from $y' = h/2$ to $y'' = h/4$ where we have $\int \frac{ds}{y} > \frac{h}{4}\frac{2}{h} = \frac{1}{2}$ since $y \leq h/2$. This is impossible because $g^*$ is a hyperbolic motion.

Therefore all points of the orbit of $z_0$ with maximal imaginary part are transformed into each other by multiples of $z + h$. Discussing the structure of $\Delta_1$ one shows as above that the boundary of $\Delta_1$ contains two vertical rays starting at some points $z^*$ and $z^* + h$

respectively. Note that points on the two rays with equal imaginary parts represent the same point of $R$.

Finally take $w = e^{\frac{2\pi}{h}iz}$ and observe that $w(z+h) = w(z)$. Hence there is some punctured disc $D' = \{w : 0 < |w| < \delta\}$ which corresponds to some domain $U$, $U \subset R$. Here the boundary point $\infty$ of $\triangle_1$ corresponds to the puncture 0 of $D'$. Therefore the boundary of $\triangle_1$ contains explicitly a puncture that can be patched.

**2.** Our next goal is the proof of

PROPOSITION 1. *Suppose the punctured disc $D' = \{z : 0 < |z| < 1\}$ is by the function $f$ holomorphically mapped into a hyperbolic Riemann surface $R$. Then either 0 is a removable singularity of $f$ or 0 corresponds to a puncture of $R$. Moreover if the puncture is patched in $R$, then 0 is once more a removable singularity.*

PROOF. Since $R$ is hyperbolic it can be represented by $R = H/G$ where $H$ is the upper half plane and $G$ is the group of cover transformations of $R$. Cut $D'$ along the interval $\{-1 < \operatorname{Re} z < 0\}$ and denote the thus obtained domain by $D^*$. Fixing the germ of $f$ into $R = H/G$ at $1/e$ and then making analytic continuation we get a function $f^* : D^* \to H$. Let $C(r) = \{re^{i\theta} : r = \text{const.}, 0 < r < 1, -\pi < \theta < \pi\}$ and denote its image under $f^*$ by $C'(r)$. Since the initial point $a$ (for $\theta \to -\pi$) and the endpoint $b$ (for $\theta \to \pi$) of $C'(r)$ correspond to the same point of $R$, there is an element $g(r) \in G$ that maps $a$ onto $b$. We claim that $g(r)$ does not depend on $r$. Because otherwise there would be an element $g(r')$ such that in any neighborhood of $r'$ one could find an element $g(r) \neq g(r')$. Hence a product $g^{-1}(r) \circ g(r')$ could be given that is arbitrarily close to the identity element of $G$. But $G$ does not contain infinitesimal transformations.

The function $h : z \to w = -i \ln z$, $h(1/e) = i$, maps the universal covering surface of $D'$ onto $H$. To the hyperbolic metric $\rho(w) = \frac{1}{\operatorname{Im} w}$ on $H$ corresponds in $D'$ the hyperbolic metric $\rho(z) = \frac{1}{|z| \ln \frac{1}{|z|}}$. We note that $C(r)$ has hyperbolic length $\ell(r) = \frac{2\pi}{\ln \frac{1}{r}}$. Using the germ of $f^* \circ h^{-1}$ at $i$ we get by analytic continuation a function $f' : H \to H$. The curve that corresponds in $H$ to $C(r)$ is mapped by $f'$ onto $C'(r)$. By the Schwarz lemma applied to $f'$ the hyperbolic length $\ell'(r)$ of $C'(r)$ satisfies the inequality $\ell'(r) \leq \ell(r)$.

Next the element $g$ obtained above cannot be hyperbolic, since $\ell(r)$ and a fortiori $\ell'(r)$ converge to 0 for $r \to 0$. Hence $g$ is either the identity element or is parabolic. In the first case $f^* : D^* \to H$ can be extended to $D'$ and is holomorphic there. Hence 0 is by the Casorati-Weierstrass theorem a removable singularity. Since $f^*(0)$ can be chosen as an interior point of some fundamental domain of $R = H/G$, the first part of Proposition 1 follows.

In the second case $g$ must be a multiple of some primitive element that may be chosen as $z+h$, $h>0$. By the Lemma and its proof there is a puncture $p$ of $R$ determined by $z+h$. Since $\ell'(r) \to 0$ for $r \to 0$ the imaginary part of all points of $C'(r)$ will be arbitrarily large for $r \to 0$. Patching the puncture $p$ by using a chart $U$ homeomorphic to $D'$ we see that some annulus $\{0<|z|<\delta\}$ is mapped by $f$ into $U$. So 0 is similarly as above a removable singularity.

REMARK 1. Proposition 1 must be attributed to M. Ohtsuka [6]. Its first part – without the statements concerning punctures – goes also back to H. Huber [3]. Simpler or alternate versions have been given by M. Heins [2], Marden, Richards and Rodin [4] and Royden [7].

**3.** Using the definition of a puncture and the procedure of patching a puncture we deduce from Proposition 1

COROLLARY 1. *Let $R$ and $M$ be Riemann surfaces and let $f : R \to M$ be holomorphic where $M$ is hyperbolic. If the sequence $\{p_n\}$ $(n = 1, 2, \ldots)$ converges to a puncture $p$ of $R$, then $\{f(p_n)\}$ converges either to a point $q$ or to a puncture $q$ of $M$. In both cases $f$ can be extended to $p$ and $f : R \cup \{p\} \to M \cup \{q\}$ is then holomorphic at $p$.*

The next two Corollaries are generalizations of the Picard Theorem.

COROLLARY 2. *Let $S$ denote the Riemann sphere and let $P$ and $Q$ be finite subsets of $m$ and $n$ points of $S$ respectively. Suppose $f : S\backslash P \to S\backslash Q$ is meromorphic. If $n>2$, then $f$ can be extended to a rational function that must be constant when $m<n$.*

PROOF. Since $S\backslash Q$ is hyperbolic, Corollary 1 can be applied to each point of $P$. Hence $f$ is rational. However if $m<n$ at most $m$ punctures of $S$ are patched with respect to $Q$. Therefore $S$ cannot be mapped onto $S$. So $f$ is constant.

REMARK 2. If $P=\{\infty\}$ and $n=3$, Corollary 2 contains the (little) Picard Theorem.

COROLLARY 3. *Let $R$ be a Riemann surface and let $p$ be a point of $R$. Suppose $f$ is meromorphic on $R\backslash\{p\}$. If there are 3 distinct points $a, b, c$ of the Riemann sphere $S$ such that $f^{-1}\{a,b,c\}$ is empty or finite, then $f$ can be extended to a function meromorphic on $R$.*

PROOF. Let $P = f^{-1}\{a,b,c\}$ and restrict $f$ to $R\backslash(P \cup \{p\})$. Since $S\backslash\{a,b,c\}$ is hyperbolic this restriction can by Corollary 1 be extended to the finitely many points of $P \cup \{p\}$.

REMARK 3. If $R$ is a domain in the complex plane and $p$ is an essential singularity of a function holomorphic on $R\backslash\{p\}$, then the contraposition to Corollary 3 contains the (great) Picard Theorem.

**4.** Finally we generalize the procedure of patching punctures to some countable sets that are then removable.

As a preparation we recall the Cantor process (e.g.[1], VIII, §4) to define for a closed set the kernel by successively peeling away isolated sets.

Let $C$ be a closed set in some topological space. The first step consists in removing all isolated points in order to get the derived set $C'$. Then remove all isolated points of $C'$ and get $C'' = (C')'$. Iterate this process ad infinitum. For $\omega$ as first limit ordinal take the intersection of all derived sets of lower order, i.e., $C^{\omega} = \bigcap_{i<\omega} C^{i}$. Then continue with $C^{\omega+1} = (C^{\omega})'$. Every time either take the derived set (for the next ordinal) or take some intersection (for a limit ordinal). If once we get the same set, $C^{\eta+1} = C^{\eta}$, we do no more generate new sets and the process stops. $C^{\eta}$ is then the kernel.

If the space satisfies the second countability axiom, then the process stops in the first or in the second number class. This shows that one has generated at most a countable number of sets. Since at each step at most countably many points have been removed, one has only countably many points removed in order to get the kernel. But the kernel itself must be a perfect set and is therefore uncountable if it is not empty.

Obviously this process works for a subset of a Riemann surface and we state

PROPOSITION 2. *Let $R$ and $M$ be Riemann surfaces, and let $P$ be a closed countable subset of $R$. If $f : R\backslash P \to M$ is holomorphic and $M$ is and stays hyperbolic (at each step of the following proof), then $f$ can be extended to a holomorphic function $f^*$, $f^* : R \to M^*$, where $M^*$, $M \subset M^*$, is some Riemann surface.*

PROOF. If $P$ is not empty but countable, it cannot be perfect and therefore contains isolated points. If $M$ is hyperbolic, we can apply Corollary 1 and extend $f$ to all those isolated points. If the patched $M$ is still hyperbolic we can continue with $P'$. If we reach a limit ordinal, as e.g. $\omega$, we get with respect to $P$ some intersection, but with respect to $M$ we add to $M$ the union of all points that in succession represent patched punctures. Observe that at each step countably many punctures could occur. Since $P$ is countable and the kernel therefore empty, all of $P$ will be removed.

Similarly as in the proof of Corollary 2 we deduce from Proposition 2

COROLLARY 4. *Let $P$ be a closed countable subset of a compact Riemann surface $R$, and let $Q$ be an uncountable subset of some compact Riemann surface $M$. If $f : R\backslash P \to M\backslash Q$ is holomorphic, then $f$ is constant.*

COROLLARY 5. *Let $P$ be a closed countable subset of a Riemann surface $R$, and let $M$ be a Riemann surface of genus $g$, $1 < g \leq \infty$. If $f : R \backslash P \to M$ is holomorphic, then there are extensions of $f$ to $f^*$ and $M$ to a Riemann surface $M^*$ respectively, such that $f^* : R \to M^*$ is holomorphic.*

REMARK 4. There are some more recent important generalizations of the Picard Theorem.

T. Nishino [5] has shown that a compact set of logarithmic capacity zero is removable for a holomorphic mapping into a compact Riemann surface of genus $g$, $g \geq 2$.

A simpler proof of this theorem has been given by M. Suzuki [9].

H. Shiga [8] has shown that an $AB$-nullset is removable for holomorphic mappings into certain Riemann surfaces. However, he gives an example of a holomorphic mapping into a compact Riemann surface where a compact set with linear measure zero is not removable.

## REFERENCES


[1] F. Hausdorff, *Grundzüge der Mengenlehre.* Leipzig, 1914; Chelsea, New York, 1949.

[2] M. Heins, *On Fuchsoid groups that contain parabolic transformations.* Contributions to Function Theory. Tata Institute, Bombay, 1960, p. 203–210.

[3] H. Huber, *Ueber analytische Abbildungen Riemannscher Flächen in sich.* Comment. Math. Helv. 27 (1953), 1–73.

[4] A. Marden, I. Richards and B. Rodin, *Analytic self-mappings of Riemann surfaces.* J. Analyse Math. 18 (1967), 197–225.

[5] T. Nishino, *Prolongements analytiques au sens de Riemann.* Bull. Soc. math. France 107 (1979), 97–112.

[6] M. Ohtsuka, *On the behavior of an analytic function about an isolated boundary point.* Nagoya Math. J. 4 (1952), 103–108.

[7] H. L. Royden, *The Picard theorem for Riemann surfaces.* Proc. Amer. Math. Soc. 90 (1984), 571–574.

[8] H. Shiga, *On the boundary behavior of holomorphic mappings of plane domains to Riemann surfaces.* J. Math. Kyoto Univ. (to appear).

[9] M. Suzuki, *Comportement des applications holomorphes autour d'un ensemble polaire.* C. R. Acad. Sci. Paris Sér. I Math. 304 (1987), 191–194.

# NONEQUIDIMENSIONAL VALUE DISTRIBUTION THEORY [1]

**Yum-Tong Siu** [2]
Department of Mathematics, Harvard University
Cambridge, Massachusetts 02138, U.S.A.

One of the many areas of mathematics which Professor Lars Ahlfors' work has greatly influenced is the theory of value distribution founded by Nevanlinna [N1, N2]. He cast value distribution theory in differential geometric terms and completed the value distribution theory of holomorphic curves and their associated curves in higher dimensional projective space [A1, A2]. The reformulation of value distribution theory in notions of differential geometry has a great impact on the subject. Shimizu [Sh] contributed also to the geometric formulation by using surface area to describe the characteristic of a meromorphic function. Cartan [C] first generalized Nevanlinna's theory to the case of a holomorphic curve in a higher dimensional projective space and obtained an important inequality on the defect. H. and J. Weyl [W-W1], unaware of Cartan's work, developed independently the theory of holomorphic curves in a higher dimensional projective space, but their theory was not complete. It was Ahlfors [A2] who finally completed the defect relation for holomorphic curves and their associated curves in higher dimensional projective space with the expected nondegeneracy assumption. In the early seventies the school of Griffiths [C-Gr, Gr-K] used the method of differential geometry to develop the equidimensional value distribution theory for holomorphic maps to compact algebraic manifolds. One of the major unsolved problems in value distribution theory is the defect relation for the nonequidimensional case, in particular, the case of holomorphic curves in higher dimensional complex projective spaces. In recent years, because of the theorem of Roth [R] and the solution of the Mordell conjecture [F], there is an ever growing interest in the relation between number theory and value distribution theory and curvature [V].

In this talk we will discuss conjectures in nonequidimensional value distribution theory and some recent results obtained by using meromorphic connections.

---

[1] This paper is in final form and no version of it will be submitted for publication elsewhere.

[2] Research partially supported by a grant from the National Science Foundation.

## 1. Description of Results

(1.1) *The classical 1–dimensional result.* To put our recent results in proper perspective, we review first the relevant known results in Nevanlinna theory. We start with Nevanlinna's original work in 1929 [N1]. Let $f: \mathbb{C} \to \mathbb{P}_1$ be holomorphic, *i.e.* $f$ is a meromorphic function. The characteristic of $f$ (according to Ahlfors-Shimizu [A1, Sh]) is

$$T_f(r) = \int_0^r \frac{d\rho}{\rho} \int_{|z|<\rho} \frac{1}{\pi} \frac{|f'|^2}{(1+|f|^2)^2} = \int_0^r \frac{d\rho}{\rho} \int_{|z|<\rho} \frac{\sqrt{-1}}{2\pi} \partial\bar{\partial} \log(1+|f|^2).$$

To simplify notations we introduce the iterated integral

$$\mathcal{J}_r(\eta) = \int_0^r \frac{d\rho}{\rho} \int_{|z|<\rho} \eta,$$

where $\eta$ is either a function or a (1,1)-current. The counting function for $f$ and $a \in \mathbb{P}_1$ is $N_f(r,a) = \mathcal{J}_r(\{f = a\})$, where the set $\{f = a\}$ is regarded as a (1,1)-current with multiplicity counted. The First Main Theorem of Nevanlinna says that $N_f(r,a) \leq T_f(r) + O(1)$. (Here and in the rest of this paper the Landau symbols $O(\cdot)$ and $o(\cdot)$ are used.) The defect for the map $f$ and the point $a$ is defined by

$$\delta_f(a) = \liminf_{r\to\infty} \frac{T_f(r) - N_f(r,a)}{T_f(r)}.$$

The classical defect relation of Nevanlinna is $\Sigma_\nu \delta_f(a_\nu) \leq 2$ for an arbitrary set of distinct points $\{a_\nu\}$ when $f$ satisfies the nondegeneracy condition that $f$ is nonconstant.

The characteristic function $T_f(r)$ can be expressed in terms of the Fubini-Study metric. The (1,1)-form $\omega$ of the Fubini-Study metric of $\mathbb{P}_1$ is $\omega = \frac{\sqrt{-1}}{2\pi} \partial\bar{\partial} \log(|w_0|^2+|w_1|^2)$, where $[w_0,w_1]$ are the homogeneous coordinates of $\mathbb{P}_1$. One can write $\frac{\sqrt{-1}}{2\pi} \partial\bar{\partial} \log(1+|f|^2) = f^*\omega$ and $T_f(r) = \mathcal{J}_r(f^*\omega)$.

(1.2) *Cartan's result.* When $\mathbb{P}_1$ is replaced by $\mathbb{P}_n$, the point $a$ is replaced by a hyperplane $H$ and the (1,1)-form of the Fubini-Study metric of $\mathbb{P}_n$ is $\omega = \frac{\sqrt{-1}}{2\pi} \partial\bar{\partial} \log(\Sigma_{\nu=0}^n |w_\nu|^2)$, where $[w_0,\cdots,w_n]$ are the homogeneous coordinates of $\mathbb{P}_n$. In this case H. Cartan [C] obtained the following defect relation $\Sigma_\nu \delta_f(H_\nu) \leq n + 1$ if $f$ satisfies the nondegeneracy condition that the image of $f$ is not contained in a hyperplane of $\mathbb{P}_n$ and if the hyperplanes $H_\nu$ are in general position. Actually H. Cartan proved the stronger inequality

$$(q-(n+1))\, T_f(r) \leq \sum_{\nu=1}^{q} N_f^{(n)}(r,H_\nu) + O(\log T_f(r) + \log r)$$

(for $r$ not belonging to some subset $I$ of $\mathbb{R}$ with $\int_I \frac{dt}{t}$ finite), where $N_f^{(n)}(r,H_\nu)$ means that when we compute the counting function $N_f(r,H_\nu)$ we truncate the multiplicity at order $n$ (*i.e.* if the multiplicity is higher than $n$, then we count it as $n$ only).

The nondegeneracy condition is very important. The difficulty that one encounters when the nondegeneracy condition fails is typified in the following simple example. The exponential map $z \to e^z$ composed with the inclusion map $\mathbb{P}_1 \hookrightarrow \mathbb{P}_2$ defines a holomorphic map $f : \mathbb{C} \to \mathbb{P}_2$. We can find four hyperplanes $H_\nu$ $(1 \leq \nu \leq 4)$ in $\mathbb{P}_2$ which intersect normally so that $H_\nu \cap f(\mathbb{C})$ is the point $0$ of $\mathbb{P}_1$ for $\nu = 1, 2$ and is the point $\infty$ of $\mathbb{P}_1$ for $\nu = 3, 4$ so that the image of $f$ is disjoint from $H_\nu$ for $1 \leq \nu \leq 4$ and the defect of $f$ for the four hyperplanes is 4 instead of the allowed limit of 3. As one sees in this simple example, the difficulty arises from the fact that the restrictions of hyperplanes in general position to the hyperplane containing the image of the map may no longer be in general position.

(1.3) *Associated curves.* Ahlfors in [A2] derived defect relations not only for a holomorphic map $f : \mathbb{C} \to \mathbb{P}_n$ but also for the associated curves of the map $f$. The associated curves are defined as follows. Let $F : \mathbb{C} \to \mathbb{C}^{n+1}$ be a holomorphic map inducing $f$. Then the $k^{th}$ associated curve of $f$ is the map $f_k : \mathbb{C} \to Gr(\mathbb{P}_k,\mathbb{P}_n)$ induced by $F \wedge F' \wedge \cdots \wedge F^{(k)}$ from $\mathbb{C}$ to $Gr(\mathbb{C}^{k+1},\mathbb{C}^{n+1})$, where $Gr(A,B)$ means the Grassmannian of $A$ in $B$. The defect relation of Ahlfors is $\Sigma_\nu\, \delta_{f_k}\left[A_\nu^{(n-k)}\right] \leq \binom{n+1}{k+1}$, where $A_\nu^{(n-k)}$ is some $\mathbb{P}_{n-k}$ in $\mathbb{P}_n$. In the definition of the defect $\delta_{f_k}\left[A_\nu^{(n-k)}\right]$ the counting function is defined by using the set $\{z \in \mathbb{C} \mid f(z) \cap A_\nu^{(n-k)} \neq \emptyset\}$ with multiplicity counted, and the characteristic function $T_r(f_k)$ is defined by using the embedding $Gr(\mathbb{P}_k,\mathbb{P}_n) \hookrightarrow \mathbb{P}_{\binom{n+1}{k+1}-1}$ given by the Plücker coordinates and using the Fubini-Study metric of $\mathbb{P}_{\binom{n+1}{k+1}-1}$. The nondegeneracy condition is that the image of $f$

is not contained in a hyperplane of $\mathbb{P}_n$ and the general position condition is that the set of all $A_\nu^{(n-k)}$ should be in general position.

There is some relationship between this result of Ahlfors' and the preceding result of Cartan's. Let $g_k$ be the map obtained by composing the associated curve $f_k : \mathbb{C} \to Gr(\mathbb{P}_k,\mathbb{P}_n)$ with the embedding $Gr(\mathbb{P}_k,\mathbb{P}_n) \hookrightarrow \mathbb{P}_{\binom{n+1}{k+1}-1}$ by Plücker coordinates. If Cartan's result is applied to the map $g_k$, one gets immediately the defect relation $\Sigma_\nu \delta_{f_k}\left[A_\nu^{(n-k)}\right] \leq \binom{n+1}{k+1}$ if one assumes the nondegeneracy condition that the image of $g_k$ is not contained in any hyperplane of $\mathbb{P}_{\binom{n+1}{k+1}-1}$. This new nondegeneracy condition in general is different from the original nondegeneracy condition that the image of $f$ is not contained in any hyperplane of $\mathbb{P}_n$. In the special case $k = 1$ the two nondegenerate conditions agree as one can see from the following simple argument. Suppose the image of $g_1$ is contained in a hyperplane $\Pi$ of $\mathbb{P}_{\binom{n+1}{k+1}-1}$ for $k = 1$. The equation of a hyperplane in $\mathbb{P}_{\binom{n+1}{k+1}-1}$ for $k = 1$ has $\binom{n+1}{2}$ coefficients. By using a projective linear transformation of $\mathbb{P}_n$ which has $(n+1)^2 - 1$ degrees of freedom, we can assume that only one coefficient of the equation of $\Pi$ is nonzero. Without loss of generality we can assume that the condition that $\Pi$ contains the image of $g_1$ is given by the vanishing of $\det \begin{bmatrix} F_0 & F_0' \\ F_1 & F_1' \end{bmatrix}$, where $(F_0,\cdots,F_n)$ are the coordinates of the map $F$. This means that $\lambda_0 F_0 + \lambda_1 F_1 \equiv 0$ for some nonzero pair of complex numbers $(\lambda_0,\lambda_1)$ and the image of $f$ is contained in some hyperplane of $\mathbb{P}_n$.

(1.4) *The equidimensional result.* The generalization of value distribution theory to holomorphic maps from $\mathbb{C}^n$ to $\mathbb{P}_n$ was first done by Stoll [St1].

The school of Griffiths used the curvature approach to look at holomorphic maps $f : \mathbb{C}^n \to \mathbb{P}_n$ and consider, instead of hyperplanes, divisors $Z$ of higher degree $d$, *i.e.* zero-sets of homogeneous polynomials of degree $d$ in $\mathbb{P}_n$. The defect relation is $\Sigma_\nu \delta_f(Z_\nu) \leq \frac{n+1}{d}$, when the rank of $df$ over $\mathbb{C}$ is $n$ at some point of $\mathbb{C}^n$ and the divisors $Z_\nu$ are all nonsingular and intersect normally. Here to define the defect the characteristic function is $d\mathcal{T}_r(f^*\omega)$ instead of just $\mathcal{T}_r(f^*\omega)$, otherwise the First Main Theorem $N_f(r,a) \leq T_f(r) + O(1)$ would fail to hold. The method of the school of Griffiths works in the more general case when $\mathbb{P}_n$ is replaced by a compact algebraic manifold M of complex dimension n. In such a case the divisors $Z_\nu$ are the zero-sets of holomorphic sections of a positive line bundle L over M and the defect is at most $\alpha$ when $\alpha\theta_L + \theta_{K_M}$ is everywhere positive, where $\theta_L$ is the curvature form of L and $K_M$ is the canonical line bundle of M. The characteristic function is defined as $\mathcal{J}_r(f^*\theta_L)$. The case $M = \mathbb{P}_n$ corresponds to $L = H^d$, where H is the hyperplane section line bundle of $\mathbb{P}_n$.

The most important assumption here is the nondegeneracy condition that the rank of df over $\mathbb{C}$ at some point of $\mathbb{C}^n$ is equal to the complex dimension n of $\mathbb{P}_n$ or M. One can also replace $\mathbb{C}^n$ by $\mathbb{C}^m$ with m different from n and, of course, necessarily with $m \geq n$. It is not known whether, analogous to the case $d = 1$, one can replace this nondegeneracy condition by the following weaker and more natural condition: the image of f is not contained in any complex hypersurface of $\mathbb{P}_n$ of degree d (or any complex hypersurface without any restriction on its degree); or in the general case of a compact algebraic manifold M, the image of f is not contained in the zero set of any holomorphic section of the line bundle L or tensor powers $L^\nu$ of L.

(1.5) *Conjectures.* One of the most important open questions in value distribution theory is how to obtain defect relations for a holomorphic map from $\mathbb{C}^m$ to $\mathbb{P}_n$ (or to a compact algebraic manifold of complex dimension n) with $m < n$ and for divisors of degree $d > 1$ in $\mathbb{P}_n$. The smaller m is, compared to n, the more difficult the question is. So the typical casc is the situation of a holomorphic map from $\mathbb{C}$ to $\mathbb{P}_n$ with divisors of degree $d > 1$.

There is a conjecture of Griffiths that for the case of a holomorphic map $f: \mathbb{C} \to \mathbb{P}_n$ and nonsingular divisors $Z_\nu$ of degree $d > 1$ in $\mathbb{P}_n$ intersecting normally the defect relation is $\Sigma_\nu\, \delta_f(Z_\nu) \leq \frac{n+1}{d}$ under an appropriate nondegeneracy condition of the map f. In the case of $d = 1$ the theorem of Cartan tells us that the appropriate nondegeneracy condition is that the image of f is not contained completely in any hyperplane of $\mathbb{P}_n$. So just as in the equidimensional case, the most natural analog of the nondegeneracy condition for f in the case of a general d should be the condition that the image of f is not contained completely in any divisor of degree d in $\mathbb{P}_n$.

Unfortunately for this nondegeneracy condition there is the following counterexample due to Biancofiore [B]. The map $f: \mathbb{C} \to \mathbb{P}_2$ is induced by the homogeneous coordinates $[1 - 2e^\zeta, e^\zeta + e^{2\zeta}, e^{3\zeta} - 2e^{2\zeta}]$, where $\zeta$ is the coordinate of $\mathbb{C}$. The degree of the divisors is 2. The defects for the three nonsingular divisors

$9x^2+36xy+8xz+16y^2=0$, $9z^2+36yz+8xz+16y^2=0$, and $xy+yz+xz=0$ which intersect normally are respectively 1, 1, and $\frac{1}{3}$, where $[x,y,z]$ are the the homogeneous coordinates of $\mathbb{P}_2$. So the sum of the defects for these three divisors exceeds $\frac{n+1}{d}=\frac{3}{2}$. The image of $f$ is not contained in any divisor of degree 2. However, it is contained in the divisor $y^3+2y^2(x+z)+(5y+x+z)xz=0$ of degree 3.

In view of this counter-example it is natural to consider the nondegeneracy condition that the image of $f$ is not contained in any complex hypersurface of $\mathbb{P}_n$ without restriction on the degree of the hypersurface. The conjecture with this nondegeneracy condition is still open.

A related conjecture is that there exists a positive integer $d(n)$ depending only on $n$ so that for any generic nonsingular divisor $Z$ of degree at least $d(n)$ in $\mathbb{P}_n$ there does not exist any nonconstant holomorphic map from $\mathbb{C}$ to $\mathbb{P}_n - Z$. Gromov told me that Grauert recently obtained a proof of this conjecture for $n = 2$.

(1.6) *Maps of exponentials of polynomials.* When each of the homogeneous coordinates of the map $f:\mathbb{C}\to\mathbb{P}_n$ is of the form $\Sigma_{j=1}^{k} e^{P_j(z)}$ with $P_j(z)$ being polynomials of $z$, the defects for divisors of degree $>1$ can be rather explicitly estimated. When the polynomials $P_j(z)$ are perturbed by lower order functions, the defects can still be rather explicitly estimated. Shiffman [S1] obtained by such explicit estimates the defect relation for maps $f:\mathbb{C}\to\mathbb{P}_n$ whose $\nu^{\text{th}}$ homogeneous coordinate is of the form $\Sigma_j\,\varphi_{\nu j}e^{P_{\nu j}}$ $(0\le\nu\le n)$ where the characteristic function $T_f(r)$ of $f$ is of order $r^\lambda$ and $P_{\nu j}$ is a polynomial of degree $\le\lambda$ and the characteristic function $T_{\varphi_{\nu j}}(r)$ of $\varphi_{\nu j}$ is of order $o(r^\lambda)$. His defect relation for divisors $D_1,\cdots,D_q$ of degree $d$ in $\mathbb{P}_n$ with no point belonging to $n+1$ of the $D_j$'s and with $f(\mathbb{C})$ not contained in $\cup_{j=1}^q D_j$ is given by $\Sigma_{j=1}^q\,\delta_f(D_j)\le 2n$. It also works when $\mathbb{C}$ is replaced by $\mathbb{C}^m$. Note that even in this special case the order of the defect is $O(1)$ instead of the expected $O(\frac{1}{d})$.

(1.7) *The use of the Veronese map.* An easy way to get a defect relation for divisors $Z$ of degree $d>1$ in $\mathbb{P}_n$ is to make the divisor linear by using the Veronese map. If we define $g:\mathbb{C}\to\mathbb{P}_{\binom{n+d}{d}-1}$ by using $[f_0^{\lambda_0}\cdots f_n^{\lambda_n}]_{\lambda_0+\cdots+\lambda_n=d}$ as the homogeneous coordinates of $g$, where $[f_0,\cdots,f_n]$ is the homogeneous coordinate of $f$, then the set

$f^{-1}(Z)$ equals the inverse image of a hyperplane in $\mathbb{P}_{\binom{n+d}{d}-1}$ under $g$. So by using the defect relation for $g$ we obtain the defect relation $\Sigma_\nu \, \delta_f(Z_\nu) \le \binom{n+d}{d}$ when the hyperplanes in $\mathbb{P}_{\binom{n+d}{d}-1}$ corresponding to the divisors $Z_\nu$ are in general position and the image of $f$ is not contained in any divisor of degree $d$ in $\mathbb{P}_n$. This defect relation obtained by using the Veronese map was first observed by Shiffman [S1]. Note that the order of the defect estimate is $d^n$ instead of the expected $\frac{1}{d}$. The number $\binom{n+d}{d}$ is large, because of the large number of terms in the defining equation for the divisor $Z$ of degree $d$ in $\mathbb{P}_n$. If most of the coefficients of the defining equation of $Z$ are zero, then we can get away with a smaller number for the defect estimate. The simplest case is the case of Fermat curves in $\mathbb{P}_2$. Let $C_a$ denote the Fermat curve defined by the equation $a_0 w_0^d + a_1 w_1^d + a_2 w_2^d = 0$ with $a = (a_0, a_1, a_2) \in \mathbb{C}^3$. Consider the map $\mathbb{P}_2 \to \mathbb{P}_2$ defined by $[w_0, w_1, w_2] \to [w_0^d, w_1^d, w_2^d]$. Unlike the case of the Veronese map the image of this map is still $\mathbb{P}_2$. Composing the map $f$ with this map, we get $g : \mathbb{C} \to \mathbb{P}_2$ with $g = [f_0^d, f_1^d, f_2^d]$. We have $\Sigma_a \, \delta_g(C_a) = \Sigma_a \, \delta_g(C_a) \le 3$ which is not of the order $\frac{1}{d}$ but is better than the order $d^2$ which one would obtain by using the Veronese map. The correct defect estimate is expected to be $\frac{6}{d}$, because of the following result of M. Green [G]: A holomorphic map from $\mathbb{C}^m$ to $\mathbb{P}_n$ omitting a Fermat variety of degree $d > n(n+1)$ must have its image contained in a linear subspace of dimension $\le \frac{n}{2}$. Moreover, this result of Green's is sharp. So for $n = 2$, $d = 6$ it means that there exists a holomorphic map $\mathbb{C} \to \mathbb{P}_2$ omitting a Fermat curve of degree 6. So the defect is 1 for that case and we should have no less than $\frac{6}{d}$ for the defect estimate.

(1.8) *Shiffman's method for Fermat curves.* Shiffman in a private communication [S2] showed me how the defect estimate $\frac{6}{d}$ can be obtained by using the result of H. Cartan in the following way. Given a set of triples $a^{(1)}, \cdots, a^{(q)} \in \mathbb{C}^3$. One adds the three triples $a^{(q+1)} = (1,0,0)$, $a^{(q+2)} = (0,1,0)$, and $a^{(q+3)} = (0,0,1)$. We now apply the result of H. Cartan to the map $g$ and the hyperplanes whose defining equations have the triples $a^{(\nu)}$ $(1 \le \nu \le q+3)$ as coefficients. Then (for $r$ not belonging to some subset $I$ of $\mathbb{R}$ with finite $\int_I \frac{dt}{t}$)

$$\{(q+3)-3\}\, T_g(r) \le \sum_{\nu=1}^{q} N_g^{(2)}(r, C_{a^{(\nu)}}) + \sum_{\nu=1}^{3} N_g^{(2)}(r, C_{a^{(q+\nu)}}) + O(\log T_g(r) + \log r).$$

Since the multiplicity of $g^{-1}(C_{a^{(q+\nu)}})$ is $d$ and in the computation of $N_g^{(2)}(r, a^{(q+\nu)})$ the multiplicity is truncated at 2, it follows that

$$N_g^{(2)}(r, C_{a^{(q+\nu)}}) \le \tfrac{2}{d} N_g(r, C_{a^{(q+\nu)}}) \le \tfrac{2}{d} T_g(r) = \tfrac{2}{d} T_f(r)$$

for $1 \le \nu \le 3$. Since the defect estimate comes from $\Sigma_{\nu=1}^{3} N_g^{(2)}(r, C_{a^{(q+\nu)}})$, we get the defect estimate $\frac{6}{d}$.

(1.9) *Results using meromorphic connections.* This method of Shiffman fails if one considers Fermat curves in different coordinate systems. Some recent results using meromorphic connections can handle such cases. Let us first state the defect relation we can get when our results are applied to Fermat curves in different coordinate systems.

**Theorem 1.** *Suppose* $f : \mathbb{C} \to \mathbb{P}_2$ *is a holomorphic map and for each* $\nu$ *the curve* $Z_\nu$ *is a Fermat curve* $\zeta_{\nu 0}^d + \zeta_{\nu 1}^d + \zeta_{\nu 2}^d = 0$ *of degree* $d$ *in some homogeneous coordinate system* $[\zeta_{\nu 0}, \zeta_{\nu 1}, \zeta_{\nu 2}]$ *of* $\mathbb{P}_2$. *When* $f$ *satisfies a suitable nondegeneracy condition and the collection* $\{Z_\nu\}$ *of Fermat curves satisfies a suitable general position condition (both to be specified in the later sketch of the proof in* §3) *we have the defect estimate* $\Sigma_\nu\, \delta_f(Z_\nu) \le \frac{1603}{d}$.

This method works in more general situations. A meromorphic connection for a compact complex manifold $M$ is a connection $\Gamma_{\alpha\beta}^{\gamma}$ for the tangent bundle of $M$ such that each component $\Gamma_{\alpha\beta}^{\gamma}$ with respcct to a local holomorphic coordinate system is meromorphic. We do not require that the connection be symmmetric, *i.e.* $\Gamma_{\alpha\beta}^{\gamma}$ may not equal $\Gamma_{\beta\alpha}^{\gamma}$. The use of meromorphic connections was introduced in [Si1, Si2] where the same single meromorphic connection is used for all the divisors in a family of divisors. For Fermat curves the method in [Si1, Si2] can be used only in the case of Fermat curves in a *fixed* coordinate system which can already be handled by Shiffman's application of the method of Cartan. To overcome this drawback we use different meromorphic connections for different divisors in the family of divisors and introduce what can very roughly be described as a value distribution theory of meromorphic connections. We are going to sketch this new method. Complete details can be found in [Si3].

We now state our result in a more general setting.

**Theorem 2.** *Let* M *be a compact complex manifold. Let* $L_{D_0}$ *be a positive line bundle such that a meromorphic connection* $D_0$ *exists for* M *which when multiplied by a non-identically-zero holomorphic section of* $L_{D_0}$ *is holomorphic. Let* L *be a line bundle and* $\{s_\nu\}$ *a collection of holomorphic sections of* L *over* M. *Let* $L_D$ *be a positive line bundle with the following property: for each* $\nu$ *a meromorphic connection* $D^{(\nu)}$ *exists for* M *such that the second fundamental form of the zero-set* $Zs_\nu$ *of* $s_\nu$ *with respect to* $D^{(\nu)}$ *is zero and the connection* $D^{(\nu)}$ *multiplied by some non-identically-zero holomorphic section* $t_\nu$ *of* $L_D$ *is holomorphic. Let* $p = \dim \Gamma(M, L_D)$ *and* $k = \dim \Gamma(M, T^*_M \otimes T^*_M \otimes T_M \otimes L_{D_0} \otimes L_D)$. *Let* F *be a nonnegative line bundle over* M *so that* $\Gamma(M, F \otimes (L_{D_0} \otimes L_D)^{\frac{n(n-1)}{2}} \otimes K_M^{-1})$ *embeds* M *into* $\mathbb{P}_{a-1}$, *where* a *is the dimension of* $\Gamma(M, F \otimes (L_{D_0} \otimes L_D)^{\frac{n(n-1)}{2}} \otimes K_M^{-1})$. *Let* $\alpha$ *be a nonnegative number such that* $\alpha\theta_L$ *minus the curvature of* $F \otimes (L_{D_0} \otimes L_D)^{\frac{n(n-1)}{2}} \otimes K_M^{-1}$ *is everywhere nonnegative on* M. *Suppose* f *is a holomorphic map from* $\mathbb{C}$ *to* M. *Assume that* f *satisfies a suitable nondegeneracy condition and* $\{Zs_\nu\}$ *satisfies a suitable general position condition specified in* §3. *Then the defect of* f *is no more than* $\alpha \binom{b}{c} + \alpha\, a$, *where* $b = \frac{n(n-1)}{2} + p + k - 1$ *and* $c = \frac{n(n-1)}{2}$.

**Corollary.** *For the special case* $M = \mathbb{P}_n$ *with* $L = H^d$ *(i.e. for divisors of degree* d*), let* $\ell$ *be the pole order of the meromorphic connections for a collection* $\{s_\nu\}$ *of holomorphic sections of* L *(i.e.* $L_D = L^\ell$*). Then the defect is no more than* $\binom{b}{c}$ *times*

$$\frac{(\ell+1)^{\frac{n(n+1)}{2}} + n + 1}{d},$$

*where* $b = \frac{n(n-1)}{2} + \binom{\ell+n}{n} + (n+1)^2(n+\ell+1)\binom{n+\ell-1}{n-1} - 1$ *and* $c = \frac{n(n-1)}{2}$.

This defect estimate in the Corollary is completely useless if $\ell$ is large. A class of divisors of degree d in $\mathbb{P}_n$ to which this Corollary can be applied to give defect estimates of order $\frac{1}{d}$ is the following. Let k be a nonnegative integer $\leq d$. The divisors in this class are defined by polynomials of the form $\Sigma_{\nu=0}^{n} P_\nu(w_0,\cdots,w_n)\, w_\nu^{d-k}$ (where each $P_\nu$ is a polynomial) with respect to some homogeneous coordinate system $[w_0,\cdots,w_n]$ of $\mathbb{P}_n$ (which may depend on the divisor). For such a divisor there is a meromorphic connection with pole order no more than $(n+1)k+4$ with respect to which the second fundamental form of the divisor is zero. The defect for such a class of divisors can be estimated by $\frac{C(k,n)}{d}$, where $C(k,n)$ is a constant depending only on k and n (computable according to the formula given in the Corollary).

## 2. Known Techniques of Value Distribution Theory

To sketch the method of proof for Theorems 1 and 2, we briefly go through the relevant major known techniques in value distribution theory and present the sketch by discussing the difficulties of applying these known techniques to the nonequidimensional case and how the difficulties are handled by using meromorphic connnections.

(2.1) *The analogous compact case.* Nevanlinna theory can be regarded as a noncompact version of the following compact situation. Let M be a compact Riemann surface and L be a holomorphic line bundle over M with a Hermitian metric. Let $\theta_L$ be the curvature form of L and s be a global holomorphic section of L. Then $\frac{\sqrt{-1}}{2\pi}\partial\bar\partial \log|s|^2 = Zs - \theta_L$, where Zs is the zero-set of s regarded as a (1,1)-current on M. By integrating this equation over M, we conclude that the number of zeros of s is equal to the integral of $\theta_L$ over M and is independent of the holomorphic section s. The integral of $\theta_L$ over M is known as the Chern class of L.

(2.2) *The basic Poincaré–Lelong equation.* Now we look at the Nevanlinna theory of a holomorphic map $f: \mathbb{C} \to M$ (with $M = \mathbb{P}_1$). We pull back the line bundle L to a line bundle $f^*L$ over $\mathbb{C}$ and pull back the section s to a section $f^*s$ of $f^*L$ over $\mathbb{C}$. We have the basic Poincaré–Lelong equation $\frac{\sqrt{-1}}{2\pi}\partial\bar\partial \log|f^*s|^2 = f^*Zs - f^*\theta_L$. Integrating by parts over $\{|z| \leq r\}$ yields

$$\frac{\sqrt{-1}}{2\pi}\int_{|z|=r} \bar\partial \log|f^*s|^2 = \int_{|z|\leq r} f^*Zs - \int_{|z|\leq r} f^*\theta_L.$$

The defect of f for the divisor Zs comes from the boundary term. The boundary term can be rewritten as an integral of the normal derivative of $\log|f^*s|^2$, because $\partial\bar\partial$ is the same as the Laplace operator up to a constant factor and the divergence theorem is equivalent to integration by parts. So an integration with respect to r yields

$$\mathcal{T}_r(f^*Zs) - \mathcal{T}_r(f^*\theta_L) = \mathcal{T}_r(\tfrac{\sqrt{-1}}{2\pi}\partial\bar\partial\log|f^*s|^2) = \tfrac{1}{4\pi}\mathcal{A}_r(\log|f^*s|^2) - \tfrac{1}{2}\log|f^*s|^2(0),$$

where the notation $\mathcal{A}_r(g)$ means $\int_{\theta=0}^{2\pi} g(re^{i\theta})$ for a function g. Since the pointwise norm of s is uniformly bounded by the compactness of M, the First Main Theorem $\mathcal{T}_r(f^*Zs) \le \mathcal{T}_r(f^*\theta_L) + O(1)$ follows immediately.

(2.3) *The modified Poincaré–Lelong equation.* To handle the boundary term, we analyze the equation $\frac{\sqrt{-1}}{2\pi}\partial\bar\partial\log|f^*s|^2 = f^*Zs - f^*\theta_L$ which we rewrite as $\frac{\sqrt{-1}}{2\pi}\partial\bar\partial\log\frac{1}{|f^*s|^2} = f^*\theta_L - f^*Zs$. The equation is a consequence of the fact that $\partial\bar\partial\log|F|^2 = 0$ for any nowhere zero holomorphic function F. The right-hand side of the above equation comes from the "twisting" of the metric of L and the zero-set of the holomorphic section s of L. If we have another term on the right-hand side with the correct sign, then we can take care of the boundary term. We try to replace $\log\frac{1}{|f^*s|^2}$ by another expression which would give essentially $f^*\theta_L - f^*Zs$ and an additional term we want. Let us forget the twisting of the metric and the zero-set of the holomorphic section. We would like to have $\partial_z\bar\partial_z\Phi(F) = \Psi(F)$ for some positive-valued functions $\Phi$ and $\Psi$ so that $\Psi$ dominates $\Phi$ in a suitable sense. A suitable candidate to use for $\Phi$ is a metric of constant negative curvature. We have

$$\partial_z\bar\partial_z\log\frac{1}{|z|^2(\log|z|^2)^2} = \frac{2}{|z|^2(\log|z|^2)^2}\,.$$

Thus when we use

$$\Phi(z) = \log\frac{1}{|z|^2(\log|z|^2)^2} \quad\text{and}\quad \Psi(z) = \frac{2}{|z|^2(\log|z|^2)^2}$$

and keep track of the twisting of the metric and the zero-set of the holomorphic section, we get the modified Poincaré–Lelong equation

(2.3.1)
$$\frac{\sqrt{-1}}{2\pi} \partial\bar{\partial} \log \frac{1}{|f^*s|^2(\log|f^*s|^2)^2}$$

$$= (f^*\theta_L - f^*Zs)\left[1 + \frac{2}{(\log|f^*s|^2)}\right] + \frac{2|f^*Ds|^2}{|f^*s|^2(\log|f^*s|^2)^2},$$

where Ds means the covariant differentiation of s. (The use of hyperbolic metrics in the derivation of the logarithmic derivative lemma which corresponds to this step was first introduced by F. Nevanlinna [Ne]. See [N2, pp. 242–243]. The formulation in the form given here is due to Griffiths.) Because of the factor $|f^*Ds|^2$ in the numerator of

$$\frac{2|f^*Ds|^2}{|f^*s|^2(\log|f^*s|^2)^2},$$

we still cannot use the integral

$$\mathscr{I}_r\left[\frac{2|f^*Ds|^2}{|f^*s|^2(\log|f^*s|^2)^2}\right]$$

to dominate

$$\mathscr{I}_r\left[\log\frac{1}{|f^*s|^2(\log|f^*s|^2)^2}\right].$$

(2.4) *The role of the anticanonical line bundle.* To take care of the factor $|f^*Ds|^2$, we have to use the equation

(2.4.1)
$$\frac{\sqrt{-1}}{2\pi} \partial\bar{\partial} \log \frac{|df|^2}{|f^*s|^2(\log|f^*s|^2)^2} = Z(df) - f^*\theta_{K_M^{-1}} +$$

$$+ (f^*\theta_L - f^*Zs)\left[1 + \frac{2}{(\log|f^*s|^2)}\right] + \frac{2|f^*Ds|^2}{|f^*s|^2(\log|f^*s|^2)^2},$$

where $Z(df)$ is the zero-set of the differential df of f and the pointwise norm $|df|$ is taken with respect to the metric of M. The curvature $\theta_{K_M^{-1}}$ is the curvature of the anticanonical line bundle $K_M^{-1}$ of M. The boundary term from

$$\frac{\sqrt{-1}}{2\pi} \partial\bar{\partial} \log \frac{|df|^2}{|f^*s|^2(\log|f^*s|^2)^2}$$

is going to be taken care of by the term

$$\frac{2|f^*Ds|^2}{|f^*s|^2(\log|f^*s|^2)^2}$$

on the right-hand side. Since the term

$$\frac{\sqrt{-1}}{2\pi}\partial\bar\partial\log\frac{|df|^2}{|f^*s|^2(\log|f^*s|^2)^2}$$

involves two differentiations and the term

$$\frac{2|f^*Ds|^2}{|f^*s|^2(\log|f^*s|^2)^2}$$

involves no differentiation and both are to be integrated twice, to compare the two terms we have to handle the problem of comparing an integrand and its integral.

(2.5) *Basic inequality from a calculus lemma and the concavity of logarithm.* To handle the problem of comparing an integrand and its integral, we observe that when the integral of a nonnegative integrand is small, the integrand must be small outside a small interval. One can express this fact more precisely in the following Calculus Lemma. Suppose $\alpha(\cdot)$ is a positive-valued function defined on the set of positive real numbers and $\psi$ is a positive-valued function, then the set I of points x with $\psi'(x) \le \alpha(\psi(x))$ is of measure $\le \int_{y=0}^{\infty}\frac{dy}{\alpha(y)}$ as is readily verified by the substitution $y = \psi(x)$. Usually this is applied to the case where $\alpha(x) = x^\lambda + 1$ with $\lambda > 1$ and $\psi$ is an indefinite integral. For any function g by the fundamental theorem of calculus we have

$$\frac{1}{r}\frac{d}{dr}\Big(\frac{1}{r}\frac{d}{dr}\mathcal{T}_r(g)\Big) = \frac{1}{r}\frac{d}{dr}\Big[\int_{|z|<r} g\Big] = \mathcal{A}_r(g).$$

By applying the Calculus Lemma twice with $x = \log r$, we get

$$\log\mathcal{A}_r(g) \le \varepsilon\log r + (1+\delta)\log\mathcal{T}_r(g) \quad \|_{\varepsilon,\delta}$$

where $\|_{\varepsilon,\delta}$ (or simply $\|$) means that the inequality holds for all r outside an open set $I_{\varepsilon,\delta}$ depending on $\varepsilon,\delta$ such that $\int_{I_{\varepsilon,\delta}}\frac{dt}{t}$ is finite. By the concavity of the logarithmic function, we have

$$\mathcal{A}_r(\log g) \le 2\pi(1+\delta)\log\mathcal{T}_r(g) + O(1) \quad \|_{\varepsilon,\delta}.$$

Since Ds is nonzero near $s = 0$, we have $|df|^2 \leq \text{constant} \cdot |f^*Ds|^2$ on the pullback under $f$ of a neighborhood of $s = 0$. Hence

$$(2.5.1) \qquad \mathcal{A}_r\left[\log \frac{|df|^2}{|f^*s|^2(\log|f^*s|^2)^2}\right]$$

$$\leq \text{constant} \cdot \log \mathcal{I}_r\left[\frac{2|f^*Ds|^2}{|f^*s|^2(\log|f^*s|^2)^2}\right] + O(1) \quad \|.$$

Note that this inequality is simply a direct consequence of the Calculus Lemma and the concavity of the logarithmic function and does not depend on any other arguments.

**(2.6)** *Final derivation of classical defect relation.* Applying the Green's theorem $\mathcal{I}_r(\partial_z\partial_{\bar{z}}\, g) = \frac{1}{4}\mathcal{A}_r(g) - \frac{\pi}{2}g(0)$ to the equation (2.4.1) we get

$$\frac{1}{4\pi}\mathcal{A}_r\left[\log \frac{|df|^2}{|f^*s|^2(\log|f^*s|^2)^2}\right]$$

$$\leq \frac{1}{2}\left[\log \frac{|df|^2}{|f^*s|^2(\log|f^*s|^2)^2}\right](0) - \mathcal{I}_r\left[f^*\theta_{K_M^{-1}}\right] +$$

$$+ \mathcal{I}_r\left[(f^*\theta_L - f^*Zs)\left[1 + \frac{2}{(\log|f^*s|^2)}\right]\right] + \mathcal{I}_r\left[\frac{2|f^*Ds|^2}{|f^*s|^2(\log|f^*s|^2)^2}\right].$$

By replacing the pointwise length $|s|$ of $s$ by its multiple with a sufficiently small positive constant, we can make the absolute value of $\dfrac{2}{(\log|f^*s|^2)}$ less than any prescribed positive number. By using the basic inequality (2.5.1) we obtain the defect

$$\liminf_{r \to \infty} \frac{\mathcal{I}_r(f^*\theta_L) - \mathcal{I}_r(f^*Zs)}{\mathcal{I}_r(f^*\theta_L)} \leq \alpha$$

for any constant $\alpha$ with the property that $\alpha\theta_L - \theta_{K_M^{-1}}$ is everywhere nonnegative.

This translates back to the classical Nevanlinna defect being 2 when one uses the $q^{th}$ tensor power of the hyperplane section line bundle as the line bundle $L$ and uses a holomorphic section $s$ of $L$ with $q$ distinct zeros.

The case of holomorphic maps from $\mathbb{C}^n$ to $\mathbb{P}_1$ is the same. The only modification is that one has to multiply both sides of (2.4.1) by $\left[\frac{\sqrt{-1}}{2\pi}\partial\bar{\partial}\log|z|^2\right]^{n-1}$ before one integrates. This is the same as restricting oneself to complex lines in $\mathbb{C}^n$ through the origin and then averaging over all such complex lines.

(2.7) *Transplantation to the higher equidimensional case.* The preceding argument can be modified in the following way to work for the case of holomorphic maps f from $\mathbb{C}^n$ to $\mathbb{P}_n$ and also the more general case of holomorphic maps from $\mathbb{C}^n$ to a compact complex manifold M of complex dimension n. For this higher equidimensional case it is crucial that one has the assumption of normal crossing for the divisors as well as the nondegeneracy condition that the complex rank of f is n. Assume that L is a holomorphic line bundle over M with positive curvature $\theta_L$ and $\{s_\nu\}$ is a collection of holomorphic sections of L over M so that each of the zero-sets $Zs_\nu$ of $s_\nu$ is nonsingular and they intersect normally. The natural analog of the equation (2.4.1) is

$$(2.7.1)\qquad \frac{\sqrt{-1}}{2\pi}\partial\bar\partial\log\prod_{\nu=1}^{q}\frac{1}{|f^*s_\nu|^2(\log|f^*s_\nu|^2)^2}$$

$$=\sum_{\nu=1}^{q}(f^*\theta_L-f^*Zs_\nu)\left[1+\frac{2}{(\log|f^*s_\nu|^2)}\right]+\sum_{\nu=1}^{q}\sqrt{-1}\,\frac{Df^*s_\nu\wedge\overline{Df^*s_\nu}}{|f^*s_\nu|^2(\log|f^*s_\nu|^2)^2}.$$

Again one multiplies the equation by $\left[\frac{\sqrt{-1}}{2\pi}\partial\bar\partial\log|z|^2\right]^{n-1}$ before one integrates.

The main difference between this higher equidimensional case and the classical 1-dimensional case is in getting the analog of the basic inequality (2.5.1) which in the 1-dimensional case was derived simply from the Calculus Lemma and the concavity of the logarithmic function. For the higher equidimensional case, in place of $|df|^2$ in the numerator of $\frac{|df|^2}{|f^*s|^2(\log|f^*s|^2)^2}$, one uses the Jacobian determinant $|Jf|^2$ of the holomorphic map f. We have to try to dominate the boundary term coming from

$$\left[\frac{\sqrt{-1}}{2\pi}\partial\bar\partial\log|z|^2\right]^{n-1}\wedge\frac{\sqrt{-1}}{2\pi}\partial\bar\partial\log\left\{|Jf|^2\prod_{\nu=1}^{q}\frac{1}{|f^*s_\nu|^2(\log|f^*s_\nu|^2)^2}\right\}$$

by the integral of

$$\left[\frac{\sqrt{-1}}{2\pi}\partial\bar\partial\log|z|^2\right]^{n-1}\wedge\sum_{\nu=1}^{q}\sqrt{-1}\,\frac{Df^*s_\nu\wedge\overline{Df^*s_\nu}}{|f^*s_\nu|^2(\log|f^*s_\nu|^2)^2}.$$

The key point is to use the domination of geometric means by arithmetic means. Since the zero-sets $Zs_\nu$ intersect normally, the expression

$$\prod_{\nu=1}^{q} \frac{1}{|f^*s_\nu|^2(\log|f^*s_\nu|^2)^2}$$

is dominated by a constant times the sum of

$$\sum_{\nu_1<\cdots<\nu_n} \prod_{i=1}^{n} \frac{1}{|f^*s_{\nu_i}|^2(\log|f^*s_{\nu_i}|^2)^2}\,.$$

Because of the domination of geometric means by arithmetic means and the assumption of normal intersection of the zero-sets of $s_{\nu_i}$, each

$$\left\{|Jf|^2 \prod_{i=1}^{n} \frac{1}{|f^*s_{\nu_i}|^2(\log|f^*s_{\nu_i}|^2)^2}\right\}^{1/n}$$

times the volume form $\left[\frac{\sqrt{-1}}{2\pi}\partial\bar\partial\log|z|^2\right]^n$ is dominated by a constant times

$$\left[\frac{\sqrt{-1}}{2\pi}\partial\bar\partial\log|z|^2\right]^{n-1} \wedge \sum_{i=1}^{n} \sqrt{-1}\,\frac{Df^*s_{\nu_i}\wedge\overline{Df^*s_{\nu_i}}}{|f^*s_{\nu_i}|^2(\log|f^*s_{\nu_i}|^2)^2}\,.$$

Now the argument used in the 1-dimensional case carries over to give the same defect estimate $\sum_{\nu=1}^{q}\delta_f(Zs_\nu)\le\alpha$ for any constant $\alpha$ with the property that $\alpha\theta_L-\theta_{K_M^{-1}}$ is everywhere nonnegative.

The case of holomorphic maps $f$ from $\mathbb{C}^m$ to a compact complex manifold $M$ of complex dimension $n\le m$ is the same if the rank of the differential of $f$ over $\mathbb{C}$ is $n$. One simply multiplies the equation (2.7.1) by $\left[\frac{\sqrt{-1}}{2\pi}\partial\bar\partial\log|z|^2\right]^{m-1}$ instead of by $\left[\frac{\sqrt{-1}}{2\pi}\partial\bar\partial\log|z|^2\right]^{n-1}$ before one integrates.

## 3. Sketch of the Method of Meromorphic Connections

(3.1) *Replacement for the basic inequality from calculus lemma and concavity of logarithm.* To make the sketch of the method of meromorphic connections more easily understood, we consider first the case of a holomorphic map $f$ from $\mathbb{C}$ to a compact complex manifold $M$ of complex dimension 2, especially when $M$ is the complex projective plane $\mathbb{P}_2$. Again we assume that we have a holomorphic line bundle $L$ over $M$ whose curvature form $\theta_L$ is positive definite and we have a holomorphic section $s$ of $L$ whose zero-set $Zs$ is smooth. One can use just in the classical case the equation (2.3.1). However, we do not have the basic inequality (2.5.1). In the present case $f^*Ds = (f_z^\alpha D_\alpha s)dz$ (summation convention being used) is the inner product of the two vectors $df = (f_z^\alpha)$ and $Ds = (D_\alpha s)$. So it is important to know at what angle $f(\mathbb{C})$ intersects the zero-set $Zs$ of $s$. According to the conjecture for this case the estimate in the defect should involve the anticanonical line bundle $K_M^{-1}$. Since the pullback of a holomorphic 2-form by $f$ is always zero, we have to consider associated curves. We introduce a connection in the tangent bundle of $M$ and consider the first and second differentials $df : T_\mathbb{C} \to T_M$ and $Ddf : T_\mathbb{C} \otimes T_\mathbb{C} \to T_M$ of $f$, where $T_M$ denotes the tangent bundle of $M$. Take a tangent vector $t$ of $\mathbb{C}$. Both $(df)(t)$ and $(Ddf)(t \otimes t)$ are in $T_M$ and their exterior product $(df)(t) \wedge (Ddf)(t)$ lies in $\wedge^2 T_M$. So we get a map $df \wedge Ddf : T\mathbb{C} \to \Lambda^2 T_M$, which is not linear along the fibers of $T_\mathbb{C}$ but of degree 3. We write $f_z \wedge D_z f_z$ for its value at $\frac{\partial}{\partial z}$.

The reason to consider the inequality (2.5.1) in the classical case is to take care of the undesirable boundary term $\mathscr{A}_r\left[\log \frac{1}{|f^*s|^2(\log|f^*s|^2)^2}\right]$ resulting from integrating the equation (2.3.1). In the classical case the price to pay is the addition of the factor $|df|^2$ to the numerator and that factor accounts for the defect. In the present case we hope to get the following inequality

$$(3.1.1) \qquad \mathscr{T}_r\left[\frac{|f_z \wedge D_z f_z|^2}{|f^*s|^2(\log|f^*s|^2)^2}\right] \leq \text{constant} \cdot \left[\mathscr{T}_r(f^*\theta_L)^c + r^c\right] \quad \|$$

(where $c$ is a positive constant) so that we have as a corollary of the Calculus Lemma and the concavity of the logarithmic function the following natural analog of the basic inequality (2.5.1)

$$\mathscr{A}_r\left[\log \frac{|f_z \wedge D_z f_z|^2}{|f^*s|^2(\log|f^*s|^2)^2}\right] \le O(\log \mathscr{T}_r(f^*\theta_L) + \log r) \quad \|.$$

This inequality as a replacement for the basic inequality (2.5.1) would give us a defect relation if $f_z \wedge D_z f_z$ is holomorphic and the defect would come from the anticanonical line bundle $K_M^{-1}$ of M.

(3.2) *Heuristic discussion of the proof of the replacement basic inequality.* Let us first heuristically discuss how one may get the replacment basic inequality (3.1.1) under suitable assumptions. The difficulty in our present case is that we cannot dominate $|df|^2$ by $|f^*Ds|^2$ to get the basic inequality (2.5.1). However, we do get from the Calculus Lemma and the concavity of the logarithmic function the weaker inequality

(3.2.1)
$$\mathscr{A}_r\left[\log \frac{|f^*Ds|^2}{|f^*s|^2(\log|f^*s|^2)^2}\right]$$
$$\le \text{constant}\cdot\log \mathscr{T}_r\left[\frac{2|f^*Ds|^2}{|f^*s|^2(\log|f^*s|^2)^2}\right] + O(1) \quad \|.$$

By integrating the equation (2.3.1) for

$$\partial\bar\partial \log \frac{1}{|f^*s|^2(\log|f^*s|^2)^2},$$

we get

$$\mathscr{T}_r\left[\frac{|f^*Ds|^2}{|f^*s|^2(\log|f^*s|^2)^2}\right] \le \text{constant}\cdot\left[\mathscr{T}_r(f^*\theta_L) + r^c\right] \quad \|.$$

We say that we have an estimate for $\mathscr{T}_r(g)$ if we have

$$\mathscr{T}_r(g) \le \text{constant}\cdot\left[\mathscr{T}_r(f^*\theta_L)^c + r^c\right] \quad \|.$$

We repeat the argument for

$$\partial\bar\partial \log \frac{1}{|f^*Ds|^2(\log|f^*Ds|^2)^2}$$

instead of

$$\partial\bar\partial \log \frac{1}{|f^*s|^2(\log|f^*s|^2)^2}$$

and get an estimate for

$$\mathscr{I}_r\left[\frac{|D(f^*Ds)|^2}{|f^*Ds|^2(\log|f^*Ds|^2)^2}\right]$$

which is almost the same as an estimate for

$$\mathscr{I}_r\left[\frac{|D(f^*Ds)|^2}{|f^*Ds|^2}\right].$$

Writing

$$\frac{|D(f^*Ds)|^2}{|f^*s|^2(\log|f^*s|^2)^2} = \frac{|D(f^*Ds)|^2}{|f^*Ds|^2}\frac{|f^*Ds|^2}{|f^*s|^2(\log|f^*s|^2)^2},$$

we combine together the estimates for

$$\mathscr{I}_r\left[\frac{|f^*Ds|^2}{|f^*s|^2(\log|f^*s|^2)^2}\right] \quad \text{and} \quad \mathscr{I}_r\left[\frac{|D(f^*Ds)|^2}{|f^*Ds|^2}\right]$$

to get an estimate for

$$\mathscr{I}_r\left[\frac{|D(f^*Ds)|^2}{|f^*s|^2(\log|f^*s|^2)^2}\right].$$

We add together the estimates for

$$\mathscr{I}_r\left[\frac{|f^*Ds|^2}{|f^*s|^2(\log|f^*s|^2)^2}\right] \quad \text{and} \quad \mathscr{I}_r\left[\frac{|D(f^*Ds)|^2}{|f^*s|^2(\log|f^*s|^2)^2}\right]$$

to get an estimate for

$$\mathscr{I}_r\left[\frac{|f^*Ds|^2 + |D(f^*Ds)|^2}{|f^*s|^2(\log|f^*s|^2)^2}\right].$$

Now

$$\begin{cases} f^*Ds = f_z^\alpha D_\alpha s, \\[2mm] D(f^*Ds) = (D_z f_z^\alpha)D_\alpha s + f_z^\alpha f_z^\beta D_\beta D_\alpha s. \end{cases}$$

If $D_\beta D_\alpha s = 0$, then

$$\begin{cases} f_z^1 D_1 s + f_z^2 D_2 s = f^*Ds, \\[2mm] (D_z f_z^1)D_1 s + (D_z f_z^2)D_2 s = D(f^*Ds). \end{cases}$$

Solving for $D_1 s$ and $D_2 s$ by Cramer's rule, we can estimate

$$(D_\nu s) \det \begin{pmatrix} f_z^1 & f_z^2 \\ D_z f_z^1 & D_z f_z^2 \end{pmatrix}$$

in terms of $f^*Ds$ and $D(f^*Ds)$ for $\nu = 1, 2$. The nonsingularity of $Zs$ gives a positive lower bound for $|D_1 s|^2 + |D_2 s|^2$ near $s = 0$. Hence we get an estimate for

$$\mathcal{T}_r\left[\frac{|f_z \wedge D_z f_z|^2}{|f^* s|^2 (\log |f^* s|^2)^2}\right]$$

and a defect relation.

(3.3) *Inaccuracies in the heuristic argument, meromorphic connections, and the second fundamental form.* We now look at the several points of inaccuracy which we gloss over in our preceding heuristic discussion. The first point is the holomorphicity of $f_z \wedge D_z f_z$. For $M = \mathbb{P}_2$ and the Levi-Civita connection, $f_z \wedge D_z f_z$ is holomorphic always for any holomorphic map $f$ because the component $R_{1\bar{1}1\bar{2}}$ of the curvature tensor vanishes for any orthonormal frame. When $Zs$ is linear, we have $D_\alpha D_\beta s = 0$. So in that case everything works. However, we do not know how to use the condition of normal crossing. When we have a finite number of divisors $Zs_\nu$ $(1 \leq \nu \leq q)$ no more than two of which interesect at a point, we can use the inequality

$$\prod_{\nu=1}^{q} \frac{1}{|f^* s_\nu|^2 (\log |f^* s_\nu|^2)^2} \leq \text{constant} \cdot \sum_{\mu \neq \nu} \frac{1}{|f^* s_\mu|^2 (\log |f^* s_\mu|^2)^2} \frac{1}{|f^* s_\nu|^2 (\log |f^* s_\nu|^2)^2}$$

and we have to use the factor $|f_z \wedge D_z f_z|^4$ instead of $|f_z \wedge D_z f_z|^2$ at the expense of losing a factor of 2 in the defect relation.

In the general case, in order to make $f_z \wedge D_z f_z$ holomorphic, the most obvious way is to introduce a holomorphic connection in the sense that $\Gamma^\gamma_{\alpha\beta}$ is holomorphic in local holomorphic coordinates (without requiring $\Gamma^\gamma_{\alpha\beta}$ to be symmetric in $\alpha$ and $\beta$). Note that the Levi-Civita connection is not holomorphic in this sense. In general no such holomorphic connection exists. However, meromorphic connections (*i.e.* connections with $\Gamma^\gamma_{\alpha\beta}$ meromorphic) always exist so that $t f_z \wedge D_z f_z$ is holomorphic for some non--identically-zero holomorphic section $t$ of some holomorphic line bundle over $M$. The presence of $t$ weakens the defect relation when we estimate $\mathcal{A}_r(\log |f_z \wedge D_z f_z|^2)$ from below.

There are some other minor inaccuracies. The first one is that in the step of getting the estimate for

$$\mathcal{I}_r\left[\frac{|D(f^*Ds)|^2}{|f^*Ds|^2(\log|f^*Ds|^2)^2}\right],$$

one should use

$$\partial\bar\partial\log\frac{1}{|f^*Ds|^2(A+(\log|f^*Ds|^2)^2)}$$

instead of

$$\partial\bar\partial\log\frac{1}{|f^*Ds|^2(\log|f^*Ds|^2)^2}$$

for a suitable positive constant $A$, because unlike $\frac{2}{(\log|f^*s|^2)}$ one cannot make $\frac{2}{(\log|f^*Ds|^2)}$ small by rescaling the norm $|\cdot|$ for the line bundle and one has to make $\frac{2}{A+(\log|f^*Ds|^2)^2}$ small by choosing $A$ large. The second one is that in the step to get the estimate for

$$\mathcal{I}_r\left[\frac{|f_z\wedge D_zf_z|^2}{|f^*s|^2(\log|f^*s|^2)^2}\right]$$

one actually gets the estimate

$$\mathcal{I}_r\left(\left[\frac{|f_z\wedge D_zf_z|^{2\lambda}}{|f^*s|^2(\log|f^*s|^2)^2}\right]^{\mu}\right)\le\text{constant}\cdot\left[\mathcal{I}_r(f^*\theta_L)^c+r^c\right]$$

for $\lambda>1$ and $\mu>0$ with both $\lambda-1$ and $\mu$ sufficiently small. The $\lambda$ is used to compensate for the inaccuracy in the step of passing from an estimate for

$$\mathcal{I}_r\left[\frac{|D(f^*Ds)|^2}{|f^*s|^2(\log|f^*s|^2)^2}\right]$$

to an estimate for

$$\mathcal{I}_r\left[\frac{|D(f^*Ds)|^2}{|f^*Ds|^2}\right].$$

The $\mu$ is used because of the use of the Hölder inequality when one uses

$$\frac{|D(f^*Ds)|^2}{|f^*s|^2(\log|f^*s|^2)^2}=\frac{|D(f^*Ds)|^2}{|f^*Ds|^2}\,\frac{|f^*Ds|^2}{|f^*s|^2(\log|f^*s|^2)^2}$$

to get the estimate for

$$\mathscr{I}_r\left[\frac{|D(f^*Ds)|^2}{|f^*Ds|^2(\log|f^*Ds|^2)^2}\right]$$

from estimates for

$$\mathscr{I}_r\left[\frac{|D(f^*Ds)|^2}{|f^*Ds|^2}\right] \quad \text{and} \quad \mathscr{I}_r\left[\frac{|f^*Ds|^2}{|f^*s|^2(\log|f^*s|^2)^2}\right].$$

The most serious difficulty is the requirement that $D_\alpha D_\beta s$ should vanish with respect to the connection. For the argument to go through what actually is needed is that $t D_\alpha D_\beta s = A_\alpha D_\beta s + B_\beta D_\alpha s + C_{\alpha\beta} s$ for some smooth vector-bundle valued tensors $A_\alpha$, $B_\beta$, and $C_{\alpha\beta}$, where $t$ is a non-identically-zero holomorphic section of some holomorphic bundle over $M$. This is used to take care of the pole set of the connection. Geometrically this means that if a tangent vector $\xi$ is tangential to $Zs$, then the covariant derivative $D\xi$ of $\xi$ is also tangential to $Zs$. In other words, the second fundamental form of $Zs$ with respect to the connection is zero. This is required to relate $D_z f_z$ to $D(f^*Ds)$ so that we can replace factors $|f^*Ds|^2$ and $|D(f^*Ds)|^2$ by the factor $|f_z \wedge D_z f_z|^2$ in our estimates.

(3.4) *Meromorphic connections for Fermat curves in the projective plane.* In the case $M = \mathbb{P}_2$ the holomorphic section $s$ of $H^d$ can be represented by a polynomial of degree $d$ on the affine part $\mathbb{C}^2$ of $\mathbb{P}_2$. We can always find rational functions $a^\gamma_{\alpha\beta}$ and $b_{\alpha\beta}$ such that $\partial_\alpha \partial_\beta s = \Sigma_\gamma a^\gamma_{\alpha\beta} \partial_\gamma s + b_{\alpha\beta} s$. We can always use the meromorphic connection $\Gamma^\gamma_{\alpha\beta} = -a^\gamma_{\alpha\beta}$. However, in general the pole order of $a^\gamma_{\alpha\beta}$ is high and is of the same order as the degree $d$, thereby making it impossible to have a useful defect relation. For special cases such as the Fermat curve $s = \zeta_1^d + \zeta_2^d - 1$ in inhomogeneous coordinates $\zeta_1$, $\zeta_2$, we have

$$\begin{cases} \partial_1 \partial_1 s = \dfrac{d-1}{\zeta_1} \partial_1 s \\ \partial_1 \partial_2 s = 0 \\ \partial_2 \partial_2 s = \dfrac{d-1}{\zeta_2} \partial_2 s \end{cases}$$

and the pole order of the connection in the affine part $\mathbb{C}^2$ is 1 no matter how big $d$ is. The pole order of the connection in all of $\mathbb{P}_2$ including the infinity line of $\mathbb{P}_2$ is 3.

From this method we get the conclusion that for a Fermat curve the defect is $\leq \frac{6}{d}$.

Suppose we have a finite number of holomorphic sections $\{s_\nu\}_{\nu=1}^q$ of $H^d$ so that each $s_\nu$ can be represented by a homogeneous polynomial of degree $d$ of the form $w_{\nu 0}^d + w_{\nu 1}^d + w_{\nu 2}^d$, where for each $1 \leq \nu \leq q$, $[w_{\nu 0}, w_{\nu 1}, w_{\nu 2}]$ is some homogeneous coordinate system of $\mathbb{P}_2$. For each $\nu$ we have a meromorphic connection $\Gamma(\nu)_{\alpha\beta}^\gamma$ which has pole order 3 in all of $\mathbb{P}_2$ such that the second fundamental form of the zero-set of $s_\nu$ with respect to the connection $\Gamma(\nu)_{\alpha\beta}^\gamma$ is zero. In other words, $t_\nu D_\alpha^{(\nu)} D_\beta^{(\nu)} s = A_{\nu\alpha} D_\beta^{(\nu)} s + B_{\nu\beta} D_\alpha^{(\nu)} s + C_{\nu\alpha\beta} s$ for some smooth tensors $A_{\nu\alpha}$, $B_{\nu\beta}$, $C_{\nu\alpha\beta}$, where $D^{(\nu)}$ denotes covariant differentiation with respect to the connection $\Gamma(\nu)_{\alpha\beta}^\gamma$ and $t_\nu$ is a non-identically-zero holomorphic section of $H^3$ to take care of the pole set of $\Gamma(\nu)_{\alpha\beta}^\gamma$. To get the defect relation we have to obtain a lower bound of $\mathcal{M}_r\left[\log \Pi_{\nu=1}^q |f_z \wedge t_\nu D_z^{(\nu)} f_z|^2\right]$ in terms of $\mathcal{T}_r(f^* \theta_H)$. To make the defect relation useful we have to have a lower bound independent of $q$.

(3.5) *Value distribution theory for meromorphic connections.* For a holomorphic map $g = (g_0, \cdots, g_N) : \mathbb{C} \to \mathbb{C}^{N+1} - 0$ and $a = (a_0, \cdots, a_N) \in \mathbb{C}^{N+1} - 0$, to have a lower bound of $\mathcal{M}_r(\log |\Sigma_{j=0}^N a_j g_j|^2)$ in terms of the characteristic function of the map $h : \mathbb{C} \to \mathbb{P}_N$ defined by $g$ means to have an estimate for the defect of $h$ and the hyperplane in $\mathbb{P}_N$ defined by $a$. When we have a finite subset $\{a^{(\nu)}\}_{\nu=1}^q$ of $\mathbb{C}^{N+1} - 0$, a lower bound for $\mathcal{M}_r(\Pi_{\nu=1}^q \log|\Sigma_{j=0}^N a_j^{(\nu)} g_j|^2)$ corresponds to an estimate of the sum of defects for the hyperplanes defined by $a^{(\nu)}$. Now we are going to use the trick of the Veronese map to make $f_z \wedge t_\nu D_z^{(\nu)} f_z$ correspond to $\Sigma_{j=0}^n a_j^{(\nu)} g_j$ so that we can get a lower bound estimate for $\mathcal{M}_r\left[\log \Pi_{\nu=1}^q |f_z \wedge t_\nu D_z^{(\nu)} f_z|^2\right]$.

The difference of two meromorphic connections after being made holomorphic by multiplication by some holomorphic section of $H^\ell$ is a section of $T_{\mathbb{P}_2}^* \otimes T_{\mathbb{P}_2}^* \otimes T_{\mathbb{P}_2} \otimes H^\ell$ and $\Gamma(\mathbb{P}_2, T_{\mathbb{P}_2}^* \otimes T_{\mathbb{P}_2}^* \otimes T_{\mathbb{P}_2} \otimes H^\ell)$ is a vector space. So the set of connections can be parametrized by a vector space. We can use the Cartan-Ahlfors result for a suitable holomorphic map $h : \mathbb{C} \to \mathbb{P}_N$ to get a lower bound of $\mathcal{M}_r\left[\log \Pi_{\nu=1}^q |f_z \wedge t_\nu D_z^{(\nu)} f_z|^2\right]$ in terms of $\mathcal{T}_r(f^* \theta_H)$. The trick of the Veronese map does not give the best bound. There is some loss. However, this loss is not related to the degree d. It relates only to the pole order of the meromorphic connections. Let us do the counting for the case of Fermat curves. Fix an initial meromorphic connection $D^{(0)}$ with pole order $\ell'$ which can be chosen to be 1. The pole order of the connections $D^{(\nu)}$ for the vanishing of the second fundamental form of the Fermat curve is $\ell$ which can be chosen to be 3. Since we take the difference of $D^{(\nu)}$ and $D^{(0)}$ we have to use $t_0 t_\nu$. The pole set is of order $\ell + \ell' = 4$. The characteristic function of the map h is $T_g(r)$ which is of the order $(\ell + \ell' + 3)\, \mathcal{T}_r(f^* \theta_H)$, where the number 3 comes from $K_{\mathbb{P}_2}^{-1} = H^3$. Next we have to consider the number of constants, *i.e.* the dimension N of $\mathbb{P}_N$ plus 1. It is equal to $p + k + 3$, where $p = \dim \Gamma(\mathbb{P}_2, H^\ell) = \binom{\ell+2}{2} = \binom{5}{2} = 10$ and $k = \dim \Gamma(\mathbb{P}_2, T^*_{\mathbb{P}_2} \otimes T^*_{\mathbb{P}_2} \otimes T_{\mathbb{P}_2} \otimes H^{\ell+\ell'}) \leq 9(\ell + \ell')(\ell + \ell' + 2) = 216$. Here p has to be used because of the freedom in the choice of $t_\nu$. The estimate for k comes from the general formula $\dim \Gamma(\mathbb{P}_n, T^*_{\mathbb{P}_n} \otimes T^*_{\mathbb{P}_n} \otimes T_{\mathbb{P}_n} \otimes H^\ell) \leq (n+1)^2 (n+\ell) \binom{n+\ell-2}{\ell-1}$ for $n \geq 2$. Also we have to add $f_0^{\ell+\ell'+3}, f_1^{\ell+\ell'+3}, f_2^{\ell+\ell'+3}$ to make sure that the homogeneous coordinates of h do not vanish simultaneously. So the defect is $10 + 216 + 3 = 229$ for h. Finally we compare $T_f(r)$, which is of the order $d\, \mathcal{T}_r(f^* \theta_H)$, with $T_h(r)$ and get $\frac{229 \times 7}{d} = \frac{1603}{d}$. The nondegeneracy condition is that $f_z \wedge D_z^{(\nu)} f_z$ is not identically zero for any $\nu$. The general position condition that the hyperplanes in $\mathbb{P}_{p+k+2}$ coming from the connections $D^{(\nu)}$ should be in general position. Note that the identical vanishing of $f_z \wedge D_z^{(\nu)} f_z$ means that the image of the map obtained by using the trick of the Veronese map is contained completely in the hyperplane in $\mathbb{P}_{p+k+2}$ corresponding to $\nu$.

(3.6) *The general case of a curve in a compact n–fold.* We use the notations of Theorem 2. We have to use

$$f_z \wedge t_\nu D_z^{(\nu)} f_z \wedge (t_\nu D^{(\nu)})^2 f_z \wedge \cdots \wedge (t_\nu D^{(\nu)})^{n-1} f_z .$$

The number of constants is equal to the number of monomials of homogeneous degree $1 + 2 + \cdots + (n-1) = \frac{n(n-1)}{2}$ in $p + k$ variables. So the number is the binomial coefficient $\binom{b}{c}$, where $b = \frac{n(n-1)}{2} + p + k - 1$ and $c = \frac{n(n-1)}{2}$. Because of

$$f_z \wedge t_\nu D_z^{(\nu)} f_z \wedge (t_\nu D^{(\nu)})^2 f_z \wedge \cdots \wedge (t_\nu D^{(\nu)})^{n-1} f_z$$

we have to use $(L_{D_0} \otimes L_D)^{\frac{n(n-1)}{2}} \otimes K_M^{-1}$ in the trick of the Veronese map. Moreover, we have to add the $a$ sections of $\Gamma(M, F \otimes (L_{D_0} \otimes L_D)^{\frac{n(n-1)}{2}} \otimes K_M^{-1})$ to make sure that we have no common zeros in the map into the complex projective space. So the map $h$ is from $\mathbb{C}$ to $\mathbb{P}_N$ with $N$ equal to $\binom{b}{c} + a - 1$, where $b = \frac{n(n-1)}{2} + p + k - 1$ and $c = \frac{n(n-1)}{2}$. The defect for $h$ is therefore no more than $\binom{b}{c} + a$. The characteristic function for $h$ is the one for $F \otimes (L_{D_0} \otimes L_D)^{\frac{n(n-1)}{2}} \otimes K_M^{-1}$. The sections $s_\nu$ are sections of $L$ and the characteristic function for $f$ comes from the one for $L$. So the defect is no more than $\alpha \binom{b}{c} + \alpha\, a$, where $\alpha$ is a nonnegative number such that $\alpha\theta_L$ minus the curvature of $F \otimes (L_{D_0} \otimes L_D)^{\frac{n(n-1)}{2}} \otimes K_M^{-1}$ is everywhere nonnegative on $M$. Again the nondegeneracy condition is that

$$f_z \wedge t_\nu D_z^{(\nu)} f_z \wedge (t_\nu D^{(\nu)})^2 f_z \wedge \cdots \wedge (t_\nu D^{(\nu)})^{n-1} f_z$$

is not identically zero for any $\nu$. The general position condition is that the hyperplanes in $\mathbb{P}_N$ coming from the connections $D^{(\nu)}$ should be in general position.

Consider now the special case $M = \mathbb{P}_n$ with $L = H^d$ (*i.e.* for divisors of degree $d$). Let $\ell$ be the pole order of the meromorphic connections for a collection $\{s_\nu\}$ of holomorphic sections of $L$ (*i.e.* $L_D = L^\ell$). We can choose a meromorphic connection $D_0$ with pole order 1. Then the defect is no more than

$$\binom{b}{c} \text{ times } \frac{(\ell+1)^{\frac{n(n+1)}{2}} + n + 1}{d},$$

where

$$c = \frac{n(n-1)}{2} \text{ and } b = \frac{n(n-1)}{2} + \binom{\ell+n}{n} + (n+1)^2(n+\ell+1)\binom{n+\ell-1}{n-1} - 1.$$

## References


[A1] L. V. Ahlfors, Beiträge zur Theorie der meromorphen Funktionen, C. R. 7$^{e}$ Congrès des Math. Scand. (Oslo, 1929), 84–88, Oslo, 1930.

[A2] L. V. Ahlfors, The theory of meromorphic curves. *Acta Soc. Sci. Fenn. Nova Ser A* **3** (4) (1941), 1–31.

[B] A. Biancofiore, A hypersurface defect relation for a class of meromorphic maps, *Trans. Amer. Math. Soc.* **270** (1982), 47–80.

[C-Gr] J. Carlson and Ph. Griffiths, Defect relation for equidimensional holomorphic mappings between algebraic varieties, *Ann. of Math.* **95** (1972), 557–584.

[C] H. Cartan, Sur les zéros des combinaisons linéaires de p fonctions holomorphes données, *Mathematica* 7 (1933), 80–103.

[Co-Gr] M. Cowen and Ph. Griffiths, Holomorphic curves and metrics of non-negative curvature, *J. Analyse Math.* **29** (1976), 93–153.

[F] G. Faltings, Endlichkeitssätze für abelsche Varietäten über Zahlkörpern, *Invent. Math.* **73** (1983), 349–366.

[G] M. Green, Some Picard theorems for holomorphic maps to algebraic varieties, *Amer. J. Math.* **97** (1975), 43–75.

[Gr-K] Ph. Griffiths and J. King, Nevanlinna theory and holomorphic mappings between algebraic varieties, *Acta Math.* **130** (1973), 145–220.

[Ne] F. Nevanlinna, Über die Anwendung einer Klasse von uniformisierenden Transzendenten zur Untersuchung der Wertverteilung analytischer Funktionen, *Acta Math.* **50** (1927), 159–188.

[N1] R. Nevanlinna, Le théorème de Picard-Borel et la théorie des fonctions méromorphes, Gauthier Villars, Paris, 1929, reprint Chelsea Publ. Co., New York, 1974.

[N2] R. Nevanlinna, Eindeutige analytische Funktionen, Berlin, 1936.

[No] J. Noguchi, Holomorphic curves in algebraic varieties, *Hiroshima Math. J.* 7 (1977), 833–853.

[R] K. F. Roth, Rational approximations to algebraic numbers, *Mathematika* **2** (1955), 1–20, Corrigendum, *ibid*, 168.

[S1] B. Shiffman, On holomorphic curves and meromorphic maps in projective spaces, *Indiana Univ. Math. J.* **28** (1979), 627–641.

[S2] B. Shiffman, Private communication, 1986.

[Sh] T. Shimizu, On the theory of meromorphic functions, *Japanese Journal of Mathematics* **6** (1929), 119–171.

[Si1] Y. T. Siu, Nonequidimensional value distribution theory and subvariety extension, Complex Analysis and Algebraic Geometry (Göttingen, 1985), 158–174, Lecture Notes in Math., 1194, Springer-Verlag, 1986.

[Si2] Y. T. Siu, Defect relations for holomorphic maps between spaces of different dimensions, *Duke Math. J.* **55** (1987), 213–251.

[Si3] Y. T. Siu, Nonequidimensional value distribution theory and meromorphic connections, preprint, 1987.

[St1] W. Stoll, Die beiden Hauptsätze der Wertverteilungstheorie bei Funktionen mehrerer komplexen Veränderlichen, I. *Acta Math.* **90** (1953), 1–115, II. *Acta*

*Math.* **92** (1954), 55–169.

[St2] W. Stoll, About the value distribution of holomorphic maps into projective space, *Acta Math.* **123** (1969), 83–114.

[V] P. Vojta, A higher dimensional Mordell conjecture, Arithmetic Geometry (Storrs, Conn., 1984), 341–353, Springer-Verlag, 1986.

[W-W1] H. Weyl and J. Weyl, Meromorphic curves, *Ann. of Math.* **39** (1938), 516–538.

[W-W2] H. Weyl and J. Weyl, Meromorphic functions and analytic curves, Ann. of Math. Stud., 12, Princeton University Press, Princeton, N.J., 1943.

# POINCARÉ DOMAINS IN THE PLANE

WAYNE SMITH and DAVID A. STEGENGA
Department of Mathematics, University of Hawaii
Honolulu, Hawaii 96822, U.S.A.

**1. Introduction.** Let $D$ denote the open unit disk and $\Omega$ a simply connected domain in the plane. Let $g : D \to \Omega$ be a Riemann mapping function mapping the origin to $w_0$. Let $\delta_\Omega(w)$ denote the Euclidean distance from $w$ to the boundary of $\Omega$ and define the pseudo-hyperbolic metric on $\Omega$ by

$$\tilde{\rho}_\Omega(w_1, w_2) = \inf\{\int_\gamma \frac{|dw|}{\delta_\Omega(w)} : \gamma \text{ is an arc in } \Omega \text{ from } w_1 \text{ to } w_2\}.$$

Then $\frac{1}{2}\rho_\Omega \leq \tilde{\rho}_\Omega \leq 2\rho_\Omega$ where $\rho_\Omega$ is the usual hyperbolic metric on $\Omega$ (see [**SS**]).

Assume that the area of $\Omega$, $A(\Omega)$, is finite. If $u$ is an integrable function on $\Omega$, then denote by $u_\Omega$ its average on $\Omega$. Let $H(\Omega)$ denote the space of holomorphic functions on $\Omega$ and $C^1(\Omega)$ the space of continuously differentiable functions on $\Omega$. Define

$$k_\Omega = \sup_{u \in C^1(\Omega)} \frac{\iint_\Omega |u - u_\Omega|^2 \, dxdy}{\iint_\Omega |\nabla u|^2 \, dxdy} \tag{1.1}$$

$$K_\Omega(w_0) = \sup_{F \in H(\Omega)} \frac{\iint_\Omega |F - F(w_0)|^2 \, dxdy}{\iint_\Omega |F'|^2 \, dxdy}. \tag{1.2}$$

If $k_\Omega$ ($K_\Omega(w_0)$) is finite, then we say that $\Omega$ is a (analytic) Poincaré domain. Hamilton [**H**] has proved that a simply connected planar domain with finite area is an analytic Poincaré domain if and only if it is a Poincaré domain. In [**AS**], Axler and Shields show that a simply connected domain $\Omega$, with finite area, is an analytic Poincaré domain (with mapping function g) if and only if $|g'|^2 \, dxdy$ is a Carleson measure for the Dirichlet space $\mathcal{D}$ of holomorphic functions on $D$ with square integrable derivative over $D$. These measures were defined in [**S**] and are characterized as follows:

---

This paper is in final form and no version of it will be submitted for publication elsewhere.
First author is supported in part by a grant from the National Science Foundation.

THEOREM 1.1 [S]. *The measure $|g'|^2\,dxdy$ is a Carleson measure for $D$ if and only if there is a finite constant $K_g$ so that*

$$\iint_{\cup S(I_j)} |g'|^2\,dxdy \le K_g/\log \frac{2}{\mathrm{Cap}(\cup I_j)} \tag{1.3}$$

*holds for arbitrary finite disjoint collections of arcs $I_j$ on the unit circle $\partial D$.*

In the theorem, we have denoted by Cap$(E)$ the logarithmic capacity of the set $E$ and by $S(I)$ the region between the arc $I$ and the hyperbolic geodesic in $D$ with the same endpoints as $I$. This geodesic is the circular arc orthogonal to the unit circle with the same endpoints as $I$. Hence a simply connected (analytic) Poincaré domain $\Omega$ is characterized by the inequality (1.3). A different but somewhat similiar characterization was given by Maz'ya [**M1**] using the notion of capacities of condensers.

Both of these characterizations are difficult to apply to examples, and our goal in this paper is to obtain a more geometric understanding of Poincaré domains. As a starting point we investigate to what extent a single arc can be used in (1.3). For a single arc $I$, (1.3) says that

$$\log \frac{2}{|I|} \cdot A(g(S(I))) \lesssim K_g. \tag{1.4}$$

We use the notation $a \lesssim b$ to mean that $a$ is less than some absolute constant multiple of $b$. Observe that $\log \frac{2}{|I|}$, where $|I|$ denotes the arc length of $I$, is comparable to the hyperbolic distance from 0 to the hyperbolic geodesic $J = \partial S(I) \cap D$. Since hyperbolic distances and geodesics are invariant under conformal mappings, (1.4) has a geometric interpretation on $\Omega$.

Fix $w_0 \in \Omega$ and let $J$ be a hyperbolic geodesic in $\Omega$ which does not contain $w_0$. For such a geodesic, define int$(J)$ to be the component of $\Omega \setminus J$ not containing $w_0$. Let

$$\rho \cdot A_\Omega(w_0) = \sup\{\tilde{\rho}_\Omega(w_0, J) A(\mathrm{int}(J)) : J \text{ hyperbolic geodesic in } \Omega\},$$

so that by the above discussion we have a lower bound for $K_g$, namely

$$\rho \cdot A_\Omega(w_0) \lesssim K_g \approx K_\Omega(w_0). \tag{1.5}$$

In the above, we used $a \approx b$ to indicate that $a$ and $b$ are comparable. The constants in (1.5) are absolute constants.

In order to compute the quantity in (1.5), we will need to eliminate certain arcs from consideration. The following theorem is useful for this purpose.

THEOREM 1.2. *Let $\Omega$ be a simply connected planar domain containing a disk $D$.*

(a) [**J**, p. 118] *If $w_1$, $w_2$ are points of $\Omega$ which belong to $D$, then the hyperbolic geodesic $J$ connecting these points is also contained in $D$.*

(b) [**P**, Theorem 10.10] *If a hyperbolic geodesic $J$ is tangent to $\partial D$, then $J$ is disjoint from $D$.*

Rephrasing our question above we ask: for what domains $\Omega$ does the finiteness of the quantity $\rho \cdot A_\Omega(w_0)$ imply that $\Omega$ is a Poincaré domain? To illustrate the potential value of this quantity we give two simple examples. Consider the rectangle $R = (-b, b) \times (-a, a)$ where $0 < a < b$ and assume that $b/a$ is large. If we take $J$ to be the hyperbolic geodesic in $R$ that contains the point $b/2$ and has a vertical tangent there, then using Theorem 1.2 we see that $J$ is disjoint from the two disks in $R$ which have maximal radius and are tangent at $b/2$. One then computes that $\tilde{\rho}_R(0, J) \approx b/a$, that $A(\text{int}(J)) \approx ab$ and so $\rho \cdot A_R(0) \gtrsim b^2$. This would appear to be the worst case, and in fact we show below that $K_R(0) \approx b^2$. Hence $\rho \cdot A_R(0) \approx K_R(0)$. Since $b/a$ is large, we get a better result than was previously known for convex or star-shaped regions (see [**GT**, Page 164] or [**AS**]).

As a second example, let $0 < a < 1$ with $b$ small and $b/a$ large. Let

$$\Omega = (-1, 0) \times (-\frac{1}{2}, \frac{1}{2}) \cup (0, b) \times (-\frac{a}{2}, \frac{a}{2}) \cup (b, 2b) \times (-\frac{b}{2}, \frac{b}{2})$$

so that $\Omega$ consists of a large room centered at $w_0 = (-1/2, 0)$, a narrow connecting corridor and a small room. If we take $J$ to be the hyperbolic geodesic in $\Omega$ which has a vertical tangent at the point $w = b/2$, then $\tilde{\rho}_\Omega(w_0, J) \approx ba^{-1}$ provided $a$ is sufficiently small. So $\rho \cdot A_\Omega(w_0) \gtrsim b^3 a^{-1}$. If we take $b_n = 2^{-n}$, $a_n = 2^{-4n}$ and attach infinitely many rooms and corridors to the main room, then clearly one can construct a Jordan domain $\Omega$, with a rectifiable boundary, so that $K_\Omega(w_0) \gtrsim \rho \cdot A_\Omega(w_0) = +\infty$. Hence $\Omega$ is not a Poincaré domain. This construction, but with a different proof, can be found in [**CH**, p. 521] or [**M2**, p. 253].

In addition to the above example of a non-Poincaré domain, there are several other examples in the literature, see [**A**], [**AS**], [**Hu**], [**N**] and [**M2**, p. 10]. In all of these examples, we can show that $\rho \cdot A_\Omega(w_0)$ is infinite. In section 5 we give several examples of non-Poincaré domains for which $\rho \cdot A_\Omega(w_0)$ is finite. In sections 2–4 we describe some geometric restrictions on $\Omega$ which imply that $\rho \cdot A_\Omega(w_0)$ is equivalent to $K_\Omega(w_0)$.

**2. Bounded Star-Shaped Domains.** Let $\Omega$ be a star-shaped domain with respect to the origin and satisfy $D(0, a) \subset \Omega \subset D(0, b)$, where $0 < a < b$ and $D(z, r)$ denotes the disk centered at $z$ of radius $r$. In [**AS**] it is proved that $\Omega$ is a Poincaré domain and their proof shows that $K_\Omega(0) \lesssim b^3/a$. An example where $K_\Omega(0)$ is large is obtained by removing $n$ radial slits starting at $r = a$, i.e., let

$$\Omega_n = D(0,b) \setminus \bigcup_{k=1}^{n} [ae^{2\pi ik/n}, be^{2\pi ik/n}].$$

One can show that for large $n$, $\rho \cdot A_{\Omega_n}(0) \gtrsim b^2 \log(1+b/a)$. Simply construct two tangent circles of maximal radius in the middle of one of the long corridors. The hyperbolic geodesic which is also tangent to these circles must be disjoint from their interiors by Theorem 1.2. The computation can now be easily done. This turns out to be the worst case as the following theorem shows.

THEOREM 2.1. *If $\Omega$ is a star-shaped domain with respect to the origin and satisfies $D(0,a) \subset \Omega \subset D(0,b)$, then*

$$K_\Omega(0) \lesssim b^2 \log(1 + \frac{b}{a}). \tag{2.1}$$

PROOF: Consider such a domain $\Omega$ and let $F \in H(\Omega)$ with $F(0) = 0$. Let $0 \le \theta < 2\pi$ be fixed. Then $\Omega \cap \{re^{i\theta} : r \ge 0\} = [0, r(\theta)e^{i\theta})$, where $a \le r(\theta) \le b$. Put $I_0 = [a/2, r(\theta)]$ and define

$$u_\theta(x) = \begin{cases} |F'(xe^{i\theta})| & x \in I_\theta \\ 0 & x \in \mathbb{R} \setminus I_\theta. \end{cases}$$

Denote the Hardy-Littlewood maximal function of $u_\theta$ by $Mu_\theta(x)$. As is easily seen, the weight function $w(x) = \min(|x|+a, b+a)$ satisfies Muckenhoupt's $A_2$ condition, namely,

$$\sup_I \left( \frac{1}{|I|} \int_I w\, dx \cdot \frac{1}{|I|} \int_I \frac{dx}{w} \right) \lesssim \log(1 + \frac{b}{a}),$$

where $I$ ranges over all intervals on the line. Hence by the weighted norm inequalities [Mu] we have

$$\begin{aligned} \int_{\frac{a}{2}}^{r(\theta)} |F(re^{i\theta}) - F(\frac{a}{2}e^{i\theta})|^2 r\, dr &\le b^2 \int_{\frac{a}{2}}^{r(\theta)} (Mu_\theta)^2 x\, dx \\ &\lesssim b^2 \log(1 + \frac{b}{a}) \int_{\frac{a}{2}}^{r(\theta)} (u_\theta)^2 x\, dx \\ &= b^2 \log(1 + \frac{b}{a}) \int_{\frac{a}{2}}^{r(\theta)} |F'(re^{i\theta})|^2 r\, dr. \end{aligned}$$

Integrating the above with respect to $d\theta$ yields that

$$\iint_{\Omega \setminus D(0,\frac{a}{2})} |F(z) - F(\frac{a}{2}\frac{z}{|z|})|^2\, dxdy \lesssim b^2 \log(1 + \frac{b}{a}) \iint_\Omega |F'|^2\, dxdy. \tag{2.2}$$

An elementary inequality shows that $\sup|F|^2$ over $D(0,a/2)$ is dominated by the integral of $|F'|^2$ over $D(0,a)$ and hence

(2.3) $$\iint_{\Omega\setminus D(0,\frac{a}{2})} |F(\frac{a}{2}\frac{z}{|z|})|^2\,dxdy \lesssim b^2 \iint_{D(0,a)} |F'|^2\,dxdy.$$

Since $D(0,1)$ is obviously an analytic Poincaré domain, we see that $K_{D(0,a/2)}(0) \approx a^2$. Combining this fact with (2.2) and (2.3) we obtain that

$$\iint_\Omega |F|^2\,dxdy \lesssim b^2 \log(1+\frac{b}{a}) \iint_\Omega |F'|^2\,dxdy,$$

which is (2.1) and the proof is complete.

THEOREM 2.2. *Suppose that $h(x)$ is an lower semicontinuous function defined on the interval $[-a,a]$ and satisfies $0 < a \le h(x) \le b$ for all $x$. If*

$$\Omega = \{(x,y) : -a < x < a; -a < y < h(x)\},$$

*then $K_\Omega(0) \lesssim b^2$.*

PROOF: As in the proof of Theorem 2.1 we have, for $F \in H(\Omega)$ with $F(0)=0$, that

$$\iint_\Omega |F(x,y)-F(x,0)|^2\,dxdy \lesssim b^2 \iint_\Omega |F'|^2\,dxdy.$$

Note that the weighted norm inequality is not needed. Thus it suffices to prove that

(2.4) $$\iint_\Omega |F(x,0)|^2\,dxdy \lesssim ba \iint_{D(0,a)} |F'|^2\,dxdy.$$

Using the well known fact that the Hardy space $H^2(D) \subset \mathcal{D}$, we see by the Riesz-Féjer Theorem [**FR**] that

(2.5) $$\int_{-1}^{1} |G(x)-G(0)|^2\,dx \lesssim \iint_D |G'|^2\,dxdy,$$

for any $G \in H(D)$. The proof is completed by observing that (2.4) follows from (2.5) by a change of variables.

COROLLARY 2.3. *Let $\Omega$ be an $a \times b$ rectangle with $0 < a \le b$. If $w_0 \in \Omega$ satisfies $\delta_\Omega(w_0) = a/2$, then $K_\Omega(w_0) \approx b^2 \approx \rho \cdot A_\Omega(w_0)$.*

**3. Corridors and Rooms.** Let $\Omega$ consist of a main room $\Omega_0 = (-1,1)\times(-1,1)$ along with countably many rectangular corridors, attached at the top of $\Omega_0$ and leading into rectangular rooms. All these rectangles have their sides parallel to the coordinate axes and their reflections about the line $y=1$ are contained in $\Omega_0$. Let $w_0$ be the center of $\Omega_0$. The region $\Omega$ is a generalization of the second example in the introduction.

THEOREM 3.1. *If $\Omega$ is as above, then $K_\Omega(w_0) \approx \rho \cdot A_\Omega(w_0)$.*

LEMMA 3.2. *Let $\Delta$ be a simply connected domain in the plane and let $F \in H(\Delta)$. Then*

$$|F(w_1) - F(w_2)|^2 \lesssim \rho_\Delta(w_1, w_2) \iint_\Delta |F'|^2 \, dxdy.$$

PROOF: Let $G$ be the Riemann mapping function mapping D onto $\Delta$ with $G(0) = w_1$ and $G(r) = w_2$, where $0 \le r < 1$. Then we see that the inequality (3.1) is equivalent to

$$|F \circ G(r) - F \circ G(0)|^2 \lesssim \log \frac{1+r}{1-r} \iint_D |(F \circ G)'|^2 \, dxdy,$$

which is well known for functions analytic on D.

PROOF OF THE THEOREM: By (1.5) it suffices to prove that $K_\Omega(w_0) \lesssim \rho \cdot A_\Omega(w_0)$. Hence we may assume that $\rho \cdot A_\Omega(w_0) < \infty$.

Let $\Delta_n$ be the union of the $n^{th}$ room $\Omega_n$ and connecting corridor $\Lambda_n$, along with their reflections $\Lambda_n^*$ and $\Omega_n^*$, about the line $y = 1$. We are assuming that $\Delta_n$ is contained in $\Omega$. By considering the left and right sides of $\Omega_n$ separately, we may assume without loss of generality that $\Omega_n$ and $\Lambda_n$ have their left boundaries lying on the same vertical line. Let the size of $\Lambda_n$ be $a_n \times b_n$ and the size of $\Omega_n$ be $c_n \times d_n$. If $d_n < a_n$, then we redefine $\Lambda_n$ and $\Omega_n$. We enlarge $\Lambda_n$ to include all points of $\Delta_n$ above $\Lambda_n$ and remove these points from $\Omega_n$.

Denote by $w_n$ the center of $\Omega_n$. We choose two disks of maximal radius in $\Omega$ which are tangent at the point $w_n$. By Theorem 1.2 there is a hyperbolic geodesic $J$ through $w_n$ which is disjoint from these disks. It is now obvious that there is a hyperbolic geodesic $J_n$ in $\Omega$, which contains the point $w_n$ and satisfies $A(\Omega_n) \approx A(\mathrm{int}(J_n))$ and $\tilde{\rho}_\Omega(w_0, w_n) \approx \tilde{\rho}_\Omega(w_0, J_n)$. This proves that

$$\tilde{\rho}(w_0, w_n) A(\Omega_n) \lesssim \rho \cdot A_\Omega(w_0). \tag{3.1}$$

Let $F \in H(\Omega)$ satisfy $F(w_0) = 0$. By the hypotheses, each room $\Omega_n$ is a rectangle with largest side smaller than 2. It follows from Corollary 2.3 that $K_{\Omega_n}(w_n) \lesssim 1$. Hence

$$\sum_n \iint_{\Omega_n} |F|^2 \, dxdy \lesssim \sum_n |F(w_n)|^2 A(\Omega_n) + \iint_{\cup \Omega_n} |F'|^2 \, dxdy. \tag{3.2}$$

Let $\gamma$ be an arc in $\Omega$ from $w_0$ to $w_n$ and let $\gamma_1$ be the portion of $\gamma$ which starts at the last exit of $\gamma$ from $\Omega_0$ and continues to $w_n$ through $\Delta_n$. Then since $\gamma_1^* \cup \gamma_1$ is an arc in $\Delta_n$ from $w_n^*$ to $w_n$ we have by symmetry that

$$\tilde{\rho}_{\Delta_n}(w_n^*, w_n) \le 2 \int_{\gamma_1} \frac{|dw|}{\delta_{\Delta_n}} \le 2 \int_\gamma \frac{|dw|}{\delta_\Omega},$$

and hence $\tilde{\rho}_{\Delta_n}(w_n^\star, w_n) \le 2\tilde{\rho}_\Omega(w_0, w_n)$. It now follows from (3.1) and Lemma 3.2 that

$$\begin{aligned}|F(w_n^\star) - F(w_n)|^2 A(\Omega_n) &\lesssim \tilde{\rho}_{\Delta_n}(w_n^\star, w_n) A(\Omega_n) \iint_{\Delta_n} |F'|^2\, dxdy \\ &\lesssim \tilde{\rho}_\Omega(w_0, w_n) A(\Omega_n) \iint_{\Delta_n} |F'|^2\, dxdy \\ &\lesssim \rho \cdot A_\Omega(w_0) \iint_{\Delta_n} |F'|^2\, dxdy.\end{aligned}$$

Summing over $n$ yields

$$\sum_n |F(w_n^\star) - F(w_n)|^2 A(\Omega_n) \lesssim \rho \cdot A_\Omega(w_0) \iint_\Omega |F'|^2\, dxdy. \tag{3.3}$$

Next we integrate the inequality $|F(w_n^\star)|^2 \lesssim |F(w) - F(w_n^\star)|^2 + |F(w)|^2$ over $\Omega_n^\star$ to obtain that

$$|F(w_n^\star)|^2 A(\Omega_n) \lesssim \iint_{\Omega_n^\star} \left(|F'|^2 + |F|^2\right)\, dxdy. \tag{3.4}$$

Combining (3.2), (3.3) and (3.4) we see that

$$\iint_{\cup\Omega_n} |F|^2\, dxdy \lesssim \rho \cdot A_\Omega(w_0) \iint_\Omega |F'|^2\, dxdy + \iint_{\Omega_0} |F|^2\, dxdy.$$

Let $\tilde{\Omega}_0$ be the union of $\Omega_0$ along with all the corridors $A_n$. By Theorem 2.2 it follows that $K_{\tilde{\Omega}_0}(w_0) \lesssim 1$ because the diameter of $\tilde{\Omega}_0$ is comparable to one. Since $F(w_0) = 0$, we finally obtain that the integral of $|F|^2$ over $\Omega$ is dominated by $\rho \cdot A_\Omega(w_0)$ times the integral of $|F'|^2$ over $\Omega$. So $K_\Omega(w_0) \lesssim \rho \cdot A_\Omega(w_0)$ and the proof is complete.

**4. Monotone Type Domains.** Suppose that $h : [0,\infty) \to [0,\infty)$ is a nonincreasing function which is continuous from the right and satisfies $h(x) = h_0$ for $0 \le x < h_0$. Define

$$\Omega = \{\, (x,y) \mid |y| < h(x), 0 \le x < \infty \,\} \cup (-h_0, 0) \times (-h_0, h_0),$$

so that $\Omega$ is an open simply connected domain which is symmetric about the real axis.

THEOREM 4.1. *If $\Omega$ is as above, then*

$$K_\Omega(0) \approx \rho \cdot A_\Omega(0). \tag{4.1}$$

PROOF: By (1.5) it suffices to prove that $K_\Omega(0) \lesssim \rho \cdot A_\Omega(0)$ . Hence we may assume that $\rho \cdot A_\Omega(0) < \infty$. In order to simplify the notation, we will assume that $h(x) > 0$ for all $x > 0$. If $h(x) = 0$ for some $x$, then $\Omega$ is bounded and only minor alterations of the proof below are needed.

Since (4.1) is invariant under the dilation $z \to tz$ $(t > 0)$ and $\rho \cdot A_{t\Omega}(0) = t^2 \rho \cdot A_\Omega(0)$, we may assume that $\rho \cdot A_\Omega(0)$ is small. Without loss of generality, we assume that $\rho \cdot A_\Omega(0) = \pi/1024$. Suppose that $J$ is the hyperbolic geodesic in $\Omega$ which has a vertical tangent at $w_1 = -3h_0/4$. By Theorem 1.2, $J$ does not intersect the interior of the disk $D(w_1, h_0/4)$. It follows that

$$h_n^2 \le h_0^2 \le \frac{32}{\pi}\rho \cdot A_\Omega(0) = \frac{1}{32}. \tag{4.2}$$

Define, for $n \ge 1$, $h_n = h(n)$ and

$$\Omega_n = (n - 4h_n, n) \times (-h_n, h_n) \cup \{ (x, y) \in \Omega \mid n \le x < n+1 \}.$$

Put $w_n = n - 2h_n$ , $\Omega_0 = \{ (x, y) \in \Omega \mid |x| \le 1 \}$ and $w_0 = 0$. Let $J_n$ be the hyperbolic geodesic in $\Omega$, which contains the point $w_n$ and has a vertical tangent there. Since $\Omega$ is symmetric with respect to the real axis, the same must be true of $J_n$ and it follows that $\rho_\Omega(0, w_n) = \rho_\Omega(0, J_n)$. Now let $D_n$ be the largest disk in $\Omega$ which has a vertical tangent at $w_n$ and whose center is to the right of $w_n$. Then by Theorem 1.2, $J_n$ is disjoint from $D_n$. Thus, $J_n$ is prevented from traveling past $D_n$ and hence an elementary argument, when combined with the above discussion, yields the lower bound

$$\begin{aligned} \tilde\rho_\Omega(w_{n-1}, w_n) A(\Omega_n) &\le 4\tilde\rho_\Omega(0, J_n) A(\Omega_n) \\ &\le 16\tilde\rho_\Omega(0, J_n) A(\mathrm{int}(J_n)) \le 16\rho \cdot A_\Omega(0), \end{aligned} \tag{4.3}$$

which holds for all $n \ge 1$.

Using (4.2), we see that

$$w_n - (n-1) = 1 - 2h_n \ge \frac{1}{2}, \qquad n \ge 1,$$

and hence the Schwarz inequality yields that

$$\frac{1}{4} \le \left( \int_{n-1}^{w_n} dx \right)^2 \le \int_{w_{n-1}}^{w_n} \frac{dx}{h} \cdot \int_{n-1}^{w_n} h\, dx \le \tilde\rho(w_{n-1}, w_n) A(\Omega_{n-1}). \tag{4.4}$$

Combining (4.3) and (4.4) we get the geometric decay,

$$A(\Omega_n) \le \frac{\pi}{16} A(\Omega_{n-1}), \tag{4.5}$$

in the areas of the $\Omega_n$'s.

Now let $F \in H(\Omega)$ satisfy $F(0)=0$. From Lemma 3.2 and (4.3) we obtain that

$$|F(w_n)-F(w_{n-1})|^2 A(\Omega_n) \lesssim \tilde{\rho}_\Omega(w_n,w_{n-1})A(\Omega_n)\iint\limits_{\Omega_{n-1}} |F'|^2\,dxdy$$
$$\lesssim \rho\cdot A_\Omega(0)\iint\limits_{\Omega_{n-1}} |F'|^2\,dxdy,$$

and hence (4.5) implies that

$$|F(w_n)|^2 A(\Omega_n) \le c\rho\cdot A_\Omega(0)\iint\limits_{\Omega_{n-1}} |F'|^2\,dxdy + 2|F(w_{n-1})|^2 A(\Omega_n)$$
$$\le c\rho\cdot A_\Omega(0)\iint\limits_{\Omega_{n-1}} |F'|^2\,dxdy + \frac{1}{2}|F(w_{n-1})|^2 A(\Omega_{n-1}),$$

for some absolute constant $c$. Summing on $n$ yields that

$$\sum_{n=1}^{\infty}|F(w_n)|^2 A(\Omega_n) \le 2c\rho\cdot A_\Omega(0)\sum_{n=1}^{\infty}\iint\limits_{\Omega_{n-1}} |F'|^2\,dxdy$$
$$\lesssim \iint\limits_{\Omega} |F'|^2\,dxdy. \tag{4.6}$$

Finally, since $K_{\Omega_n}(w_n)\lesssim 1$ by Theorem 2.2, we use (4.6) to obtain that

$$\iint\limits_{\Omega}|F|^2\,dxdy \lesssim \iint\limits_{\Omega_0}|F'|^2\,dxdy + \sum_{n=1}^{\infty}\left(\iint\limits_{\Omega_n}|F'|^2\,dxdy + |F(w_n)|^2A(\Omega_n)\right)$$
$$\lesssim \iint\limits_{\Omega}|F'|^2\,dxdy.$$

Thus $K_\Omega(0)\lesssim 1$. Since $\rho\cdot A_\Omega(0)\gtrsim 1$, the proof is complete.

We do not know the best constants in the comparability relation (4.1).

5. **Two Examples.** In the previous sections, the quantity $\rho\cdot A_\Omega(w_0)$ determined whether the region $\Omega$ satisfied the Poincaré inequality or not. This is due to the simplified geometry of the regions involved for in this section we construct two domains, one bounded and the other starshaped, such that each domain satisfies $\rho\cdot A_\Omega(w_0)<\infty$ and $K_\Omega(w_0)=+\infty$.

Fix an integer $n$ and let $\Omega$ be the rectangle $(0,n)\times(0,1)$ with the following parts removed. For any integer $m$, satisfying $2\le m\le n-1$, and any

$$y=\frac{1}{2}\pm\frac{1}{2^2}\pm\cdots\pm\frac{1}{2^m},$$

we remove the rectangle $[m,n]\times[y-2^{-n-1},y+2^{-n-1}]$. We do the same thing for $m=1$ with $y=1/2$, so that a total of $2^{n-1}-1$ rectangles are removed. Thus, $\Omega$ is a Jordan region and it follows from Theorem 2.2 that $K_\Omega(w_0)\lesssim n^2$, where $w_0=(1/2,1/2)$. By considering the function $F(w)=w$, we see that $K_\Omega(w_0)\approx n^2$.

THEOREM 5.1. *Suppose that for each positive integer $n$, we construct the region $\Omega_n$ as above. Then*

$$K_{\Omega_n}(w_0) \approx n^2 \quad \text{and} \quad \rho \cdot A_{\Omega_n}(w_0) \approx n.$$

PROOF: Fix an integer $n$ and let $\Omega = \Omega_n$. We must prove that $\rho \cdot A_{\Omega_n}(w_0) \approx n$. If $m$ is an integer, with $0 \le m \le n-1$, then the portion of $\Omega$ with $m < x < m+1$ consists of $2^m$ rectanglar rooms of size $1 \times 2\delta_m$, where $\delta_m = (2^{-m} - 2^{-n})/2$. Denote these rooms by $\Omega_j^m$, where $0 \le m \le n-1$ and $0 \le j \le 2^m - 1$. We order the rooms according to their heights with $\Omega_0^m$ being the bottom most room. Let $w_j^m$ be the center of $\Omega_j^m$ and let $J_j^m$ be the hyperbolic geodesic in $\Omega$, which has a vertical tangent at $w_j^m$. By Theorem 1.2, the hyperbolic geodesic $J_j^m$ does not intersect the disks of radius $\delta_m$ in $\Omega$, which are tangent at $w_j^m$. We easily deduce that

$$\tilde{\rho}_\Omega(w_0, J_j^m) \approx 2^m \quad \text{and} \quad A(\text{int}(J_j^m)) \approx (n-m)2^{-m}.$$

Thus, $\rho \cdot A_\Omega(w_0) \gtrsim n$.

Fix a geodesic $J$ with endpoints $a$, $b \in \partial\Omega$. Then $a$ lies on the boundary of some room $\Omega_{j_a}^{m_a}$ and similiarly, $b$ determines a room $\Omega_{j_b}^{m_b}$. We may assume that $m_a \le m_b$. Let $\Omega_j^m$ be the unique room with minimal $m$ and $\Omega_j^m \cap J$ nonempty.

First consider the case that $m+1 < m_b$. This means that there is a point $w_1 \in J \cap \Omega_j^m$ and a successor room $\Omega_{j'}^{m+1}$ to $\Omega_j^m$ so that $J$ must pass through $\Omega_{j'}^{m+1}$ in order to connect $w_1$ to $b$. Let $D = D(w_{j'}^{m+1}, \delta_{m+1})$ and $J(w_1, a)$ be the subarc of $J$, from the point $w_1$ to the endpoint $a$. It follows from Theorem 1.2 that $J(w_1, a)$ is disjoint from D. Let $c^+, c^-$ be defined by $w_{j'}^{m+1} \pm \delta_{m+1} i$, then we clearly have that $J$ separates $c^+$ and $c^-$ in $\Omega$. It now follows from [**P**, Corollary 10.3] (see also problem 1 [**P**, Page 318]) that

$$\delta_\Omega(w) \approx 2^{-m}, \qquad w \in J \cap D.$$

We easily calculate that $\tilde{\rho}_\Omega(w_0, J) \approx 2^m$, $A(\text{int}(J)) \lesssim (n-m)2^{-m}$ and hence that

$$\tilde{\rho}_\Omega(w_0, J) \cdot A(\text{int}(J)) \lesssim n. \tag{5.1}$$

Next consider the case that $m_b \le m+1$. Let $\tilde{\Omega}_j^m$ be the union of $\Omega_j^m$ along with its two successor rooms. Since $m \le m_a \le m_b \le m+1$, we see that $a, b \in \partial\tilde{\Omega}_j^m$. If $a$ and $b$ belong to different components of $\partial\Omega \cap \partial\tilde{\Omega}_j^m$, then we obviously have that $\tilde{\rho}(w_j^m, J) \lesssim 2^m$ and so (5.1) holds.

Finally, assume that $a, b$ belong to the same component of $\partial\Omega \cap \partial\tilde{\Omega}_j^m$. Let $\delta = \max\{\delta(w) \mid w \in J\}$ and observe that $\delta \le \delta_m$ by the minimality. Again, if $\delta \approx \delta_m$, then we have (5.1) as before, so we may assume that $\delta$ is small compared to $\delta_m$. Let

$I \subset \partial\Omega \cap \partial\tilde{\Omega}_j^m$ be the arc determined by $a, b$. Then $|I| \leq 2 + 2^{-n}$. By placing two disks of radius $\delta$ and tangent to $\partial\Omega$ on either side of $I$, we see by Theorem 1.2 that

$$A(\text{int}(J)) \lesssim \delta \cdot d_\Omega(a,b) \lesssim \delta, \tag{5.2}$$

where $d_\Omega(a,b)$ is the infimum of the lengths of all curves in $\Omega$ with endpoints $a$ and $b$.

An easy calculation shows that

$$\tilde{\rho}_\Omega(w_j^m, J) \lesssim 2^m + \log\frac{1}{\delta} \tag{5.3}$$

and hence (5.2) and (5.3) imply that (5.1) holds with righthand side equal to one. This proves that $\rho \cdot A_\Omega(w_0) \lesssim n$. We have already proved the other assertions, so the proof is complete.

COROLLARY 5.2. *There exists an lower semicontinuous function $h(x)$, defined on $(-1,1)$ and satisfying $h(x) \geq 1$ for all $x$, which has the following property: If we define the simply connected domain by*

$$\Omega = \{(x,y) \mid -1 < x < 1; -1 < y < h(x)\},$$

*then $K_\Omega(0) = \infty$ and $\rho \cdot A_\Omega(0) < \infty$.*

PROOF: For an integer $k$, define

$$a_k = \frac{1}{4^k} \quad \text{and} \quad n_k = 8^k, \qquad k \geq 1.$$

Let $R_0$ be the square $(-1,1)\times(-1,1)$ and attach to the top of $R_0$ a sequence of rectangles $R_k$ of size $a_k \times n_k a_k$. We assume that the rectangles are placed in order of decreasing width as we move from left to right and that the space between $R_k$ and $R_{k+1}$ is $a_k$. This can be done since $\sum a_k = 1/3 < 1$.

Now modify each $R_k$ so that it is a dilation of the region $\Omega_{n_k}$ in Theorem 5.1. We assume that the subdividing of $R_k$ takes place as one travels upward into $R_k$. Denote this modified region as $\Omega$. Then $\Omega$ is a simply connected region which lies under the graph of an lower semicontinuous function. However, $\Omega$ is no longer a Jordan domain.

Let $w_0 = 0$ and $F(w) = w$. Then one calculates that

$$\iint_\Omega |F|^2\, dxdy \approx \sum_{k=1}^{\infty} (n_k a_k)^3 a_k = +\infty, \quad \text{and}$$

$$\iint_\Omega |F'|^2\, dxdy = A(\Omega) = 4 + \sum_{k=1}^{\infty} n_k a_k^2 < \infty.$$

Hence $K_\Omega(0) = \infty$.

We must prove that $\rho \cdot A_\Omega(w_0) < \infty$. Fix $k$ and let $R = R_k$. If we define $J_j^m$ to be the hyperbolic geodesic of $\Omega$, with horizontal tangent at the center of the room $R_j^m$ (indexed as in Theorem 5.1), then

$$\tilde{\rho}_\Omega(w_0, J_j^m) \approx \log \frac{1}{a_k} + 2^m \quad \text{and}$$
$$A(\text{int}(J_j^m)) \approx (n_k - m)a_k \cdot \frac{a_k}{2^m}$$

so that

$$\tilde{\rho}_\Omega(w_0, J_j^m) \cdot A(\text{int}(J_j^m)) \lesssim n_k a_k^2 \log \frac{1}{a_k} \lesssim 1. \tag{5.4}$$

In order to prove that (5.4) holds for all hyperbolic geodesics in $\Omega$ we must use the techniques employed in the proof of Theorem 5.1. For example, the case that both endpoints of the geodesic $J$ are contained in some $R_k$ is straightforward, since placing a disk of radius $a_k$ at the opening into $R_k$ from $R_0$ will prevent $J$ from traveling very far into $R_0$. So the analysis is the same as in the proof of Theorem 5.1.

Consider the case that the geodesic $J$ has one endpoint $a$ in $R_k$ and the other endpoint $b$ in $R_l$, where $k < l$. Let $\delta$ be the maximal value of $\delta(w)$ on $J$. If $\delta \approx 1$, then $\tilde{\rho}_\Omega(w_0, J) \approx 1$ and there is nothing to prove. So we assume that $\delta$ is small. We place two disks of radius $\delta$, which are tangent to the top boundary of $R_0$, with one just to left of $R_k$ and the other just to the right of $R_l$. By Theorem 1.2, $J$ must lie in the region bounded by the two semicircles and the line $y = 1 - \delta$.

Since the $a_k$'s are decreasing geometrically, we see that the distance between the centers of these disks is comparable to $a_k$. Using [**P**, Corollary 10.3] again, we see that $\delta \gtrsim a_k$ and it follows that

$$\tilde{\rho}_\Omega(w_0, J) \approx \log \frac{1}{\delta} \lesssim \log \frac{1}{a_k} \quad \text{and}$$
$$A(\text{int}(J)) \approx a_k\delta + \sum_{k'=k}^{l} n_{k'} a_{k'}^2 \approx n_k a_k^2.$$

Thus, (5.4) holds in this case.

The remaining cases follow in a similiar manner and the rest of the proof is left to the reader.

COROLLARY 5.3. *There is a non-Poincaré domain $\Omega$ which is starshaped about the origin and for which $\rho \cdot A_\Omega(0)$ is finite.*

PROOF: We modify the construction in the proof of Corollary 5.2. We start with the unit disk and attach the sequence of rectangular configurations $R_k$ described above so that the line of symmetry lies on a radius from the origin. The result will be a non-Poincaré domain $\Omega$ for which $\rho \cdot A_\Omega(0) < \infty$. But $\Omega$ is not starshaped. Fix an integer $k$. The first subroom of $R_k$ branches into two succeeding rooms. We rotate each of these succeeding rooms (along with all of their succeesor rooms) so that the centerline of each room lies on a radius through the origin. The new "Y" shaped attachment is starshaped through two levels of subrooms. Continuing we modify the attachment $R_k$ so that it is starshaped through all of its subrooms.

Observe that the modification to one attachment does not overlap modifications to another attachments since these modified regions lie in the sector determined by their initial opening doorway. In addition, the verification that this modified region is not a Poincaré domain and satisfies the $\rho \cdot A$ condition follows immediately from the proof of Corollary 5.2. The proof is therefore complete.

We now construct a bounded domain with the property that $\rho \cdot A_\Omega$ is finite and yet $\Omega$ is not a Poincaré domain. First we subdivide a fixed rectangle with the subdivision depending on $n$, then we construct the domain by attaching an infinite sequence of dilated versions onto a single square.

Fix a positive integer $n$ and let $\Omega$ be the rectangle $(-1,2) \times (0,1)$ with the following line segments removed. In the vertical direction we remove the lines $x = 2^{-m}$, where $0 \leq m \leq n$. If $m$ is an integer, with $1 \leq m \leq n$, and if

$$y = \frac{1}{2} \pm \frac{1}{2^2} \pm \cdots \pm \frac{1}{2^{m+1}},$$

then we remove the line segment from $(-1, y)$ to $(2^{-m}, y)$. We do the same thing when $m = 0$ with $y = 1/2$, so that $\Omega$ consists of $2^{n+1} - 1$ square rooms and $2^{n+1}$ rectangular rooms of size $(1 + 2^{-n}) \times 2^{-n-1}$. For $0 \leq m \leq n$, we label the $2^m$ square rooms of side $2^{-m}$ by $\Omega_j^m$, where $1 \leq j \leq 2^m$. Similiarly, we denote by $\Omega_j^{n+1}$ the rectangular rooms. Let $w_j^m$ denote the center of the room $\Omega_j^m$. To create a simply connected region, we open a doorway of size $2^{-2^m}$, symmetrically placed on the right side of each room $\Omega_j^m$, where $1 \leq m \leq n+1$ and $1 \leq j \leq 2^m$.

Let $J_j^m$ be the hyperbolic geodesic of $\Omega$ which has a vertical tangent at the point $w_j^m$. Put $w_0 = w_1^0$. Then using our techniques, we see that

$$\tilde{\rho}_\Omega(w_0, J_j^m) \lesssim 2^m \quad \text{and} \quad A(\mathrm{int}(J_j^m)) \approx \frac{1}{2^m}.$$

The above computation combined with an argument similiar to the proof of Theorem 5.1 shows that $\rho \cdot A_\Omega(w_0) \lesssim 1$.

THEOREM 5.4. *Suppose that for each positive integer $n$, we construct the region $\Omega_n$ as above, then*

$$k_{\Omega_n} \gtrsim n \quad \textit{and} \quad \rho \cdot A_{\Omega_n}(w_0) \lesssim 1$$

PROOF: Fix an integer $n$ and let $\Omega = \Omega_n$. We already know that $\rho \cdot A_\Omega(w_0) \lesssim 1$, so we must prove that $k_\Omega \gtrsim n$. It follows from (1.1) that

$$\iint\limits_{\Omega} |u - u_\Omega|^2 \, dxdy \lesssim k_\Omega \iint\limits_{\Omega} |\nabla u|^2 \, dxdy \tag{5.5}$$

provided that $u(x,y)$ is a piecewise smooth function and $u_\Omega$ is the average of $u$ over $\Omega$.

Hence it suffices to produce such a function $u(x,y)$ for which $u = 0$ on $\Omega_1^0$, $u = 1$ on $\cup \Omega_j^{n+1}$ and such that

$$\iint\limits_{\Omega} |\nabla u|^2 \, dxdy \lesssim \frac{1}{n}. \tag{5.6}$$

For if (5.6) holds, then it follows from (5.5) that

$$1 \lesssim \iint\limits_{\Omega} |u - u_\Omega|^2 \, dxdy \lesssim \frac{k_\Omega}{n},$$

which yields the desired lower bound.

For $1 \leq m \leq n$ and $1 \leq j \leq 2^m$, we construct a function $u_j^m$ which is zero everywhere except on $\Omega_j^m$ and all succeeding rooms. In addition, $u_j^m$ is one in all succeeding rooms to $\Omega_j^m$. Let $a = w_j^m - 2^{-m-1}$, so that $a$ is the center of the doorway leading into $\Omega_j^m$. We define $u_j^m = 1$ on $\Omega_j^m \setminus D(a, 2^{-m-1})$, $u_j^m = 0$ on $D(a, 2^{-2^m-1})$, and for all other points $w \in \Omega_j^m$ we define

$$u_j^m(w) = 1 - \frac{\log 2^{m+1}|w - a|}{\log 2^{m-2^m}}.$$

It follows immediately that

$$\iint\limits_{\Omega} |\nabla u_j^m|^2 \, dxdy \approx \frac{1}{2^m}.$$

Put $u = \frac{1}{n} \sum_{m,j} u_j^m$, so that $u = 1$ on $\cup_j \Omega_j^{n+1}$ and $u = 0$ on $\Omega_1^0$. Finally, we compute

$$\iint\limits_{\Omega} |\nabla u|^2 \, dxdy = \frac{1}{n^2} \sum_{m,j} \iint\limits_{\Omega_j^m} |\nabla u_j^m|^2 \, dxdy \approx \frac{1}{n^2} \sum_{m,j} \frac{1}{2^m} = \frac{1}{n},$$

and hence (5.6) holds. This completes the proof.

COROLLARY 5.5. *There is a bounded simply connected non-Poincaré domain $\Omega$ for which $\rho \cdot A_\Omega(w_0)$ is finite.*

PROOF: Let $R_0$ and $w_0$ be as in the proof of Corollary 5.2 and let $a_k = 2^{-k}$. Attach a sequence of rectangles $R_k$, for $k \geq 1$, of size $a_k \times 3a_k$ to the top of $R_0$, with a space of

$a_k$ between $R_k$ and $R_{k+1}$. We modify each $R_k$ so that it is a dilation of the region $\Omega_{n_k}$ in Theorem 5.4, where $a_k^2 n_k$ tends to infinity. Let $\Omega$ be this modified region.

The proof that $\rho \cdot A_\Omega(w_0) \lesssim 1$ is similiar to the proof given in Corollary 5.2. Let $u_k = 0$ except on $R_k$, where it is the dilation of the function $u(x, y)$ in the proof of Theorem 5.4. We have that

$$\iint_\Omega |u_k - (u_k)_\Omega|^2 \, dxdy \gtrsim A(R_k) \quad \text{and} \quad \iint_\Omega |\nabla u_k|^2 \, dxdy \lesssim \frac{1}{n_k}$$

for all $k \geq 1$. Hence $a_k^2 n_k \lesssim k_\Omega$ for all $k \geq 1$. This means that $\Omega$ is not a Poincaré domain and the proof is complete.

We remark that Corollary 5.5 when combined with Hamilton's result [**H**] gives a new example of a measure, namely $|g'|^2 \, dxdy$ for which (1.3) holds for intervals but not in general. Such a measure was constructed in [S], but the measure was not given by a conformal mapping as above.

## REFERENCES


[**A**] Amick, C. J., *Some remarks on Rellich's theorem and the Poincaré inequality*, J. London Math. Soc. **18** (1978), 81–93.

[**AS**] Axler, S. and Shields A., *Univalent multipliers of the Dirichlet space*, Michigan Math. J. **32** (1985), 65–80.

[**CH**] Courant, R. and Hilbert, D., "Methoden der mathematischen Physik, Vol. II," Springer, Berlin, 1937.

[**FR**] Fejér, L. and Riesz, F., *Über einige funktionentheoretische Ungleichungen*, Math. Z. **11** (1921), 305–314.

[**GT**] Gilbarg, D. and Trudinger, N. S., "Elliptic Partial Differential Equations of Second Order," Springer-Verlag, Berlin Heidelberg, 1983.

[**H**] Hamilton, D. H., *On the Poincaré inequality*, Complex Variables Theory Appl. **5** (1986), 265–270.

[**Hu**] Hummel, J., *Counterexample to the Poincaré inequality*, Proc. Amer. Math. Soc. **8** (1957), 207–210.

[**J**] Jørgensen, V., *On an inequality for the hyperbolic metric and its application to the theory of functions*, Math. Scand. **4** (1956), 113–124.

[**M1**] Maz'ya, V. G., *On Neumann's problem in domains with nonregular boundaries*, Siberian Math. J. **9** (1968), 990–1012.

[**M2**] Maz'ya, V. G., "Sobolev Spaces," Springer-Verlag, Berlin Heidelberg, 1985.

[**Mu**] Muckenhoupt, B., *Weighted norm inequalities for the Hardy maximal function*, Trans. Amer. Math. Soc. **165** (1972), 207–226.

[**N**] Nikodým, O., *Sur une classe de fonctions considérées dans le problème de Dirichlet*, Fundamenta Math. **21** (1933), 129–150.

[P] Pommerenke, Ch., "Univalent Functions," Vanderhoeck & Ruprecht, Göttingen, 1975.

[S] Stegenga, D. A., *Multipliers of the Dirichlet space*, Illinois J. Math. **24** (1980), 113–139.

[SS] Smith, W., and Stegenga, D. A., *A geometric characterization of Hölder domains*, J. London Math. Soc. **35** (1987), 471–480.

# CONCERNING THE GROSS STAR THEOREM

KENNETH STEPHENSON
University of Tennessee
Knoxville, TN 37996, U.S.A.

According to the Gross Star Theorem, if you stand at any point on the image surface of an entire function, you can see infinity in almost any direction. What about inner functions in the unit disc, which are in some ways the hyperbolic analogues of entire functions? Can you stand on the image surface of an inner function and see the unit circle in almost every direction? The answer is a resounding no; we construct an inner function whose image surface contains a full open set of points from which one can see none of the unit circle!

Inner functions in the unit disc $\mathbf{D}$ are in some ways the hyperbolic analogues of entire functions in the plane $\mathbf{C}$. Here we investigate certain similarities and differences concerning the behavior of their inverse mappings. The similarities occur when one considers continuation of inverses along Brownian paths, while the differences show up in continuation along rays. In this last regard, we consider the famous Gross Star Theorem for entire functions and show how utterly the analogue fails for inner functions. Our methods are geometric and involve the construction of Riemann surfaces with appropriate properties. We close with some preliminary results concerning the sharpness of the Gross Star Theorem for entire functions. My thanks to Carl Sundberg for valuable discussions on the material presented here.

## PRELIMINARIES

Let $g$ be an analytic function on a simply connected domain $\Gamma$. The image surface $\mathcal{R}_g$ is the Riemann surface serving as the natural domain of the inverse function $g^{-1}$. This is simply connected and may be pictured as a surface spread over the range of $g$ and having a projection $\pi$ to that range. When we speak of "geometry" of the image surface we are referring to this setting. For instance, a "ray" on the surface is a curve whose projection is a ray in $\mathbf{C}$; a "disc" refers to a connected component of the open subset of the surface lying over a disc in $\mathbf{C}$.

Traveling along a curve in $\mathcal{R}_g$, one can encounter only two types of singularities:

The author gratefully acknowledges support of the National Science Foundation.

namely, branch points of $\pi$ or the ideal boundary $\partial_\infty \mathcal{R}_g$. In the first case, the preimage of the curve in $\Gamma$ (i.e., its image under $g^{-1}$) will encounter a branch point of $g$; while in the second, the preimage will go to $\partial\Gamma$. There are several useful notions of "boundary" for Riemann surfaces. We will use the notation $\partial\mathcal{R}_g$ for the *accessible* boundary points; that is, the boundary obtained when we compactify $\mathcal{R}_g$ to make $\pi$ continuous. Each boundary point, then, is determined by a curve (actually an equivalence class of curves) in $\mathcal{R}_g$ which goes to the ideal boundary and along which $\pi$ has an asymptotic value in $\mathbf{C}^\infty$. Traveling to the boundary along one of these curves is the equivalent of "falling off the edge" of $\mathcal{R}_g$ over the corresponding value in $\mathbf{C}$.

Brownian motion is frequently useful in the study of properties of $\mathcal{R}_g$ and hence those of the function $g$ itself (see Kakutani [6] [7]). First, one can use it to determine whether the surface is the plane or the disc. In the latter case, exit distributions can be used to determine whether there is harmonic measure supported on $\partial\mathcal{R}_g$ and if so, where. For instance, if $g$ is a bounded analytic function on $\mathbf{D}$, then harmonic measure resides on $\partial R_g$ in such a way that the measure it induces on $\mathbf{C}$ under $\pi$ is the same as that which normalized Lebesgue measure on the unit circle induces under the (boundary) function $g$. Finally, recall that by Lévy's theorem analytic functions map Brownian motion to Brownian motion, so one can track Brownian motion on $\mathcal{R}_g$ by lifting it from $\mathbf{C}$. See Davis [3] for background and applications.

## Entire and Inner Functions

The reader is certainly familiar with entire functions, analytic functions $F : \mathbf{C} \longrightarrow \mathbf{C}$. An analytic function $I : \mathbf{D} \longrightarrow \mathbf{D}$ is an *inner function* if it has radial limits $I(e^{i\theta})$ lie in the unit circle $\mathbf{T}$ for almost all $\theta \in [0, 2\pi]$. Therefore, $\mathcal{R}_I$ is spread over $\mathbf{D}$ in such a way that its harmonic measure is supported on the points of $\partial\mathcal{R}_I$ over $\mathbf{T}$. Inner functions have long been important in function theory in their own right, but they are also related to entire functions in the following way: Suppose $f$ is entire and $\Delta = \Delta(a, r)$ is a disc of radius $r > 0$ centered at $a \in \mathbf{C}$. Let $\Omega$ be a disc in $\mathcal{R}_g$ above $\Delta$ (a component of $f^{-1}(\Delta)$). By the maximum principle, $\Omega$ is simply connected. Let $\rho : \mathbf{D} \longrightarrow \Omega$ be a conformal mapping, $\psi : \Delta \longrightarrow \mathbf{D}$ a conformal mapping, and define $I \equiv \psi \circ f \circ \rho : \mathbf{D} \longrightarrow \mathbf{D}$. Then $I$ is an inner function. In the terminology of Heins [5], both inner functions and entire functions are of "type-Bl"; we will refer to $I$ as a "localization" of $f$ to $\Delta$.

Consider an inner function $I$ and assume for convenience that $I(0) = 0$ and $I'(0) \neq 0$. Let $Z_t$ denote standard Brownian motion starting at $z = 0$ and stopped at $\mathbf{T}$ and let $\mathcal{B}_z$ denote its (continuous) sample paths. Define the process $W_t \equiv I(Z_t)$ with paths $\mathcal{B}_w \equiv I(\mathcal{B}_z)$. We claim that $W_t$ is also Brownian motion stopped at $\mathbf{T}$. To see this, note that almost every path in $\mathcal{B}_z$ is stopped at a point of $\mathbf{T}$ and, as a bounded analytic function, $I$ has asymptotic values along almost all of these. Moreover, as an inner function, these asymptotic values are almost all in $\mathbf{T}$. Applying Lévy's theorem, we see that $W_t$ is Brownian motion in the image disc started at 0 and stopped at $\mathbf{T}$, differing from standard Brownian motion only in speed. If $I^{-1}$ denotes the branch of the inverse function defined in a neighborhood of $w = 0$ by $I^{-1}(0) = 0$, then, *a fortiori*, $I^{-1}$ continues (analytically) on every path in $\mathcal{B}_w$. We conclude:

PROPOSITION 1. *Let $I : \mathbf{D} \longrightarrow \mathbf{D}$ be an inner function, and $I^{-1}$ a branch of the inverse function defined in a neighborhood of $a \in \mathbf{D}$. Then $I^{-1}$ continues analytically on almost every Brownian path starting at $a$ until it reaches $\mathbf{T}$.*

Using localizations, we can study the behavior of inverses of entire functions. Let $F$ be entire and let $F^{-1}$ be a branch of its inverse defined in a neighborhood of $a \in \mathbf{C}$. Given $r > 0$, let $I$ be the localization of $F$ to $\Delta = \Delta(a, r)$ as defined earlier. By the last proposition, $I^{-1}$ continues on almost every Brownian path until it reaches $\mathbf{T}$. Unwrapping the definition of $I$, we see that $F^{-1}$ continues on almost every Brownian path starting at $a$ until it first reaches $\partial\Delta$. Letting $r$ go to infinity through a countable sequence, we reach this conclusion:

PROPOSITION 2. *Let $F$ be entire, $F^{-1}$ a branch of the inverse function defined in a neighborhood of $a \in \mathbf{C}$. Then $F^{-1}$ continues analytically along almost every Brownian path in $\mathbf{C}$ starting at $a$.*

We will return to these results briefly at the end of the paper to point out an easy converse to Proposition 1 whose analogue fails for entire functions.

## THE COMPARISONS

To summarize the feature common to entire and inner functions, we give these results a geometric interpretation: Consider yourself as a Brownian traveler on the image surface $\mathcal{R}_f$ of some analytic function $f$, starting at a point $x$ of $\mathcal{R}_f$. You may encounter a branch point in $\mathcal{R}_f$, which prevents you from going on; however, since there are at most countably many of these, this event has probability zero. The main danger in your travels is the boundary of $\mathcal{R}_f$ — you may fall off the edge! According to Proposition 2, if $f$ is entire, this will almost never happen. On the other hand, if $f$ is an inner function on $\mathbf{D}$, then you will almost surely fall off the edge, but it will almost surely be at a point lying over the unit circle. Summing up: *The singularities of the inverse of an entire function are so thin that the Brownian traveler almost never encounters them. Likewise, those singularities of the inverse of an inner function which lie over the interior of the unit disc are so thin that the Brownian traveler almost never encounters them.*

To see where differences arise, we go back to a surprising result of W. Gross [4] established in 1918 — this was one of the first uses of the important length-area method. (See [8] [9] for details and statements applying to more general settings.)

GROSS STAR THEOREM (GST). *Let $F$ be entire, $F^{-1}$ a branch of the inverse function defined in a neighborhood of $a \in \mathbf{C}$. Then $F^{-1}$ continues analytically on the ray $\{a + re^{i\theta} : r > 0\}$ for almost every $\theta \in [0, 2\pi]$ with respect to Lebesgue measure.*

In our geometric language, this may be paraphrased: The singularites of the inverse of an entire function are so thin that, standing at any point of $\mathcal{R}_F$, one can "see infinity" in almost every direction. What would be the analogous statement for an inner function $I$? First, the Euclidean ray should be replaced by the hyperbolic ray in $\mathbf{D}$; that is, by the arc of a circle through the point in question which is orthogonal to the unit circle. (From a point over the origin, these are just Euclidean radial segments.) Standing at any point of $\mathcal{R}_I$, can one "see" the unit circle along almost every hyperbolic ray? The answer is no. Indeed, there is an inner function with a full open set of points from which one can see no further than a fixed finite hyperbolic distance.

THEOREM. *There exists an inner function $I$, an open set $\Omega \subseteq \mathbf{D}$, and a branch $g$ of $I^{-1}$ defined on $\Omega$ with the following property: Along any hyperbolic ray emanating from $\Omega$, the analytic continuation of $g$ fails within a fixed finite hyperbolic distance.*

Thus, though one encounters the boundary of $\mathcal{R}_I$ over the interior of the disc along virtually every ray, still the Brownian traveler will almost surely miss that boundary and reach the unit circle. In the proof, we obtain $I$ by constructing its image surface first. Its properties follow because we can manipulate the geometry of that surface to cut off rays while leaving the harmonic measure over the unit circle.

## THE CONSTRUCTION

To construct $\mathcal{R}$, we use the method informally known as "cutting and pasting": We start with a collection of simply connected surfaces, each having a natural projection to $\mathbf{D}$ (i.e., we picture them as spread over $\mathbf{D}$). Starting with one surface, we successively attach others along common simply connected edges. The final surface is simply connected by Van Kampen's theorem and it inherits a projection $\pi$ to $\mathbf{D}$. Letting $\phi : \mathbf{D} \longrightarrow \mathcal{R}$ be a conformal mapping, we define $I \equiv \pi \circ \phi$. Then $\mathcal{R}_I = \mathcal{R}$, and to conclude that $I$ is an inner function, it suffices to prove that the harmonic measure of $\mathcal{R}$ is supported on the part of $\partial\mathcal{R}$ lying over $\mathbf{T}$.

In our case, the pieces of $\mathcal{R}$ are either pieces of the unit disc or of the universal covering surface of the unit disc with two points removed. The corresponding projections to $\mathbf{D}$ will always be denoted $\pi$ — though technically there is a different projection for each fragment, they agree along identified edges. We want to construct $\mathcal{R}$ so that any curve starting in the disc $\Omega = \Delta(0, \frac{1}{8})$ whose projection lies along a hyperbolic ray in **D** will encounter $\partial\mathcal{R}$ before leaving $\Delta(0, \frac{1}{2})$, yet so that almost every Brownian traveler will encounter $\partial\mathcal{R}$ at some point over the unit circle.

For ease of exposition, we first construct a surface for which rays starting at 0 have the desired property; later we indicate the modifications necessary to accommodate rays from arbitrary points of $\Omega$. The initial piece $R_0$ in our construction will be the disc $\Delta(0, \frac{1}{4})$. The other pieces will consist of the "corridors" and "wings" which we describe now.

For $n$ a positive integer, define

$$r_n = \frac{1}{4} + \sum_{j=1}^{n-1} (\frac{1}{2})^{j+2};$$

so $r_1 = \frac{1}{4}$ and $\lim_{n\to\infty} r_n = \frac{1}{2}$. At the $n$th stage the corridors will be sets of the form

$$C_{n,k} = \left\{z : r_n < |z| < r_{n+1},\ \frac{2\pi(k-1)}{2^N} < \arg z < \frac{2\pi k}{2^N}\right\},$$

where $N = N(n)$ is to be chosen later, and $k = 1, 2, \ldots, 2^N$. Figure 1 illustrates the placement of the corridors.

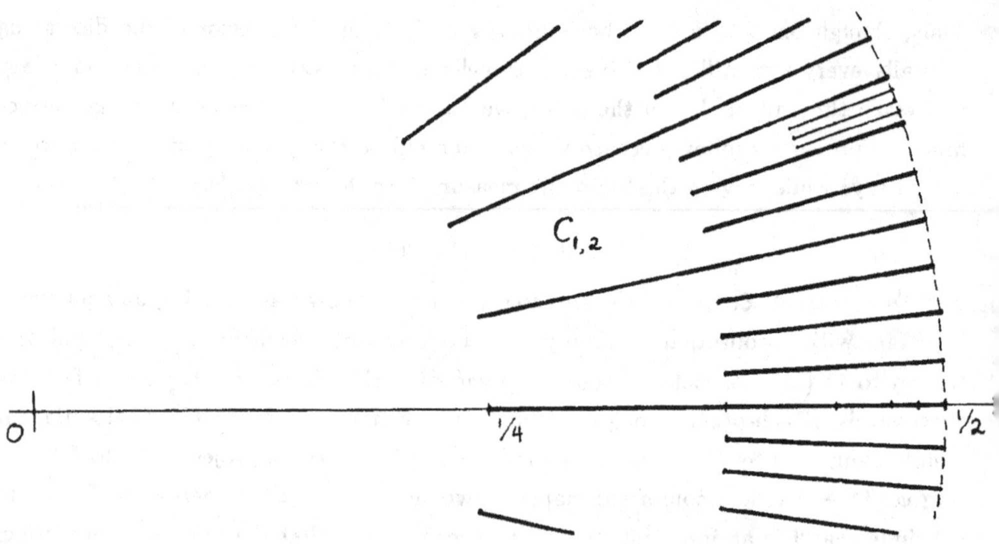


Figure 1.

Figure 2 shows a typical corridor, call it $C$, which has a threshold, an exit, and two sides. On those sides we will attach left and right wings. Let points $a$ and $b$ be the endpoints of the left side, as indicated in Figure 2. The universal covering surface of $\mathbf{D}\setminus\{a,b\}$ can be broken in two by a crosscut (on one sheet only) along the segment $[a,b]$. Thinking of this as a directed segment, discard the piece on the right of the cut and attach the piece on the left to the left side of the corridor — this is the left wing of $C$. The right wing is described in a similar fashion, using the endpoints $a'$, $b'$.

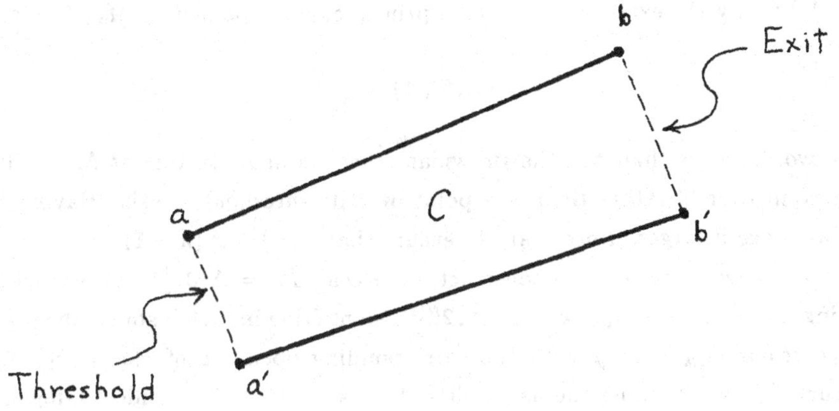


Figure 2.

The structure thus obtained will be denoted $A = A_{n,k}$ and will be termed an "assembly", and with its natural projection $\pi$, is a covering surface of the disc. On $A$, we will denote by $M = M_{n,k}$ the points lying over the circle $\{w : |w| = \frac{r_n + r_{n+1}}{2}\}$ which bisects $[a,b]$. The boundary of $A$ consists of the threshold and exit (of $C \subseteq A$) and points over the unit circle (in the boundaries of the wings). Almost every Brownian traveler on $A$ starting at a point of $M$ will fall off $A$ at some point of one of these. Our next task is to choose $N(n)$, which determines the width of the corridor, so that falling off over $\mathbf{T}$ occurs more frequently. Note that $M$ separates the threshold of $A$ from the exit.

For fixed $n$, we now choose $N(n)$; the assemblies $A_{n,k}$ for different $k$ are rotations of one another about the origin, so it suffices to work with $k = 1$. Let $E$ denote the circular arcs which form the threshold and exit of $C = C_{n,1}$, and let $G$ denote the circle $\{z : |z| = \frac{r_n + r_{n-1}}{2}\}$. It is clear that the harmonic measure $\omega(z, E, \mathbf{D}\setminus E)$ goes to zero uniformly for $z \in G$ as the length of the arcs in $E$ goes to zero. These arcs have length

$\frac{2\pi r_n}{2^N}$ and $\frac{2\pi r_{n+1}}{2^N}$, so we may choose $N$ sufficiently large that

$$\omega(z, \mathbf{T}, \mathbf{D}\backslash E) > \frac{1}{2}, \quad z \in G.$$

If $W$ is the universal covering surface of $\mathbf{D}\backslash E$, then this inequality implies that more than $\frac{1}{2}$ of the Brownian travelers starting at any point of $W$ lying over $G$ will exit $\partial W$ at a point over the unit circle. A moment's reflection will show that $W$ is embedded as an open subsurface of $A = A_{n,1}$ and that the restriction $\pi|_W$ is the universal covering map of $\mathbf{D}\backslash E$. By the extension of domain principle, every point $w \in M_{n,1} \subseteq A$ satisfies

$$(*) \qquad \omega(w, \mathbf{T}, A) > \frac{1}{2}.$$

In otherwords, more than half the Brownian travelers in $A$ starting at $M_{n,1}$ will leave $A$ at a point over $\mathbf{T}$ rather than at a point over its threshold or exit. Having chosen $N(n)$, we make it larger, if necessary to ensure that $N(n) > N(n-1)$.

We are now in position to construct $\mathcal{R}$. Recall $R_0 = \Delta(0, \frac{1}{4})$. Construct $R_1$ by attaching the assemblies $A_{1,k}$, $k = 1, \ldots, 2^{N(1)}$, identifying in each instance the threshold of the corridor $C_{1,k} \subseteq A_{1,k}$ with the corresponding open arc of $\{|z| = r_1\}$ in $\partial R_0$. Construct $R_2$ by attaching the assemblies $A_{2,k}$, $k = 1, \ldots, 2^{N(2)}$, identifying in each instance the threshold of the corridor $C_{2,k}$ with the corresponding open arc of $\{|z| = r_2\}$ which lies in the exit of one of the corridors $C_{1,j} \subseteq R_1$. Continue in this way, attaching the thresholds of each generation of assemblies to the exits of the previous. The resulting surfaces $R_0, R_1, \ldots$ are nested and lead (in the inductive limit) to the surface $\mathcal{R}$ we want. $\mathcal{R}$ is simply connected and inherits a natural projection $\pi$ to $\mathbf{D}$. It remains to check that it satisfies the geometric properties we intended.

It is not difficult to check that $\partial\mathcal{R}$ consists of points lying over one of the sets $B_1, B_2$, or $B_3$, where: $B_1$ consists of the endpoints of the thresholds of all the corridors $C_{n,k}$; $B_2$ is the circle $\{|z| = \frac{1}{2}\}$; and $B_3 = \mathbf{T}$. First, let us consider the part of $\partial\mathcal{R}$ which an observer can see while standing at the origin (actually the point of $R_0 \subseteq \mathcal{R}$ above the origin); that is, which boundary points are accessible along radial segments? Such a ray either travels down an infinite succession of corridors $C_{n,k}$ or hits one of the threshold endpoints. In the former case, since $r_n \to \frac{1}{2}$, the ray leaves $\mathcal{R}$ at a point over $B_2$; in the latter case, at a point over $B_1 \subseteq \{|z| < \frac{1}{2}\}$. Next, we need to consider the Brownian traveler on $\mathcal{R}$ starting over the origin: Since $B_1$ is countable, almost no Brownian travelers will exit from the surface at points over $B_1$. Almost no Brownian travelers will exit at points over $B_2$ either, but this is where the details of our construction enter. First, observe that no points on the boundary of any single assembly $A = A_{n,k}$ will lie over $B_2$ — indeed, the boundary of $A$ consists of its threshold and exit, which lie in

$\{|z| < \frac{1}{2}\}$, along with points over the unit circle. Therefore, any traveler who would hit the boundary of $\mathcal{R}$ at a point over $B_2$ must travel through assemblies belonging to infinitely many generations; in particular, given any integer $n > 0$, there exist integers $J > n$ and $k$, $1 \leq k \leq 2^N(J)$, so that the traveler reaches some point $w_0$ of $A_{J,k}$. Since $\mathcal{R}$ is simply connected, this entails that the traveler pass through a unique sequence of assemblies in generations $1, 2, \ldots, n$. Denote these by $A_{1,k_1}, A_{2,k_2}, \ldots, A_{n,k_n}$. By "passing through" the assembly $A_{j,k_j}$, we mean that the traveler reaches a point of $M_{j,k_j} \subseteq A_{j,k_j}$ at some time $t_j$ and reaches the exit of $A_{j,k_j}$ at some later time $\tau_j$. These are stopping times for the Brownian motion, and it is clear that we can define them inductively so that $0 < t_1 < \tau_1 < t_2 < \tau_2 < \cdots < t_n < \tau_n$. But recall inequality (*): The Brownian traveler who reaches $M_{j,k_j}$ has probability less than one half of reaching the exit of $A_{j,k_j}$ before falling off over $\mathbf{T}$. These events are independent, so the probability that a traveler will successfully pass through $n$ assemblies is no bigger than $(\frac{1}{2})^n$. Therefore, almost no Brownian travelers will leave $\mathcal{R}$ at points over $B_2$. We conclude that almost all Brownian travelers leave $\mathcal{R}$ at points over $B_3 = \mathbf{T}$; equivalently, the harmonic measure of $\mathcal{R}$ resides at boundary points over the unit circle. This completes the construction in case we want to prevent someone standing at the origin from seeing any of the unit circle.

Next, we indicate how to modify the construction to obscure the view from a whole open set $\Omega$, which we take as $\Delta(0, \frac{1}{8})$. The basic idea is the same, $\mathcal{R}$ being built with repeated generations of corridors having narrowing thresholds and attached wings. Again, each corridor is accessible only by passing through a unique succession of corridors ("ancestors") from previous generations. Now, however, the shapes of the corridors will need to be adjusted and their projections will, of necessity, overlap. The thresholds and exits for the $n$th generation will remain over the circles $\{|z| = r_n\}$ and $\{|z| = r_{n+1}\}$, respectively, with every exit at one stage being a disjoint union of thresholds for the next. The shapes of the corridors $C_{n,k}$ are determined inductively using (hyperbolic) rays from $\Omega$ as follows: At the first stage, let $T$ be the open subarc of $\{|z| = r_1 = \frac{1}{4}\}$ which forms the threshold of a corridor $C = C_{1,k}$. Consider the union, $H$, of rays from points of $\Omega$ through $T$. Then define $C = H \cap \{r_1 < |z| < r_2\}$. For the inductive step, assume $C'$ has been determined at the $(n-1)$st stage using a collection $H'$ of rays and let $T$ be the subarc of the exit of $C'$ which will be the threshold of $C$. If $H$ is the union of all rays of $H'$ which pass through $T$, then define $C = H \cap \{r_n < |z| < r_{n+1}\}$. Note that every corridor $C$ lies in the union of rays passing through its ancestors.

In Figure 3 two corridors of the first generation have been drawn for illustration. Note that these and subsequent corridors may overlap; this causes no difficulty —

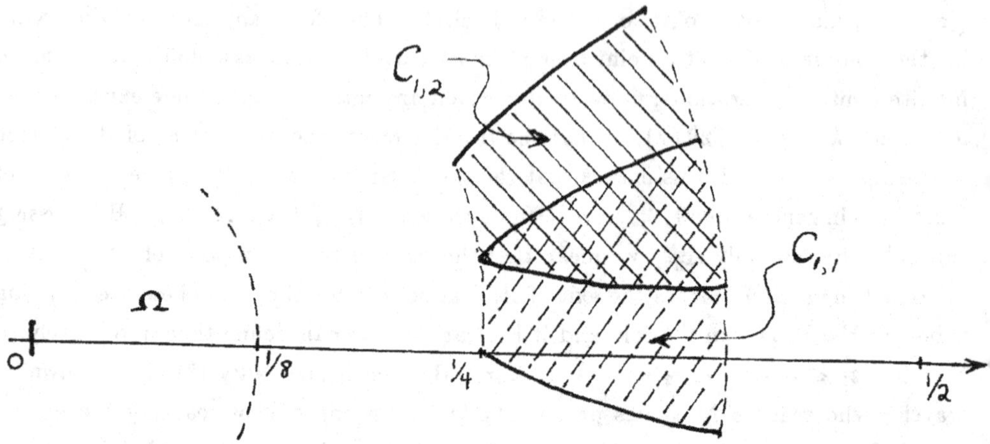


Figure 3.

attachments will still be along the thresholds, so overlaps just represent different sheets of the resulting surface. The wings of the corridors are exactly as before. Unlike our earlier construction, the length of the exit of a given corridor will not necessarily go to zero as the length of the threshold (which is what we control) goes to zero. This suggests problems with the estimate (*). Fortunately, for our Brownian motion arguments we need (*) only for some subsequence of indices. Thus we can remedy things by choosing the threshold sizes with a little more forethought: Assume by induction that we have constructed all corridors of the $n_0$ generation and wish to ensure that (*) holds for assemblies at the $m = n_0 + 3$ stage. Rays which define the shape of a corridor $C = C_{m,k}$ must pass through thresholds at the $m-2$ and $m-1$ stages. Knowing ahead of time how small the threshold and exit arcs of $C$ (what we denoted as $E$ before) need to be to obtain (*), we can adjust the threshold sizes at these two earlier stages to make the rays defining $C$ nearly parallel, then adjust the size of the threshold at the $m$th stage to make $E$ sufficiently small.

The remaining construction and its properties are essentially unchanged. From a point of $\Omega$, every ray either hits one of a countable number of endpoints of thresholds or goes down an infinite number of succesively narrower corridors, approaching a point over $\{|z| = \frac{1}{2}\}$. As before, almost every Brownian traveler will wander onto a wing of some assembly and end up falling off over the unit circle. The details are left to the interested reader.

As a final comment on the constuction, note the available flexibility: We could replace $\Omega$ by nearly any open set in $\mathbf{D}$ having closure in $\mathbf{D}$; the hyperbolic rays could be replaced by other trajectories, for instance the level lines of some analytic function; one could work with generalizations of inner functions in domains (or surfaces) other than $\mathbf{D}$; and so forth.

## SHARPNESS IN THE GST

We conclude with brief comments on the use of construction techniques in studying entire functions; for instance, in studying sharpness in the GST. For $F$ entire, fix some point $x$ of $\mathcal{R}_F$ over the origin and let $E_F \subseteq [0, 2\pi]$ denote the directions of rays from $x$ which encounter $\partial\mathcal{R}_F$ in the finite plane. $E_F$ is an $F_\sigma$-set and, according to the GST, has Lebesgue measure zero. Is it the case that this "exceptional" set is actually much smaller? Hausdorff dimension zero? (log) capacity zero? This is the sharpness question raised in [1, Problem 1.32] by A. E. Eremenko. The length-area methods which prove the GST will not help, so there has been an interest in finding examples. Here we indicate some preliminary results using constructive techniques.

First, note that these techniques require much more subtlety in the entire function setting. To illustrate, go back to Proposition 1. This has an easily proven converse: *If $\mathcal{R}$ is a simply Riemann surface lying over $\mathbf{D}$ and if almost every Brownian traveler starting at a point of $\mathcal{R}$ falls off the boundary at a point over $\mathbf{T}$, then $\mathcal{R}$ is the image surface of an inner function.*

On the other hand, the analogous result for entire functions (the converse of Proposition 2) is false: Consider $\mathcal{R}$ to be the universal covering surface of $\mathbf{C}$ punctured at $\{-1, 1\}$. Almost every Brownian traveler will miss these two points (and infinity), and hence will not fall off $\mathcal{R}$. Nonetheless, since $\mathcal{R}$ is the disc, it is the image surface of a function on $\mathbf{D}$ rather than an entire function. This is related to the classical "type problem" and reflects the fact that image surfaces which are not the disc can be very difficult to build! To illustrate, we consider a very restricted class of surfaces. Given a compact subset $E \subseteq [0, 2\pi]$, construct $\mathcal{R}$ as follows: Let $\{z_j\}$ be an enumeration of the points $e^{i\theta}$, where $\theta \in E$ is a left or right endpoint of a complementary open interval. Let $\mathcal{R}_0$ be a copy of $\mathbf{C}$ which is slit along each of the rays $\gamma_j = \{rz_j : r \geq 1\}$. If $z_j$ is a left endpoint, then take the universal covering surface of $\mathbf{C} \backslash z_j$, slit it along $\gamma_j$, throw away the right half, and attach the left half to its opposite on $\mathcal{R}_0$. If $z_j$ is a right endpoint, proceed in the symmetric fashion. Doing this for all $z_j$ gives a simply connected surface $\mathcal{R}$. If $\mathcal{R}$ is the plane, the corresponding function $F$ is entire.

QUESTION. *For which sets $E$ will $\mathcal{R}$ be the plane?*

A necessary condition is that the capacity of $E$ be zero; this follows from a result of Beurling (see [2, page 56]). Whether this is sufficient remains open. If $E$ is finite, then a result of Nevanlinna [8, Chp. XI] implies that $F$ is entire. One can use this result and Brownian motion arguments to obtain, first, countable sets $E$ and then certain perfect (hence uncountable) sets $E$. Details of these arguments will be provided elsewhere; the key ingredient is this criterion due to Kakutani [7]: Suppose $\mathcal{R}$ is a simple connected open Riemann surface, $K \subseteq \mathcal{R}$ is a compact set with nonempty interior, and $p$ is a point in the unbounded component of the complement of $K$. If almost every Brownian traveler starting at $p$ hits $K$, then $\mathcal{R}$ is conformally equivalent to the plane.

## REFERENCES

[1] K. F. Barth, D. A. Brannan and W. K. Hayman, *Research problems in complex analysis*, Bull. London Math. Soc. **16** (1984), 490–517.

[2] E. F. Collingwood and A. J. Lohwater, "The Theory of Cluster Sets," Cambridge University Press, Cambridge, 1966.

[3] Burgess Davis, *Brownian motion and analytic functions*, Ann. Probab. **7** (1979), 913–932.

[4] W. Gross, *Über die Singularitäten analytischer Funktionen*, Mh. Math. Phys. **29** (1918).

[5] Maurice Heins, *On the Lindelöf principle*, Ann. of Math. **61** (1955), 440–473.

[6] S. Kakutani, *Two dimensional Brownian motion and harmonic functions*, Proc. Japan Acad. **20** (1944), 706–714.

[7] S. Kakutani, *Two dimensional Brownian motion and the type problem of Riemann surfaces*, Proc. Japan Acad. **21** (1945), 138–140.

[8] Rolf Nevanlinna, "Analytic Functions," Springer-Verlag, New York-Heidelberg-Berlin, 1970.

[9] K. Noshiro, "Cluster Sets," Springer-Verlag, Berlin-Göttingen-Heidelberg, 1960.

# ON HARMONIC MAJORIZATION [1]

**Nobuyuki Suita**
Department of Mathematics, Tokyo Institute of Technology
Tokyo, Japan

**1. Introduction.** It was shown by Bañuelos and Wolff [1] that every holomorphic function $f(z)$ on the unit disc $D$ has two subdomains $D_j$ $(j = 1,2)$ such that the restriction of $f$ onto each $D_j$ belongs to $H^2(D_j)$ with $D = D_1 \cup D_2$. Here $H^2(D_j)$ denotes the Hardy class of index 2 on $D_j$. This result has been extended by Nakai and Tada [2] for an arbitrary open Riemann surface and positive index.

Since the fact that $f \in H^p(D_j)$ means that $|f|^p$ has a harmonic majorant on $D_j$, such a problem reduces to the existence of a harmonic majorant. In the present note we show that every upper semicontinuous function $f(z)$ on an open Riemann surface $R$ has two subdomains $D_j$ $(j = 1,2)$ with $R = D_1 \cup D_2$, on each of which its restriction has a harmonic majorant. The method of construction of subdomains $D_j$ is similar as in Bañuelos and Wolff [1].

In an oral communication with Nakai and Tada they said that they realized this extended result. However the proof given here is much simpler than theirs especially for subharmonic functions. We present the result in the widest form and its proof in the simplest.

**2. Statement of results.** We show

**Theorem.** *Let* $R$ *be an open Riemann surface and* $f(z)$ *be an upper semicontinuous function on* $R$. *Then there exist two subdomains* $D_j$ $(j = 1,2)$ *with* $R = D_1 \cup D_2$ *such that the restriction of* $f$ *onto each* $D_j$ *has a harmonic majorant on* $D_j$.

Before proving Theorem we prepare a lemma which is unnecessary for subharmonic functions.

**Lemma.** *Let* $B_t$ *be a rectangle,* $0 < \mathrm{Re}\, z < 1$ *and* $0 < \mathrm{Im}\, z < t$, *and let* $\omega_t(z)$ *be the harmonic measure of the union of the three edges other than* $[0, it]$ *with*

[1] This paper is in final form and no version of it will be submitted for publication elsewhere.

*respect to* $B_t$. *Then the (inner) normal derivative* $\partial\omega_t/\partial n = \partial\omega_t/\partial x$ *tends to infinity uniformly on the open segment* $(0, it)$ *as* $t$ *tends to zero.*

*Proof.* $\omega_t(z)$ is greater than

$$u_t(z) = \frac{2}{\pi}\left(\arg\frac{e^{-\pi z/t}-1}{e^{-\pi z/t}+1}-\frac{\pi}{2}\right).$$

We have

$$\frac{\partial\omega_t(z)}{\partial n} \geq \frac{\partial u_t(z)}{\partial x}$$

on $(0, it)$. The assertion is easily verified by direct calculation.

**3. Proof of Theorem.** Take an exaustion $\{R_n\}_{n=1}^{\infty}$ of $R$ with the following properties:

i) $R_n$ is a relatively compact subregion of $R$ with a finite number of boundary components consisting of closed analytic curves,

ii) $R - R_n$ has no relatively compact components each of which has only one boundary contour and

iii) $\overline{R}_n \subset R_{n+1}$ and $R = \cup_{n=1}^{\infty} R_n$.

Here $\overline{R}_n$ denotes the closure of $R_n$ in $R$.

Set $U_n = R_{2n+1} - \overline{R}_{2n}$, $n = 0,1,\dots$ with $R_0 = \emptyset$. Then $U_n$ is open, consisting of at most a finite number of components $U_{nj}$, $j = 1,2,\dots,N_n$. Starting from $U_0 = R_1$, we connect $U_{n-1}$ with $U_n$ by narrow canals $B_{nj}$, $j = 1,2,\dots,N_n$ in $R_{2n} - \overline{R}_{2n-1}$. For simplicity's sake we take mutually disjoint $B_{nj}$'s and each $B_{nj}$ is a quadrilateral whose two opposite sides are on $\partial U_n$ and $\partial U_{n-1}$. Let $B_n$ be the union of $B_{nj}$ with respect to $j$. Then, on letting $A^i$ stand for the interior of $A$, the set $D_1$ will be given by

$$D_1 = \bigcup_{n=0}^{\infty} (\overline{U_n \cup B_{n+1} \cup U_{n+1}})^i.$$

Let $f(z)$ be an upper semicontinuous function on $R$. Taking $\max(f(z), 0)$ if necessary, we may suppose that $f$ is nonnegative. Set

$$L_n = n + \max_{z \in R_n} f(z).$$

We define $u_1(z) = L_1$ on $U_0$. Clearly $f(z) \leq u_1(z)$ on $U_0$.

Let $R_1'$ be a slightly larger regular subregion of $R_2$ containig $R_1$ such that $R_1' - \overline{R}_1$ consists of doubly connected domains and that $f(z) < L_1$ holds on $R_1'$. We choose the union of canals in the following way. The set $\partial U_0 \cap \partial B_1$ is so small that

$$L_3 \omega(a; \partial U_0 \cap \partial B_1, U_0) < 1/2 \tag{1}$$

for a fixed $a \in U_0$ and that

$$L_3 \omega(z; \partial U_0 \cap \partial B_1, (\overline{B_1 \cup U_1})^i) < 1 \tag{2}$$

on the set $(B_1 \cup U_1) - R_1'$. Here the $\omega$ in (2) denotes the harmonic measure of the set $\partial U_0 \cap \partial B_0$ in the open set $(\overline{B_1 \cup U_1})^i$. These can be done by making the set $\partial U_0 \cap \partial B_1$ sufficiently small.

Let $D_{11}$ be the domain $(\overline{U_0 \cup B_1 \cup U_1})^i$. Let $u_2(z)$ be the solution of the Dirichlet problem on $D_{11}$ with the data:

$$u_2 = \begin{cases} L_1 & \text{on } \partial D_{11} \cap \partial U_0, \\ L_3 & \text{on } \partial D_{11} \cap \partial (B_1 \cup U_1). \end{cases}$$

In order to prove $u_1 < u_2$ on $U_0$ we consider an auxiliary function $v_2(z)$ on $(\overline{B_1 \cup U_1})^i$ which is the solution of the Dirichlet problem with the data:

$$v_2 = \begin{cases} L_1 & \text{on } \partial D_{11} \cap \partial U_0, \\ L_3 & \text{on } \partial D_{11} \cap \partial (B_1 \cup U_1). \end{cases}$$

Then the function $\hat{u}_2(z)$ defined by

$$\hat{u}_2(z) = \begin{cases} u_1(z) & \text{on } U_0, \\ v_2(z) & \text{otherwise on } D_{11}, \end{cases}$$

is subharmonic on $D_{11}$, since $\hat{u}_2(z)$ is continuous on $D_{11}$ and inequality

$$\frac{\partial u_1}{\partial n} < \frac{\partial v_2}{\partial n} \tag{3}$$

is automatically satisfied on $\partial U_0 \cap \partial B_1 \cap D_{11}$. Since $u_2$ is the harmonic function with the same boundary values as $\hat{u}_2$, we have

$$u_1(z) = \hat{u}_2(z) < u_2(z).$$

**4. Continued.** Let $D_{1n}$ be $(\overline{D_{1n-1} \cup B_n \cup U_n})^i$. After $u_{n-1}$ is defined on $D_{1n-1}$, we choose $B_n$ so that the following three conditions are satisfied;

$$L_{2n+1}\omega(a; \partial B_n \cap \partial D_{1n-1}, D_{1n-1}) < 2^{-n} \tag{4}$$

and

$$L_{2n+1}\omega(a; \partial B_n \cap \partial D_{1n-1}, (\overline{B_n \cup U_n})^i) < 1 \tag{5}$$

for every $z$ in the outside of a large domain $R'_{2n-1}$ between $R_{2n-1}$ and $R_{2n}$, where $f(z) < L_{2n-1}$ holds. To give the third condition let $v_n(z)$ be the solution of the Dirichlet problem on the open set $(\overline{B_n \cup U_n})^i$ with data;

$$v_n = \begin{cases} L_{2n-1} & \text{on } \partial D_{1n-1} \cap \partial B_n, \\ L_{2n+1} & \text{otherwise on } \partial(B_n \cup U_n). \end{cases}$$

Then the third condition is given by

$$\frac{\partial u_{n-1}}{\partial n} < \frac{\partial v_n}{\partial n} \text{ on } \partial D_{1n-1} \cap \partial B_n \cap D_{1n}. \tag{6}$$

By making use of Lemma, the condition (6) is verified for a sufficiently small $\partial B_n \cap \partial D_{1n-1}$.

Let $u_n(z)$ be inductively defined as the solution of the Dirichlet problem with the data:

$$u_n = \begin{cases} L_{2n+1} & \text{on } \partial(B_n \cup U_n) \cap \partial D_{1n} \\ L_{2n-1} & \text{otherwise on } \partial D_{1n}. \end{cases}$$

Then as in no. 3 we have $u_{n-1} < u_n$ in $D_{1n-1}$. Clearly $f \leq u_n$ in $D_{1n}$. In fact we deduce form (5) that $L_{2n+1} - u_n(z) < 1$ outside of $R'_{2n-1}$ in $(\overline{B_n \cup U_n})^i$ while $f < L_{2n-1}$ in $R'_{2n-1} \cap (\overline{B_n \cup U_n})^i$. Since $\{u_n\}$ is increasing and

$$u_n(a) - u_{n-1}(a) < L_{2n+1}\omega(a; \partial B_n \cap \partial D_{1n-1}, D_{1n-1}) < 2^{-n},$$

$\{u_n\}$ converges to a harmonic function $u(z)$ uniformly on every compact subset of $D_1$. It is easy to see that $u(z)$ dominates $f(z)$ on $D_1$.

To construct $D_2$, we take slightly larger domain $R'_{2n}$ between $R_{2n}$ and $R_{2n+1}$ and smaller domains $R^*_{2n-1}$ between $R_{2n-1}$ and $R_{2n-2}$. Set

$V_n = R'_{2n} - \overline{R^*_{2n-1}}$, $n = 1,2,\ldots$. We connect $V_n$ with $V_{n-1}$ by a set of canals $C_n$. $D_2$ is given by

$$D_2 = \bigcup_{n=1}^{\infty} \overline{(V_n \cup C_{n+1} \cup V_{n+1})}^i.$$

The same argument guarantees the existence of a harmonic majorant $v(z)$ on $D_2$. It is clear that $R = D_1 \cup D_2$.

**5. Remarks.** If $f(z)$ is subharmonic on $R$, we may assume $f \geq 0$. In this case, we can take $u_n$ as the harmonization of $f$ on $D_{1n}$ under the choice of $B_n$ satisfying only (4). Here $L_{2n+1}$ can be replaced by

$$M_{2n+1} = \max_{z \in R_{2n+1}} f(z).$$

$\{u_n\}$ is increasing and the condition (4) guarantees the convergence of $\{u_n\}$ to a harmonic majorant.

It is obvious that such a harmonic majoration is valid for higher dimensional cases.

## References


[1] Bañuelos, R. and Wolff, T., Note on $H^2$ on planar domains. Proc. Amer. Math. Soc. 95 (1985), 217–218.

[2] Nakai, M. and Tada, T., Note on $H^p$ on Riemann surfaces. J. Math. Soc. Japan 39 (1987), 663–666.

# HOMEOMORPHIC CONJUGATES OF FUCHSIAN GROUPS: AN OUTLINE

**Pekka Tukia**
University of Helsinki, Department of Mathematics
Hallituskatu 15, SF-00100 Helsinki, Finland

This is an outline of my paper with the same title. The following notation is used:

$S = \{x \in R^2 : |x| = 1\}$,
$D = \{x \in R^2 : |x| < 1\}$,
$H$ = the group of homeomorphisms of $S$, and
$M$ = the group of Möbius transformations of $S$.

By a Fuchsian group we mean a discrete subgroup of $M$. It may contain also orientation reversing elements. Usually a Fuchsian group is thought to act on $\bar{D}$ but this is not a problem since the action on $S$ can be uniquely extended to an action on $\bar{D}$ and later we might also consider a Fuchsian group to act on $\bar{D}$, depending on the situation.

We are here concerned with the

**Problem.** *When is a group $G$ of homeomorphisms of $S$ topologically conjugate to a Fuchsian group?*

We propose that this happens if and only if $G$ is a *convergence group* of $S$ which means that it satisfies the *convergence property*. This means that whenever $g_i \in G$ are distinct, we can pass to a subsequence in such a way that, for some $a, b \in S$,

$$g_i | S \setminus \{a\} \to b, \qquad \text{(CON)}$$

locally uniformly. This notion is due to Gehring and Martin [GM] and corresponds to their definition of a discrete convergence group.

We may also speak of convergence groups of other spaces than $S$ by which we mean that (CON) is true with $S$ replaced by the space in question.

Every Fuchsian group is a convergence group as follows, for instance, by a normal family argument, and Gehring and Martin showed that many of the topological properties of Fuchsian groups generalize to convergence groups. For instance, one can define the *limit set* L(G) of G as the complement of the set where G acts discontinuously and then L(G) seems to have all the topological properties of the limit set of a Fuchsian group. Thus we adopt the same terminology for convergence groups as for Fuchsian groups and speak of *elementary* groups and of the groups of the *first* and *second* kind.

Although we have not been able to prove that every convergence group is a topological conjugate of a Fuchsian group, we have the

**Theorem.** *Let* G *be a convergence group of* S. *Then either* G *is topologically conjugate to a Fuchsian group or* G *has a semitriangle subgroup of finite index.*

A *semitriangle* group is a non-elementary group which is generated by u and v such that, for some p, q, r > 1,

$$u^p = v^q = (v^{-1}u^{-1})^r = 1.$$

Thus such a group is a factor group of a triangle group. If the group is Fuchsian, this is about equivalent to the fact that it is a triangle group but we do not allow parabolic punctures (that is, $p, q, r < \infty$).

We outline the proof. For simplicity we assume from now on that all the elements are orientation preserving .

The proof of this theorem starts from the fact that single elements of convergence groups are conjugate to Möbius transformations. If $g \in G\setminus\{id\}$ has finite order, then it is easily seen to be conjugate to an elliptic element of M. Otherwise $g^n$, $n \in Z$, are distinct and we can find by the convergence property $n_i$ and $a, b \in S$ such that

$$g^{n_i}|S\setminus\{a\} \to b \tag{1}$$

uniformly on compact subsets of $S\setminus\{a\}$. If

$$X = \{x \in S : g(x) = x\},$$

then X can contain at most two points in view of (1). If X contains two points, let I and J be the intervals of $S\setminus X$. Since g is monotone on I and J, the maps $g|I$ and $g|J$ are conjugate to the translation $z \mapsto z+1$ of R. Using the convergence property, one sees that the directions of the translations $g|I$ and $g|J$ are so compatible that g is conjugate to a hyperbolic map of M.

Similarly, if X contains one point, g is conjugate to a parabolic Möbius transformation, and if X is empty, g is conjugate to an elliptic element of M of finite order. Thus the theorem is true for one-generator groups and we can call elements of a convergence group *hyperbolic*, *parabolic* or *elliptic*.

The treatment of the general case is based on the notion of a *simple axis*. An axis is abstractly defined to be a point-pair $(a,b)$ of distinct elements of $S$ though sometimes it is more convenient to identify it with the hyperbolic line of $D$ with end-points $a$ and $b$. An axis is basically a directed axis though we may sometimes forget this and identify $(a,b)$ and $(b,a)$. Two axes *intersect* if and only if the corresponding hyperbolic lines are distinct and intersect. If $g \in G$, then $gA = (g(a),g(b))$ and $A = (a,b)$ is *simple* (with respect to $G$) if it satisfies

$1^\circ$ $A$ and $gA$ do not intersect for any $g \in G$, and

$2^\circ$ if $A_i = g_i A$ are the axes conjugate to $A$ in $G$ then, given $\varepsilon > 0$, $|g_i(a) - g_i(b)| > \varepsilon$ for only finitely many $i$.

We assume that there is a simple axis $A$ and then prove that $G$ is topologically conjugate to a Fuchsian group. If $A$ is simple, we can form the 1-complex

$$X_A = S \cup (\cup_{g \in G} gA)$$

(where $gA$ is interpreted as a hyperbolic line). Since $A$ is simple, it is easy to extend the action of $G$ to $X_A$ in such a way that we obtain a convergence group of $X_A$.

Let $C$ be a component of $D \setminus X_A$. Then $\partial C$ is homeomorphic to $S$ and

$$G_C = \{g | \partial C : g \in G \text{ and } g(\partial C) = \partial C\} \tag{2}$$

is a convergence group of $\partial C$. Since $L(G) \subset S$, $G_C$ is of the second kind and it is not difficult to show that $G_C$ has a simple axis (see below). Hence we can repeat this process, first in $G_C$ and then in all components of $D \setminus X_A$ conjugate in $G$ to $C$ (that is, their boundary is conjugate to $\partial C$) in such a way that we obtain a convergence group of a 1-complex $Y \supset X_A$.

We can repeat this process (perhaps infinitely many times) and find a 1-complex $X$ on which $G$ acts as a convergence group such that, if $C$ is a component of $D \setminus X$, then $G_C$ is very simple (conjugate to a Fuchsian elementary group). At this stage we can extend $G$ to the whole $\bar{D}$ to obtain a convergence group of $\bar{D}$.

However, a convergence group of $\bar{D}$ can be shown to be conjugate (as a group of $\bar{D}$) to a Fuchsian group [MT]. In our situation we can observe that we can assume that the extension to $\bar{D}$ is constructed in such a way that there is a G-invariant triangulation of $D' = \bar{D} \setminus L(G)$. It is easy to define a conformal structure on $D'$ so that $G$ is conformal in this structure. One can show that, with respect to this structure, $G | \bar{D} \setminus L(G)$ is conformally conjugate to the action of a Fuchsian group $F$ in $\bar{D} \setminus L(F)$

and that the conjugacy extends to $\bar{D}$.

### The existence of a simple axis

Thus it is important to know whether there are simple axes. Not all groups have simple axes. For instance, if G is a Fuchsian triangle group with compact orbit space, then there are no simple axes. The proof of our theorem is based on a procedure which finds either a simple axis or a semitriangle subgroup.

We start from the fact that in any case there are *regular axes* and A is such an axis if it satisfies $2^\circ$ and if its *intersection number* $I_A$ is finite. If the endpoints of A are not the fixed points of a hyperbolic $g \in G$, then A is *non–hyperbolic* and we define

$$I_A = \text{number of axes conjugate to A in G and intersecting A.}$$

If the endpoints of A are the fixed points of a hyperbolic $g \in G$, then A is a *hyperbolic* axis and, setting

$$G_A = \{g \in G : g \text{ fixes endpoints of } A\},$$

the intersection number $I_A$ of A is defined to be the number of $G_A$-conjugacy classes of axes conjugate to A in G and which intersect A.

Each hyperbolic $g \in G$ defines an axis

$$A_g = A = (a,b)$$

such that a is the repelling and b the attractive fixed point of g. It can be shown that such an axis is a regular axis and we also denote its intersection number by

$$I_g = I_A .$$

If G is non-elementary, that is, the limit set of G contains more than two points, there are hyperbolics. This is a fact that is valid for any non-elementary convergence group of $S^n$ (see [GM]) since in every neighbourhood of $(a,b) \in L(G)$ there are endpoints of some hyperbolic axis of G. We by-pass the proof of this topological fact which is a not difficult consequence of the convergence property.

If G is elementary, then a straightforward reasoning, not unlike the one for one-generator convergence groups, shows that G is conjugate to a subgroup of M. If G is non-elementary, we can find a hyperbolic axis A and, if A is not simple, try to reduce its intersection number.

An axis B conjugate to A in G is called an A-*axis.* If $I_A > 0$, then there is an A-axis B intersecting A. Thus $B = uA$ for some $u \in G$. Here u is not unique but we may replace u by hu where $h \in G_B$.

Assume that A and B intersect positively, as in Figures 1a and 1b, and let $C = uB = u^2A$. In the figures we have drawn the axes as arrows so that the end marked with < corresponds to the positive endpoint of the axis. We denote the positive endpoint of an axis A by $P_A$ and the negative endpoint by $N_A$ so that $A = (N_A, P_A)$.

Choosing u right, we can obtain that we have the situation in Figure 1a or 1b and that if N is the number of A-axes which intersect A and have one endpoint in the interval $(P_C, N_A)$ (=the interval of $S\backslash\{P_C, N_A\}$ which does not contain any of the endpoints of the axes A, B, and C), then

$$N < I_A . \tag{3}$$

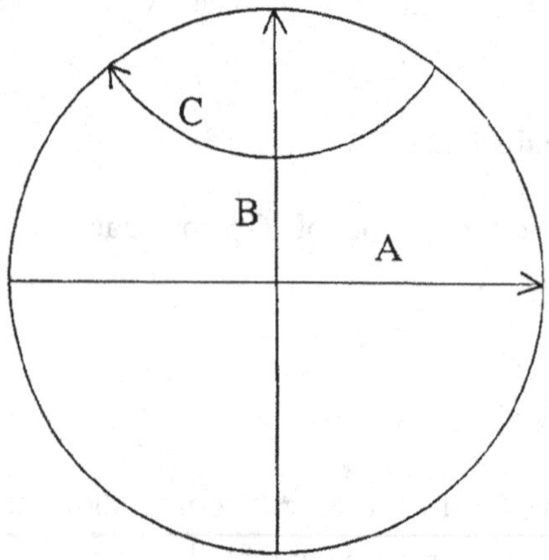


Figure 1a

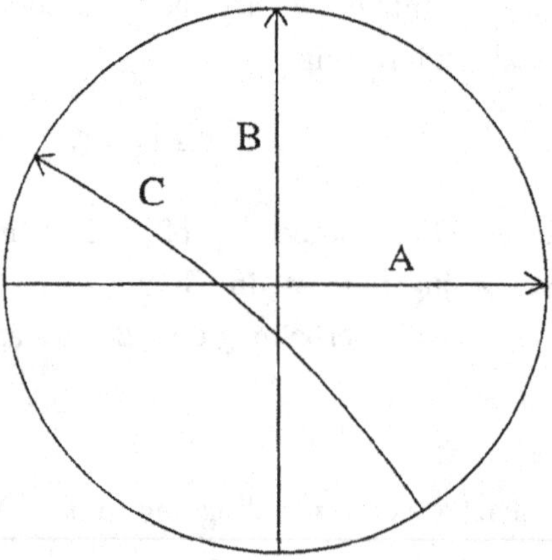


Figure 1b

Suppose that (3) is true. Then (A,B,C) is called a *canonical* G-*triple.* The situation of Figure 1a is simpler and therefore, if we have the situation of Figure 1a, (A,B,C) is called a *good canonical* G-*triple.*

Thus given A and an A-axis B intersecting A, there is a uniquely determined A-axis C such that (A,B,C) is a canonical G-triple. One can show that, given A, there is an A-axis B intersecting A such that the canonical G-triple determined by A and B is good.

So, if we have non-elementary G, there is a good canonical G-triple (A,B,C) $= (A, uA, u^2A)$. If u is hyperbolic, then

$$I_u < I_A .$$

This can be seen as follows. If $A' = A_u$ and $B'$ is an $A'$-axis intersecting $A'$, it is possible to associate to $B'$ an A-axis $B''$ intersecting $A$ with one endpoint in $(P_C, N_A)$ which depends only on the $G_{A'}$-conjugacy class of $B'$. The map $B' \mapsto B''$ is injective, and hence $I_u \leq N < I_A$ by (3).

If g is torsionless, it is clear that the process we have described finds either a simple hyperbolic axis or a parabolic element in G. However, parabolic elements are about as good as simple hyperbolic axes.

If G contains elliptic elements, then u may also be elliptic and we are no better off. But in this situation we can consider also the maps

$$v = g_B^{-1}u, \text{ and } w = v^{-1}u^{-1}.$$

Figure 2 illustrates the situation. (The intersection pattern may not always be as shown, D may intersect A and also other differences are possible. But this is the basic situation which occurs, for instance, if all the elements u, v, w are hyperbolic.)

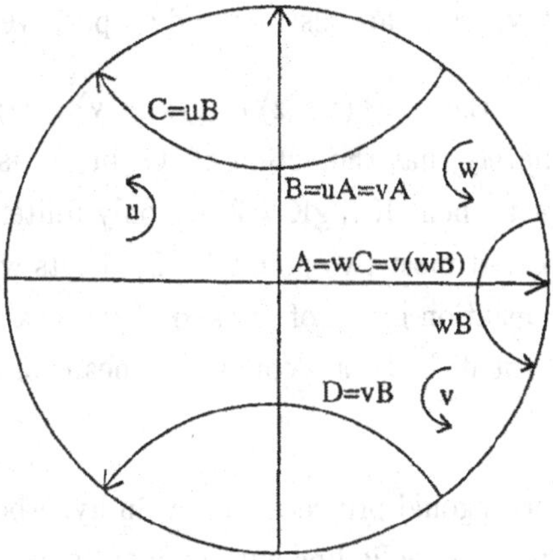


Figure 2

It is possible to show that if h is any of the maps u, v, w and h is hyperbolic, then $I_h < I_A$. If all of them are elliptic then the group generated by them is a semitriangle group. Consequently, either there is a parabolic or a simple hyperbolic element or then there is a semitriangle subgroup.

There need not be regular non-hyperbolic axes (a Fuchsian triangle group with a compact orbit space is a counter-example) but if there are such axes, then there are simple axes. The treatment of this case resembles the hyperbolic case. We only say that if A is non-hyperbolic and not simple, and $I \subset S$ is an open interval such that one endpoint of I is an endpoint of A and the other is an endpoint of an A-axis intersecting A, but that I contains no endpoints of A-axes intersecting A, then the axis with the same endpoints as I has smaller intersection number than A.

If the endpoints of A either ordinary points or parabolic fixed points, then A is a regular axis. Thus there is a simple axis if the above described process finds a parabolic element. Since the group $G_C$ in (2) is of the second kind, all limit points being on S, it is clear that we can always find a simple axis of $G_C$ and hence the above outlined process for the extension of G to $\bar{D}$ can be carried out as soon as there is a simple axis.

## The 3-manifold

To conclude that a semitriangle subgroup is of finite index, we need to introduce the space of triples

$$T = \{(x,y,z) : x, y, z \in S \text{ are distinct and on positive order on } S\}.$$

The group G acts in T by the rule $(x,y,z) \mapsto (g(x),g(y),g(z))$. The convergence property implies almost immediately that the action of G in T is properly discontinuous, that is, if $K \subset T$ is compact, then $K \cap gK \neq \emptyset$ for only finitely many $g \in G$. Furthermore, since $g \in G\setminus\{id\}$ fixes at most two points in S, G acts without fixed points in T and hence G is a cover translation group of T and T/G is a 3-manifold.

The space T and the disk D are connected means of the projection $p : T \to D$ defined by

$$p(x,y,z) = \text{the orthogonal projection of } z \text{ (in hyperbolic geometry) on the hyperbolic line with endpoints } y \text{ and } x.$$

One easily sees that $p^{-1}(w)$ is homeomorphic to a circle and that p is compatible with Möbius transformations (i.e. $pg = gp$ for $g \in M$). It follows that if F is Fuchsian, D/F is compact if and only if T/F is compact.

One can use the space T to show that $T/G_0$ is compact if $G_0$ is a semi-triangle group. This starts from the theorem that if $T/G_0$ is non-compact, then there is a simple axis and hence $G_0$ is conjugate to a Fuchsian group. If $T/G_0$ were non-compact, we could thus conjugate $G_0$ to a Fuchsian triangle group F which does not have parabolics. Hence $D/F$, and consequently $T/F$, is compact which is a contradiction. Thus $T/G_0$ must be compact which implies that the index of a semitriangle subgroup of $G$ is finite.

The fact that $T/G$ is a 3-manifold would make it possible to approach the conjugacy problem via 3-manifold theory (cf. Kerckhoff [K, p. 260] and Scott [S, p. 484]). If $T/G$ is Haken, then known 3-manifold theory can be used to conclude that $G$ is conjugate to a Fuchsian group.

It is also of interest to note that a subgroup $G$ of orientation preserving elements of $H$ is a convergence group if and only if it acts properly discontinuously on $T$.

**The Nielsen realization problem**

Our theorem is closely connected to the Nielsen realization problem (NRP) to which Kerckhoff [K] has recently given a complete solution. If we could prove that every convergence group is conjugate to a Fuchsian group, we would obtain a new proof of the NRP. We explain the connection of the NRP to our theorem.

Let $N$ be a compact Riemann surface and let $\{f_i\}$ be a finite set of homeomorphisms of $N$. It is assumed that $\{f_i\}$ is a group on the level of homotopy. Thus for instance $f_i \circ f_j$ is homotopic to some $f_k$, etc. The NRP asks whether it is possible to find $h_i$ homotopic to $f_i$ such that $\{h_i\}$ is already a group on the level of homeomorphisms.

To obtain the connection to our theorem, we represent $N$ by means of a Fuchsian group $F$ so that $N=D/F$. For each $f_i$ there are countably many lifts $\tilde{f}_{ij}$ to a homeomorphism $f$ of $D$ such that

$$p\tilde{f}_{ij} = f_i p$$

when $p : D \to N$ is the canonical projection. A theorem which goes back to Nielsen asserts that each $\tilde{f}_{ij}$ can be extended to a homeomorphism of $\bar{D}$. It is also true that $\tilde{f}_{ij}|S$ depends only on the automorphism of $F$ it induces which does not change if we perform a homotopy respecting the group action. Since $\{f_i\}$ is a group on the homotopy level, it follows that

$$G = \{\tilde{f}_{ij}|S\} \tag{4}$$

is a group. It contains the Fuchsian group $F$ as a normal subgroup of finite index.

Fuchsian groups are convergence groups and hence so is G.

Suppose that there is $f \in H$ such that $fGf^{-1}$ is Fuchsian. It is a theorem of Nielsen that we can extend f to a homeomorphism of $\bar{D}$ in such a way that $fFf^{-1}$ is a Fuchsian group of $\bar{D}$. Thus we can replace F by $fFf^{-1}$ and then G defined by (4) is Fuchsian, that is $\tilde{f}_{ij}|S \in M$. Let $\tilde{h}_{ij}$ be the Möbius transformation of $\bar{D}$ which coincides with $\tilde{f}_{ij}$ in S. Then

$$\{\tilde{h}_{ij}\}$$

is a Fuchsian group of $\bar{D}$ which contains F as a normal subgroup of finite index when we consider F to act in $\bar{D}$. If $h_i$ is the homeomorphism of N defined by $\tilde{h}_{ij}$, then $h_i$ is homotopic to $f_i$ and $\{h_i\}$ is a group.

We remark that if one combines our theorem with Kerkchoff's, one obtains the following generalization of the NRP:

*Let* G *be a group of homeomorphisms of* S *which contains a Fuchsian group as a subgroup of finite index. Then* G *is topologically conjugate to a Fuchsian group.*

To see this, let $F \subset G$ be a Fuchsian subgroup of finite index. Since F is a convergence group, so is G. Now, $D/F$ is compact if and only if $T/F$ is compact. If $T/F$ is compact, then F and G are finitely generated and hence we can replace F by a smaller subgroup of finite index which is normal in G and thus we can apply Kerkchoff's theorem. Otherwise G has a simple axis as we have seen and we can carry out the above outlined process and have the conjugacy.

We remark that from the perspective of the NRP our proof is especially close to Zieschang's proof [Z] (which is in turn close to Nielsen's proof [N]). From our point of view, Zieschang proved our theorem for such convergence groups which are finitely generated and have a Fuchsian subgroup of finite index.

**Final remarks**

The theorem, as formulated above, gives our basic result. We can also give some other conditions guaranteeing the topological conjugacy of G to a Fuchsian group. As we have seen this follows as soon as there is a simple axis and actually a weaker condition "there is an axis with at most binary intersection" would suffice.

Other conditions which imply the topological conjugacy are that G is torsionless, G is of the second kind, G has parabolics, G is infinitely generated, the manifold

$T/G_0$ is non-compact or virtually Haken ($G_0$ is the orientation preserving subgroup). The last result assumes Kerkchoff's theorem (but if $T/G_0$ is already Haken, it is not needed).

## References

[GM] F. W. Gehring and G. J. Martin, Discrete quasiconformal groups I. Proc. London Math. Soc. (3) 55 (1987), 331–358.

[K] S. P. Kerkchoff, The Nielsen realization problem. Ann. of Math. (2) 117 (1983), 235–265.

[MT] G. J. Martin and P. Tukia, Convergence and Möbius groups. To appear in the proceedings of the Geometric Function Theory workshop at the MSRI 1986.

[N] J. Nielsen, Abbildungsklassen endlicher Ordnung. Acta Math. 75 (1942), 23–115.

[S] P. Scott, The geometries of 3-manifolds. Bull. London Math. Soc. 15 (1983), 401–487.

[Z] H. Zieschang, Finite groups of mapping classes of surfaces. Lecture Notes in Math., 875, Springer-Verlag, Berlin, 1981.

# KONFORME UND QUASIKONFORME ABBILDUNGEN UND DAS KRÜMMUNGSVERHALTEN VON KURVENSCHAREN

**Alfred Wohlhauser**
Federal Institute of Technology
DMA, CH-1015 Lausanne, Switzerland

**Abstract.** Let $z_0$ be a point of the complex plane. We consider the family of curves S, whose members intersect at $z_0$, all having the same curvature there. We show that a mapping f is conformal at $z_0$ if and only if the centers of curvature (calculated from the point $f(z_0)$) constitute a conic section. In addition, introducing the notion of a quasi-ellipse and its diameters, we characterize the quasi-conformity of f at $z_0$.

## 1. Der Dilatationsquotient

Sei $z_0$ ein Punkt in der komplexen Ebene und $f(z) = u(x,y) + iv(x,y)$ eine topologische und stetig differenzierbare Abbildung einer Umgebung von $z_0$; wir nehmen also an, dass die Funktionaldeterminante $J = u_x v_y - u_y v_x$ in $z_0$ von Null verschieden sei. Der Quotient

$$D(z_0) := \frac{\max|df/dz|_{z=z_0}}{\min|df/dz|_{z=z_0}}$$

heisst der *Dilatationsquotient* von f in $z_0$; er ist definitionsgemäss $\geq 1$. Sind in $z_0$ die Cauchy-Riemannschen Differentialgleichungen erfüllt, d.h. ist $u_x = v_y$ und $u_y = -v_x$, so ist $df/dz$ in $z_0$ richtungsunabhängig und somit $D = 1$. Ist umgekehrt $D = 1$, so folgen daraus die Cauchy-Riemannschen Differentialgleichungen. Die Abbildung f ist also im Punkt $z_0$ genau dann *konform,* d.h. f ist dort komplex differenzierbar und $f'(z_0)$ verschieden von Null, wenn $D(z_0) = 1$ ist. Die Abbildung f heisst in $z_0$ K-*quasikonform*, wenn $D(z_0) = K$ ist.

## 2. Die konforme Abbildung

Wir betrachten nun die Kurvenschar S, die aus denjenigen Kurven besteht, welche durch $z_0$ gehen und in diesem Punkt die Krümmung $1/r$ besitzen. Die Krümmungsmittelpunkte dieser Kurven liegen also auf dem Kreis vom Radius r mit Zentrum $z_0$. Im folgenden sei f zweimal stetig differenzierbar. Es gilt der

**Satz.** *Die Abbildung* f *ist in* $z_0$ *genau dann konform, wenn die zum Punkt* $f(z_0)$ *gehörigen Krümmungsmittelpunkte der Bildkurven der Schar* S *auf einem Kegelschnitt liegen.*

Zum Beweis. Durch geeignete Ähnlichkeitstransformationen wird erreicht, dass $z_0 = 0$ und

$$f(z) = f(x + iy) = Dx + iy + o(z); \qquad (1)$$

D ist der Dilatationsquotient von f im Nullpunkt.

Sei nun $z(t) = x(t) + iy(t)$ eine Kurve der Schar S; $z(0) = 0$, $|z'(0)| \neq 0$. Es ist also

$$\left.\frac{x'y'' - x''y'}{(x'^2 + y'^2)^{3/2}}\right|_{t=0} = \frac{1}{r}. \qquad (2)$$

$\alpha$ bezeichne den Winkel, den $z(t)$ im Nullpunkt mit der positiven x-Achse bildet. $\omega(\alpha)$ sei entsprechend der Winkel, den $f(z(t))$, immer im Nullpunkt, mit der positiven u-Achse bildet. Um $\omega(\alpha)$ zu bestimmen, kommt es auf dasselbe heraus, ob wir uns längs der Kurve $z(t)$ oder auf ihrer Tangente im Nullpunkt dem Nullpunkt nähern. Vermöge (1) gilt für jeden Punkt z, genügend nahe bei $z_0 = 0$, eines Strahls $F_\alpha = \{z \mid z = re^{i\alpha}, r > 0\}$, dass

$$\arg f(z) = \arg(Dx + iy + o(z)) = \arg(D \cos \alpha + i \sin \alpha) + \arg(1 + \varepsilon(z)),$$

mit $\varepsilon(z) \to 0$ für $z \to 0$ (vrgl. [3], pg. 11).

Man hat also

$$\omega(\alpha) = \arg(D \cos \alpha + i \sin \alpha) = \operatorname{arc\,tg}\left(\frac{\operatorname{tg} \alpha}{D}\right),$$

und somit

$$\frac{y'}{x'} = \operatorname{tg} \alpha = D \operatorname{tg} \omega(\alpha) = D \operatorname{ctg} \varphi,$$

wobei $\varphi = \frac{\pi}{2} - \omega(\alpha)$ (siehe Fig. 1).

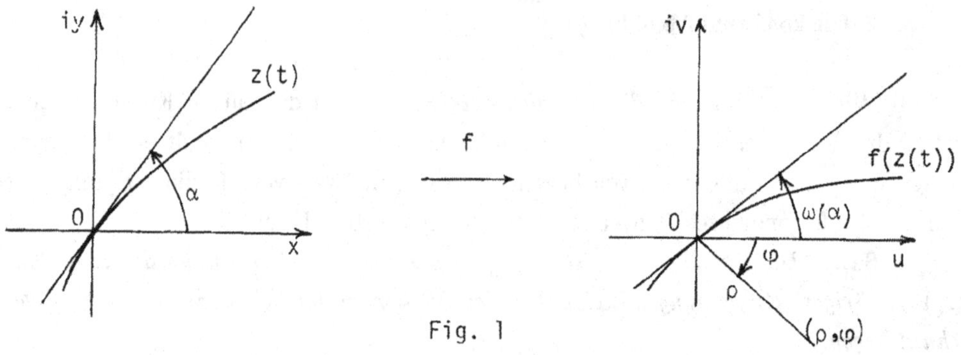


Fig. 1

Vermöge (1) gilt für die Krümmung $1/\rho$ der Bildkurve $f(z(t))$ im Nullpunkt:

$$\frac{1}{\rho}=\frac{D(x'y''-x''y')+Dx'(v_{xx}x'^2+v_{yy}y'^2+2v_{xy}x'y')-y'(u_{xx}x'^2+u_{yy}y'^2+2u_{xy}x'y')}{((Dx')^2+y'^2)^{3/2}}. \quad (4)$$

Jetzt formt man den 1. Term der rechten Seite von (4) unter Berücksichtigung von (3) folgendermassen um:

$$\begin{aligned}\frac{D(x'y''-x''y')}{((Dx')^2+y'^2)^{3/2}}&=\frac{D(x'y''-x''y')}{(x'^2+y'^2)^{3/2}}\cdot\frac{(x'^2+y'^2)^{3/2}}{(D^2x'^2+y'^2)^{3/2}}\\&=\frac{D}{r}\left[\frac{1+\mathrm{tg}^2\alpha}{D^2+\mathrm{tg}^2\alpha}\right]^{3/2}=\frac{D}{r}\left[\frac{1+D^2\mathrm{ctg}^2\varphi}{D^2(1+\mathrm{ctg}^2\varphi)}\right]^{3/2}\\&=\frac{1}{rD^2}\left[\frac{v^2+D^2u^2}{v^2+u^2}\right]^{3/2}\end{aligned}$$

(bei der letzten Umformung hat man verwendet, dass $\mathrm{ctg}\,\varphi=u/v$ ist). Schreibt man erneut $x, y$ an Stelle von $u, v$, so wird aus (4), da $\rho=\sqrt{x^2+y^2}$ und

$$y'^2=\sin^2\alpha=\frac{D^2\mathrm{ctg}^2\varphi}{1+D^2\mathrm{ctg}^2\varphi}=\frac{D^2u^2}{v^2+D^2u^2},\quad x'^2=\cos^2\alpha=\frac{v^2}{v^2+D^2u^2} \quad (5)$$

(ohne Einschränkung hat man $x'^2+y'^2=1$ gesetzt):

$$\frac{1}{\sqrt{x^2+y^2}}=\frac{\frac{1}{rD^2}(D^2x^2+y^2)^{3/2}+\frac{1}{D^3}\{\dots\}}{(x^2+y^2)^{3/2}},$$

mit

$$\{\ldots\} = \{-D^3 u_{yy} x^3 + D v_{xx} y^3 + D^2(D v_{yy} - 2u_{xy}) x^2 y + D(2D v_{xy} - u_{xx}) x y^2\}. \tag{6}$$

Daraus folgt für den Ort der Krümmungsmittelpunkte der Bildkurven der Schar S:

$$1 = \frac{\frac{1}{rD^2}(D^2x^2+y^2)^{3/2} + \frac{1}{D^3}\{\ldots\}}{x^2+y^2}, \tag{7}$$

wobei $\{\ldots\} = (6)$.

Sei nun f in $z_0$ konform. Dann hat man $D = 1$ und die Cauchy-Riemannschen Differentialgleichungen. Aus (7) erhält man damit:

$$x^2 + y^2 + r^2(u_{xx}x + v_{xx}y - 1)^2 = 0, \tag{8}$$

und das ist ein Kegelschnitt (vrgl. [2] , pg. 57–60).

Umgekehrt nimmt man an, dass der Ort der zum Punkt $f(z_0)$ gehörigen Krümmungsmittelpunkte der Bildkurven der Schar S ein Kegelschnitt sei. Die Gleichung

$$\frac{1}{rD^2}(D^2x^2 + y^2)^{3/2} + (\alpha x^3 + \beta y^3 + \gamma x^2 y + \delta x y^2) - (x^2 + y^2) = 0 \tag{9}$$

(siehe (7)) definiere also einen reelle Kegelschnitt, d.h. (abgesehen eventuell vom Punkt (0,0)) eine Punktmenge, die einer Gleichung der Form

$$ax^2 + by^2 + cxy + dx + ey + f = 0 \tag{10}$$

genügt. Es ist $r > 0$ und $D \geq 1$.

Zur Abkürzung setze man

$$A(k) = \alpha + \beta k^3 + \gamma k + \delta k^2 \text{ und}$$
$$B(k) = a + bk^2 + ck, \quad k \in \mathbb{R}.$$

Wir schneiden nun den Kegelschnitt mit den Geraden $y = kx$, $k \in \mathbb{R}$ (vrgl. dazu [1], pg. 8). Für die Schnittpunkte mit (9) gilt

$$x_{1,2} = \frac{1 + k^2}{\pm \frac{1}{rD^2}(D^2 + k^2)^{3/2} + A(k)}, \text{ d.h.}$$

$$x_1 + x_2 = \frac{2A(k)(1 + k^2)}{A(k)^2 - \frac{1}{r^2D^4}(D^2 + k^2)^3} \text{ und} \tag{11}$$

$$x_1 \cdot x_2 = \frac{(1+k^2)^2}{A(k)^2 - \frac{1}{r^2D^4}(D^2+k^2)^3}.$$

Für die Schnittpunkte mit (10) hat man

$$x_1 + x_2 = -\frac{d + kc}{B(k)} \quad \text{und} \quad x_1 \cdot x_2 = \frac{f}{B(k)}. \tag{12}$$

Aus (11) und (12) folgen durch Gleichsetzen die Identitäten

$$2(1+k^2)A(k)B(k) + (d+ek)\left((A(k)^2 - \frac{1}{r^2D^4}(D^2+k^2)^3\right) = 0, \tag{13}$$

$$(1+k^2)^2B(k) - f\left(A(k)^2 - \frac{1}{r^2D^4}(D^2+k^2)^3\right) = 0. \tag{14}$$

Aus (14) erhält man, da $f \neq 0$, nach (11) und (12):

$$A(k)^2 - \frac{1}{r^2D^4}(D^2+k^2)^3 = \frac{1}{f}(1+k^2)^2B(k).$$

Setzt man dies in (13) ein, so ergibt sich

$$2(1+k^2)A(k)B(k) + (\tfrac{d}{f} + \tfrac{e}{f}k)(1+k^2)^2B(k) = 0, \text{ d.h.}$$

$$B(k)[2A(k) + (\tfrac{d}{f} + \tfrac{e}{f}k)(1+k^2)] = 0.$$

Hieraus folgt

$$B(k) = 0 \;\; \forall k \in \mathbb{R} \text{ oder}$$

$$2A(k) + (\tfrac{d}{f} + \tfrac{e}{f}k)(1+k^2) = 0 \;\; \forall k \in \mathbb{R}.$$

Die erste Möglichkeit ist auszuschliessen: Aus $B(k) = 0$ $(\forall k \in \mathbb{R})$ würde nach (11) und (12) $A(k)^2 - \frac{1}{r^2D^4}(D^2+k^2)^3 = 0$ $(\forall k \in \mathbb{R})$ folgen, d.h. durch Annullieren der Koeffizienten von $k^0, k^1, k^3, k^6$:

$$\beta^2 = \frac{1}{r^2D^4}, \quad \beta\alpha + \gamma\delta = 0, \quad \gamma\alpha = 0, \quad \alpha^2 = \frac{D^2}{r^2};$$

dies impliziert infolge $D \geq 1$

$$\alpha \neq 0, \quad \gamma = 0, \quad \beta = 0, \quad \beta = \pm\frac{1}{rD^2},$$

was nicht möglich ist. Folglich gilt die Identität

$$2A(k) + (\tfrac{d}{f} + \tfrac{e}{f}k)(1 + k^2) = 0. \tag{15}$$

Durch Annullieren der Koeffizienten von $k^0, k^1, k^2, k^3$ in (15) resultiert

$$\tfrac{e}{f} = -2\beta = -2\gamma \text{ und } \tfrac{d}{f} = -2\alpha = -2\delta,$$

also insbesondere

$$\beta = \gamma \text{ und } \alpha = \delta. \tag{16}$$

Annullieren der Koeffizienten von $k^0$, $k^2$ und $k^6$ in (14) ergibt weiter die 3 Gleichungen

$$\tfrac{a}{f} = \alpha^2 - D^6, \tag{17}$$

$$\tfrac{2a}{f} + \tfrac{b}{f} = \gamma^2 + 2\alpha\delta - 3D^4, \tag{18}$$

$$\tfrac{b}{f} = \beta^2 - 1. \tag{19}$$

Setzt man (17) und (19) in (18) ein und berücksichtigt man (16), so folgt schliesslich die Bedingung

$$2D^6 - 3D^4 + 1 = 0.$$

Da $D \geq 1$, kommt nur $D = 1$ in Frage, d.h. f muss in $z_0$ konform sein. Damit ist der Satz bewiesen.

## 2. Die quasikonforme Abbildung

Wir betrachten wiederum die Kurvenschar S, die aus denjenigen Kurven besteht, welche durch $z_0$ gehen und in diesem Punkt die Krümmung $1/r$ besitzen.

Durch geeignete Ähnlichkeitstransformationen wird erreicht, dass $z_0 = 0$ und $f(z) = f(x + iy) = Dx + iy + o(z)$, wobei D der Dilatationsquotient von f im Nullpunkt ist.

Sei nun vorerst f linear, d.h. von der Form $f(z) = f(x + iy) = Dx + iy$. Dann wird der Ort der zum Punkt $f(z_0) = f(0) = 0$ gehörigen Krümmungsmittelpunkte der Bildkurven der Schar S durch folgende Beziehung beschrieben (siehe (7)):

$$1 = \frac{\frac{1}{rD^2}(D^2x^2 + y^2)^{3/2}}{x^2 + y^2}. \tag{20}$$

Die durch (20) bestimmte Kurve soll die zu $f(z_0)$ gehörige r-*Quasiellipse* (von f) heissen. Ihre Gestalt ist aus Fig. 2 ersichtlich.

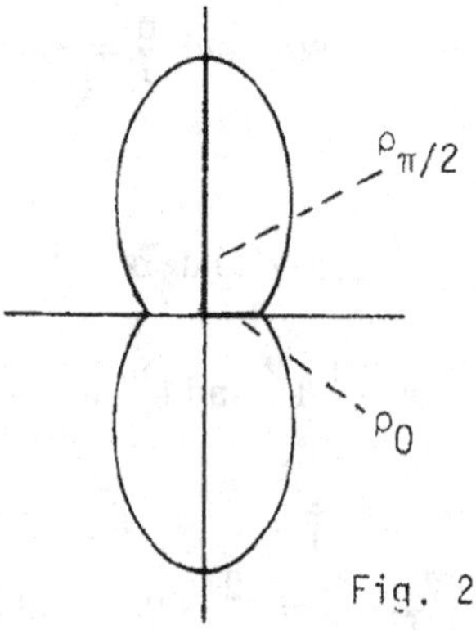


Fig. 2

Wie man leicht zeigt, ist eine r-Quasiellipse ($r > 0$ beliebig) genau dann ein Kreis, wenn $D = 1$ ist.

In Polarkoordinaten schreibt sich die r-Quasiellipse (20) wie folgt:

$$\rho = rD^2 \left[ \frac{1}{1 + (D^2 - 1)\cos^2\varphi} \right]^{3/2} \tag{21}$$

Für $\varphi = 0$ hat man $\rho_0(r) := \rho(r, \varphi = 0) = r/D$ und für $\varphi = \pi/2$ wird $\rho_{\pi/2}(r) := \rho(r, \varphi = \pi/2) = rD^2$. $\rho_0(r)$ und $\rho_{\pi/2}(r)$ heissen die *Halbmesser* der zu $f(z_0)$ gehörigen r-Quasiellipse (20) (siehe Fig. 2).

Man hat nun unmittelbar den

**Satz.** *Sei* f *linear. Dann ist* f *in* $z_0$ *genau dann* K-*quasikonform, wenn*

$$\frac{\rho_{\pi/2}(r)}{\rho_0(r)} = K^3$$

*ist, wo* $\rho_{\pi/2}(r)$ *und* $\rho_0(r)$ *die beiden Halbmesser einer beliebigen zu* $f(z_0)$ *gehörigen* r-*Quasiellipse sind.*

Für den allgemeinen Fall, d.h. f braucht nicht linear zu sein, ergibt sich der

**Satz.** *Die Abbildung* f *ist in* $z_0$ *genau dann* K*-quasikonform, wenn gilt:*

$$\lim_{r \to 0} \frac{\rho_{\pi/2}(r)}{\rho_0(r)} = K^3.$$

Anwendungen obiger Resultate in den Ingenieurwissenschaften werden später mitgeteilt.

## Literaturverzeichnis

[1] Kunz, E., Kommutative Algebra und algebraische Geometrie, Vieweg 1980.

[2] Ringleb, Fr., Über das Verhalten der Krümmung ebener Kurven bei konformer Abbildung, Jahresbericht der deutschen Math. Ver. 45, 1935.

[3] Taari, O., Charakterisierung der Quasikonformität mit Hilfe der Winkelverzerrung, Ann. Acad. Sci. Fenn. 390, 1966.

# CARLESON MEASURE AND BMO INTERPOLATION [1]

**Yao Bi-yun** [2]
Hangzhou University, Department of Mathematics
Hangzhou, China

## 1. Introduction

Let $\{z_n\}$ be a sequence in the unit disk $D = \{z : |z| < 1\}$. If given any bounded sequence $\{\alpha_n\}$ there exists a function $f \in H^\infty$ such that $f(z_n) = \alpha_n$ $(n = 1,2,\ldots)$, then $\{z_n\}$ is called an interpolating sequence. Carleson [1] proved that $\{z_n\}$ is an interpolating sequence if and only if $\{z_n\}$ is uniformly separated, i.e. there exists a positive number $\delta$ such that

$$\sup_k \prod_{\substack{j=1 \\ j\neq k}}^{\infty} \left| \frac{z_j - z_k}{1 - \bar{z}_j z_k} \right| \geq \delta. \tag{1}$$

Condition (1) is equivalent to the following condition: there is a $\delta_1 > 0$ such that

$$\left| \frac{z_j - z_k}{1 - \bar{z}_j z_k} \right| \geq \delta_1 \quad (\forall\, j, k,\ j \neq k), \tag{2}$$

and the measure

$$\mu = \sum_{n=1}^{\infty} (1 - |z_n|^2) \delta_{z_n} \tag{3}$$

is a Carleson measure, where $\delta_{z_n}$ is Kronecker's symbol.

The following result is well-known for $H^p$ interpolations [2].

**Theorem A.** *A sequence $\{z_n\}$ in $D$ is an interpolating sequence if and only if*

---

[1] This paper is in final form and no version of it will be submitted for publication elsewhere.

[2] Project supported by the National Natural Science Foundation of China (NSFC).

$$T_p(H^p) = \ell^p \tag{4}$$

*for some* p $(0 < p \leq \infty)$, *where* $T_p$ *is the linear operator on* $H^p$ *defined by* $T_p(f) = \{f(z_n)(1-|z_n|^2)^{1/p}\}$ $(0 < p < \infty)$ *and* $T_\infty(f) = \{f(z_n)\}$.

The case $1 \leq p \leq \infty$ of Theorem A is due to Shapiro and Shields [3] and the case $0 < p < 1$ is due to Kabaela [4].

It is shown in [5] that the relation (4) in Theorem A can be replaced by

$$T_p(H^p) \supset \ell^p \tag{5}$$

for $1 < p < \infty$, or by

$$T_\infty(\mathrm{BMOA}) \supset \ell^\infty \tag{6}$$

for $p = \infty$.

The formula (5) is presented in the form of a weighted interpolation in $H^p$. It can be easily turned into the form of a (simple) interpolation.

**Theorem B.** *A sequence* $\{z_n\}$ *in* D *is an interpolating sequence if and only if for given a sequence* $\{w_n\}$ *satisfying*

$$\sum_{n=1}^{\infty} |w_n|^p(1-|z_n|^2) < \infty \tag{7}$$

*for some* p $(1 < p < \infty)$ *there exists an* $f \in H^p$ *such that* $f(z_n) = w_n$ $(n = 1,2,\ldots)$.

It is natural to ask how do we find the condition (7). This is due to the characteristic of Carleson measure, which is usually stated as Carleson's theorem.

**Theorem C.** *Let* $\mu$ *be a positive finite measure, then* $\mu$ *is a Carleson measure if and only if there exists a constant* A *such that*

$$\int_D |f|^p \, d\mu \leq A\|f\|_p \tag{8}$$

*for some* p $(0 < p < \infty)$ *and all* $f \in H^p$.

Evidently, if $\mu$ satisfies (3), then (8) gives (7), where $f(z_n)$ is replaced by $w_n$.

Recently, Sundberg [6] considered the BMO interpolation. With a careful construction he proved

**Theorem D.** *Let* $\{z_n\}$ *be an interpolating sequence and* $\{\alpha_n\}$ *be a sequence. Then the following conditions are equivalent.*

(a) *There is an* $f \in BMOA$ *such that* $f(z_n) = \alpha_n$ $(n = 1,2,...)$.

(b) *There is a harmonic extension* $f(z)$ *of a* BMO *function* $f(e^{i\theta})$ *on* $\partial D$ *such that* $f(z_n) = \alpha_n$ $(n = 1,2,...)$.

(c) *There is a positive number* $\lambda$ *and a function* $\beta(z)$ *on* D *such that*

$$\sup_{z \in D} \sum_{n=1}^{\infty} e^{\lambda|\alpha_n - \beta(z)|} [1 - \rho(z,z_n)^2] < \infty, \tag{9}$$

*where* $\rho(z,w) = |z - w|/|1 - \bar{w}z|$.

(d) *There is a positive number* $\lambda$ *such that*

$$\sup_{z \in D} \sum_{n=1}^{\infty} e^{\lambda|\alpha_n - \alpha(z)|} [1 - \rho(z,z_n)^2] < \infty, \tag{10}$$

*where*

$$\alpha(z) = \frac{\sum \alpha_n [1 - \rho(z,z_n)^2]}{\sum [1 - \rho(z,z_n)^2]}.$$

What is the analogue in BMOA, which is closely related to (9) like (8) to (7)? What is the analogue of Theorem A or B in BMOA? This note is devoted to answer these questions.

## 2. A characteristic of Carleson measure

**Theorem 1.** *Let* $\mu$ *be a positive finite measure on* D, *then* $\mu$ *is a Carleson measure if and only if for given* $f \in BMOA$ *there exists a* $\lambda > 0$ *such that*

$$\sup_{|\zeta|<1} \int_D e^{\lambda|f(z) - f(\zeta)|} \frac{1 - |\zeta|^2}{|1 - \bar{\zeta}z|^2} d\mu(z) \le C, \tag{11}$$

*where* C *is a constant depending only on* $\mu$.

*Proof.* The only if part. If $f \in BMOA$, there is a $\lambda > 0$ and an $M > 0$ such that

$$\sup_{|\zeta|<1} \sup_{0<r<1} \int_0^{2\pi} e^{\lambda|f_\zeta(re^{i\theta})|} d\theta \le M, \tag{12}$$

where

$$f_\zeta(z) = f\Big(\frac{z+\zeta}{1+\bar\zeta z}\Big) - f(\zeta)$$

and M may be taken as a universal constant if $\lambda$ is sufficiently small [7]. Hereafter we consider $\lambda$ sufficiently small. By the hypothesis $\mu$ is a Carleson measure, and

$$e^{\lambda[f(z)-f(\zeta)]}\frac{1-|\zeta|^2}{(1-\bar\zeta z)^2} \in H^1.$$

From Theorem C it follows

$$\int_D \left|e^{\lambda[f(z)-f(\zeta)]}\right| \frac{1-|\zeta|^2}{|1-\bar\zeta z|^2}\, d\mu(z) \le A\int_{\partial D} \left|e^{\lambda[f(z)-f(\zeta)]}\right| \frac{1-|\zeta|^2}{|1-\bar\zeta z|^2}|dz|$$
$$\le A\int_{\partial D} \left|e^{\lambda f_\zeta(z)}\right| |dz|.$$

Taking the supremum over all $\zeta \in D$ and using (12) we obtain

$$\sup_{|\zeta|<1}\int_D \left|e^{\lambda[f(z)-f(\zeta)]}\right| \frac{1-|\zeta|^2}{|1-\bar\zeta z|^2}\, d\mu(z) \le A \sup_{|\zeta|<1}\int_{\partial D} \left|e^{\lambda f_\zeta(z)}\right| |dz|$$
$$\le A \sup_{|\zeta|<1}\int_{\partial D} e^{\lambda|f_\zeta(z)|}\, |dz| \le AM. \tag{13}$$

Substituting f into (13) with $-f$ and $\mp if$ respectively, it implies that

$$\sup_{|\zeta|<1}\int_D e^{-\lambda \mathrm{Re}[f(z)-f(\zeta)]}\frac{1-|\zeta|^2}{|1-\bar\zeta z|^2}\, d\mu(z) \le A \sup_{|\zeta|<1}\left\|e^{\lambda|f_\zeta|}\right\|_1$$

and

$$\sup_{|\zeta|<1}\int_D e^{\pm\lambda \mathrm{Im}[f(z)-f(\zeta)]}\frac{1-|\zeta|^2}{|1-\bar\zeta z|^2}\, d\mu(z) \le A \sup_{|\zeta|<1}\left\|e^{\lambda|f_\zeta|}\right\|_1$$

respectively. According to these inequalities and the following inequalities:

$$e^{\lambda|\mathrm{Re}[f(z)-f(\zeta)]|} \le e^{\lambda \mathrm{Re}[f(z)-f(\zeta)]} + e^{-\lambda \mathrm{Re}[f(z)-f(\zeta)]}$$

and

$$e^{\lambda|\mathrm{Im}[f(z)-f(\zeta)]|} \le e^{\lambda \mathrm{Im}[f(z)-f(\zeta)]} + e^{-\lambda \mathrm{Im}[f(z)-f(\zeta)]},$$

using the Hölder inequality we obtain

$$\int_D e^{\lambda|f(z)-f(\zeta)|}\frac{1-|\zeta|^2}{|1-\bar{\zeta}z|^2}\,d\mu(z)$$

$$\leq \int_D e^{\lambda\{|\mathrm{Re}[f(z)-f(\zeta)]|+|\mathrm{Im}[f(z)-f(\zeta)]|\}}\frac{1-|\zeta|^2}{|1-\bar{\zeta}z|^2}\,d\mu(z)$$

$$\leq \left\{\int_D e^{2\lambda|\mathrm{Re}[f(z)-f(\zeta)]|}\frac{1-|\zeta|^2}{|1-\bar{\zeta}z|^2}\,d\mu(z)\right\}^{1/2}\cdot$$

$$\left\{\int_D e^{2\lambda|\mathrm{Im}[f(z)-f(\zeta)]|}\frac{1-|\zeta|^2}{|1-\bar{\zeta}z|^2}\,d\mu(z)\right\}^{1/2}$$

$$\leq 2A\left\|e^{2\lambda|f_\zeta|}\right\|_1.$$

Consequently, there exists a $\lambda > 0$ such that

$$\sup_{|\zeta|<1}\int_D e^{\lambda|f(z)-f(\zeta)|}\frac{1-|\zeta|^2}{|1-\bar{\zeta}z|^2}\,d\mu(z) \leq 2A\sup_{|\zeta|<1}\left\|e^{2\lambda|f_\zeta|}\right\|_1 \leq 2AM = C.$$

Hence (11) is proved.

The if part. Suppose that (11) holds for $\mu$ and for every $f \in \mathrm{BMOA}$. Set

$$f(z) = \log(1-\bar{z}_0 z)^{-1} \quad (z_0 \text{ is fixed with } |z_0| < 1).$$

Denote $S_h = \{re^{i\theta} : \theta_0 \leq \theta \leq \theta_0 + h,\ 1-h \leq r < 1\}$. Without loss of generality we can choose $\theta_0 = -h/2$ and $z_0 = 1-2h$; for the general case of $\theta_0$ a change of the argument of $z_0$ is needed.

When $\zeta = z_0$,

$$f(z) - f(\zeta) = \log\frac{1}{1-\bar{z}_0 z} - \log\frac{1}{1-|z_0|^2} = \log\frac{1-|z_0|^2}{1-\bar{z}_0 z}$$

for $z \in S_h$. If $h$ is sufficiently small, then

$$|1-\bar{z}_0 z| \leq \{(\tfrac{h}{2})^2 + [1-(1-2h)(1-h)]^2\}^{1/2} < \tfrac{\sqrt{37}}{2}h.$$

Therefore

$$e^{\lambda|f(z)-f(z_0)|} = \exp\left(\lambda\left|\log\frac{1-|z_0|^2}{1-\bar z_0 z}\right|\right) \geq \exp\left(\lambda\log\frac{1-|z_0|^2}{|1-\bar z_0 z|}\right)$$

$$= \left(\frac{1-|z_0|^2}{|1-\bar z_0 z|}\right)^{\lambda} \geq \left(\frac{1-(1-2h)^2}{\sqrt{37}\,h/2}\right)^{\lambda} \geq K,$$

where $K$ is a constant. This implies that

$$\sup_{|\zeta|<1}\int_D e^{\lambda|f(z)-f(\zeta)|}\frac{1-|\zeta|^2}{|1-\bar\zeta z|^2}\,d\mu(z)$$

$$\geq \int_{S_h} e^{\lambda|f(z)-f(z_0)|}\frac{1-|z_0|^2}{|1-\bar z_0 z|^2}\,d\mu(z)$$

$$\geq K\cdot\frac{1-(1-2h)^2}{37h^2/4}\cdot\mu(S_h) \geq \frac{K'}{h}\,\mu(S_h). \tag{14}$$

Finally (14) and (11) imply that

$$\frac{K'}{h}\,\mu(S_h) \leq C$$

and $\mu$ is a Carleson measure.

Evidently, if $\{z_n\}$ is an interpolating sequence in $D$, then $\mu = \Sigma\,(1-|z_n|^2)\delta_{z_n}$ is a Carleson measure. For given $f \in BMOA$ there is a $\lambda > 0$ such that (11) holds, that is,

$$\sup_{|\zeta|<1}\sum_{n=1}^{\infty} e^{\lambda|f(z_n)-f(\zeta)|}\frac{1-|\zeta|^2}{|1-\bar\zeta z_n|^2}(1-|z_n|^2) < \infty. \tag{15}$$

This is the same as (9) in Theorem D, where $\alpha_n$ is replaced by $f(z_n)$, and $\beta(z)$ is replaced by $f(z)$.

### 3. A new characteristic of interpolating sequences

We now state a necessary and sufficient condition for a sequence to be an interpolating sequence by means of BMOA interpolation, which is an analogue of Theorem B in the space BMOA.

**Theorem 2.** *Let* $\{z_n\}$ *be a sequence in* D. *Then* $\{z_n\}$ *is an interpolating sequence if and only if for a given sequence* $\{\alpha_n\}$, *if there is a* $\lambda > 0$ *and a function* $\beta(z)$ *on* D *such that*

$$\sup_{|z|<1} \sum_{n=1}^{\infty} e^{\lambda|\alpha_n - \beta(z)|}\,[1 - \rho(z,z_n)^2] < \infty, \tag{16}$$

*then there exists an* $f \in$ BMOA *such that* $f(z_n) = \alpha_n$ $(n = 1,2,\ldots)$.

*Proof.* The only if part is proved by Theorem D. We now prove the if part. From (16) we observe that $\{z_n\}$ satisfies the inequality

$$\sup_{|z|<1} \sum_{n=1}^{\infty} [1 - \rho(z,z_n)^2] < \infty. \tag{17}$$

Suppose that $\{\alpha_n\}$ is a bounded sequence. Let

$$\alpha(z) = \frac{\sum \alpha_n[1 - \rho(z,z_n)^2]}{\sum [1 - \rho(z,z_n)^2]}.$$

It is easy to check that

$$\sup_{|z|<1} \sum_{n=1}^{\infty} e^{\lambda|\alpha_n - \alpha(z)|}\,[1 - \rho(z,z_n)^2] < \infty$$

for every $\lambda > 0$ (here (17) is used). So (16) holds for $\beta(z) = \alpha(z)$. By the hypothesis, there exists an $f \in$ BMOA such that $f(z_n) = \alpha_n$ $(n = 1,2,\ldots)$. This means that (6) holds. So, we know from [5] that $\{z_n\}$ is an interpolating sequence.

## 4. Remarks

**Remark 1.** In [8], we got another form of the characteristic of Carleson measure through BMOA functions.

**Theorem E.** *Let* $\mu$ *be a positive finite measure on* D, *then* $\mu$ *is a Carleson measure if and only if*

$$\sup_{|\zeta|<1}\left\{\int_D |f(z)-f(\zeta)|^p \frac{1-|\zeta|^2}{|1-\bar{\zeta}z|^2}\, d\mu(z)\right\}^{1/p} \le C_p\|f\|_* \tag{18}$$

*for every* $f \in$ BMOA *and some* $p$ $(0 < p < \infty)$, *where* $C_p$ *is a constant and* $\|\cdot\|_*$ *is the* BMO($\partial D$) *norm.*

The formula (18) is simpler than (11), so we may ask whether it is possible to replace (9) by a simpler form.

**Remark 2.** Theorem 1 admits an extension to higher dimensional spaces.

Suppose $f \in L_{loc}(R^n)$ and

$$\int_{R^n} |f(x)|(1+|x|^2)^{-\frac{n+1}{2}}\, dx < \infty.$$

The Poisson integral of $f$ is denoted by $u(x,y)$, and the Hardy-Littlewood maximal function of $f$ is

$$Mf(x) = \sup_{q \ni x} \frac{1}{|q|}\int_q |f(x)|\, dx,$$

$q$ is the cube in $R^n$ with side parallel to the axes.

**Theorem 3.** *Let* $\sigma$ *be a positive measure on* $R_+^{n+1}$. *Then* $\sigma$ *is a Carleson measure if and only if for any* $f \in \text{BMOA}(R^n)$ *with* $Mf \not\equiv 0$, *there exists a* $\lambda > 0$ *such that*

$$\sup_{(\xi,\eta)\in R_+^{n+1}} \int_{R_+^{n+1}} e^{\lambda|u(x,y)-u(\xi,\eta)|} \frac{\eta\, d\sigma}{[|x-\xi|^2 + (y+\eta)^2]^{\frac{n+1}{2}}} \le C, \tag{19}$$

*where* $C$ *is a constant depending only on* $\sigma$.

The proof of Theorem 3 is similar but simpler than that of Theorem 1 in [9]. We omit it. One can now ask about the analogues of Theorem D and Theorem 2 in $R_+^{n+1}$.

## References


[1] Garnett, J., Bounded analytic functions. Academic Press, 1981.

[2] Duren, P., Theory of $H^p$ spaces. Academic Press, 1970.

[3] Shapiro, H. and Shields, A., On some interpolation problems for analytic func-

tions. Amer. J. Math. 83 (1961), 513–532.

[4] Kabaela, V., Interpolation sequences for the $H_p$ classes in the case $p < 1$. (Russian) Litovsk. Mat. Sb. 3:1 (1963), 141–147.

[5] Garnett, J., Harmonic interpolating sequences, $L^p$ and BMO. Ann. Inst. Fourier (Grenoble) 28:4 (1978), 215–228.

[6] Sundberg, C., Values of BMOA functions on interpolating sequences. Michigan Math. J. 31 (1984), 21–30.

[7] Baernstein, A. II, Analytic functions of bounded mean oscillation. Aspects of Contemporary Complex Analysis (Durham, 1979), 3–36, Academic Press, 1980.

[8] Yao Bi-yun, Carleson measures and BMO functions. (Chinese) J. Hangzhou Univ. 14:2 (1987), 132–140.

[9] Yao Bi-yun, The characteristic of Carleson measure on $R_+^{n+1}$. To appear in Chinese Annals, Series A.

# EXTENDING PLURISUBHARMONIC AND HOLOMORPHIC FUNCTIONS [1]

**M. L. Ben Yattou**
Department of Mathematics, Faculty of Sciences of Tunis
Tunis 1060, Tunisia

## 1. Results on strongly convex functions

**Definition 1.** A continuous function $\varphi$ with real values defined on a complex manifold $M$ of $\dim_{\mathbb{C}} M = m$ is said to be *$\rho$-strongly convex* if there is a real continuous exterior form $\omega$ of type (1,1) on $M$ satisfying the following properties:

i) at any point $z_0$ of $M$ there exists a complex coordinate system centered at $z_0$ such that if $\omega$ has the local expression $i \sum \omega_{jk}\, dz_j \wedge d\bar{z}_k$ then the Hermitian matrix $\omega_{jk}$ has at least $m - \rho + 1$ positive eigenvalues on the domain of the coordinate system;

ii) the current $i\partial\bar{\partial}\varphi - \omega$ is positive on $M$.

**Examples.** 1. If $\varphi$ is a function of class $C^2$ on a domain $M$ of $\mathbb{C}^m$ such that the Hessian of $\varphi$ has $m - \rho + 1$ positive eigenvalues at each point $z_0$ of $M$, then the conditions of Definition 1 will be satisfied by taking $\omega = i\partial\bar{\partial}\varphi$.

2. We consider a continuous plurisubharmonic function $u$ on a domain $M$ of $\mathbb{C}^m$ and we define the function $\varphi$ by

$$\varphi(z) = \sum_{j=1}^{k} x_j^2 + u(z),$$

where $z_j = x_j + iy_j$, $1 \le j \le m$. Then $\varphi$ is a $(m - k + 1)$-strongly convex function on $M$.

3. We consider a domain $M$ of $\mathbb{C}^{m+1}$ of the form $D \times G$, where $D$ is a domain of $\mathbb{C}$ and $G$ a domain of $\mathbb{C}^m$. Let $u$ be a continuous strongly subharmonic func-

---

[1] This paper is in final form and no version of it will be submitted for publication elsewhere.

tion on D and v a continuous $\rho$-strongly convex function on G. Then the function $\varphi$ defined by

$$\varphi(z) = u(z_1) + v(z_2,\dots,z_{m+1}),$$

is a $\rho$-strongly convex function on M.

4. Complete real submanifolds of $\mathbb{C}^m$ can be defined by 1-strongly convex functions, see [1] or [2].

The following result is due to El Mir, see [3] for the proof. Note that $[\varphi = 0] = \{z \in M \mid \varphi(z) = 0\}$.

**Theorem 1.** *Let $\varphi$ be a non-negative continuous $\rho$-strongly convex function defined on a complex manifold M of $\dim_{\mathbb{C}} M = m \geq \rho + 1$. Then any closed positive current of bidimension (p,p), defined on $M \setminus [\varphi = 0]$ and satisfying the condition $p \geq \rho$ has a locally finite mass all along $[\varphi = 0]$.*

For the following proposition, cf. [4].

**Proposition 1.** *Let $\varphi$ be a continuous $\rho$-strongly convex function defined on a complex manifold M of $\dim_{\mathbb{C}} M = m$. Then any complex analytic subset of M contained in $[\varphi = 0]$ has complex dimension smaller than $\rho - 1$.*

**Remark.** In Proposition 1 we do not assume $\varphi$ to be non-negative. When $\varphi$ is of class $C^2$ this proposition is already known, see [5], p. 119.

## 2. An extension theorem for plurisubharmonic functions

Our first result is the following:

**Theorem 2.** *Let M be a domain of $\mathbb{C}^m$. Let $\varphi$ be a continuous non-negative $\rho$-strongly convex function on M such that $m \geq \rho + 1$ and the set $[\varphi = 0]$ is polar. Then each plurisubharmonic function u on $M \setminus [\varphi = 0]$ has a unique plurisubharmonic extension to M.*

We need the following preliminaries:

**Definition 2.** Let h be a non-negative harmonic function defined on a domain M of $\mathbb{C}^m$. We say that h is *quasibounded* provided there exists a nondecreasing sequence of non-negative bounded harmonic functions on M converging to h.

Define the function $\tilde{\varphi}$ on $\mathbb{R}$ by $\tilde{\varphi}(x) = (x^+)^2$ where $x^+ = \max(0,x)$. Then $\tilde{\varphi}$ is a non-negative nondecreasing convex function on $\mathbb{R}$. The following theorem is originally a theorem of Privalov, extended to a more general setting by Solomentsev. The proof given by Heins [6] in the framework of Riemann surfaces ([6], Theorem 2) remains valid for the present setting:

**Theorem 3.** *Given a subharmonic function* u *defined on a domain* M *of* $\mathbb{C}^m$, *suppose that* $\tilde{\varphi} \circ u$ *has a harmonic majorant. Then* $u^+ = \max(0,u)$ *has a harmonic majorant and the least harmonic majorant* $M(u^+)$ *of* $u^+$ *is quasibounded on* M.

The following argument has been used by Järvi in [7].

*Proof of Theorem* 2. $\tilde{u} = \tilde{\varphi} \circ u$ is a plurisubharmonic function on $M \setminus [\varphi = 0]$. The current $T = i\partial\bar{\partial}\tilde{u}$ of bidimension $(m-1; m-1)$ satisfies the conditions of Theorem 1 so that it has locally finite mass all along $[\varphi = 0]$. Restricting eventually M we can suppose that it has finite mass on $M \setminus [\varphi = 0]$. We deduce that $\Delta\tilde{u}$ has also finite mass on $M \setminus [\varphi = 0]$. Introduce the Green's function G of M and define the potential $\Psi$ by

$$\Psi(z) = \int_{M \setminus [\varphi=0]} G(z,\xi)\ \Delta\tilde{u}(\xi).$$

Since $\Delta\tilde{u}$ has finite mass, $\Psi$ is a superharmonic function $\not\equiv +\infty$. On the other hand, $\tilde{u} + \Psi$ is harmonic on $M \setminus [\varphi = 0]$, so that $\tilde{\varphi} \circ u$ has a harmonic majorant.

By Theorem 3, $u^+$ has a quasibounded majorant $M(u^+)$. Since $[\varphi = 0]$ is polar, $M(u)^+$ extends to a harmonic function on M and so $u\ (\leq u^+)$ is locally bounded from above on $M \setminus [\varphi = 0]$. By Lelong's extension theorem in [8], u extends to a unique plurisubharmonic function on M.

**Corollary 1.** *With the same hypothesis as in Theorem* 2, *any pluriharmonic function* h *defined on* $M \setminus [\varphi = 0]$ *extends to a unique pluriharmonic function on* M.

*Proof.* Let $h_1$ and $h_2$ be the unique plurisubharmonic extensions of h and $-h$, respectively. $0 \equiv h - h$ has $h_1 + h_2$ and 0 as extensions. Therefore $h_1 \equiv -h_2$ is harmonic.

Our second result is:

**Theorem 4.** *Let* $\varphi$ *be a non-negative continuous* $\rho$*-strongly convex function on a domain* M *of* $\mathbb{C}^m$ *such that* $m - \rho \geq 1$ *and that the set* $[\varphi = 0]$ *is polar. Then each holomorphic function* f *on* $M \setminus [\varphi = 0]$ *extends to a holomorphic function on* M.

*Proof.* For a more general statement using Hausdorff measures, see [4]. We may work locally assuming that M is simply connected. Let $\lambda \geq 0$ be a strict majorant of $|f|$ on M. This number exists because $|f|$ has a plurisubharmonic extension by Theorem 2. Then $\log|\lambda - f| = h$ is a pluriharmonic function on $M \setminus [\varphi = 0]$ which extends to a pluriharmonic function $\tilde{h}$ on $M$. Therefore $\tilde{h} = \mathrm{Re}\, g$, where $g$ is holomorphic on $M$. We deduce that $|\lambda - f| = |\exp(g)|$ on $M \setminus [\varphi = 0]$. But $M \setminus [\varphi = 0]$ is connected so that there exists a complex number $c$ with modulus one satisfying $\lambda - f = c \exp(g)$ on $M \setminus [\varphi = 0]$. Then $\lambda - c\exp(g)$ is the holomorphic extension of $f$ to $M$.

### 3. A potential theoretic result in $\mathbb{C}^m$

We prove the following proposition:

**Proposition 2.** *Given a non-negative plurisubharmonic function* $u$ *defined on a domain* $M$ *of* $\mathbb{C}^m$, *there exist two measurable subsets* $M_1$ *and* $M_2$ *of* $[u = 0]$ *such that*

i) $\Delta u(M_1) = 0$;

ii) $[u = 0] = M_1 \cup M_2$ *and the Lebesgue measure in* $\mathbb{C}^m$ *of* $M_2$ *is zero.*

*Proof.* For $m = 1$ this proposition is contained in [12]. Suppose $m \geq 2$ and consider the Radon-Nikodym decomposition of $\Delta u$ on $M$. Without loss of generality, let $M$ be a polydisc.

Now, $\Delta u = (\Delta u)_a + (\Delta u)_s$, where $(\Delta u)_a$ is completely continuous and $(\Delta u)_s$ is singular with respect to Lebesgue measure $d\mathscr{L}^{2m}$. We know that there exist two measurable subsets $A$ and $B$ of $M$ such that $M = A \cup B$, $\emptyset = A \cap B$ and $(\Delta u)_s(A) = 0$. $\mathscr{L}^{2m}(B) = 0$ so that $(\Delta u)_s$ is supported by a subset $B$ of Lebesgue measure zero. Denote $[u = 0]' = [u = 0] \cap A$. Then

$$\int_M \chi_{[u=0]'}\, \Delta u = \int_M \chi_{[u=0]'}\, (\Delta u)_a + \int_M \chi_{[u=0]'}\, (\Delta u)_s = \int_M \chi_{[u=0]'}\, (\Delta u)_a\,,$$

since $\mathrm{supp}(\Delta u)_s \cap [u = 0]' = \emptyset$. By Fubini's theorem this is equal to

$$\int_{\Delta^{m-1}} d\mathscr{L}^{2(m-1)} \left( \int_{P \cap [u=0]'} (\Delta u)_a \right),$$

where we have identified $(\Delta u)_a$ with its density and where $P$ is the complex line defined by the coordinate $z_1$ when the other coordinates are fixed. Call $L_p$ the linear projection of $\mathbb{C}^m$ over $P$. Now

$$\int_{P\cap[u=0]'} (\Delta u)_a = \int_P \chi_{[u=0]'} (\Delta u)_a = \int_P \chi_{[u=0]'} \Delta u$$

$$= \int_P \chi_{[u=0]'} \Delta u \circ L_p = \int_P \chi_{[u=0]'} \Delta(u \circ L_p)$$

$$= \int_{P\cap[u=0]'} \Delta(u \circ L_p),$$

because of the condition on the support of $(\Delta u)_s$. For almost all P, i.e. $z' \in \Delta^{m-1}$, $\Delta(u \circ L_p) | P\cap[u=0]'$ is completely continuous with respect to Lebesgue measure over P by Fubini's theorem. In fact this is a consequence of

$$\Delta(u \circ L_p) = \Delta u \circ L_p = (\Delta u)_a \circ L_p + (\Delta u)_s \circ L_p$$

and

$$\operatorname{supp}((\Delta u)_s \circ L_p) \cap P \cap [u = 0]' = \emptyset.$$

Using the same notations in the one-dimensional case, since $u \circ L_p$ is $\geq 0$ and subharmonic, we have

$$0 \leq \int_{[u=0]'\cap P} \Delta(u \circ L_p) = \int_{[u\circ L_p=0]'\cap P} \Delta(u \circ L_p) = 0$$

by the result of Erëmenko and Sodin. We deduce that $\int_{[u=0]'} \Delta u = 0$, and the assertion follows.

**Corollary 2.** *For a strictly plurisubharmonic and non-negative function* u *defined on a domain* M *of* $\mathbb{C}^m$, *we have* $\mathcal{L}^{2m}([u = 0]) = 0$.

*Proof.* The Lebesgue measure is (locally) absolutely continuous with respect to $\Delta u$ and therefore $\mathcal{L}^{2m}(M_1) = 0$.

**Proposition 3.** *Let* $\varphi$ *be a non-negative continuous* $\rho$*-strongly convex function defined on a domain* M *of* $\mathbb{C}^m$ *satisfying the condition* $m - \rho \geq 0$. *Then* $\mathcal{L}^{2m}([\varphi = 0]) = 0$.

*Proof.* The proof is local; so choose a neighborhood of $Z_0 \in [\varphi = 0]$ of the form $\Delta \times \Delta^{m-1}$ such that for any $w \in \Delta^{m-1}$, $\varphi_w : z \in \Delta \to \varphi(z,w)$ is a non-negative strictly subharmonic function, see [4], Lemma 4.

By Corollary 2, $\mathcal{L}^2([\varphi_w = 0]) = 0$. On the other hand,

$$\int_{[\varphi=0]\cap\Delta\times\Delta^{m-1}} d\mathscr{L}^{2m} = \int_{w\in\Delta^{m-1}} \left[ \int_{[\varphi_w=0]\cap\Delta} d\mathscr{L}^2 \right] d\mathscr{L}(w)^{2(m-1)} = 0$$

by Fubini's theorem.

## 4. More extension results

Here we study an example of general importance. We consider a domain in $\mathbb{C}^2$ of the form $D_1 \times D_2$ where $D_1$ contains the unit circle C. We choose two functions u and v defined on $D_1$ and $D_2$, respectively. We suppose v to be strictly subharmonic on $D_2$, u subharmonic on $D_1$ and u, v to be continuous and non-negative. Define the function $\varphi$ by

$$\varphi(z_1;z_2) = (|z_1|^2 - 1)^2 + u(z_1) + v(z_2).$$

The function $\varphi$ is seen to be 1-strongly convex on a neighborhood of $C \times D_2$. On the other hand, $[\varphi = 0] \subset C \times [v = 0]$. By [11], Theorem 2.10.45, there exists a $\lambda > 0$ such that

$$\mathscr{H}_3([\varphi = 0]) \leq \mathscr{H}_3(C \times [v = 0]) = \lambda \mathscr{H}_1(C) \times \mathscr{H}_2([v = 0]) = 0$$

(since $\mathscr{H}_2([v = 0]) = 0$ by Corollary 2 and $0 < \mathscr{H}_1(C) < +\infty$). By [4], Proposition 6, any holomorphic function f defined on $D_1{\times}D_2 \setminus [\varphi = 0]$ extends holomorphically to $D_1 \times D_2$. This can also be proved using Corollary 2 and the Levy extension theorem.

Next we cancel the quantity $(|z_1|^2 - 1)^2$ and the hypothesis on the circle by considering the function $\tilde{\varphi}(z) = u(z_1) + v(z_2)$. Let now u be strictly subharmonic too (this was the case of $(|z_1|^2 - 1)^2 + u(z_1)$ in a neighborhood of C).We claim that any holomorphic function f defined on $D_1{\times}D_2 \setminus [\tilde{\varphi} = 0]$ extends holomorphically to $D_1 \times D_2$. The proof makes use of a recent result of Shiffman [13], Corollary 1, on the meromorphic almost everywhere extension of separately holomorphic functions.

*Proof of the claim.* The sets $B_1 = [u = 0]$ and $B_2 = [v = 0]$ have Lebesgue measure zero by Corollary 2. Also, $[\tilde{\varphi} = 0] = B_1 \times B_2$ has Lebesgue measure zero. Put $B = B_1 \cup B_2$ and suppose without loss of generality that $D_1 = D_2$. Clearly, $\mathscr{L}^2(B) = 0$.

i) For each $z_2 \in D_2 \setminus B$, $B_1^{z_2} = \{w \in \mathbb{C} \mid (w,z_2) \in [\tilde{\varphi} = 0]\} = \emptyset$ so that $\mathscr{L}^2(B_1^{z_2}) = 0$ and the function $f_1^{z_2}$ defined by $f_1^{z_2} : z_1 \to f(z_1,z_2)$ is holomorphic on $D_1$. Also, $B_2^{z_2} = \{w \in \mathbb{C} \mid (z_2,w) \in [\tilde{\varphi} = 0]\} = \emptyset$, $\mathscr{L}^2(B_2^{z_2}) = 0$ and $f_2^{z_2} : w \to f(z_2,w)$ is holomorphic on $D_2$.

ii) For each $z_1 \in D_1 \setminus B$, $B_1^{z_1} = \{w \in C \mid (w,z_1) \in [\tilde{\varphi} = 0]\} = \emptyset$ so that $\mathscr{L}^2(B_1^{z_1}) = 0$ and $f_1^{z_1} : w \to f(w,z_1)$ is holomorphic on $D_1$. Also, $B_2^{z_1} = \{w \in C \mid (z_1,w) \in [\tilde{\varphi} = 0]\} = \emptyset$, $\mathscr{L}^2(B_2^{z_1}) = 0$ and $f_2^{z_1} : w \to f(z_1,w)$ is holomorphic on $D_2$. By [13], Corollary 1, there now exists a meromorphic function $g$ on $D_1 \times D_2$ equal to $f$ almost everywhere on $D_1 \times D_2$. Clearly, we have $f \equiv g$ on $D_1 \times D_2 \setminus [\tilde{\varphi} = 0]$ so that $g$ is a meromorphic extension of $f$ to $D_1 \times D_2$. But the subset of $D_1 \times D_2$, where $g$ is not holomorphic, is contained in $[\tilde{\varphi} = 0]$. By Proposition 1 this complex analytic subset has dimension $\leq 1 - 1 = 0$. Therefore, $f$ has a holomorphic extension to $D_1 \times D_2$.

## References


[1] El Mir, H., Sur le prolongement des courants positifs fermés. Acta Math. 153 (1984), 1–45.

[2] Henkin, M. G. and Leiterer, J., Theory of functions on complex manifolds. Birkhäuser, Basel, 1984.

[3] El Mir, H., Prolongement des courants positifs fermés et fonctions k convexes. Proceedings of the Conference on Complex Analysis and Applications (Varna, 1985), 199–211.

[4] Ben Yattou, M. L., Extension of analytical objects across $\rho$-strongly convex subsets. To appear in the Proceedings of the Conference on Complex Analysis and Applications (Varna, 1987).

[5] Siu, Y.-T. and Trautmann, G., Gapsheaves and extension of coherent analytic subsheaves. Lecture Notes in Math., 172, Springer-Verlag, Berlin, 1971.

[6] Heins, M., Hardy classes on Riemann surfaces. Lecture Notes in Math., 98, Springer-Verlag, Berlin, 1969.

[7] Järvi, P., Removable singularities for $H^p$-functions. Proc. Amer. Math. Soc. 86 (1982), 596–598.

[8] Harvey, R. and Polking, J. G., Extending analytic objects. Comm. Pure Appl. Math. 28 (1975), 701–727.

[9] Shiffman, B., Extension of positive line bundles and meromorphic maps. Invent. Math. 15 (1972), 332–347.

[10] Shiffman, B., A course in real and complex analysis (1978/79) at the Johns

Hopkins University (unpublished).

[11] Federer, H., Geometric measure theory. Grundlehren der mathematischen. Wissenschaften, 153, Springer-Verlag, Berlin, 1969.

[12] Erëmenko, A. E. and Sodin, M. L., A hypothesis of Littlewood and the distribution of values of entire functions. (Russian) Funktsional. Anal. i Prilozhen. 20 (1986), 71–72.

[13] Shiffman, B., Complete characterization of holomorphic chains of codimension one. Math. Ann. 274 (1986), 233–256.

# LECTURE NOTES IN MATHEMATICS

Edited by A. Dold and B. Eckmann

**Some general remarks on the publication of proceedings of congresses and symposia**

Lecture Notes aim to report new developments - quickly, informally and at a high level. The following describes criteria and procedures which apply to proceedings volumes. The editors of a volume are strongly advised to inform contributors about these points at an early stage.

§1. One (or more) expert participant(s) of the meeting should act as the responsible editor(s) of the proceedings. They select the papers which are suitable (cf. §§ 2, 3) for inclusion in the proceedings, and have them individually refereed (as for a journal). It should not be assumed that the published proceedings must reflect conference events faithfully and in their entirety. Contributions to the meeting which are not included in the proceedings can be listed by title. The series editors will normally not interfere with the editing of a particular proceedings volume - except in fairly obvious cases, or on technical matters, such as described in §§ 2, 3. The names of the responsible editors appear on the title page of the volume.

§2. The proceedings should be reasonably homogeneous (concerned with a limited area). For instance, the proceedings of a congress on "Analysis" or "Mathematics in Wonderland" would normally not be sufficiently homogeneous.

One or two longer survey articles on recent developments in the field are often very useful additions to such proceedings - even if they do not correspond to actual lectures at the congress. An extensive introduction on the subject of the congress would be desirable.

§3. The contributions should be of a high mathematical standard and of current interest. Research articles should present new material and not duplicate other papers already published or due to be published. They should contain sufficient information and motivation and they should present proofs, or at least outlines of such, in sufficient detail to enable an expert to complete them. Thus resumes and mere announcements of papers appearing elsewhere cannot be included, although more detailed versions of a contribution may well be published in other places later.

Surveys, if included, should cover a sufficiently broad topic, and should in general not simply review the author's own recent research. In the case of surveys, exceptionally, proofs of results may not be necessary.

"Mathematical Reviews" and "Zentralblatt für Mathematik" require that papers in proceedings volumes carry an explicit statement that they are in final form and that no similar paper has been or is being submitted elsewhere, if these papers are to be considered for a review. Normally, papers that satisfy the criteria of the Lecture Notes in Mathematics series also satisfy this

.../...

requirement, but we would strongly recommend that the contributing authors be asked to give this guarantee explicitly at the beginning or end of their paper. There will occasionally be cases where this does not apply but where, for special reasons, the paper is still acceptable for LNM.

§4. Proceedings should appear soon after the meeeting. The publisher should, therefore, receive the complete manuscript within nine months of the date of the meeting at the latest.

§5. Plans or proposals for proceedings volumes should be sent to one of the editors of the series or to Springer-Verlag Heidelberg. They should give sufficient information on the conference or symposium, and on the proposed proceedings. In particular, they should contain a list of the expected contributions with their prospective length. Abstracts or early versions (drafts) of some of the contributions are very helpful.

§6. Lecture Notes are printed by photo-offset from camera-ready typed copy provided by the editors. For this purpose Springer-Verlag provides editors with technical instructions for the preparation of manuscripts and these should be distributed to all contributing authors. Springer-Verlag can also, on request, supply stationery on which the prescribed typing area is outlined. Some homogeneity in the presentation of the contributions is desirable.

Careful preparation of manuscripts will help keep production time short and ensure a satisfactory appearance of the finished book. The actual production of a Lecture Notes volume normally takes 6 -8 weeks.

Manuscripts should be at least 100 pages long. The final version should include a table of contents and as far as applicable a subject index.

§7. Editors receive a total of 50 free copies of their volume for distribution to the contributing authors, but no royalties. (Unfortunately, no reprints of individual contributions can be supplied.) They are entitled to purchase further copies of their book for their personal use at a discount of 33.3 %, other Springer mathematics books at a discount of 20 % directly from Springer-Verlag. Contributing authors may purchase the volume in which their article appears at a discount of 33.3 %.

Commitment to publish is made by letter of intent rather than by signing a formal contract. Springer-Verlag secures the copyright for each volume.

LECTURE NOTES

ESSENTIALS FOR THE PREPARATION OF CAMERA-READY MANUSCRIPTS

Springer

Springer-Verlag
Berlin Heidelberg New York
London Paris Tokyo Hong Kong

The preparation of manuscripts which are to be reproduced by photo-offset require special care. Manuscripts which are submitted in technically unsuitable form will be returned to the author for retyping. There is normally no possibility of carrying out further corrections after a manuscript is given to production. Hence it is crucial that the following instructions be adhered to closely. If in doubt, please send us 1 - 2 sample pages for examination.

General. The characters must be uniformly black both within a single character and down the page. Original manuscripts are required: photocopies are acceptable only if they are sharp and without smudges.

On request, Springer-Verlag will supply special paper with the text area outlined. The standard TEXT AREA (OUTPUT SIZE if you are using a 14 point font) is 18 x 26.5 cm (7.5 x 11 inches). This will be scale-reduced to 75% in the printing process. If you are using computer typesetting, please see also the following page.

Make sure the TEXT AREA IS COMPLETELY FILLED. Set the margins so that they precisely match the outline and type right from the top to the bottom line. (Note that the page number will lie outside this area). Lines of text should not end more than three spaces inside or outside the right margin (see example on page 4).

Type on one side of the paper only.

Spacing and Headings (Monographs). Use ONE-AND-A-HALF line spacing in the text. Please leave sufficient space for the title to stand out clearly and do NOT use a new page for the beginning of subdivisons of chapters. Leave THREE LINES blank above and TWO below headings of such subdivisions.

Spacing and Headings (Proceedings). Use ONE-AND-A-HALF line spacing in the text. Do not use a new page for the beginning of subdivisons of a single paper. Leave THREE LINES blank above and TWO below headings of such subdivisions. Make sure headings of equal importance are in the same form.

The first page of each contribution should be prepared in the same way. The title should stand out clearly. We therefore recommend that the editor prepare a sample page and pass it on to the authors together with these instructions. Please take the following as an example. Begin heading 2 cm below upper edge of text area.

MATHEMATICAL STRUCTURE IN QUANTUM FIELD THEORY

John E. Robert
Mathematisches Institut, Universität Heidelberg
Im Neuenheimer Feld 288, D-6900 Heidelberg

Please leave THREE LINES blank below heading and address of the author, then continue with the actual text on the same page.

Footnotes. These should preferable be avoided. If necessary, type them in SINGLE LINE SPACING to finish exactly on the outline, and separate them from the preceding main text by a line.

**Symbols**. Anything which cannot be typed may be entered by hand in BLACK AND ONLY BLACK ink. (A fine-tipped rapidograph is suitable for this purpose; a good black ball-point will do, but a pencil will not). Do not draw straight lines by hand without a ruler (not even in fractions).

**Literature References.** These should be placed at the end of each paper or chapter, or at the end of the work, as desired. Type them with single line spacing and start each reference on a new line. Follow "Zentralblatt für Mathematik"/"Mathematical Reviews" for abbreviated titles of mathematical journals and "Bibliographic Guide for Editors and Authors (BGEA)" for chemical, biological, and physics journals. Please ensure that all references are COMPLETE and ACCURATE.

**IMPORTANT**

**Pagination.** For typescript, number pages in the upper right-hand corner in LIGHT BLUE OR GREEN PENCIL ONLY. The printers will insert the final page numbers. For computer type, you may insert page numbers (1 cm above outer edge of text area).

It is safer to number pages AFTER the text has been typed and corrected. Page 1 (Arabic) should be THE FIRST PAGE OF THE ACTUAL TEXT. The Roman pagination (table of contents, preface, abstract, acknowledgements, brief introductions, etc.) will be done by Springer-Verlag.

If including running heads, these should be aligned with the inside edge of the text area while the page number is aligned with the outside edge noting that right-hand pages are odd-numbered. Running heads and page numbers appear on the same line. Normally, the running head on the left-hand page is the chapter heading and that on the right-hand page is the section heading. Running heads should not be included in proceedings contributions unless this is being done consistently by all authors.

**Corrections.** When corrections have to be made, cut the new text to fit and paste it over the old. White correction fluid may also be used.

Never make corrections or insertions in the text by hand.

If the typescript has to be marked for any reason, e.g. for provisional page numbers or to mark corrections for the typist, this can be done VERY FAINTLY with BLUE or GREEN PENCIL but NO OTHER COLOR: these colors do not appear after reproduction.

**COMPUTER-TYPESETTING.** Further, to the above instructions, please note with respect to your printout that
- the characters should be sharp and sufficiently black;
- it is not strictly necessary to use Springer's special typing paper. Any white paper of reasonable quality is acceptable.

If you are using a significantly different font size, you should modify the output size correspondingly, keeping length to breadth ratio 1 : 0.68, so that scaling down to 10 point font size, yields a text area of 13.5 x 20 cm (5 3/8 x 8 in), e.g.

Differential equations.: use output size 13.5 x 20 cm.

Differential equations.: use output size 16 x 23.5 cm.

Differential equations.: use output size 18 x 26.5 cm.

Interline spacing: 5.5 mm base-to-base for 14 point characters (standard format of 18 x 26.5 cm).
If in any doubt, please send us 1 - 2 sample pages for examination. We will be glad to give advice.

Vol. 1201: Curvature and Topology of Riemannian Manifolds. Proceedings, 1985. Edited by K. Shiohama, T. Sakai and T. Sunada. VII, 336 pages. 1986.

Vol. 1202: A. Dür, Möbius Functions, Incidence Algebras and Power Series Representations. XI, 134 pages. 1986.

Vol. 1203: Stochastic Processes and Their Applications. Proceedings, 1985. Edited by K. Itô and T. Hida. VI, 222 pages. 1986.

Vol. 1204: Séminaire de Probabilités XX, 1984/85. Proceedings. Edité par J. Azéma et M. Yor. V, 639 pages. 1986.

Vol. 1205: B.Z. Moroz, Analytic Arithmetic in Algebraic Number Fields. VII, 177 pages. 1986.

Vol. 1206: Probability and Analysis, Varenna (Como) 1985. Seminar. Edited by G. Letta and M. Pratelli. VIII, 280 pages. 1986.

Vol. 1207: P.H. Bérard, Spectral Geometry: Direct and Inverse Problems. With an Appendix by G. Besson. XIII, 272 pages. 1986.

Vol. 1208: S. Kaijser, J.W. Pelletier, Interpolation Functors and Duality. IV, 167 pages. 1986.

Vol. 1209: Differential Geometry, Peñíscola 1985. Proceedings. Edited by A.M. Naveira, A. Ferrández and F. Mascaró. VIII, 306 pages. 1986.

Vol. 1210: Probability Measures on Groups VIII. Proceedings, 1985. Edited by H. Heyer. X, 386 pages. 1986.

Vol. 1211: M.B. Sevryuk, Reversible Systems. V, 319 pages. 1986.

Vol. 1212: Stochastic Spatial Processes. Proceedings, 1984. Edited by P. Tautu. VIII, 311 pages. 1986.

Vol. 1213: L.G. Lewis, Jr., J.P. May, M. Steinberger, Equivariant Stable Homotopy Theory. IX, 538 pages. 1986.

Vol. 1214: Global Analysis – Studies and Applications II. Edited by Yu.G. Borisovich and Yu.E. Gliklikh. V, 275 pages. 1986.

Vol. 1215: Lectures in Probability and Statistics. Edited by G. del Pino and R. Rebolledo. V, 491 pages. 1986.

Vol. 1216: J. Kogan, Bifurcation of Extremals in Optimal Control. VIII, 106 pages. 1986.

Vol. 1217: Transformation Groups. Proceedings, 1985. Edited by S. Jackowski and K. Pawalowski. X, 396 pages. 1986.

Vol. 1218: Schrödinger Operators, Aarhus 1985. Seminar. Edited by E. Balslev. V, 222 pages. 1986.

Vol. 1219: R. Weissauer, Stabile Modulformen und Eisensteinreihen. III, 147 Seiten. 1986.

Vol. 1220: Séminaire d'Algèbre Paul Dubreil et Marie-Paule Malliavin. Proceedings, 1985. Edité par M.-P. Malliavin. IV, 200 pages. 1986.

Vol. 1221: Probability and Banach Spaces. Proceedings, 1985. Edited by J. Bastero and M. San Miguel. XI, 222 pages. 1986.

Vol. 1222: A. Katok, J.-M. Strelcyn, with the collaboration of F. Ledrappier and F. Przytycki, Invariant Manifolds, Entropy and Billiards; Smooth Maps with Singularities. VIII, 283 pages. 1986.

Vol. 1223: Differential Equations in Banach Spaces. Proceedings, 1985. Edited by A. Favini and E. Obrecht. VIII, 299 pages. 1986.

Vol. 1224: Nonlinear Diffusion Problems, Montecatini Terme 1985. Seminar. Edited by A. Fasano and M. Primicerio. VIII, 188 pages. 1986.

Vol. 1225: Inverse Problems, Montecatini Terme 1986. Seminar. Edited by G. Talenti. VIII, 204 pages. 1986.

Vol. 1226: A. Buium, Differential Function Fields and Moduli of Algebraic Varieties. IX, 146 pages. 1986.

Vol. 1227: H. Helson, The Spectral Theorem. VI, 104 pages. 1986.

Vol. 1228: Multigrid Methods II. Proceedings, 1985. Edited by W. Hackbusch and U. Trottenberg. VI, 336 pages. 1986.

Vol. 1229: O. Bratteli, Derivations, Dissipations and Group Actions on C*-algebras. IV, 277 pages. 1986.

Vol. 1230: Numerical Analysis. Proceedings, 1984. Edited by J.-P. Hennart. X, 234 pages. 1986.

Vol. 1231: E.-U. Gekeler, Drinfeld Modular Curves. XIV, 107 pages. 1986.

Vol. 1232: P.C. Schuur, Asymptotic Analysis of Soliton Problems. VIII, 180 pages. 1986.

Vol. 1233: Stability Problems for Stochastic Models. Proceedings, 1985. Edited by V.V. Kalashnikov, B. Penkov and V.M. Zolotarev. VI, 223 pages. 1986.

Vol. 1234: Combinatoire énumérative. Proceedings, 1985. Edité par G. Labelle et P. Leroux. XIV, 387 pages. 1986.

Vol. 1235: Séminaire de Théorie du Potentiel, Paris, No. 8. Directeurs: M. Brelot, G. Choquet et J. Deny. Rédacteurs: F. Hirsch et G. Mokobodzki. III, 209 pages. 1987.

Vol. 1236: Stochastic Partial Differential Equations and Applications. Proceedings, 1985. Edited by G. Da Prato and L. Tubaro. V, 257 pages. 1987.

Vol. 1237: Rational Approximation and its Applications in Mathematics and Physics. Proceedings, 1985. Edited by J. Gilewicz, M. Pindor and W. Siemaszko. XII, 350 pages. 1987.

Vol. 1238: M. Holz, K.-P. Podewski and K. Steffens, Injective Choice Functions. VI, 183 pages. 1987.

Vol. 1239: P. Vojta, Diophantine Approximations and Value Distribution Theory. X, 132 pages. 1987.

Vol. 1240: Number Theory, New York 1984–85. Seminar. Edited by D.V. Chudnovsky, G.V. Chudnovsky, H. Cohn and M.B. Nathanson. V, 324 pages. 1987.

Vol. 1241: L. Gårding, Singularities in Linear Wave Propagation. III, 125 pages. 1987.

Vol. 1242: Functional Analysis II, with Contributions by J. Hoffmann-Jørgensen et al. Edited by S. Kurepa, H. Kraljević and D. Butković. VII, 432 pages. 1987.

Vol. 1243: Non Commutative Harmonic Analysis and Lie Groups. Proceedings, 1985. Edited by J. Carmona, P. Delorme and M. Vergne. V, 309 pages. 1987.

Vol. 1244: W. Müller, Manifolds with Cusps of Rank One. XI, 158 pages. 1987.

Vol. 1245: S. Rallis, L-Functions and the Oscillator Representation. XVI, 239 pages. 1987.

Vol. 1246: Hodge Theory. Proceedings, 1985. Edited by E. Cattani, F. Guillén, A. Kaplan and F. Puerta. VII, 175 pages. 1987.

Vol. 1247: Séminaire de Probabilités XXI. Proceedings. Edité par J. Azéma, P.A. Meyer et M. Yor. IV, 579 pages. 1987.

Vol. 1248: Nonlinear Semigroups, Partial Differential Equations and Attractors. Proceedings, 1985. Edited by T.L. Gill and W.W. Zachary. IX, 185 pages. 1987.

Vol. 1249: I. van den Berg, Nonstandard Asymptotic Analysis. IX, 187 pages. 1987.

Vol. 1250: Stochastic Processes – Mathematics and Physics II. Proceedings 1985. Edited by S. Albeverio, Ph. Blanchard and L. Streit. VI, 359 pages. 1987.

Vol. 1251: Differential Geometric Methods in Mathematical Physics. Proceedings, 1985. Edited by P.L. García and A. Pérez-Rendón. VII, 300 pages. 1987.

Vol. 1252: T. Kaise, Représentations de Weil et $GL_2$ Algèbres de division et $GL_n$. VII, 203 pages. 1987.

Vol. 1253: J. Fischer, An Approach to the Selberg Trace Formula via the Selberg Zeta-Function. III, 184 pages. 1987.

Vol. 1254: S. Gelbart, I. Piatetski-Shapiro, S. Rallis. Explicit Constructions of Automorphic L-Functions. VI, 152 pages. 1987.

Vol. 1255: Differential Geometry and Differential Equations. Proceedings, 1985. Edited by C. Gu, M. Berger and R.L. Bryant. XII, 243 pages. 1987.

Vol. 1256: Pseudo-Differential Operators. Proceedings, 1986. Edited by H.O. Cordes, B. Gramsch and H. Widom. X, 479 pages. 1987.

Vol. 1257: X. Wang, On the C*-Algebras of Foliations in the Plane. V, 165 pages. 1987.

Vol. 1258: J. Weidmann, Spectral Theory of Ordinary Differential Operators. VI, 303 pages. 1987.

Vol. 1259: F. Cano Torres, Desingularization Strategies for Three-Dimensional Vector Fields. IX, 189 pages. 1987.

Vol. 1260: N.H. Pavel, Nonlinear Evolution Operators and Semigroups. VI, 285 pages. 1987.

Vol. 1261: H. Abels, Finite Presentability of S-Arithmetic Groups. Compact Presentability of Solvable Groups. VI, 178 pages. 1987.

Vol. 1262: E. Hlawka (Hrsg.), Zahlentheoretische Analysis II. Seminar, 1984–86. V, 158 Seiten. 1987.

Vol. 1263: V.L. Hansen (Ed.), Differential Geometry. Proceedings, 1985. XI, 288 pages. 1987.

Vol. 1264: Wu Wen-tsün, Rational Homotopy Type. VIII, 219 pages. 1987.

Vol. 1265: W. Van Assche, Asymptotics for Orthogonal Polynomials. VI, 201 pages. 1987.

Vol. 1266: F. Ghione, C. Peskine, E. Sernesi (Eds.), Space Curves. Proceedings, 1985. VI, 272 pages. 1987.

Vol. 1267: J. Lindenstrauss, V.D. Milman (Eds.), Geometrical Aspects of Functional Analysis. Seminar. VII, 212 pages. 1987.

Vol. 1268: S.G. Krantz (Ed.), Complex Analysis. Seminar, 1986. VII, 195 pages. 1987.

Vol. 1269: M. Shiota, Nash Manifolds. VI, 223 pages. 1987.

Vol. 1270: C. Carasso, P.-A. Raviart, D. Serre (Eds.), Nonlinear Hyperbolic Problems. Proceedings, 1986. XV, 341 pages. 1987.

Vol. 1271: A.M. Cohen, W.H. Hesselink, W.L.J. van der Kallen, J.R. Strooker (Eds.), Algebraic Groups Utrecht 1986. Proceedings. XII, 284 pages. 1987.

Vol. 1272: M.S. Livšic, L.L. Waksman, Commuting Nonselfadjoint Operators in Hilbert Space. III, 115 pages. 1987.

Vol. 1273: G.-M. Greuel, G. Trautmann (Eds.), Singularities, Representation of Algebras, and Vector Bundles. Proceedings, 1985. XIV, 383 pages. 1987.

Vol. 1274: N. C. Phillips, Equivariant K-Theory and Freeness of Group Actions on C*-Algebras. VIII, 371 pages. 1987.

Vol. 1275: C.A. Berenstein (Ed.), Complex Analysis I. Proceedings, 1985–86. XV, 331 pages. 1987.

Vol. 1276: C.A. Berenstein (Ed.), Complex Analysis II. Proceedings, 1985–86. IX, 320 pages. 1987.

Vol. 1277: C.A. Berenstein (Ed.), Complex Analysis III. Proceedings, 1985–86. X, 350 pages. 1987.

Vol. 1278: S.S. Koh (Ed.), Invariant Theory. Proceedings, 1985. V, 102 pages. 1987.

Vol. 1279: D. Ieşan, Saint-Venant's Problem. VIII, 162 Seiten. 1987.

Vol. 1280: E. Neher, Jordan Triple Systems by the Grid Approach. XII, 193 pages. 1987.

Vol. 1281: O.H. Kegel, F. Menegazzo, G. Zacher (Eds.), Group Theory. Proceedings, 1986. VII, 179 pages. 1987.

Vol. 1282: D.E. Handelman, Positive Polynomials, Convex Integral Polytopes, and a Random Walk Problem. XI, 136 pages. 1987.

Vol. 1283: S. Mardešić, J. Segal (Eds.), Geometric Topology and Shape Theory. Proceedings, 1986. V, 261 pages. 1987.

Vol. 1284: B.H. Matzat, Konstruktive Galoistheorie. X, 286 pages. 1987.

Vol. 1285: I.W. Knowles, Y. Saitō (Eds.), Differential Equations and Mathematical Physics. Proceedings, 1986. XVI, 499 pages. 1987.

Vol. 1286: H.R. Miller, D.C. Ravenel (Eds.), Algebraic Topology. Proceedings, 1986. VII, 341 pages. 1987.

Vol. 1287: E.B. Saff (Ed.), Approximation Theory, Tampa. Proceedings, 1985–1986. V, 228 pages. 1987.

Vol. 1288: Yu. L. Rodin, Generalized Analytic Functions on Riemann Surfaces. V, 128 pages, 1987.

Vol. 1289: Yu. I. Manin (Ed.), K-Theory, Arithmetic and Geometry. Seminar, 1984–1986. V, 399 pages. 1987.

Vol. 1290: G. Wüstholz (Ed.), Diophantine Approximation and Transcendence Theory. Seminar, 1985. V, 243 pages. 1987.

Vol. 1291: C. Mœglin, M.-F. Vignéras, J.-L. Waldspurger, Correspondances de Howe sur un Corps p-adique. VII, 163 pages. 1987

Vol. 1292: J.T. Baldwin (Ed.), Classification Theory. Proceedings, 1985. VI, 500 pages. 1987.

Vol. 1293: W. Ebeling, The Monodromy Groups of Isolated Singularities of Complete Intersections. XIV, 153 pages. 1987.

Vol. 1294: M. Queffélec, Substitution Dynamical Systems – Spectral Analysis. XIII, 240 pages. 1987.

Vol. 1295: P. Lelong, P. Dolbeault, H. Skoda (Réd.), Séminaire d'Analyse P. Lelong – P. Dolbeault – H. Skoda. Seminar, 1985/1986. VII, 283 pages. 1987.

Vol. 1296: M.-P. Malliavin (Ed.), Séminaire d'Algèbre Paul Dubreil et Marie-Paule Malliavin. Proceedings, 1986. IV, 324 pages. 1987.

Vol. 1297: Zhu Y.-l., Guo B.-y. (Eds.), Numerical Methods for Partial Differential Equations. Proceedings. XI, 244 pages. 1987.

Vol. 1298: J. Aguadé, R. Kane (Eds.), Algebraic Topology, Barcelona 1986. Proceedings. X, 255 pages. 1987.

Vol. 1299: S. Watanabe, Yu.V. Prokhorov (Eds.), Probability Theory and Mathematical Statistics. Proceedings, 1986. VIII, 589 pages. 1988.

Vol. 1300: G.B. Seligman, Constructions of Lie Algebras and their Modules. VI, 190 pages. 1988.

Vol. 1301: N. Schappacher, Periods of Hecke Characters. XV, 160 pages. 1988.

Vol. 1302: M. Cwikel, J. Peetre, Y. Sagher, H. Wallin (Eds.), Function Spaces and Applications. Proceedings, 1986. VI, 445 pages. 1988.

Vol. 1303: L. Accardi, W. von Waldenfels (Eds.), Quantum Probability and Applications III. Proceedings, 1987. VI, 373 pages. 1988.

Vol. 1304: F.Q. Gouvêa, Arithmetic of $p$-adic Modular Forms. VIII, 121 pages. 1988.

Vol. 1305: D.S. Lubinsky, E.B. Saff, Strong Asymptotics for Extremal Polynomials Associated with Weights on $\mathbb{R}$. VII, 153 pages. 1988.

Vol. 1306: S.S. Chern (Ed.), Partial Differential Equations. Proceedings, 1986. VI, 294 pages. 1988.

Vol. 1307: T. Murai, A Real Variable Method for the Cauchy Transform, and Analytic Capacity. VIII, 133 pages. 1988.

Vol. 1308: P. Imkeller, Two-Parameter Martingales and Their Quadratic Variation. IV, 177 pages. 1988.

Vol. 1309: B. Fiedler, Global Bifurcation of Periodic Solutions with Symmetry. VIII, 144 pages. 1988.

Vol. 1310: O.A. Laudal, G. Pfister, Local Moduli and Singularities. V, 117 pages. 1988.

Vol. 1311: A. Holme, R. Speiser (Eds.), Algebraic Geometry, Sundance 1986. Proceedings. VI, 320 pages. 1988.

Vol. 1312: N.A. Shirokov, Analytic Functions Smooth up to the Boundary. III, 213 pages. 1988.

Vol. 1313: F. Colonius, Optimal Periodic Control. VI, 177 pages. 1988.

Vol. 1314: A. Futaki, Kähler-Einstein Metrics and Integral Invariants. IV, 140 pages. 1988.

Vol. 1315: R.A. McCoy, I. Ntantu, Topological Properties of Spaces of Continuous Functions. IV, 124 pages. 1988.

Vol. 1316: H. Korezlioglu, A.S. Ustunel (Eds.), Stochastic Analysis and Related Topics. Proceedings, 1986. V, 371 pages. 1988.

Vol. 1317: J. Lindenstrauss, V.D. Milman (Eds.), Geometric Aspects of Functional Analysis. Seminar, 1986–87. VII, 289 pages. 1988.

Vol. 1318: Y. Felix (Ed.), Algebraic Topology – Rational Homotopy. Proceedings, 1986. VIII, 245 pages. 1988

Vol. 1319: M. Vuorinen, Conformal Geometry and Quasiregular Mappings. XIX, 209 pages. 1988.